T0290710

Statistical Methods for Stochastic Differential Equations

MONOGRAPHS ON STATISTICS AND APPLIED PROBABILITY

General Editors

F. Bunea, V. Isham, N. Keiding, T. Louis, R. L. Smith, and H. Tong

Monographs on Statistics and Applied Probability 124

Statistical Methods for Stochastic Differential Equations

Edited by
Mathieu Kessler
Alexander Lindner
Michael Sørensen

CRC Press
Taylor & Francis Group
Boca Raton London New York

CRC Press is an imprint of the
Taylor & Francis Group, an informa business

A CHAPMAN & HALL BOOK

CRC Press
Taylor & Francis Group
6000 Broken Sound Parkway NW, Suite 300
Boca Raton, FL 33487-2742

© 2012 by Taylor & Francis Group, LLC
CRC Press is an imprint of Taylor & Francis Group, an Informa business

No claim to original U.S. Government works

Printed in the United States of America on acid-free paper
Version Date: 20120326

International Standard Book Number: 978-1-4398-4940-8 (Hardback)

Library of Congress Cataloging-in-Publication Data

Statistical methods for stochastic differential equations / [edited by] Mathieu Kessler, Alexander Lindner, Michael Sørensen.
    p. cm. -- (Monographs on statistics and applied probability ; 124)
    Includes bibliographical references and index.
    ISBN 978-1-4398-4940-8 (hardback)
    1. Stochastic differential equations--Statistical methods. I. Kessler, Mathieu. II. Lindner, Alexander, 1973- III. Sørensen, Michael.

QA274.23.S75 2012
515'.35--dc23
                                                                2012005166

Visit the Taylor & Francis Web site at
http://www.taylorandfrancis.com

and the CRC Press Web site at
http://www.crcpress.com

Contents

CONTENTS

Preface

The chapters of this volume represent the revised versions of the main papers given at the seventh Séminaire Européen de Statistique on "Statistics for Stochastic Differential Equations Models," held at La Manga del Mar Menor, Cartagena, Spain, May 7th–12th, 2007. The aim of the Séminaire Européen de Statistique is to provide talented young researchers with an opportunity to get quickly to the forefront of knowledge and research in areas of statistical science which are of major current interest. As a consequence, this volume is tutorial, following the tradition of the books based on the previous seminars in the series entitled:

- *Networks and Chaos – Statistical and Probabilistic Aspects*
- *Time Series Models in Econometrics, Finance and Other Fields*
- *Stochastic Geometry: Likelihood and Computation*
- *Complex Stochastic Systems*
- *Extreme Values in Finance, Telecommunications and the Environment*
- *Statistics of Spatio-Temporal Systems*

About 40 young scientists from 15 different nationalities mainly from European countries participated. More than half presented their recent work in short communications; an additional poster session was organized, all contributions being of high quality.

The importance of stochastic differential equations as the modeling basis for phenomena ranging from finance to neurosciences has increased dramatically in recent years. Effective and well behaved statistical methods for these models are therefore of great interest. However, the mathematical complexity of the involved objects raises theoretical but also computational challenges. The Séminaire and the present book present recent developments that address, on one hand, properties of the statistical structure of the corresponding models and, on the other hand, relevant implementation issues, thus providing a valuable and updated overview of the field.

The first chapter of the book, written by Michael Sørensen, describes the application of estimating functions to diffusion-type models. Estimating functions are a comparatively recent tool to estimate parameters of discretely observed stochastic processes. They generalize the method of maximum likelihood estimation by searching for the roots of a so-called estimating equation, but have the advantage that these equations can typically be calculated and solved more easily than the likelihood equations, which often require extensive calculations. The idea is to approximate the likelihood equations, and in certain situations, estimating functions provide fully efficient estimators. Maximum likelihood estimation is discussed as a particular case.

The second chapter, written by Per Mykland and Lan Zhang addresses the modeling of high-frequency data of financial prices. The considered model is assumed to be a semimartingale plus an additive error, the so-called microstructure noise. This noise causes difficulty in estimation, since its impact on the estimators may be higher than that of the relevant model parameters. An approach is presented to overcome these difficulties, using multiscale realized volatility. It is shown how statistical data can be used in a sensible way for the trading of options, hence combining the probabilistic part of mathematical finance with statistical issues that may arise by misspecification of the model or other errors.

In Chapter 3, Jean Jacod treats inference for general jump diffusion processes based on high-frequency data. This means that one observes a stochastic process at equidistant time points between time 0 and time T, where the corresponding interval between two consecutive observation times is small and in the limit tends to zero. Such models have many applications, in particular in finance, where one is interested in estimating the integrated volatility. A number of estimation techniques for such general processes are presented, mainly based on variants of the quadratic variation, and the corresponding limit theorems are explained. This allows in particular the development of tests to distinguish whether processes have jumps or not.

Chapter 4, written by Omiros Papaspiliopoulos and Gareth Roberts, focuses on computational methods for the implementation of likelihood-based inference procedures for diffusion models. After a detailed overview on various simulation techniques for diffusions, the exact simulation method is presented with particular emphasis on the simulation of conditioned diffusions. Rather than using an Euler approximation scheme, these methods simulate the path of a (conditioned) diffusion exactly, without any discretization error. The exact simulation method can then be combined with Monte Carlo techniques to compute efficiently maximum likelihood and Bayesian estimators for diffusions.

Chapter 5, written by Fabienne Comte, Valentine Genon-Catalot and Yves Rozenholc, gives insight on non-parametric methods for stochastic differential equations models, several methods being presented and the corresponding convergence rates investigated. Several examples are used to illustrate the behavior of the suggested procedures.

In Chapter 6, Peter Brockwell and Alexander Lindner discuss some recent stochastic volatility models where the driving process is a Lévy process with jumps. After presenting the motivations for such models and their properties, estimation methods are described.

Finally, in Chapter 7, written by Grigorios Pavliotis, Yvo Pokern, and Andrew Stuart, the modeling of the multiscale characteristic that may be present in data is addressed and the procedures that can be used to find a useful diffusion approximation are described. Some examples from physics and molecular dynamics are presented.

The Séminaire Européen de Statistique is an activity of the European Regional Committee of the Bernoulli Society for Mathematical Statistics and Probability. The scientific organization for the seventh Séminaire Européen was in the hands of Ole E. Barndorff-Nielsen, Aarhus University; Bärbel Finkenstädt, University of Warwick; Leonhard Held, University of Zürich; Ingrid van Keilegom, Université Catholique de Louvain; Alexander Lindner, Technical Braunschweig University; Enno Mammen, University of Mannheim; Gesine Reinert, University of Oxford; Michael Sørensen, University of Copenhagen and Aad van der Vaart, University of Amsterdam. The seventh Séminaire was part of the second series of the European Mathematical Society Summer Schools and Conferences in pure and applied mathematics and, as such, was funded by the European Commission as a Marie Curie Conference and Training Course under EU Contract MSCF-CT-2005-029473. The organization committee is grateful to the Technical University of Cartagena and, in particular, to its Department of Applied Mathematics and Statistics for providing support to the Conference.

Contributors

Peter J. Brockwell
Department of Statistics
Colorado State University
Fort Collins, Colorado, USA

Fabienne Comte
UFR de Mathématiques et Informatique
Université Paris Descartes – Paris 5
Paris, France

Valentine Genon-Catalot
UFR de Mathématiques et Informatique
Université Paris Descartes – Paris 5
Paris, France

Jean Jacod
Institut de mathématiques de Jussieu
Université Pierre et Marie Curie (Paris-6) and CNRS, UMR 7586
Paris, France

Mathieu Kessler
Departamento de Matemática Aplicada y Estadistica
Universidad Politécnica de Cartagena
Cartagena, Spain

Alexander Lindner
Institut für Mathematische Stochastik
Technische Universität Braunschweig
Braunschweig, Germany

Per A. Mykland
Department of Statistics
University of Chicago
Chicago, Illinois, USA

Omiros Papaspiliopoulos
Department of Economics
Universitat Pompeu Fabra
Barcelona, Spain

Grigorios A. Pavliotis
Department of Mathematics
Imperial College London
South Kensington Campus, London, UK

Yvo Pokern
Department of Statistical Science
University College London
London, UK

Gareth Roberts
Department of Statistics
University of Warwick
Coventry, UK

Yves Rozenholc
UFR de Mathématiques et Informatique
Université Paris Descartes – Paris 5
Paris, France

Michael Sørensen
Department of Mathematical Sciences
University of Copenhagen
Copenhagen, Denmark

Andrew M. Stuart
Mathematics Department
University of Warwick
Coventry, UK

Lan Zhang
Department of Finance
University of Illinois at Chicago
Chicago, Illinois, USA

CHAPTER 1

Estimating functions for diffusion-type processes

Michael Sørensen

Department of Mathematical Sciences, University of Copenhagen
Universitetsparken 5, DK-2100 Copenhagen Ø, Denmark

1.1 Introduction

In this chapter we consider parametric inference based on discrete time observations $X_0, X_{t_1}, \ldots, X_{t_n}$ from a d-dimensional stochastic process. In most of the chapter the statistical model for the data will be a diffusion model given by a stochastic differential equation. We shall, however, also consider some examples of non-Markovian models, where we typically assume that the data are partial observations of a multivariate stochastic differential equation. We assume that the statistical model is indexed by a p-dimensional parameter θ.

The focus will be on estimating functions. An *estimating function* is a p-dimensional function of the parameter θ and the data:

$$G_n(\theta; X_0, X_{t_1}, \ldots, X_{t_n}).$$

Usually we suppress the dependence on the observations in the notation and write $G_n(\theta)$. We obtain an estimator by solving the equation

$$G_n(\theta) = 0. \tag{1.1}$$

Estimating functions provide a general framework for finding estimators and studying their properties in many different kinds of statistical models. The estimating function approach has turned out to be very useful for discretely sampled parametric diffusion-type models, where the likelihood function is usually not explicitly known. Estimating functions are typically constructed by combining relationships (dependent on the unknown parameter) between an obser-

1

vation and one or more of the previous observations that are informative about the parameters.

As an example, suppose the statistical model for the data $X_0, X_\Delta, X_{2\Delta}, \dots,$ $X_{n\Delta}$ is the one-dimensional stochastic differential equation

$$dX_t = -\theta \tan(X_t)dt + dW_t,$$

where $\theta > 0$ and W is a Wiener process. The state-space is $(-\pi/2, \pi/2)$. This model will be considered in more detail in Subsection 1.3.6. For this process Kessler and Sørensen (1999) proposed the estimating function

$$G_n(\theta) = \sum_{i=1}^{n} \sin(X_{(i-1)\Delta}) \left[\sin(X_{i\Delta}) - e^{-(\theta+\frac{1}{2})\Delta} \sin(X_{(i-1)\Delta}) \right],$$

which can be shown to be a martingale, when θ is the true parameter. For such martingale estimating functions, asymptotic properties of the estimators as the number of observations tends to infinity can be studied by means of martingale limit theory, see Subsection 1.3.1. An explicit estimator $\hat{\theta}_n$ of the parameter θ is obtained by solving the estimating equation (1.1):

$$\hat{\theta}_n = \Delta^{-1} \log \left(\frac{\sum_{i=1}^{n} \sin(X_{(i-1)\Delta}) \sin(X_{i\Delta})}{\sum_{i=1}^{n} \sin(X_{(i-1)\Delta})^2} \right) - \frac{1}{2},$$

provided that

$$\sum_{i=1}^{n} \sin(X_{(i-1)\Delta}) \sin(X_{i\Delta}) > 0. \tag{1.2}$$

If this condition is not satisfied, the estimating equation (1.1) has no solution, but fortunately it can be shown that the probability that (1.2) holds tends to one as n tends to infinity. As illustrated by this example, it is quite possible that the estimating equation (1.1) has no solution. We shall give general conditions that ensure the existence of a unique solution as the number of observations tend to infinity.

The idea of using estimating equations is an old one and goes back at least to Karl Pearson's introduction of the method of moments. The term estimating function may have been coined by Kimball (1946). In the econometric literature, the method was introduced by Hansen (1982, 1985), who generalized the theory to a broad stochastic process setting. In this literature it is known as the generalized method of moments (GMM). A discussion of links between the econometrics and statistics literature can be found in Hansen (2001).

A general asymptotic theory for estimating functions is presented in Section 1.10, and Section 1.11 reviews the theory of optimal estimating functions. Given a collection of relations between observations at different time points that can be used for estimation, this theory clarifies how to combine the relations in an optimal way, i.e. in such a way that the most efficient estimator is obtained. In Section 1.2 we present conditions ensuring that estimators from estimating functions are consistent and asymptotically normal under the so-called low-frequency asymptotics, which is the same as usual large sample asymptotics. In Section 1.3 we present martingale estimating functions for diffusion models including asymptotics and two optimality criteria. One of these criteria, small Δ-optimality, is particular of diffusion models. Likelihood inference is included as a particular case. We also discuss how to implement martingale estimating functions. There are several methods available for calculating approximations to the likelihood function. These are briefly reviewed in Section 1.4, where a particular expansion approach is presented in detail. Non-martingale estimating functions are considered in Section 1.5. An important aspect of the statistical theory of diffusion processes is that a number of alternative asymptotic scenarios particular to diffusions are available to supplement the traditional large sample asymptotics. High frequency asymptotics, high frequency asymptotics in a fixed time-interval, and small-diffusion asymptotics are presented in the Sections 1.6, 1.7, and 1.8. A number of non-Markovian models are considered in Section 1.9, including observations with measurement errors, integrated diffusions, sums of diffusions, stochastic volatility models, and compartment models. A general tool for these models are prediction-based estimating functions, which generalize the martingale estimating functions and share some of their convenient features.

1.2 Low-frequency asymptotics

In this section, we assume that observations have been made at the equidistant time points $i\Delta$, $i = 1, \ldots, n$, and consider the classical large sample asymptotic scenario, where the time between observations, Δ, is fixed, and the number of observations, n, goes to infinity. Since Δ is fixed, we will generally suppress Δ in the notation in this section. We assume that the statistical model is indexed by a p-dimensional parameter $\theta \in \Theta$, which we want to estimate. The corresponding probability measures are denoted by P_θ. The distribution of the data is given by the true probability measure, which we denote by P.

Under the true probability measure, P, it is assumed that $\{X_{i\Delta}\}$ is a stationary process with state space $D \subseteq \mathbb{R}^d$. We study the asymptotic properties of an estimator, $\hat{\theta}_n$, obtained by solving the estimating equation (1.1) when G_n is an estimating function of the general form

$$G_n(\theta) = \frac{1}{n} \sum_{i=r}^{n} g(X_{(i-r+1)\Delta}, \ldots, X_{i\Delta}; \theta), \qquad (1.3)$$

where r is a fixed integer smaller than n, and g is a suitable function with values in $\mathbb{R}^p$. All estimators discussed in this chapter can be represented in this way. We shall present several useful examples of how g can be chosen in the subsequent sections. *A priori* there is no guarantee that a unique solution to (1.1) exists, but conditions ensuring this for large sample sizes are given below. By a G_n–estimator, we mean an estimator, $\hat{\theta}_n$, which solves (1.1) when the data belongs to a subset $A_n \subseteq D^n$, and is otherwise given a value $\delta \notin \Theta$. We give results ensuring that, as $n \to \infty$, the probability of A_n tends to one.

Let Q denote the joint distribution of $(X_\Delta, \ldots, X_{r\Delta})$, and $Q(f)$ the expectation of $f(X_\Delta, \ldots, X_{r\Delta})$ for a function $f : D^r \mapsto \mathbb{R}$. To obtain our asymptotic results about G_n–estimators, we need to assume that a law of large numbers (an ergodic theorem) as well as a central limit theorem hold. Specifically, we assume that as $n \to \infty$

$$\frac{1}{n} \sum_{i=r}^{n} f(X_{(i-r+1)\Delta}, \ldots, X_{i\Delta}) \xrightarrow{P} Q(f) \tag{1.4}$$

for any function $f : D^r \mapsto \mathbb{R}$ such that $Q(|f|) < \infty$, and that the estimating function (1.3) satisfies

$$\frac{1}{\sqrt{n}} \sum_{i=r}^{n} g(X_{(i-r+1)\Delta}, \ldots, X_{i\Delta}; \theta) \xrightarrow{D} N(0, V(\theta)) \tag{1.5}$$

under P for any $\theta \in \Theta$ for which $Q(g(\theta)) = 0$. Here $V(\theta)$ is a positive definite $p \times p$-matrix. Moreover, $g(\theta)$ denotes the function $(x_1, \ldots, x_r) \mapsto g(x_1, \ldots, x_r; \theta)$, convergence in probability under P is indicated by $\xrightarrow{P}$, and $\xrightarrow{D}$ denotes convergence in distribution.

The following condition ensures the existence of a consistent G_n–estimator. We denote transposition of matrices by ^{T}, and $\partial_{\theta^T} G_n(\theta)$ denotes the $p \times p$-matrix, where the ijth entry is $\partial_{\theta_j} G_n(\theta)_i$.

Condition 1.1 *A parameter value $\bar{\theta} \in \text{int}\,\Theta$ and a neighbourhood N of $\bar{\theta}$ in Θ exist such that:*

(1) The function $g(\theta) : (x_1, \ldots, x_r) \mapsto g(x_1, \ldots, x_r; \theta)$ is integrable with respect to Q for all $\theta \in N$, and

$$Q\left(g(\bar{\theta})\right) = 0. \tag{1.6}$$

(2) The function $\theta \mapsto g(x_1, \ldots, x_r; \theta)$ is continuously differentiable on N for all $(x_1, \ldots, x_r) \in D^r$.

(3) The function $(x_1, \ldots, x_r) \mapsto \|\partial_{\theta^T} g(x_1, \ldots, x_r; \theta)\|$ is dominated for all $\theta \in N$ by a function which is integrable with respect to Q.

(4) The $p \times p$ matrix

$$W = Q\left(\partial_{\theta^T} g(\bar{\theta})\right) \tag{1.7}$$

is invertible.

Here and later $Q(g(\theta))$ denotes the vector $(Q(g_j(\theta)))_{j=1,...,p}$, where g_j is the jth coordinate of g, and $Q(\partial_{\theta^T} g(\theta))$ is the matrix $\{Q(\partial_{\theta_j} g_i(\theta))\}_{i,j=1,...,p}$.

To formulate the uniqueness result in the following theorem, we need the concept of locally dominated integrability. A function $f : D^r \times \Theta \mapsto \mathbb{R}^q$ is called *locally dominated integrable* with respect to Q if for each $\theta' \in \Theta$ there exists a neighbourhood $U_{\theta'}$ of θ' and a non-negative Q-integrable function $h_{\theta'} : D^r \mapsto \mathbb{R}$ such that $|f(x_1,\ldots,x_r;\theta)| \leq h_{\theta'}(x_1,\ldots,x_r)$ for all $(x_1,\ldots,x_r,\theta) \in D^r \times U_{\theta'}$.

Theorem 1.2 *Assume that Condition 1.1 and the central limit result (1.5) hold. Then a $\bar{\theta}$-consistent G_n–estimator, $\hat{\theta}_n$, exists, and*

$$\sqrt{n}(\hat{\theta}_n - \bar{\theta}) \xrightarrow{D} N_p\left(0, W^{-1}VW^{T-1}\right) \qquad (1.8)$$

under P, where $V = V(\bar{\theta})$. If, moreover, the function $g(x_1,\ldots,x_r;\theta)$ is locally dominated integrable with respect to Q and

$$Q(g(\theta)) \neq 0 \text{ for all } \theta \neq \bar{\theta}, \qquad (1.9)$$

then the estimator $\hat{\theta}_n$ is the unique G_n–estimator on any bounded subset of Θ containing $\bar{\theta}$ with probability approaching one as $n \to \infty$.

Remark: By a $\bar{\theta}$-consistent estimator is meant that $\hat{\theta}_n \xrightarrow{P} \bar{\theta}$ as $n \to \infty$. If the true model belongs to the statistical model, i.e. if $P = P_{\theta_0}$ for some $\theta_0 \in \Theta$, then the estimator $\hat{\theta}_n$ is most useful if Theorem 1.2 holds with $\bar{\theta} = \theta_0$. Note that because $\bar{\theta} \in \text{int } \Theta$, a $\bar{\theta}$-consistent G_n–estimator $\hat{\theta}_n$ will satisfy $G_n(\hat{\theta}_n) = 0$ with probability approaching one as $n \to \infty$.

In order to prove Theorem 1.2, we need the following uniform law of large numbers.

Lemma 1.3 *Consider a function $f : D^r \times K \mapsto \mathbb{R}^q$, where K is a compact subset of Θ. Suppose f is a continuous function of θ for all $(x_1,\ldots,x_r) \in D^r$, and that there exists a Q-integrable function $h : D^r \mapsto \mathbb{R}$ such that $\|f(x_1,\ldots,x_r;\theta)\| \leq h(x_1,\ldots,x_r)$ for all $\theta \in K$. Then the function $\theta \mapsto Q(f(\theta))$ is continuous, and*

$$\sup_{\theta \in K} \| \frac{1}{n} \sum_{i=r}^{n} f(X_{(i-r+1)\Delta},\ldots,X_{i\Delta};\theta) - Q(f(\theta)) \| \xrightarrow{P} 0. \qquad (1.10)$$

Proof: That $Q(f(\theta))$ is continuous follows from the dominated convergence theorem. To prove (1.10), define for $\eta > 0$:

$$k(\eta; x_1,\ldots,x_r) = \sup_{\theta,\theta' \in K: \|\theta'-\theta\| \leq \eta} \|f(x_1,\ldots,x_r;\theta') - f(x_1,\ldots,x_r;\theta)\|,$$

and let $k(\eta)$ denote the function $(x_1, \ldots, x_r) \mapsto k(\eta; x_1, \ldots, x_r)$. Since $k(\eta)$ $\leq 2h$, it follows from the dominated convergence theorem that $Q(k(\eta)) \to 0$ as $\eta \to 0$. Moreover, $Q(f(\theta))$ is uniformly continuous on the compact set K. Hence for any given $\epsilon > 0$, we can find $\eta > 0$ such that $Q(k(\eta)) \leq \epsilon$ and such that $\|\theta - \theta'\| < \eta$ implies that $\|Q(f(\theta)) - Q(f(\theta'))\| \leq \epsilon$ for $\theta, \theta' \in K$. Define the balls $B_\eta(\theta) = \{\theta' : \|\theta - \theta'\| < \eta\}$. Since K is compact, there exists a finite covering

$$K \subseteq \bigcup_{j=1}^{m} B_\eta(\theta_j),$$

where $\theta_1, \ldots, \theta_m \in K$, so for every $\theta \in K$ we can find $\theta_\ell, \ell \in \{1, \ldots, m\}$, such that $\theta \in B_\eta(\theta_\ell)$. Thus with

$$F_n(\theta) = \frac{1}{n} \sum_{i=r}^{n} f(X_{(i-r+1)\Delta}, \ldots, X_{i\Delta}; \theta)$$

we have

$\|F_n(\theta) - Q(f(\theta))\|$

$\leq \quad \|F_n(\theta) - F_n(\theta_\ell)\| + \|F_n(\theta_\ell) - Q(f(\theta_\ell))\| + \|Q(f(\theta_\ell)) - Q(f(\theta))\|$

$\leq \quad \dfrac{1}{n} \sum_{\nu=r}^{n} k(\eta; X_{(\nu-r+1)\Delta}, \ldots, X_{\nu\Delta}) + \|F_n(\theta_\ell) - Q(f(\theta_\ell))\| + \epsilon$

$\leq \quad \left| \dfrac{1}{n} \sum_{\nu=r}^{n} k(\eta; X_{(\nu-r+1)\Delta}, \ldots, X_{\nu\Delta}) - Q(k(\eta)) \right|$

$$+ Q(k(\eta)) + \|F_n(\theta_\ell) - Q(f(\theta_\ell))\| + \epsilon$$

$\leq \quad Z_n + 2\epsilon,$

where

$$Z_n = \left| \frac{1}{n} \sum_{\nu=r}^{n} k(\eta; X_{(\nu-r+1)\Delta}, \ldots, X_{\nu\Delta}) - Q(k(\eta)) \right|$$

$$+ \max_{1 \leq \ell \leq m} \|F_n(\theta_\ell) - Q(f(\theta_\ell))\|.$$

By (1.4), $P(Z_n > \epsilon) \to 0$ as $n \to \infty$, so

$$P\left(\sup_{\theta \in K} \|F_n(\theta) - Q(f(\theta))\| > 3\epsilon \right) \to 0$$

for all $\epsilon > 0$. $\qquad\qquad\qquad\qquad\qquad\qquad\qquad\qquad\qquad\qquad\qquad\qquad\square$

Proof (of Theorem 1.2): The existence of a $\bar{\theta}$-consistent G_n–estimator $\hat{\theta}_n$ follows from Theorem 1.58. Condition (i) follows from (1.4) and (1.6). Define the function $W(\theta) = Q\left(\partial_{\theta^T} g(\theta)\right)$. Then condition (iii) in Theorem 1.58 is equal to Condition 1.1 (4). Finally, let M be a compact subset of N containing $\bar{\theta}$. Then the conditions of Lemma 1.3 are satisfied for $f = \partial_{\theta^T} g$, so (1.159) is satisfied. The asymptotic normality, (1.20), follows from Theorem 1.60 and (1.5).

In order to prove the last statement, let K be a compact subset of Θ containing $\bar{\theta}$. By the finite covering property of a compact set, it follows from the local dominated integrability of g that g satisfies the conditions of Lemma 1.3. Hence (1.160) holds with $G(\theta) = Q(g(\theta))$ and $M = K$. From the local dominated integrability of g and the dominated convergence theorem it follows that $G(\theta)$ is a continuous function, so (1.9) implies that

$$\inf_{K \setminus \bar{B}_\epsilon(\bar{\theta})} |G(\theta)| > 0,$$

for all $\epsilon > 0$, where $\bar{B}_\epsilon(\bar{\theta})$ is the closed ball with radius ϵ centered at $\bar{\theta}$. By Theorem 1.59 it follows that (1.162) holds with $M = K$ for every $\epsilon > 0$. Let $\hat{\theta}'_n$ be a G_n–estimator, and define a G_n–estimator by $\hat{\theta}''_n = \hat{\theta}'_n 1\{\hat{\theta}'_n \in K\} + \hat{\theta}_n 1\{\hat{\theta}'_n \notin K\}$, where 1 denotes an indicator function, and $\hat{\theta}_n$ is the consistent G_n–estimator we know exists. By (1.162) the estimator $\hat{\theta}''_n$ is consistent, so by Theorem 1.58, $P(\hat{\theta}_n \neq \hat{\theta}''_n) \to 0$ as $n \to \infty$. Hence $\hat{\theta}_n$ is eventually the unique G_n–estimator on K.

$\square$

1.3 Martingale estimating functions

In this section we consider observations $X_0, X_{t_1}, \ldots, X_{t_n}$ of a d-dimensional diffusion process given by the stochastic differential equation

$$dX_t = b(X_t; \theta)dt + \sigma(X_t; \theta)dW_t, \qquad (1.11)$$

where σ is a $d \times d$-matrix and W a d-dimensional standard Wiener process. Here and elsewhere in this chapter, the observation times t_i are non-random. We denote the state space of X by D. When $d = 1$, the state space is an interval (ℓ, r), where ℓ could possibly be $-\infty$, and r might be ∞. The drift b and the diffusion matrix σ depend on a parameter θ which varies in a subset Θ of $\mathbb{R}^p$. The equation (1.11) is assumed to have a weak solution, and the coefficients b and σ are assumed to be smooth enough to ensure, for every $\theta \in \Theta$, the uniqueness of the law of the solution, which we denote by P_θ. We denote the true parameter value by θ_0.

We suppose that the transition distribution has a density $y \mapsto p(\Delta, x, y; \theta)$ with respect to the Lebesgue measure on D, and that $p(\Delta, x, y; \theta) > 0$ for all

$y \in D$. The transition density is the conditional density under P_θ of $X_{t+\Delta}$ given that $X_t = x$.

We shall, in this section, be concerned with statistical inference based on estimating functions of the form

$$G_n(\theta) = \sum_{i=1}^{n} g(\Delta_i, X_{t_{i-1}}, X_{t_i}; \theta). \tag{1.12}$$

where g is a p-dimensional function which satisfies that

$$\int_D g(\Delta, x, y; \theta) p(\Delta, x, y; \theta) dy = 0 \tag{1.13}$$

for all $\Delta > 0$, $x \in D$ and $\theta \in \Theta$. Thus, by the Markov property, the stochastic process $\{G_n(\theta)\}_{n \in \mathbb{N}}$ is a martingale with respect to $\{\mathcal{F}_n\}_{n \in \mathbb{N}}$ under P_θ. Here and later $\mathcal{F}_n = \sigma(X_{t_i} : i \leq n)$. An estimating function with this property is called a *martingale estimating function*.

1.3.1 Asymptotics

In this subsection we give asymptotic results for estimators obtained from martingale estimating functions as the number of observations goes to infinity. To simplify the exposition, the observation time points are assumed to be equidistant, i.e., $t_i = i\Delta$, $i = 0, 1, \ldots, n$. Since Δ is fixed, we will in most cases suppress Δ in the notation and write for example $p(x, y; \theta)$ and $g(x, y; \theta)$.

It is assumed that the diffusion given by (1.11) is ergodic, that its invariant probability measure has density function μ_θ for all $\theta \in \Theta$, and that $X_0 \sim \mu_\theta$ under P_θ. Thus the diffusion is stationary.

When the observed process, X, is a one-dimensional diffusion, the following simple conditions ensure *ergodicity*, and an explicit expression exists for the density of the invariant probability measure. The *scale measure* of X has Lebesgue density

$$s(x; \theta) = \exp\left(-2 \int_{x^\#}^{x} \frac{b(y; \theta)}{\sigma^2(y; \theta)} dy\right), \quad x \in (\ell, r), \tag{1.14}$$

where $x^\# \in (\ell, r)$ is arbitrary.

Condition 1.4 *The following holds for all $\theta \in \Theta$:*

$$\int_{x^\#}^{r} s(x; \theta) dx = \int_{\ell}^{x^\#} s(x; \theta) dx = \infty$$

and

$$\int_{\ell}^{r} [s(x; \theta) \sigma^2(x; \theta)]^{-1} dx = A(\theta) < \infty.$$

Under Condition 1.4 the process X is ergodic with an invariant probability measure with Lebesgue density

$$\mu_\theta(x) = [A(\theta)s(x;\theta)\sigma^2(x;\theta)]^{-1}, \quad x \in (\ell, r); \quad (1.15)$$

for details see e.g. Skorokhod (1989). For general one-dimensional diffusions, the measure with Lebesgue density proportional to $[s(x;\theta)\sigma^2(x;\theta)]^{-1}$ is called the speed measure.

Let Q_θ denote the probability measure on D^2 given by

$$Q_\theta(dx, dy) = \mu_\theta(x)p(\Delta, x, y; \theta)dxdy. \quad (1.16)$$

This is the distribution of two consecutive observations $(X_{\Delta(i-1)}, X_{\Delta i})$. Under the assumption of ergodicity the law of large numbers (1.4) is satisfied for any function $f : D^2 \mapsto \mathbb{R}$ such that $Q(|f|) < \infty$, see e.g. Skorokhod (1989).

We impose the following condition on the function g in the estimating function (1.12)

$$Q_\theta\left(g(\theta)^T g(\theta)\right) = \quad (1.17)$$
$$\int_{D^2} g(y, x; \theta)^T g(y, x; \theta)\mu_\theta(x)p(x, y; \theta)dydx < \infty,$$

for all $\theta \in \Theta$. By (1.4),

$$\frac{1}{n}\sum_{i=1}^{n} g(X_{\Delta i}, X_{\Delta(i-1)}; \theta') \xrightarrow{P_\theta} Q_\theta(g(\theta')), \quad (1.18)$$

for all $\theta, \theta' \in \Theta$. Since the estimating function $G_n(\theta)$ is a square integrable martingale under P_θ, the asymptotic normality in (1.5) follows without further conditions from the central limit theorem for martingales, see P. Hall and Heyde (1980). This result for stationary processes goes back to Billingsley (1961). In the martingale case the asymptotic covariance matrix $V(\theta)$ in (1.5) is given by

$$V(\theta) = Q_{\theta_0}\left(g(\theta)g(\theta)^T\right). \quad (1.19)$$

Thus we have the following particular case of Theorem 1.2.

Theorem 1.5 *Assume Condition 1.1 is satisfied with $r = 2$, $\bar{\theta} = \theta_0$, and $Q = Q_{\theta_0}$, where θ_0 is the true parameter value, and that (1.17) holds for $\theta = \theta_0$. Then a θ_0-consistent G_n–estimator $\hat{\theta}_n$ exists, and*

$$\sqrt{n}(\hat{\theta}_n - \theta_0) \xrightarrow{D} N_p\left(0, W^{-1}VW^{T^{-1}}\right) \quad (1.20)$$

under P_{θ_0}, where W is given by (1.7) with $\bar{\theta} = \theta_0$, and $V = V(\theta_0)$ with $V(\theta)$ given by (1.19). If, moreover, the function $g(x, y; \theta)$ is locally dominated integrable with respect to Q_{θ_0} and

$$Q_{\theta_0}(g(\theta)) \neq 0 \text{ for all } \theta \neq \theta_0, \quad (1.21)$$

then the estimator $\hat{\theta}_n$ is the unique G_n–estimator on any bounded subset of Θ containing θ_0 with probability approaching one as $n \to \infty$.

In practice we do not know the value of θ_0, so it is necessary to check that the conditions of Theorem 1.5 hold for any value of $\theta_0 \in \text{int } \Theta$.

The asymptotic covariance matrix of the estimator $\hat{\theta}_n$ can be estimated consistently by means of the following theorem.

Theorem 1.6   *Under Condition 1.1 (2) – (4) (with $r = 2$, $\bar{\theta} = \theta_0$, and $Q = Q_{\theta_0}$),*

$$W_n = \frac{1}{n} \sum_{i=1}^{n} \partial_{\theta^T} g(X_{(i-1)\Delta}, X_{i\Delta}; \hat{\theta}_n) \xrightarrow{P_{\theta_0}} W, \qquad (1.22)$$

where $\hat{\theta}_n$ is a θ_0-consistent estimator. The probability that W_n is invertible approaches one as $n \to \infty$. If, moreover, the function $(x, y) \mapsto \|g(x, y; \theta)\|$ is dominated for all $\theta \in N$ by a function which is square integrable with respect to Q_{θ_0}, then

$$V_n = \frac{1}{n} \sum_{i=1}^{n} g(X_{(i-1)\Delta}, X_{i\Delta}; \hat{\theta}_n) g(X_{(i-1)\Delta}, X_{i\Delta}; \hat{\theta}_n)^T \xrightarrow{P_{\theta_0}} V. \quad (1.23)$$

Proof: Let C be a compact subset of N such that $\theta_0 \in \text{int } C$. By Lemma 1.3, $\frac{1}{n} \sum_{i=1}^{n} \partial_{\theta^T} g(X_{(i-1)\Delta}, X_{i\Delta}; \theta)$ converges to $Q_{\theta_0}(\partial_{\theta^T} g(\theta))$ in probability uniformly for $\theta \in C$. This implies (1.22) because $\hat{\theta}_n$ converges in probability to θ_0. The result about invertibility follows because W is invertible. Also the uniform convergence in probability for $\theta \in C$ of $\frac{1}{n} \sum_{i=1}^{n} g(X_{(i-1)\Delta}, X_{i\Delta}; \theta) g(X_{(i-1)\Delta}, X_{i\Delta}; \theta)^T$ to $Q_{\theta_0}(g(\theta) g(\theta)^T)$ follows from Lemma 1.3.

□

1.3.2 Likelihood inference

The diffusion process X is a Markov process, so the likelihood function based on the observations $X_0, X_{t_1}, \cdots, X_{t_n}$, conditional on X_0, is

$$L_n(\theta) = \prod_{i=1}^{n} p(t_i - t_{i-1}, X_{t_{i-1}}, X_{t_i}; \theta), \qquad (1.24)$$

where $y \mapsto p(s, x, y; \theta)$ is the transition density and $t_0 = 0$. Under weak regularity conditions the maximum likelihood estimator is efficient, i.e. it has the smallest asymptotic variance among all estimators. The transition density is only rarely explicitly known, but several numerical approaches and accurate approximations make likelihood inference feasible for diffusion models. We shall return to the problem of calculating the likelihood function in Subsection 1.4.

The vector of partial derivatives of the log-likelihood function with respect to the coordinates of θ,

$$U_n(\theta) = \partial_\theta \log L_n(\theta) = \sum_{i=1}^{n} \partial_\theta \log p(\Delta_i, X_{t_{i-1}}, X_{t_i}; \theta), \qquad (1.25)$$

where $\Delta_i = t_i - t_{i-1}$, is called the *score function* (or score vector). Here it is obviously assumed that the transition density is a differentiable function of θ. The maximum likelihood estimator usually solves the estimating equation $U_n(\theta) = 0$. The score function is a martingale with respect to $\{\mathcal{F}_n\}_{n \in \mathbb{N}}$ under P_θ, which is easily seen provided that the following interchange of differentiation and integration is allowed:

$$E_\theta \left(\partial_\theta \log p(\Delta_i, X_{t_{i-1}}, X_{t_i}; \theta) \mid X_{t_1}, \dots, X_{t_{i-1}} \right)$$

$$= \int_D \frac{\partial_\theta p(\Delta_i, X_{t_{i-1}}, y; \theta)}{p(\Delta_i, X_{t_{i-1}}, y; \theta)} p(\Delta_i, X_{t_{i-1}}, y, \theta) dy$$

$$= \partial_\theta \int_D p(\Delta_i, X_{t_{i-1}}, y; \theta) dy = 0.$$

Since the score function is a martingale estimating function, the asymptotic results in the previous subsection applies to the maximum likelihood estimator. Asymptotic results for the maximum likelihood estimator in the low-frequency (fixed Δ) asymptotic scenario considered in that subsection were established by Dacunha-Castelle and Florens-Zmirou (1986). Asymptotic results when the observations are made at random time points were obtained by Aït–Sahalia and Mykland (2003).

In the case of likelihood inference, the function $Q_{\theta_0}(g(\theta))$ appearing in the identifiability condition (1.21) is related to the Kullback-Leibler divergence between the models. Specifically, if the following interchange of differentiation and integration is allowed,

$$Q_{\theta_0}(\partial_\theta \log p(x, y, \theta)) = \partial_\theta Q_{\theta_0}(\log p(x, y, \theta)) = -\partial_\theta \bar{K}(\theta, \theta_0),$$

where $\bar{K}(\theta, \theta_0)$ is the average Kullback-Leibler divergence between the transition distributions under P_{θ_0} and P_θ given by

$$\bar{K}(\theta, \theta_0) = \int_D K(\theta, \theta_0; x) \, \mu_{\theta_0}(dx),$$

with

$$K(\theta, \theta_0; x) = \int_D \log[p(x, y; \theta_0)/p(x, y; \theta)] p(x, y; \theta_0) \, dy.$$

Thus the identifiability condition can be written in the form $\partial_\theta \bar{K}(\theta, \theta_0) \neq 0$

for all $\theta \neq \theta_0$. The quantity $\bar{K}(\theta, \theta_0)$ is sometimes referred to as the Kullback-Leibler divergence between the two Markov chain models for the observed process $\{X_{i\Delta}\}$ under P_{θ_0} and P_{θ}.

A simple approximation to the likelihood function is obtained by approximating the transition density by a Gaussian density with the correct first and second conditional moments. This approximation is conventionally called a *quasi-likelihood*. For a one-dimensional diffusion we get

$$p(\Delta, x, y; \theta) \approx q(\Delta, x, y; \theta) = \frac{1}{\sqrt{2\pi\phi(\Delta, x; \theta)}} \exp\left[-\frac{(y - F(\Delta, x; \theta))^2}{2\phi(\Delta, x; \theta)} \right]$$

where

$$F(\Delta, x; \theta) = \mathrm{E}_\theta(X_\Delta | X_0 = x) = \int_\ell^r yp(\Delta, x, y; \theta)dy. \qquad (1.26)$$

and

$$\phi(\Delta, x; \theta) = \qquad (1.27)$$
$$\mathrm{Var}_\theta(X_\Delta | X_0 = x) = \int_\ell^r [y - F(\Delta, x; \theta)]^2 p(\Delta, x, y; \theta)dy.$$

In this way we obtain the quasi-likelihood

$$L_n(\theta) \approx Q_n(\theta) = \prod_{i=1}^n q(\Delta_i, X_{t_{i-1}}, X_{t_i}; \theta),$$

and by differentiation with respect to the parameter vector, we obtain the quasi-score function

$$\partial_\theta \log Q_n(\theta) = \sum_{i=1}^n \left\{ \frac{\partial_\theta F(\Delta_i, X_{t_{i-1}}; \theta)}{\phi(\Delta_i, X_{t_{i-1}}; \theta)} [X_{t_i} - F(\Delta_i, X_{t_{i-1}}; \theta)] \qquad (1.28) \right.$$
$$\left. + \frac{\partial_\theta \phi(\Delta_i, X_{t_{i-1}}; \theta)}{2\phi(\Delta_i, X_{t_{i-1}}; \theta)^2} \left[(X_{t_i} - F(\Delta_i, X_{t_{i-1}}; \theta))^2 - \phi(\Delta_i, X_{t_{i-1}}; \theta) \right] \right\}.$$

It is clear from (1.26) and (1.27) that $\{\partial_\theta \log Q_n(\theta)\}_{n \in \mathbb{N}}$ is a martingale with respect to $\{\mathcal{F}_n\}_{n \in \mathbb{N}}$ under P_θ. This quasi-score function is a particular case of the quadratic martingale estimating functions considered by Bibby and Sørensen (1995, 1996). Maximum quasi-likelihood estimation for diffusions was considered by Bollerslev and Wooldridge (1992).

1.3.3 Godambe–Heyde optimality

In this section we present a general way of approximating the score function by means of martingales of a similar form. Suppose we have a collection of real valued functions $h_j(x, y, ; \theta)$, $j = 1, \ldots, N$, satisfying

$$\int_D h_j(x, y; \theta) p(x, y; \theta) dy = 0 \qquad (1.29)$$

for all $x \in D$ and $\theta \in \Theta$. Each of the functions h_j could be used separately to define an estimating function of the form (1.3) with $g = h_j$, but a better approximation to the score function, and hence a more efficient estimator, is obtained by combining them in an optimal way. Therefore we consider estimating functions of the form

$$G_n(\theta) = \sum_{i=1}^{n} a(X_{(i-1)\Delta}, \theta) h(X_{(i-1)\Delta}, X_{i\Delta}; \theta), \qquad (1.30)$$

where $h = (h_1, \ldots, h_N)^T$, and the $p \times N$ weight matrix $a(x, \theta)$ is a function of x such that (1.30) is P_θ-integrable. It follows from (1.29) that $G_n(\theta)$ is a martingale estimating function, i.e., it is a martingale under P_θ for all $\theta \in \Theta$.

The matrix a determines how much weight is given to each of the h_js in the estimation procedure. This weight matrix can be chosen in an optimal way using Godambe and Heyde's theory of optimal estimating functions, which is reviewed in Section 1.11. The *optimal weight matrix*, a^*, gives the estimating function of the form (1.30) that provides the best possible approximation to the score function (1.25) in a mean square sense. Moreover, the optimal $g^*(x, y; \theta) = a^*(x; \theta) h(x, y; \theta)$ is obtained from $\partial_\theta \log p(x, y; \theta)$ by projection in a certain space of square integrable functions; for details see Section 1.11.

The choice of the functions h_j, on the other hand, is an art rather than a science. The ability to tailor these functions to a given model or to particular parameters of interest is a considerable strength of the estimating functions methodology. It is, however, also a source of weakness, since it is not always clear how best to choose the h_js. In the following and in the Subsections 1.3.6 and 1.3.7, we shall present ways of choosing these functions that usually work well in practice. Also the theory in Subsection 1.3.4 and Section 1.6 casts interesting light on this problem.

Example 1.7 The martingale estimating function (1.28) is of the type (1.30) with $N = 2$ and

$$\begin{aligned} h_1(x, y; \theta) &= y - F(\Delta, x; \theta), \\ h_2(x, y; \theta) &= (y - F(\Delta, x; \theta))^2 - \phi(\Delta, x, \theta), \end{aligned}$$

where F and ϕ are given by (1.26) and (1.27). The weight matrix is

$$\left(\frac{\partial_\theta F(\Delta, x; \theta)}{\phi(\Delta, x; \theta)}, \frac{\partial_\theta \phi(\Delta, x; \theta)}{2\phi^2(\Delta, x; \theta)\Delta} \right), \qquad (1.31)$$

which we shall see is an approximation to the optimal weight matrix. □

We will now find the optimal estimating function $G_n^*(\theta)$, i.e. the estimating functions of the form (1.30) with the optimal weight matrix. We assume that the functions h_j satisfy the following condition.

Condition 1.8

(1) The functions $y \mapsto h_j(x, y; \theta)$, $j = 1, \ldots N$, are linearly independent for all $x \in D$ and $\theta \in \Theta$.

(2) The functions $y \mapsto h_j(x, y; \theta)$, $j = 1, \ldots N$, are square integrable with respect to $p(x, y; \theta)$ for all $x \in D$ and $\theta \in \Theta$.

(3) $h(x, y; \theta)$ is differentiable with respect to θ.

(4) The functions $y \mapsto \partial_{\theta_i} h_j(x, y; \theta)$ are integrable with respect to $p(x, y; \theta)$ for all $x \in D$ and $\theta \in \Theta$.

The class of estimating functions considered here is a particular case of the class treated in detail in Example 1.64. By (1.183), the optimal choice of the weight matrix a is given by

$$a^*(x; \theta) = B_h(x; \theta) \, V_h(x; \theta)^{-1}, \tag{1.32}$$

where

$$B_h(x; \theta) = \int_D \partial_\theta h(x, y; \theta)^T p(x, y; \theta) dy \tag{1.33}$$

and

$$V_h(x; \theta) = \int_D h(x, y; \theta) h(x, y; \theta)^T p(x, y; \theta) dy. \tag{1.34}$$

The matrix $V_h(x; \theta)$ is invertible because the functions h_j, $j = 1, \ldots N$, are linearly independent. Compared to (1.183), we have omitted a minus here. This can be done because an optimal estimating function multiplied by an invertible $p \times p$-matrix is also an optimal estimating function and yields the same estimator.

The asymptotic variance of an optimal estimator, i.e. a G_n^*–estimator, is simpler than the general expression in (1.20) because in this case the matrices W and V given by (1.7) and (1.19) are identical and given by (1.35). This is a general property of optimal estimating functions as discussed in Section 1.11. The result can easily be verified under the assumption that $a^*(x; \theta)$ is a differentiable function of θ. By (1.29)

$$\int_D [\partial_{\theta_i} a^*(x; \theta)] \, h(x, y; \theta) p(x, y; \theta) dy = 0,$$

so that

$$\begin{aligned} W &= \int_{D^2} \partial_{\theta^T} [a^*(x; \theta_0) h(x, y; \theta_0)] Q_{\theta_0}(dx, dy) \\ &= \mu_{\theta_0} (a^*(\theta_0) B_h(\theta_0)^T) = \mu_{\theta_0} \left(B_h(\theta_0) V_h(\theta_0)^{-1} B_h(\theta_0)^T \right), \end{aligned}$$

and by direct calculation

$$
\begin{aligned}
V &= Q_{\theta_0}(a^*(\theta_0)h(\theta_0)h(\theta_0)^T a^*(\theta_0)^T) \\
&- \mu_{\theta_0}\left(B_h(\theta_0)V_h(\theta_0)^{-1}B_h(\theta_0)^T\right).
\end{aligned} \tag{1.35}
$$

Thus provided that $g^*(x, y, \theta) = a^*(x; \theta)h(x, y; \theta)$ satisfies the conditions of Theorem 1.5, we can, as a corollary to that theorem, conclude that the sequence $\hat{\theta}_n$ of consistent G_n^*-estimators has the asymptotic distribution

$$
\sqrt{n}(\hat{\theta}_n - \theta_0) \xrightarrow{\mathcal{D}} N_p\left(0, V^{-1}\right). \tag{1.36}
$$

In the econometrics literature, a popular way of using functions like $h_j(x, y, ; \theta)$, $j = 1, \ldots, N$, to estimate the parameter θ is the *generalized method of moments* (GMM) of Hansen (1982). In practice, the method is often implemented as follows, see e.g. Campbell, Lo, and MacKinlay (1997). Consider

$$
F_n(\theta) = \frac{1}{n}\sum_{i=1}^{n} h(X_{(i-1)\Delta}, X_{i\Delta}; \theta).
$$

Under weak conditions, cf. Theorem 1.6, a consistent estimator of the asymptotic covariance matrix M of $\sqrt{n}F_n(\theta_0)$ is

$$
M_n = \frac{1}{n}\sum_{i=1}^{n} h(X_{(i-1)\Delta}, X_{i\Delta}; \tilde{\theta}_n)h(X_{(i-1)\Delta}, X_{i\Delta}; \tilde{\theta}_n)^T,
$$

where $\tilde{\theta}_n$ is a θ_0-consistent estimator (for instance obtained by minimizing $F_n(\theta)^T F_n(\theta)$). The GMM-estimator is obtained by minimizing the function

$$
H_n(\theta) = F_n(\theta)^T M_n^{-1} F_n(\theta).
$$

The corresponding estimating function is obtained by differentiation with respect to θ

$$
\partial_\theta H_n(\theta) = D_n(\theta)M_n^{-1}F_n(\theta),
$$

where by (1.4)

$$
D_n(\theta) = \frac{1}{n}\sum_{i=1}^{n} \partial_\theta h(X_{(i-1)\Delta}, X_{i\Delta}; \theta)^T \xrightarrow{P_{\theta_0}} Q_{\theta_0}\left(\partial_\theta h(\theta)^T\right).
$$

Hence the estimating function $\partial_\theta H_n(\theta)$ is asymptotically equivalent to an estimating function of the form (1.30) with weight matrix

$$
a(x, \theta) = Q_{\theta_0}\left(\partial_\theta h(\theta)^T\right) M^{-1}.
$$

We see that, in this sense, the GMM-estimators described here are covered by the theory of optimal martingale estimating functions presented in this section. The weight matrix $Q_{\theta_0}\left(\partial_\theta h(\theta)^T\right) M^{-1}$ does not depend on the data, so this type of GMM-estimator will in general be less efficient than the estimator obtained from the optimal martingale estimating function, because the optimal weight matrix (1.32) typically does depend on the data. Note, however, that the widely used GMM-estimators described here are only a particular case of a very general type of estimators, and that the theory in Hansen (1985) covers the optimal martingale estimating functions.

Example 1.9 Consider the martingale estimating function of form (1.30) with $N = 2$ and with h_1 and h_2 as in Example 1.7, where the diffusion is one-dimensional. The optimal weight matrix has columns given by

$$a_1^*(x;\theta) = \frac{\partial_\theta \phi(x;\theta)\eta(x;\theta) - \partial_\theta F(x;\theta)\psi(x;\theta)}{\phi(x;\theta)\psi(x;\theta) - \eta(x;\theta)^2}$$

$$a_2^*(x;\theta) = \frac{\partial_\theta F(x;\theta)\eta(x;\theta) - \partial_\theta \phi(x;\theta)\phi(x;\theta)}{\phi(x;\theta)\psi(x;\theta) - \eta(x;\theta)^2},$$

where

$$\eta(x;\theta) = E_\theta([X_\Delta - F(x;\theta)]^3 | X_0 = x)$$

and

$$\psi(x;\theta) = E_\theta([X_\Delta - F(x;\theta)]^4 | X_0 = x) - \phi(x;\theta)^2.$$

For the square-root diffusion (the a.k.a. the CIR-model)

$$dX_t = -\beta(X_t - \alpha)dt + \tau\sqrt{X_t}dW_t, \quad X_0 > 0, \qquad (1.37)$$

where $\alpha, \beta, \tau > 0$, the optimal weights can be found explicitly. For this model

$$F(x;\theta) = xe^{-\beta\Delta} + \alpha(1 - e^{-\beta\Delta})$$

$$\phi(x;\theta) = \frac{\tau^2}{\beta}\left((\tfrac{1}{2}\alpha - x)e^{-2\beta\Delta} - (\alpha - x)e^{-\beta\Delta} + \tfrac{1}{2}\alpha\right)$$

$$\eta(x;\theta) = \frac{\tau^4}{2\beta^2}\left(\alpha - 3(\alpha - x)e^{-\beta\Delta} + 3(\alpha - 2x)e^{-2\beta\Delta}\right.$$

$$\left. - (\alpha - 3x)e^{-3\beta\Delta}\right)$$

$$\psi(x;\theta) = \frac{3\tau^6}{4\beta^3}\left((\alpha - 4x)e^{-4\beta\Delta} - 4(\alpha - 3x)e^{-3\beta\Delta}\right.$$

$$\left. + 6(\alpha - 2x)e^{-2\beta\Delta} - 4(\alpha - x)e^{-\beta\Delta} + \alpha\right) + 2\phi(x;\theta)^2.$$

We give a method to derive these expression in Subsection 1.3.6.

The expressions for a_1^* and a_2^* can for general diffusions be simplified by the approximations

$$\eta(t,x;\theta) \approx 0 \quad \text{and} \quad \psi(t,x;\theta) \approx 2\phi(t,x;\theta)^2, \qquad (1.38)$$

which would be exactly true if the transition density were a Gaussian density function. If we insert these Gaussian approximations into the expressions for a_1^* and a_2^*, we obtain the weight functions in (1.28). When Δ is not large this can be justified, because the transition distribution is not far from Gaussian. In

Section 1.4 we present a data transformation after which the transition distribution is close to a normal distribution. □

In Subsections 1.3.6 and 1.3.7 we present martingale estimating functions for which the matrices $B_h(x;\theta)$ and $V_h(x;\theta)$ can be found explicitly, but for most models these matrices must be found by simulation, a problem considered in Subsection 1.3.5. In situations where a^* must be determined by a relatively time consuming numerical method, it might be preferable to use the estimating function

$$G_n^\bullet(\theta) = \sum_{i=1}^n a^*(X_{(i-1)\Delta};\tilde{\theta}_n)h(X_{(i-1)\Delta},X_{i\Delta};\theta), \tag{1.39}$$

where $\tilde{\theta}_n$ is a θ_0-consistent estimator, for instance obtained by some simple choice of the weight matrix a. In this way a^* needs to be calculated only once per observation point, whereas the weight matrix must be recalculated for every call of $G_n^*(\theta)$ given by (1.30) with $a = a^*$. Typically, $G_n^*(\theta)$ will be called many times in a numerical search procedure in order to find the G_n^*-estimator. Under weak regularity conditions, the $G_n^\bullet$-estimator has the same efficiency as the optimal G_n^*-estimator; see e.g. Jacod and Sørensen (2012).

Most martingale estimating functions proposed in the literature are of the form (1.30) with

$$h_j(x,y;\theta) = f_j(y;\theta) - \pi_\Delta^\theta(f_j(\theta))(x), \tag{1.40}$$

or more specifically,

$$G_n(\theta) = \sum_{i=1}^n a(X_{(i-1)\Delta},\theta)\left[f(X_{i\Delta};\theta) - \pi_\Delta^\theta(f(\theta))(X_{(i-1)\Delta})\right]. \tag{1.41}$$

Here $f = (f_1,\ldots,f_N)^T$ maps $D \times \Theta$ into $\mathbb{R}^N$, and π_Δ^θ denotes the *transition operator*

$$\pi_s^\theta(f)(x) = \int_D f(y)p(s,x,y;\theta)dy = E_\theta(f(X_s)\,|\,X_0 = x), \tag{1.42}$$

applied to each coordinate of f. The polynomial estimating functions given by $f_j(y) = y^j$, $j = 1,\ldots,N$, are an example. For martingale estimating functions of the special form (1.41), the expression for the optimal weight matrix simplifies a bit because

$$B_h(x;\theta)_{ij} = \pi_\Delta^\theta(\partial_{\theta_i}f_j(\theta))(x) - \partial_{\theta_i}\pi_\Delta^\theta(f_j(\theta))(x), \tag{1.43}$$

$i = 1,\ldots p, j = 1,\ldots,N$, and

$$V_h(x;\theta)_{ij} = \pi_\Delta^\theta(f_i(\theta)f_j(\theta))(x) - \pi_\Delta^\theta(f_i(\theta))(x)\pi_\Delta^\theta(f_j(\theta))(x), \tag{1.44}$$

$i,j = 1,\ldots,N$. If the functions f_j are chosen to be independent of θ, then (1.43) simplifies to

$$B_h(x;\theta)_{ij} = -\partial_{\theta_i}\pi_\Delta^\theta(f_j)(x). \tag{1.45}$$

A useful *approximation to the optimal weight matrix* can be obtained by applying the expansion of conditional moments given in the following lemma. The expansion is expressed in terms of the *generator* of the diffusion, which is defined by

$$A_\theta f(x) = \sum_{k=1}^{d} b_k(x; \theta) \partial_{x_k} f(x) + \tfrac{1}{2} \sum_{k,\ell=1}^{d} C_{k\ell}(x; \theta) \partial^2_{x_k x_\ell} f(x), \qquad (1.46)$$

where $C = \sigma\sigma^T$. By A_θ^i we mean i-fold application of A_θ, and in particular, $A_\theta^0 f = f$. For an ergodic diffusion with invariant measure with Lebesgue density μ_θ, let Φ_θ be the class of real functions f defined on the state space D that are twice continuously differentiable, square integrable with respect to μ_θ, and satisfy that

$$\int_D (A_\theta f(x))^2 \mu_\theta(x) dx < \infty$$

$$\sum_{i,j=1}^{d} \int_D \partial_{x_i} f(x) \partial_{x_j} f(x) C_{i,j}(x; \theta) \mu_\theta(x) dx < \infty.$$

Lemma 1.10 *Suppose that the diffusion process (1.11) is ergodic with invariant measure μ_θ, that f is $2(k+1)$ times continuously differentiable, and that $A_\theta^i f \in \Phi_\theta$, $i = 0, \ldots, k$. Then*

$$\pi_t^\theta(f)(x) = \sum_{i=0}^{k} \frac{t^i}{i!} A_\theta^i f(x) + O(t^{k+1}). \qquad (1.47)$$

Proof: We sketch the proof of (1.47), and consider only $d = 1$ to simplify the exposition. First consider $k = 1$. By Ito's formula

$$f(X_t) = f(X_0) + \int_0^t A_\theta f(X_s) ds + \int_0^t f'(X_s)\sigma(X_s; \theta) dW_s$$

$$A_\theta f(X_s) = A_\theta f(X_0) + \int_0^s A_\theta^2 f(X_u) du + \int_0^s \partial_x A_\theta f(X_u)\sigma(X_u; \theta) dW_u,$$

and by inserting the expression for $A_\theta f(X_s)$ given by the second equation into the Lebesgue integral in the first equation, we find that

$$f(X_t) = f(X_0) + t A_\theta f(X_0) + \int_0^t \int_0^s A_\theta^2 f(X_u) du\, ds \qquad (1.48)$$

$$+ \int_0^t \int_0^s \partial_x A_\theta f(X_u)\sigma(X_u; \theta) dW_u\, ds + \int_0^t f'(X_s)\sigma(X_s; \theta) dW_s.$$

Because $A_\theta^i f \in \Phi_\theta$, $i = 0, 1$, the Ito-integrals are proper P_θ-martingales. Hence by taking the conditional expectation given $X_0 = x$, we obtain

$$\pi_t^\theta(f)(x) = f(x) + t A_\theta f(x) + O(t^2).$$

The result for $k = 2$ is obtained by applying Ito's formula to $A_\theta^2 f(X_t)$, inserting the result into the first Lebesgue integral in (1.48), and finally taking the conditional expectation given $X_0 = x$. The result for $k \geq 3$ is obtained similarly by applying Ito's formula to $A_\theta^i f(X_t)$, $i = 3, \ldots, k$. $\qquad \square$

Note that (1.47) is an expansion result, so the corresponding power series does not necessarily converge. For a fixed k, the sum is a good approximation to the conditional expectation when t is small. The remainder term depends on k and θ. The explicit sufficient conditions in Lemma 1.10 for (1.47) to hold for ergodic diffusions was given in Jacobsen (2001). The expansion holds under mild regularity conditions for non-ergodic diffusions too. In a proof similar to that of Lemma 1.10, such conditions must essentially ensure that Ito-integrals are proper martingales and that the remainder term can be controlled.

It is often enough to use the approximation $\pi_\Delta^\theta(f_j)(x) \approx f_j(x) + \Delta A_\theta f_j(x)$. When f does not depend on θ this implies that for $d = 1$

$$B_h(x; \theta) \approx \Delta \left[\partial_\theta b(x; \theta) f'(x) + \tfrac{1}{2} \partial_\theta \sigma^2(x; \theta) f''(x) \right] \qquad (1.49)$$

and for $d = 1$ and $N = 1$

$$V_h(x; \theta) \approx \Delta \left[A_\theta(f^2)(x) - 2f(x) A_\theta f(x) \right] = \Delta \sigma^2(x; \theta) f'(x)^2. \quad (1.50)$$

We will refer to estimating functions obtained by approximating the optimal weight-matrix a^* in this way as *approximately optimal estimating functions*. Use of this approximation will save computer time and improve the numerical performance of the estimation procedure. The approximation will not affect the consistency of the estimators, and if Δ is not too large, it will just lead to a relatively minor loss of efficiency. The magnitude of this loss of efficiency can be calculated by means of (1.47).

Example 1.11 If we simplify the optimal weight matrix found in Example 1.9 by the expansion (1.47) and the Gaussian approximation (1.38), we obtain the approximately optimal quadratic martingale estimating function

$$G_n^\circ(\theta) = \sum_{i=1}^n \left\{ \frac{\partial_\theta b(X_{(i-1)\Delta}; \theta)}{\sigma^2(X_{(i-1)\Delta}; \theta)} [X_{i\Delta} - F(X_{(i-1)\Delta}; \theta)] \right. \qquad (1.51)$$

$$\left. + \frac{\partial_\theta \sigma^2(X_{(i-1)\Delta}; \theta)}{2\sigma^4(X_{(i-1)\Delta}; \theta)\Delta} \left[(X_{i\Delta} - F(X_{(i-1)\Delta}; \theta))^2 - \phi(X_{(i-1)\Delta}; \theta) \right] \right\}.$$

As in Example 1.9 the diffusion is assumed to be one-dimensional.

Consider a diffusion with *linear drift*, $b(x; \theta) = -\beta(x - \alpha)$. Diffusion models with linear drift and a given marginal distribution were studied in Bibby, Skovgaard, and Sørensen (2005). If $\int \sigma^2(x; \theta) \mu_\theta(x) dx < \infty$, then the Ito-integral in

$$X_t = X_0 - \int_0^t \beta(X_s - \alpha) ds + \int_0^t \sigma(X_s; \theta) dW_s$$

is a proper martingale with mean zero, so the function $f(t) = E_\theta(X_t \mid X_0 = x)$ satisfies that

$$f(t) = x - \beta \int_0^t f(s)ds + \beta\alpha t$$

or

$$f'(t) = -\beta f(t) + \beta\alpha, \quad f(0) = x.$$

Hence

$$f(t) = xe^{-\beta t} + \alpha(1 - e^{-\beta t})$$

or

$$F(x; \alpha, \beta) = xe^{-\beta\Delta} + \alpha(1 - e^{-\beta\Delta}).$$

If only estimates of drift parameters are needed, we can use the linear martingale estimating function of the form (1.30) with $N = 1$ and $h_1(x, y; \theta) = y - F(\Delta, x; \theta)$. If $\sigma(x; \theta) = \tau\kappa(x)$ for $\tau > 0$ and κ a positive function, then the approximately optimal estimating function of this form is

$$G_n^\circ(\alpha, \beta) = \left(\begin{array}{c} \displaystyle\sum_{i=1}^n \frac{1}{\kappa^2(X_{(i-1)\Delta})} \left[X_{i\Delta} - X_{(i-1)\Delta}e^{-\beta\Delta} - \alpha(1 - e^{-\beta\Delta}) \right] \\[2em] \displaystyle\sum_{i=1}^n \frac{X_{(i-1)\Delta}}{\kappa^2(X_{(i-1)\Delta})} \left[X_{i\Delta} - X_{(i-1)\Delta}e^{-\beta\Delta} - \alpha(1 - e^{-\beta\Delta}) \right] \end{array} \right),$$

where multiplicative constants have been omitted. To solve the estimating equation $G_n^\circ(\alpha, \beta) = 0$ we introduce the weights

$$w_i^\kappa = \kappa(X_{(i-1)\Delta})^{-2} / \sum_{j=1}^n \kappa(X_{(j-1)\Delta})^{-2},$$

and define $\bar{X}^\kappa = \sum_{i=1}^n w_i^\kappa X_{i\Delta}$ and $\bar{X}_{-1}^\kappa = \sum_{i=1}^n w_i^\kappa X_{(i-1)\Delta}$. These two quantities are conditional precision weighted sample averages of $X_{i\Delta}$ and $X_{(i-1)\Delta}$, respectively. The equation $G_n^\circ(\alpha, \beta) = 0$ has a unique explicit solution provided that the weighted sample autocorrelation

$$r_n^\kappa = \frac{\sum_{i=1}^n w_i^\kappa (X_{i\Delta} - \bar{X}^\kappa)(X_{(i-1)\Delta} - \bar{X}_{-1}^\kappa)}{\sum_{i=1}^n w_i^\kappa (X_{(i-1)\Delta} - \bar{X}_{-1}^\kappa)^2}$$

is positive. By the law of large numbers (1.4) for ergodic processes, the probability that $r_n^\kappa > 0$ tends to one as n tends to infinity. Specifically, we obtain the explicit estimators

$$\hat{\alpha}_n = \frac{\bar{X}^\kappa - r_n^\kappa \bar{X}_{-1}^\kappa}{1 - r_n^\kappa}$$

$$\hat{\beta}_n = -\frac{1}{\Delta} \log(r_n^\kappa),$$

see Christensen and Sørensen (2008). A slightly simpler and asymptotically equivalent estimator may be obtained by substituting $\bar{X}^\kappa$ for $\bar{X}^\kappa_{-1}$ everywhere, in which case α is estimated by the precision weighted sample average $\bar{X}^\kappa$. For the square-root process (CIR-model) given by (1.37), where $\kappa(x) = \sqrt{x}$, a simulation study and an investigation of the asymptotic variance of these estimators in Bibby and Sørensen (1995) show that they are not much less efficient than the estimators from the optimal estimating function. See also the simulation study in Overbeck and Rydén (1997), who find that these estimators are surprisingly efficient, a finding that can be explained by the results in Section 1.6.

To obtain an explicit approximately optimal quadratic estimating function, we need an expression for the conditional variance $\phi(x; \theta)$. As we saw in Example 1.9, $\phi(x; \theta)$ is explicitly known for the *square-root process (CIR-model)* given by (1.37). For this model the approximately optimal quadratic martingale estimating function is

$$
\left(
\begin{aligned}
& \sum_{i=1}^{n} \frac{1}{X_{(i-1)\Delta}} \left[X_{i\Delta} - X_{(i-1)\Delta} e^{-\beta\Delta} - \alpha(1 - e^{-\beta\Delta}) \right] \\
& \sum_{i=1}^{n} \left[X_{i\Delta} - X_{(i-1)\Delta} e^{-\beta\Delta} - \alpha(1 - e^{-\beta\Delta}) \right] \\
& \sum_{i=1}^{n} \frac{1}{X_{(i-1)\Delta}} \left[\left(X_{i\Delta} - X_{(i-1)\Delta} e^{-\beta\Delta} - \alpha(1 - e^{-\beta\Delta}) \right)^2 \right. \\
& \qquad \left. - \frac{\tau^2}{\beta} \left\{ (\alpha/2 - X_{(i-1)\Delta}) e^{-2\beta\Delta} - (\alpha - X_{(i-1)\Delta}) e^{-\beta\Delta} + \alpha/2 \right\} \right]
\end{aligned}
\right).
$$

This expression is obtained from (1.51) after multiplication by an invertible non-random matrix to obtain a simpler expression. This does not change the estimator. From this estimating function explicit estimators can easily be obtained:

$$
\hat{\alpha}_n = \frac{1}{n} \sum_{i=1}^{n} X_{i\Delta} + \frac{e^{-\hat{\beta}_n\Delta}}{n\left(1 - e^{-\hat{\beta}_n\Delta}\right)} (X_{n\Delta} - X_0),
$$

essentially the sample mean when n is large, and

$$
e^{-\hat{\beta}_n\Delta} = \frac{n \sum_{i=1}^{n} X_{i\Delta}/X_{(i-1)\Delta} - (\sum_{i=1}^{n} X_{i\Delta})(\sum_{i=1}^{n} X_{(i-1)\Delta}^{-1})}{n^2 - (\sum_{i=1}^{n} X_{(i-1)\Delta})(\sum_{i=1}^{n} X_{(i-1)\Delta}^{-1})}
$$

$$
\hat{\tau}_n^2 = \frac{\sum_{i=1}^{n} X_{(i-1)\Delta}^{-1} \left(X_{i\Delta} - X_{(i-1)\Delta} e^{-\hat{\beta}_n\Delta} - \hat{\alpha}_n(1 - e^{-\hat{\beta}_n}) \right)^2}{\sum_{i=1}^{n} X_{(i-1)\Delta}^{-1} \psi(X_{(i-1)\Delta}; \hat{\alpha}_n, \hat{\beta}_n)},
$$

where

$$
\psi(x; \alpha, \beta) = \left((\tfrac{1}{2}\alpha - x) e^{-2\beta\Delta} - (\alpha - x) e^{-\beta\Delta} + \tfrac{1}{2}\alpha \right) / \beta.
$$

It is obviously necessary for this solution to the estimating equation to exist that the expression for $e^{-\hat{\beta}_n \Delta}$ is strictly positive, an event that happens with a probability tending to one as $n \to \infty$. Again this follows from the law of large numbers (1.4) for ergodic processes. □

When the optimal weight matrix is approximated by means of (1.47), there is a certain loss of efficiency, which as in the previous example is often quite small; see Bibby and Sørensen (1995) and Section 1.6 on high frequency asymptotics below. Therefore the relatively simple estimating function (1.51) is often a good choice in practice.

It is tempting to go on to approximate $\pi_\Delta^\theta(f_j(\theta))(x)$ in (1.41) by (1.47) in order to obtain an explicit estimating function, but as will be demonstrated in Subsection 1.5.3, this can be a dangerous procedure. In general, the conditional expectation in π_Δ^θ should therefore be approximated by simulations. Fortunately, Kessler and Paredes (2002) have established that, provided the simulation is done with sufficient accuracy, this does not cause any bias, only a minor loss of efficiency that can be made arbitrarily small; see Subsection 1.3.5. Moreover, as we shall see in the Subsections 1.3.6 and 1.3.7, $\pi_\Delta^\theta(f_j(\theta))(x)$ can be found explicitly for a quite flexible class of diffusions.

1.3.4 Small Δ-optimality

The Godambe–Heyde optimal estimating functions discussed above are optimal within a certain class of estimating functions. In this subsection we present the concept of small Δ-optimality, introduced and studied by Jacobsen (2001, 2002). A small Δ-optimal estimating function is optimal among all estimating functions satisfying weak regularity conditions, but only for high sampling frequencies, i.e. when the time between observations is small. Thus the advantage of the concept of small Δ-optimality is that the optimality is global, while the advantage of the concept of Godambe–Heyde optimality is that the optimality holds for all sampling frequencies. Fortunately, we do not have to choose between the two, because it turns out that Godambe–Heyde optimal martingale estimating functions of the form (1.30) and (1.40) are small Δ-optimal.

Small Δ-optimality was originally introduced for general estimating functions for multivariate diffusion models, but to simplify the exposition we will concentrate on martingale estimating functions and on one-dimensional diffusions of the form

$$dX_t = b(X_t; \alpha)dt + \sigma(X_t; \beta)dW_t, \tag{1.52}$$

where $\theta = (\alpha, \beta) \in \Theta \subseteq \mathbb{R}^2$. This is the simplest model type for which the essential features of the theory appear. Note that the drift and the diffusion coefficient depend on different parameters. It is assumed that the diffusion is ergodic, that its invariant probability measure has density function μ_θ for all $\theta \in \Theta$, and that $X_0 \sim \mu_\theta$ under P_θ. Thus the diffusion is stationary.

Throughout this subsection, we shall assume that the observation times are equidistant, i.e. $t_i = i\Delta, i = 0, 1, \ldots, n$, where Δ is fixed, and that the martingale estimating function (1.12) satisfies the conditions of Theorem 1.5, so that we know that (eventually) a G_n-estimator $\hat{\theta}_n$ exists, which is asymptotically normal with covariance matrix $M(g) = W^{-1}VW^{T-1}$, where W is given by (1.7) with $\bar{\theta} = \theta_0$ and $V = V(\theta_0)$ with $V(\theta)$ given by (1.19).

The main idea of small Δ-optimality is to expand the asymptotic covariance matrix in powers of Δ

$$M(g) = \frac{1}{\Delta}v_{-1}(g) + v_0(g) + o(1). \tag{1.53}$$

Small Δ-optimal estimating functions minimize the leading term in (1.53). Jacobsen (2001) obtained (1.53) by Ito-Taylor expansions, see Kloeden and Platen (1999), of the random matrices that appear in the expressions for W and V under regularity conditions that will be given below. A similar expansion was used in Aït–Sahalia and Mykland (2003, 2004).

To formulate the conditions, we define the differential operator $\mathcal{A}_\theta, \theta \in \Theta$. Its domain, Γ, is the set of continuous real-valued functions $(s, x, y) \mapsto \varphi(s, x, y)$ of $s \geq 0$ and $(x, y) \in (\ell, r)^2$ that are continuously differentiable with respect to s and twice continuously differentiable with respect to y. The operator $\mathcal{A}_\theta$ is given by

$$\mathcal{A}_\theta \varphi(s, x, y) = \partial_s \varphi(s, x, y) + A_\theta \varphi(s, x, y), \tag{1.54}$$

where A_θ is the generator (1.46), which for fixed s and x is applied to the function $y \mapsto \varphi(s, x, y)$. The operator $\mathcal{A}_\theta$ acting on functions in Γ that do not depend on x is the generator of the space-time process $(t, X_t)_{t \geq 0}$. We also need the probability measure Q_θ^Δ given by (1.16). Note that in this section the dependence on Δ is explicit in the notation.

Condition 1.12 *The function φ belongs to Γ and satisfies that*

$$\int_\ell^r \int_\ell^r \varphi^2(s, x, y) Q_{\theta_0}^s(dx, dy) < \infty$$

$$\int_\ell^r \int_\ell^r (\mathcal{A}_{\theta_0} \varphi(s, x, y))^2 Q_{\theta_0}^s(dx, dy) < \infty$$

$$\int_\ell^r \int_\ell^r (\partial_y \varphi(s, x, y))^2 \sigma^2(y; \beta_0) Q_{\theta_0}^s(dx, dy) < \infty$$

for all $s \geq 0$.

As usual $\theta_0 = (\alpha_0, \beta_0)$ denotes the true parameter value. We will say that a function with values in $\mathbb{R}^k$ or $\mathbb{R}^{k \times \ell}$ satisfies Condition 1.12 if each component of the functions satisfies this condition.

Suppose φ satisfies Condition 1.12. Then by Ito's formula

$$\varphi(t, X_0, X_t) = \varphi(0, X_0, X_0) + \int_0^t \mathcal{A}_{\theta_0}\varphi(s, X_0, X_s)ds \qquad (1.55)$$

$$+ \int_0^t \partial_y\varphi(s, X_0, X_s)\sigma(X_s; \beta_0)dW_s$$

under P_{θ_0}. A significant consequence of Condition 1.12 is that the Ito-integral in (1.55) is a true P_{θ_0}-martingale, and thus has expectation zero under P_{θ_0}. If the function $\mathcal{A}_{\theta_0}\varphi$ satisfies Condition 1.12, a similar result holds for this function, which we can insert in the Lebesgue integral in (1.55). By doing so and then taking the conditional expectation given $X_0 = x$ on both sides of (1.55), we obtain

$$\pi_t^{\theta_0}(\varphi)(t, x) = \varphi(0, x, x) + t\mathcal{A}_{\theta_0}\varphi(0, x, x) + O(t^2), \qquad (1.56)$$

where

$$\pi_t^{\theta}(\varphi)(t, x) = \mathrm{E}_\theta\left(\varphi(t, X_0, X_t)|X_0 = x\right).$$

If the functions $\mathcal{A}_{\theta_0}^i\varphi$, $i = 0, \ldots, k$ satisfy Condition 1.12, where $\mathcal{A}_{\theta_0}^i$ denotes i-fold application of the operator $\mathcal{A}_{\theta_0}$, we obtain by similar arguments that

$$\pi_t^{\theta_0}(\varphi)(t, x) = \sum_{i=0}^k \frac{s^i}{i!}\mathcal{A}_{\theta_0}^i\varphi(0, x, x) + O(t^{k+1}). \qquad (1.57)$$

Note that $\mathcal{A}_{\theta}^0$ is the identity operator, $\mathcal{A}_{\theta}^0\varphi = \varphi$. The previously used expansion (1.47) is a particular case of (1.57). In the case where φ does not depend on x (or y) the integrals in Condition 1.12 are with respect to the invariant measure μ_{θ_0}. If, moreover, φ does not depend on time s, the conditions do not depend on s.

Theorem 1.13 *Suppose that the function $g(\Delta, x, y; \theta_0)$ in (1.12) is such that g, $\partial_{\theta^T}g$, gg^T and $\mathcal{A}_{\theta_0}g$ satisfy Condition 1.12. Assume, moreover, that we have the expansion*

$$g(\Delta, x, y; \theta_0) = g(0, x, y; \theta_0) + \Delta\partial_\Delta g(0, x, y; \theta_0) + o_{\theta_0, x, y}(\Delta).$$

If the matrix

$$S = \int_\ell^r B_{\theta_0}(x)\mu_{\theta_0}(x)dx \qquad (1.58)$$

is invertible, where

$$B_\theta(x) = \qquad (1.59)$$

$$\begin{pmatrix} \partial_\alpha b(x; \alpha)\partial_y g_1(0, x, x; \theta) & \frac{1}{2}\partial_\beta\sigma^2(x; \beta)\partial_y^2 g_1(0, x, x; \theta) \\ \partial_\alpha b(x; \alpha)\partial_y g_2(0, x, x; \theta) & \frac{1}{2}\partial_\beta\sigma^2(x; \beta)\partial_y^2 g_2(0, x, x; \theta) \end{pmatrix},$$

then (1.53) holds with

$$v_{-1}(g) \geq \begin{pmatrix} \left(\int_{\ell}^{r} (\partial_{\alpha} b(x; \alpha_0))^2 / \sigma^2(x; \beta_0) \mu_{\theta_0}(x) dx \right)^{-1} & 0 \\ & \\ 0 & 0 \end{pmatrix}. \qquad (1.60)$$

There is equality in (1.60) if

$$\partial_y g_1(0, x, x; \theta_0) = \partial_{\alpha} b(x; \alpha_0)/\sigma^2(x; \beta_0), \qquad (1.61)$$
$$\partial_y g_2(0, x, x; \theta_0) = 0 \qquad (1.62)$$

for all $x \in (\ell, r)$. In this case, the second term in (1.53) satisfies that

$$v_0(g)_{22} \geq 2 \left(\int_{\ell}^{r} \left(\partial_{\beta} \sigma^2(x; \beta_0) \right)^2 / \sigma^4(x; \beta_0) \mu_{\theta_0}(x) dx \right)^{-1}$$

with equality if

$$\partial_y^2 g_2(0, x, x; \theta_0) = \partial_{\beta} \sigma^2(x; \beta_0)/\sigma^4(x; \beta_0), \qquad (1.63)$$

for all $x \in (\ell, r)$.

By $\partial_y g_i(0, x, x; \theta)$ we mean $\partial_y g_i(0, y, x; \theta)$ evaluated at $y = x$, and similarly for second order partial derivatives. Thus the conditions for small Δ-optimality are (1.61), (1.62) and (1.63). For a proof of Theorem 1.13, see Jacobsen (2001). The condition (1.62) ensures that all entries of $v_{-1}(g)$ involving the diffusion coefficient parameter, β, are zero. Since $v_{-1}(g)$ is the Δ^{-1}-order term in the expansion (1.53) of the asymptotic covariance matrix, this dramatically decreases the asymptotic variance of the estimator of β when Δ is small. We refer to the condition (1.62) as *Jacobsen's condition*.

The reader is reminded of the trivial fact that for any non-singular 2×2 matrix, M_n, the estimating functions $M_n G_n(\theta)$ and $G_n(\theta)$ give exactly the same estimator. We call them *versions* of the same estimating function. The matrix M_n may depend on Δ_n. Therefore a given version of an estimating function needs not satisfy (1.61) – (1.63). The point is that a version must exist which satisfies these conditions.

Example 1.14 Consider a quadratic martingale estimating function of the form

$$g(\Delta, y, x; \theta) = \begin{pmatrix} a_1(x, \Delta; \theta)[y - F(\Delta, x; \theta)] \\ a_2(x, \Delta; \theta) \left[(y - F(\Delta, x; \theta))^2 - \phi(\Delta, x; \theta) \right] \end{pmatrix}, \qquad (1.64)$$

where F and ϕ are given by (1.26) and (1.27). By (1.47), $F(\Delta, x; \theta) = x + O(\Delta)$ and $\phi(\Delta, x; \theta) = O(\Delta)$, so

$$g(0, y, x; \theta) = \begin{pmatrix} a_1(x, 0; \theta)(y - x) \\ a_2(x, 0; \theta)(y - x)^2 \end{pmatrix}. \qquad (1.65)$$

Since $\partial_y g_2(0, y, x; \theta) = 2a_2(x, \Delta; \theta)(y - x)$, the Jacobsen condition (1.62) is satisfied for all quadratic martingale estimating functions. Using again (1.47), it is not difficult to see that the two other conditions (1.61) and (1.63) are satisfied in three particular cases: the optimal estimating function given in Example 1.9 and the approximations (1.28) and (1.51).   □

The following theorem gives conditions ensuring, for given functions $f_1, \ldots,$ f_N, that a small Δ-optimal estimating function of the form (1.30) and (1.40) exists. This is not always the case. We assume that the functions $f_1(\cdot; \theta), \ldots,$ $f_N(\cdot; \theta)$ are of full affine rank for all θ, i.e., for any $\theta \in \Theta$, the identity

$$\sum_{j=1}^{N} a_j^\theta f_j(x; \theta) + a_0^\theta = 0, \quad x \in (\ell, r),$$

for constants a_j^θ, implies that $a_0^\theta = a_1^\theta = \cdots = a_N^\theta = 0$.

Theorem 1.15 *Suppose that $N \geq 2$, that the functions f_j are twice continuously differentiable and satisfies that the matrix*

$$D(x) = \begin{pmatrix} \partial_x f_1(x; \theta) & \partial_x^2 f_1(x; \theta) \\ \partial_x f_2(x; \theta) & \partial_x^2 f_2(x; \theta) \end{pmatrix} \tag{1.66}$$

is invertible for μ_θ-almost all x. Moreover, assume that the coefficients b and σ are continuously differentiable with respect to the parameter. Then a specification of the weight matrix $a(x; \theta)$ exists such that the estimating function (1.30) satisfies the conditions (1.62), (1.61) and (1.63). When $N = 2$, these conditions are satisfied for

$$a(x; \theta) = \begin{pmatrix} \partial_\alpha b(x; \alpha)/v(x; \beta) & c(x; \theta) \\ 0 & \partial_\beta v(x; \beta)/v(x; \beta)^2 \end{pmatrix} D(x)^{-1} \tag{1.67}$$

for any function $c(x; \theta)$.

For a proof of Theorem 1.15, see Jacobsen (2002). In Section 1.6, we shall see that the Godambe–Heyde optimal choice (1.32) of the weight-matrix in (1.30) gives an estimating function which has a version that satisfies the conditions for small Δ-optimality, (1.61) – (1.63).

We have focused on one-dimensional diffusions to simplify the exposition. The situation becomes more complicated for multi-dimensional diffusions, as we shall now briefly describe. Details can be found in Jacobsen (2002). For a d-dimensional diffusion, $b(x; \alpha)$ is d-dimensional and $v(x; \beta) = \sigma(x; \beta)\sigma(x; \beta)^T$ is a $d \times d$-matrix. The function $\partial_y g_2(0, x, x; \theta_0)$ is now a d-dimensional vector, but otherwise the Jacobsen condition is unchanged. The other two conditions for small Δ-optimality are

$$\partial_y g_1(0, x, x; \theta_0) = \partial_\alpha b(x; \alpha_0)^T v(x; \beta_0)^{-1}$$

and

$$\text{vec}\left(\partial_y^2 g_2(0, x, x; \theta_0)\right) = \text{vec}\left(\partial_\beta v(x; \beta_0)\right)\left(v^{\otimes 2}(x; \beta_0)\right)^{-1}.$$

In the latter equation, $\text{vec}(M)$ denotes, for a $d \times d$ matrix M, the d^2-dimensional row vector consisting of the rows of M placed one after the other, and $M^{\otimes 2}$ is the $d^2 \times d^2$-matrix with $(i', j'), (ij)$th entry equal to $M_{i'i}M_{j'j}$. Thus if $M = \partial_\beta v(x; \beta)$ and $M^\bullet = (v^{\otimes 2}(x; \beta))^{-1}$, then the (i, j)th coordinate of $\text{vec}(M) M^\bullet$ is $\sum_{i'j'} M_{i'j'} M^\bullet_{(i'j'),(i,j)}$.

For a d-dimensional diffusion process, the conditions analogous to those in Theorem 1.15 ensuring the existence of a small Δ-optimal estimating function of the form (1.30) is that $N \geq d(d+3)/2$, and that the $N \times (d + d^2)$-matrix

$$\left(\begin{array}{cc} \partial_{x^T} f(x; \theta) & \partial_{x^T}^2 f(x; \theta) \end{array}\right)$$

has full rank $d(d+3)/2$. The rank of this matrix is at most $d(d+3)/2$, because the fact that $\partial^2_{x_i x_j} = \partial^2_{x_j x_i}$ implies that the rank of $\partial^2_{x^T} f$ is at most $d + d(d-1)/2$, and the rank of $\partial_{x^T} f$ is to most d.

1.3.5 Simulated martingale estimating functions

The conditional moments that appear in the martingale estimating functions can for most diffusion models not be calculated explicitly. For a versatile class of one-dimensional diffusions, optimal martingale estimating functions can be found explicitly; see Subsections 1.3.6 and 1.3.7. Estimation and inference is dramatically simplified by using a model for which an explicit optimal martingale estimating function is available. However, if for some reason a diffusion from this class is not a suitable model, the conditional moments must be determined by simulation.

The conditional moment $\pi_\theta^\Delta f(x) = E_\theta(f(X_\Delta) | X_0 = x)$ can be found straightforwardly. Simply fix θ and simulate numerically M independent trajectories $X^{(i)}$, $i = 1, \ldots, M$ of $\{X_t : t \in [0, \Delta]\}$ with $X_0 = x$. Of course, a trajectory cannot be simulated continuously for all $t \in [0, \Delta]$. In practice values of $\{X_{j\delta} : j = 0, \ldots, K\}$ are simulated, where K is a large integer and $\delta = \Delta/K$. By the law of large numbers,

$$\pi_\theta^\Delta f(x) \doteq \frac{1}{M} \sum_{i=1}^{M} f(X_\Delta^{(i)}).$$

The variance of the error can be estimated in the traditional way, and by the cental limit theorem, the error is approximately normal distributed. This simple approach can be improved by applying variance reduction methods, for instance methods that take advantage of the fact that $\pi_\theta^\Delta f(x)$ can be approximated by (1.47). Methods for numerical simulation of diffusion models can be found in Chapter 4 and Kloeden and Platen (1999).

The approach just described is sufficient when calculating the conditional expectation appearing in (1.40). Note, however, that when using a search algorithm to find a solution to the estimating equation, it is important to use the same random numbers (the same seed) when calculating the estimating functions for different values of the parameter θ. More care is needed if the optimal weight functions are calculated numerically. The problem is that the optimal weight matrix typically contains derivatives with respect to θ of functions that must be determined numerically, see e.g. Example 1.9. Pedersen (1994) proposed a procedure for determining $\partial_\theta \pi_\theta^\Delta f(x; \theta)$ by simulations based on results in Friedman (1975). However, it is often preferable to use an approximation to the optimal weight matrix obtained by using (1.47), possibly supplemented by Gaussian approximations, as explained in Subsection 1.3.3. This is not only much simpler, but also avoids potentially serious problems of numerical instability, and by results in Section 1.6 the loss of efficiency is often very small. The approach outlined here, where martingale estimating functions are approximated by simulation, is closely related to the simulated method of moments, see Duffie and Singleton (1993) and Clement (1997).

One might be worried that when approximating a martingale estimating function by simulation of conditional moments, the resulting estimator might have considerably smaller efficiency or even be inconsistent. The asymptotic properties of the estimators obtained when the conditional moments are approximated by simulation were investigated by Kessler and Paredes (2002), who found that if the simulations are done with sufficient care, there is no need to worry. However, their results also show that care is needed: if the discretization used in the simulation method is too crude, the estimator behaves badly. Kessler and Paredes (2002) considered martingale estimating functions of the general form

$$G_n(\theta) = \sum_{i=1}^{n} \left[f(X_{i\Delta}, X_{(i-1)\Delta}; \theta) - F(X_{(i-1)\Delta}; \theta) \right], \qquad (1.68)$$

where f is a p-dimensional function, and

$$F(x; \theta) = E_\theta(f(X_\Delta, x; \theta))|X_0 = x).$$

As previously, X is the unique solution of the stochastic differential equation (1.11). For simplicity X is assumed to be one-dimensional, but Kessler and Paredes (2002) point out that similar results hold for multivariate diffusions. Below the dependence of X on the initial value $X_0 = x$ and θ is, when needed, emphasized in the notation by writing $X(x, \theta)$.

Let $Y(\delta, \theta, x)$ be an approximation to the solution $X(\theta, x)$, which is calculated at discrete time points with a step size δ that is much smaller than Δ, and which satisfies that $Y_0(\delta, \theta, x) = x$. A simple example is the Euler scheme

$$Y_{i\delta} = Y_{(i-1)\delta} + b(Y_{(i-1)\delta}; \theta)\delta + \sigma(Y_{(i-1)\delta}; \theta)Z_i, \qquad Y_0 = x, \qquad (1.69)$$

where the Z_is are independent and $Z_i \sim N(0, \delta)$.

If the conditional expectation $F(x; \theta)$ is approximated by the simple method described above, we obtain the following approximation to the estimating function (1.68)

$$G_n^{M,\delta}(\theta) = \qquad\qquad\qquad\qquad\qquad\qquad\qquad\qquad\qquad\qquad (1.70)$$

$$\sum_{i=1}^{n} \left[f(X_{i\Delta}, X_{(i-1)\Delta}; \theta) - \frac{1}{M} \sum_{j=1}^{M} f(Y_\Delta^{(j)}(\delta, \theta, X_{(i-1)\Delta}), X_{(i-1)\Delta}; \theta) \right],$$

where $Y^{(j)}(\delta, \theta, x)$, $j = 1, \dots, M$, are independent copies of $Y(\delta, \theta, x)$.

Kessler and Paredes (2002) assume that the approximation scheme $Y(\delta, \theta, x)$ is of weak order $\beta > 0$ in the sense that

$$|E_\theta(g(X_\Delta(x, \theta), x; \theta)) - E(g(Y_\Delta(\delta, \theta, x), x; \theta))| \le R(x; \theta)\delta^\beta \qquad (1.71)$$

for all $\theta \in \Theta$, for all x in the state space of X, and for δ sufficiently small. Here $R(x; \theta)$ is of polynomial growth in x uniformly for θ in compact sets, i.e., for any compact subset $K \subseteq \Theta$, there exist constants $C_1, C_2 > 0$ such that $\sup_{\theta \in K} |R(x; \theta)| \le C_1(1 + |x|^{C_2})$ for all x in the state space of the diffusion. The inequality (1.71) is assumed to hold for any function $g(y, x; \theta)$ which is $2(\beta+1)$ times differentiable with respect to x, and satisfies that g and its partial derivatives with respect to x up to order $2(\beta+1)$ are of polynomial growth in x uniformly for θ in compact sets. This definition of weak order is stronger than the definition in Kloeden and Platen (1999) in that control of the polynomial order with respect to the initial value x is added, but Kessler and Paredes (2002) point out that theorems in Kloeden and Platen (1999) that give the order of approximation schemes can be modified in a tedious, but straightforward, way to ensure that the schemes satisfy the stronger condition (1.71). In particular, the Euler scheme (1.69) is of weak order one if the coefficients of the stochastic differential equation (1.11) are sufficiently smooth.

Under a number of further regularity conditions, Kessler and Paredes (2002) showed the following results about a $G_n^{M,\delta}$-estimator, $\hat{\theta}_n^{M,\delta}$, with $G_n^{M,\delta}$ given by (1.70). We shall not go into these rather technical conditions. Not surprisingly, they include conditions that ensure the eventual existence of a consistent and asymptotically normal G_n-estimator, cf. Theorem 1.5. If δ goes to zero sufficiently fast that $\sqrt{n}\delta^\beta \to 0$ as $n \to \infty$, then

$$\sqrt{n}\left(\hat{\theta}_n^{M,\delta} - \theta_0\right) \xrightarrow{\mathcal{D}} N\left(0, (1 + M^{-1})\Sigma\right),$$

where Σ denotes the asymptotic covariance matrix of a G_n-estimator, see Theorem 1.5. Thus for δ sufficiently small and M sufficiently large, it does not matter much that the conditional moment $F(x; \theta)$ has been determined by simulation in (1.70). Moreover, we can control the loss of efficiency by our choice of M. However, when $0 < \lim_{n \to \infty} \sqrt{n} \delta^\beta < \infty$,

$$\sqrt{n} \left(\hat{\theta}_n^{M,\delta} - \theta_0 \right) \xrightarrow{\mathcal{D}} N \left(m(\theta_0), (1 + M^{-1}) \Sigma \right),$$

and when $\sqrt{n} \delta^\beta \to \infty$,

$$\delta^{-\beta} \left(\hat{\theta}_n^{N,\delta} - \theta_0 \right) \to m(\theta_0)$$

in probability. Here the p-dimensional vector $m(\theta_0)$ depends on f and is generally different from zero. Thus it is essential that a sufficiently small value of δ is used. The discretization problems caused by the choice of δ can be avoided by using the exact simulation methods introduced by Beskos and Roberts (2005) and Beskos, Papaspiliopoulos, and Roberts (2006), see Chapter 4.

1.3.6 Explicit martingale estimating functions

In this section we consider one-dimensional diffusion models for which estimation is particularly easy because an explicit martingale estimating function exists.

Kessler and Sørensen (1999) proposed estimating functions of the form (1.41) where the functions f_j, $i = 1, \ldots, N$ are *eigenfunctions* for the generator (1.46), i.e.

$$A_\theta f_j(x; \theta) = -\lambda_j(\theta) f_j(x; \theta),$$

where the real number $\lambda_j(\theta) \geq 0$ is called the *eigenvalue* corresponding to $f_j(x; \theta)$. Under weak regularity conditions, f_j is also an eigenfunction for the transition operator π_t^θ defined by (1.42), i.e.

$$\pi_t^\theta (f_j(\theta))(x) = e^{-\lambda_j(\theta) t} f_j(x; \theta).$$

for all $t > 0$. Thus the function h_j given by (1.40) is explicit.

The following result holds for ergodic diffusions. The density of the stationary distribution is, as usual, denoted by μ_θ.

Theorem 1.16 *Let $\phi(x; \theta)$ be an eigenfunction for the generator (1.46) with eigenvalue $\lambda(\theta)$, and suppose that*

$$\int_\ell^r [\partial_x \phi(x; \theta) \sigma(x; \theta)]^2 \mu_\theta(dx) < \infty \tag{1.72}$$

for all $t > 0$. Then

$$\pi_t^\theta (\phi(\theta))(x) = e^{-\lambda(\theta) t} \phi(x; \theta). \tag{1.73}$$

for all $t > 0$.

Proof: Define $Y_t = e^{\lambda t}\phi(X_t)$. We suppress θ in the notation. By Ito's formula

$$\begin{aligned} Y_t &= Y_0 + \int_0^t e^{\lambda s}[A\phi(X_s) + \lambda\phi(X_s)]ds + \int_0^t e^{\lambda s}\phi'(X_s)\sigma(X_s)dW_s \\ &= Y_0 + \int_0^t e^{\lambda s}\phi'(X_s)\sigma(X_s)dW_s, \end{aligned}$$

so by (1.72), Y is a true martingale, which implies (1.73).   □

Note that if $\sigma(x;\theta)$ and $\partial_x\phi(x;\theta)$ are bounded functions of $x \in (\ell, r)$, then (1.72) holds. If ϕ is a polynomial of order k and $\sigma(x) \le C(1 + x^m)$, then (1.72) holds if the $2(k + m - 1)$'th moment of the invariant distribution μ_θ is finite.

Example 1.17 For the square-root model (CIR-model) defined by (1.37) with $\alpha > 0$, $\beta > 0$, and $\tau > 0$, the eigenfunctions are $\phi_i(x) = L_i^{(\nu)}(2\beta x\tau^{-2})$ with $\nu = 2\alpha\beta\tau^{-2} - 1$, where $L_i^{(\nu)}$ is the ith order Laguerre polynomial

$$L_i^{(\nu)}(x) = \sum_{m=0}^{i}(-1)^m \left(\begin{array}{c} i + \nu \\ i - m \end{array} \right) \frac{x^m}{m!},$$

and the eigenvalues are $\{i\beta \ : \ i = 0, 1, \cdots\}$. It is seen by direct calculation that $L_i^{(\nu)}$ solves the differential equation

$$\tfrac{1}{2}\tau^2 x f''(x) - \beta(x - \alpha)f'(x) + i\beta f(x) = 0.$$

By Theorem 1.16, (1.73) holds, so we can calculate all conditional polynomial moments, of which the first four were given in Example 1.9. Thus all polynomial martingale estimating functions are explicit for the square-root model.

□

Example 1.18 The diffusion given as the solution of

$$dX_t = -\theta\tan(X_t)dt + dW_t, \tag{1.74}$$

is an ergodic diffusion on the interval $(-\pi/2, \pi/2)$ provided that $\theta \ge 1/2$, which implies that Condition 1.4 is satisfied. This process was introduced by Kessler and Sørensen (1999), who called it an Ornstein-Uhlenbeck process on $(-\pi/2, \pi/2)$ because $\tan x \sim x$ near zero. The generalization to other finite intervals is obvious. The invariant measure has a density proportional to $\cos(x)^{2\theta}$.

The eigenfunctions are

$$\phi_i(x;\theta) = C_i^\theta(\sin(x)), \quad i = 1, 2, \ldots,$$

where C_i^θ is the Gegenbauer polynomial of order i, and the eigenvalues are

$i(\theta + i/2)$, $i = 1, 2, \ldots$. This follows because the Gegenbauer polynomial C_i^θ solves the differential equation

$$f''(y) + \frac{(2\theta + 1)y}{y^2 - 1} f'(y) - \frac{i(2\theta + i)}{y^2 - 1} f(y) = 0,$$

so that $\phi_i(x; \theta)$ solves the equation

$$\frac{1}{2} \phi_i''(x; \theta) - \theta \tan(x) \phi_i'(x; \theta) = -i(\theta + i/2) \phi_i(x; \theta).$$

Hence ϕ_i is an eigenfunction for the generator of the model with eigenvalue $i(\theta + i/2)$. From equation 8.934-2 in Gradshteyn and Ryzhik (1965) it follows that

$$\phi_i(x; \theta) = \sum_{m=0}^{i} \begin{pmatrix} \theta - 1 + m \\ m \end{pmatrix} \begin{pmatrix} \theta - 1 + i - m \\ i - m \end{pmatrix} \cos[(2m-i)(\pi/2 - x)].$$

Condition (1.72) in Theorem 1.16 is obviously satisfied because the state space is bounded, so (1.73) holds.

The first non-trivial eigenfunction is $\sin(x)$ (a constant is omitted) with eigenvalue $\theta + 1/2$. From the martingale estimating function

$$\check{G}_n(\theta) = \sum_{i=1}^{n} \sin(X_{(i-1)\Delta})[\sin(X_{i\Delta})) - e^{-(\theta+1/2)\Delta} \sin(X_{(i-1)\Delta}))], \quad (1.75)$$

we obtain the simple estimator for θ

$$\check{\theta}_n = -\Delta^{-1} \log \left(\frac{\sum_{i=1}^{n} \sin(X_{(i-1)\Delta}) \sin(X_{i\Delta})}{\sum_{i=1}^{n} \sin^2(X_{(i-1)\Delta})} \right) - 1/2, \quad (1.76)$$

which is defined when the numerator is positive.

An asymmetric generalization of (1.74) was proposed in Larsen and Sørensen (2007) as a model of the logarithm of an exchange rate in a target zone. The diffusion solves the equation

$$dX_t = -\rho \frac{\sin \left(\frac{1}{2}\pi(X_t - m)/z \right) - \varphi}{\cos \left(\frac{1}{2}\pi(X_t - m)/z \right)} dt + \sigma dW_t,$$

where $\rho > 0$, $\varphi \in (-1, 1)$, $\sigma > 0$ $z > 0$, $m \in \mathbb{R}$. The process (1.74) is obtained when $\varphi = 0$, $m = 0$, $z = \pi/2$, and $\sigma = 1$. The state space is $(m - z, m + z)$, and the process is ergodic if $\rho \geq \frac{1}{2}\sigma^2$ and $-1 + \sigma^2/(2\rho) \leq \varphi \leq 1 - \sigma^2/(2\rho)$. The eigenfunctions are

$$\phi_i(x; \rho, \varphi, \sigma, m, z) = P_i^{(\rho(1-\varphi)\sigma^{-2} - \frac{1}{2}, \rho(1+\varphi)\sigma^{-2} - \frac{1}{2})} \left(\sin(\frac{1}{2}\pi x/z - m) \right),$$

with eigenvalues $\lambda_i(\rho, \varphi, \sigma) = i \left(\rho + \frac{1}{2}n\sigma^2 \right)$, $i = 1, 2, \ldots$. Here $P_i^{(a,b)}(x)$ denotes the Jacobi polynomial of order i.   □

For most diffusion models where explicit expressions for eigenfunctions can be found, including the examples above, the eigenfunctions are of the form

$$\phi_i(y; \theta) = \sum_{j=0}^{i} a_{i,j}(\theta) \kappa(y)^j \tag{1.77}$$

where κ is a real function defined on the state space and is independent of θ. For martingale estimating functions based on eigenfunctions of this form, the optimal weight matrix (1.32) can be found explicitly too.

Theorem 1.19 *Suppose $2N$ eigenfunctions are of the form (1.77) for $i = 1, \ldots, 2N$, where the coefficients $a_{i,j}(\theta)$ are differentiable with respect to θ. If a martingale estimating function is defined by (1.40) using the first N eigenfunctions, then*

$$B_h(x, \theta)_{ij} = \sum_{k=0}^{j} \partial_{\theta_i} a_{j,k}(\theta) \nu_k(x; \theta) - \partial_{\theta_i}[e^{-\lambda_j(\theta)\Delta} \phi_j(x; \theta)] \tag{1.78}$$

and

$$V_h(x, \theta)_{i,j} = \tag{1.79}$$

$$\sum_{r=0}^{i} \sum_{s=0}^{j} a_{i,r}(\theta) a_{j,s}(\theta) \nu_{r+s}(x; \theta) - e^{-[\lambda_i(\theta)+\lambda_j(\theta)]\Delta} \phi_i(x; \theta) \phi_j(x; \theta),$$

where $\nu_i(x; \theta) = \pi_\Delta^\theta(\kappa^i)(x)$, $i = 1, \ldots, 2N$, solve the following triangular system of linear equations

$$e^{-\lambda_i(\theta)\Delta} \phi_i(x; \theta) = \sum_{j=0}^{i} a_{i,j}(\theta) \nu_j(x; \theta) \quad i = 1, \ldots, 2N, \tag{1.80}$$

with $\nu_0(x; \theta) = 1$.

Proof: The expressions for B_h and V_h follow from (1.43) and (1.44) when the eigenfunctions are of the form (1.77), and (1.80) follows by applying the transition operator π_Δ^θ to both sides of (1.77).

$\square$

Example 1.20 Consider again the diffusion (1.74) in Example 1.18. We will find the optimal martingale estimating function based on the first non-trivial eigenfunction, $\sin(x)$ (where we have omitted a non-essential multiplicative function of θ) with eigenvalue $\theta + 1/2$. It follows from (1.45) that

$$B_h(x; \theta) = \Delta e^{-(\theta+1/2)\Delta} \sin(x)$$

because $\sin(x)$ does not depend on θ. To find V_h we need Theorem 1.19. The second non-trivial eigenfunction is $2(\theta + 1)\sin^2(x) - 1$ with eigenvalue $2(\theta + 1)$, so

$$\nu_2(x; \theta) = e^{-2(\theta+1)\Delta}[\sin^2(x) - \frac{1}{2}(\theta + 1)^{-1}] + \frac{1}{2}(\theta + 1)^{-1}.$$

Hence the optimal estimating function is

$$G_n^*(\theta) = \sum_{i=1}^{n} \frac{\sin(X_{(i-1)\Delta})[\sin(X_{i\Delta}) - e^{-(\theta+\frac{1}{2})\Delta}\sin(X_{(i-1)\Delta})]}{\frac{1}{2}(e^{2(\theta+1)\Delta} - 1)/(\theta + 1) - (e^{\Delta} - 1)\sin^2(X_{(i-1)\Delta})}$$

where a constant has been omitted. When Δ is small, it is a good idea to multiply $G_n^*(\theta)$ by Δ because the denominator is then of order Δ.

Note that when Δ is sufficiently small, we can expand the exponential function in the numerator to obtain (after multiplication by Δ) the approximately optimal estimating function

$$\tilde{G}_n(\theta) = \sum_{i=1}^{n} \frac{\sin(X_{(i-1)\Delta})[\sin(X_{i\Delta}) - e^{-(\theta+\frac{1}{2})\Delta}\sin(X_{(i-1)\Delta})]}{\cos^2(X_{(i-1)\Delta})},$$

which has the explicit solution

$$\tilde{\theta}_n = -\Delta^{-1}\log\left(\frac{\sum_{i=1}^{n}\tan(X_{(i-1)\Delta})\sin(X_{i\Delta}))/\cos(X_{(i-1)\Delta})}{\sum_{i=1}^{n}\tan^2(X_{(i-1)\Delta})}\right) - \frac{1}{2}.$$

The explicit estimator $\tilde{\theta}$ can, for instance, be used as a starting value when finding the optimal estimator by solving $G_n^*(\theta) = 0$ numerically. Note, however, that for $\tilde{G}_n$ the square integrability (1.17) under Q_{θ_0} required in Theorem 1.5 (to ensure the central limit theorem) is only satisfied when $\theta_0 > 1.5$. This problem can be avoided by replacing $\cos^2(X_{(i-1)\Delta})$ in the numerator by 1, to which it is close when the process is not near the boundaries. In that way we arrive at the simple estimating function (1.75), which is thus approximately optimal too.

□

1.3.7 Pearson diffusions

A widely applicable class of diffusion models for which explicit polynomial eigenfunctions are available is the class of Pearson diffusions, see Wong (1964) and Forman and Sørensen (2008). A Pearson diffusion is a stationary solution to a stochastic differential equation of the form

$$dX_t = -\beta(X_t - \alpha)dt + \sqrt{2\beta(aX_t^2 + bX_t + c)}dW_t, \qquad (1.81)$$

where $\beta > 0$, and a, b and c are such that the square root is well defined when X_t is in the state space. The parameter $\beta > 0$ is a scaling of time that determines how fast the diffusion moves. The parameters α, a, b, and c determine the state space of the diffusion as well as the shape of the invariant distribution. In particular, α is the expectation of the invariant distribution. We define $\theta = (\alpha, \beta, a, b, c)$.

In the context of martingale estimating functions, an important property of the Pearson diffusions is that the generator (1.46) maps polynomials into polynomials. It is therefore easy to find eigenfunctions among the polynomials

$$p_n(x) = \sum_{j=0}^{n} p_{n,j} x^j.$$

Specifically, the polynomial $p_n(x)$ is an eigenfunction if an eigenvalue $\lambda_n > 0$ exists satisfying that

$$\beta(ax^2 + bx + c)p_n''(x) - \beta(x - \alpha)p_n'(x) = -\lambda_n p_n(x),$$

or

$$\sum_{j=0}^{n}\{\lambda_n - a_j\}p_{n,j}x^j + \sum_{j=0}^{n-1} b_{j+1}p_{n,j+1}x^j + \sum_{j=0}^{n-2} c_{j+2}p_{n,j+2}x^j = 0,$$

where $a_j = j\{1 - (j - 1)a\}\beta$, $b_j = j\{\alpha + (j - 1)b\}\beta$, and $c_j = j(j - 1)c\beta$ for $j = 0, 1, 2, \ldots$. Without loss of generality, we assume $p_{n,n} = 1$. Thus by equating the coefficients, we find that the eigenvalue is given by

$$\lambda_n = a_n = n\{1 - (n - 1)a\}\beta. \tag{1.82}$$

If we define $p_{n,n+1} = 0$, then the coefficients $\{p_{n,j}\}_{j=0,\ldots,n-1}$ solve the linear system of equations

$$(a_j - a_n)p_{n,j} = b_{j+1}p_{n,j+1} + c_{j+2}p_{n,j+2}. \tag{1.83}$$

Equation (1.83) is equivalent to a simple recursive formula if $a_n - a_j \neq 0$ for all $j = 0, 1, \ldots, n - 1$. Note that $a_n - a_j = 0$ if and only if there exists an integer $n - 1 \leq m < 2n - 1$ such that $a = m^{-1}$ and $j = m - n + 1$. In particular, $a_n - a_j = 0$ cannot occur if $a < (2n - 1)^{-1}$. It is important to notice that λ_n is positive if and only if $a < (n - 1)^{-1}$. We shall see below that this is exactly the condition ensuring that $p_n(x)$ is integrable with respect to the invariant distribution. If the stronger condition $a < (2n - 1)^{-1}$ is satisfied, the first n eigenfunctions belong to the space of functions that are square integrable with respect to the invariant distribution, and they are orthogonal with respect to the usual inner product in this space. The space of functions that are square integrable with respect to the invariant distribution (or a subset of this space) is often taken as the domain of the generator. Obviously, the eigenfunction $p_n(x)$ satisfies the condition (1.72) if $p_n(x)$ is square integrable with respect to the invariant distribution, i.e. if $a < (2n - 1)^{-1}$. By Theorem 1.16 this implies that the transition operator satisfies (1.73), so that $p_n(x)$ can be used to construct *explicit optimal martingale estimating functions* as explained in Subsection 1.3.6. For Pearson diffusions with $a \leq 0$, $a < (2n - 1)^{-1}$ is automatically satisfied, so there are infinitely many polynomial eigenfunctions. In these cases, the eigenfunctions are well-known families of orthogonal polynomials. When $a > 0$, there are only finitely many square integrable polynomial eigenfunctions. In these cases, more complicated eigenfunctions defined in terms of special functions exist too, see Wong (1964). It is of some historical interest that Hildebrandt (1931) derived the polynomials above from the viewpoint of Gram-Charlier expansions associated with the Pearson system. Some special cases had previously been derived by Romanovsky (1924).

From a modeling point of view, it is important that the class of stationary distributions equals the full Pearson system of distributions. Thus a very wide spectrum of marginal distributions is available ranging from distributions with compact support to very heavy-tailed distributions with tails of the Pareto-type. To see that the invariant distributions belong to the Pearson system, note that the scale measure has density

$$s(x) = \exp\left(\int_{x_0}^{x} \frac{u - \alpha}{au^2 + bu + c} du \right),$$

where x_0 is a point such that $ax_0^2 + bx_0 + c > 0$, cf. (1.14). Since the density of the invariant probability measure is given by

$$\mu_\theta(x) \propto \frac{1}{s(x)(ax^2 + bx + c)},$$

cf. (1.15), it follows that

$$\mu_\theta'(x) = -\frac{(2a + 1)x - \alpha + b}{ax^2 + bx + c} \mu_\theta(x).$$

The Pearson system is defined as the class of probability densities obtained by solving a differential equation of this form, see Pearson (1895).

In the following we present a full classification of the ergodic Pearson diffusions, which shows that all distributions in the Pearson system can be obtained as invariant distributions for a model in the class of Pearson diffusions. We consider six cases according to whether the squared diffusion coefficient is constant, linear, a convex parabola with either zero, one or two roots, or a concave parabola with two roots. The classification problem can be reduced by first noting that the Pearson class of diffusions is closed under location and scale-transformations. To be specific, if X is an ergodic Pearson diffusion, then so is $\tilde{X}$ where $\tilde{X}_t = \gamma X_t + \delta$. The parameters of the stochastic differential equation (1.81) for $\tilde{X}$ are $\tilde{a} = a$, $\tilde{b} = b\gamma - 2a\delta$, $\tilde{c} = c\gamma^2 - b\gamma\delta + a\delta^2$, $\tilde{\beta} = \beta$, and $\tilde{\alpha} = \gamma\alpha + \delta$. Hence, up to transformations of location and scale, the ergodic Pearson diffusions can take the following forms. Note that we consider scale transformations in a general sense where multiplication by a negative real number is allowed, so to each case of a diffusion with state space $(0, \infty)$ there corresponds a diffusion with state space $(-\infty, 0)$.

Case 1: $\sigma^2(x) = 2\beta.$ The solution to (1.81) is an Ornstein Uhlenbeck process. The state space is $\mathbb{R}$, and the invariant distribution is the *normal distribution* with mean α and variance 1. The eigenfunctions are the Hermite polynomials.

Case 2: $\sigma^2(x) = 2\beta x.$ The solution to (1.81) is the square root process (CIR process) (1.37) with state space $(0, \infty)$. Condition 1.4 that ensures ergodicity is satisfied if and only if $\alpha > 1$. If $0 < \alpha \leq 1$, the boundary 0 can with positive probability be reached at a finite time point, but if the boundary is made instantaneously reflecting, we obtain a stationary process. The invariant distribution is the *gamma distribution* with scale parameter 1 and shape parameter α. The eigenfunctions are the Laguerre polynomials.

Case 3: $a > 0$ and $\sigma^2(x) = 2\beta a(x^2 + 1).$ The state space is the real line, and the scale density is given by $s(x) = (x^2 + 1)^{\frac{1}{2a}} \exp(-\frac{\alpha}{a} \tan^{-1} x)$. By Condition 1.4, the solution is ergodic for all $a > 0$ and all $\alpha \in \mathbb{R}$. The invariant density is given by $\mu_\theta(x) \propto (x^2 + 1)^{-\frac{1}{2a}-1} \exp(\frac{\alpha}{a} \tan^{-1} x)$ If $\alpha = 0$ the invariant distribution is a scaled *t-distribution* with $\nu = 1 + a^{-1}$ degrees of freedom and scale parameter $\nu^{-\frac{1}{2}}$. If $\alpha \neq 0$ the invariant distribution is skew and has tails decaying at the same rate as the t-distribution with $1+a^{-1}$ degrees of freedom. A fitting name for this distribution is the *skew t-distribution*. It is also known as *Pearson's type IV distribution*. In either case, the mean is α and the invariant distribution has moments of order k for $k < 1 + a^{-1}$. Because of its skew and heavy tailed marginal distribution, the class of diffusions with $\alpha \neq 0$ is potentially very useful in many applications, e.g. finance. It was studied and fitted to financial data by Nagahara (1996) using the local linearization method of Ozaki (1985). We consider this process in more detail below.

Case 4: $a > 0$ and $\sigma^2(x) = 2\beta a x^2.$ The state space is $(0, \infty)$ and the scale density is $s(x) = x^{\frac{1}{a}} \exp(\frac{\alpha}{ax})$. Condition 1.4 holds if and only if $\alpha > 0$. The invariant distribution is given by $\mu_\theta(x) \propto x^{-\frac{1}{a}-2} \exp(-\frac{\alpha}{ax})$, and is thus an *inverse gamma distribution* with shape parameter $1 + \frac{1}{a}$ and scale parameter $\frac{\alpha}{a}$. The invariant distribution has moments of order k for $k < 1 + a^{-1}$. This process is sometimes referred to as the GARCH diffusion model. The polynomial eigenfunctions are known as the Bessel polynomials.

Case 5: $a > 0$ and $\sigma^2(x) = 2\beta a x(x + 1).$ The state space is $(0, \infty)$ and the scale density is $s(x) = (1 + x)^{\frac{\alpha+1}{a}} x^{-\frac{\alpha}{a}}$. The ergodicity Condition 1.4 holds if and only if $\frac{\alpha}{a} \geq 1$. Hence for all $a > 0$ and all $\mu \geq a$, a unique ergodic solution to (1.81) exists. If $0 < \alpha < 1$, the boundary 0 can be reached at a finite time point with positive probability, but if the boundary is made instantaneously reflecting, a stationary process is obtained. The density of the invariant distribution is given by $\mu_\theta(x) \propto (1+x)^{-\frac{\alpha+1}{a}-1} x^{\frac{\alpha}{a}-1}$. This is a scaled

F-distribution with $\frac{2\alpha}{a}$ and $\frac{2}{a}+2$ degrees of freedom and scale parameter $\frac{\alpha}{1+a}$. The invariant distribution has moments of order k for $k < 1 + a^{-1}$.

Case 6: $a < 0$ and $\sigma^2(x) = 2\beta a x(x - 1)$. The state space is $(0, \infty)$ and the scale density is $s(x) = (1 - x)^{\frac{1-\alpha}{a}} x^{\frac{\alpha}{a}}$. Condition 1.4 holds if and only if $\frac{\alpha}{a} \leq -1$ and $\frac{1-\alpha}{a} \leq -1$. Hence for all $a < 0$ and all $\alpha > 0$ such that $\min(\alpha, 1-\alpha) \geq -a$, a unique ergodic solution to (1.81) exists. If $0 < \alpha < -a$, the boundary 0 can be reached at a finite time point with positive probability, but if the boundary is made instantaneously reflecting, a stationary process is obtained. Similar remarks apply to the boundary 1 when $0 < 1-\alpha < -a$. The invariant distribution is given by $\mu_\theta(x) \propto (1-x)^{-\frac{1-\alpha}{a}-1} x^{-\frac{\alpha}{a}-1}$ and is thus the *Beta distribution* with shape parameters $\frac{\alpha}{-a}$ and $\frac{1-\alpha}{-a}$. This class of diffusions will be discussed in more detail below. It is often referred to as the *Jacobi diffusions* because the related eigenfunctions are Jacobi polynomials. Multivariate Jacobi diffusions were considered by Gourieroux and Jasiak (2006).

Example 1.21 The *skew t-distribution* with mean zero, ν degrees of freedom, and skewness parameter ρ has (unnormalized) density

$$f(z) \propto$$
$$\left\{(z/\sqrt{\nu} + \rho)^2 + 1\right\}^{-(\nu+1)/2} \exp\left\{\rho(\nu - 1)\tan^{-1}\left(z/\sqrt{\nu} + \rho\right)\right\},$$

which is the invariant density of the diffusion $Z_t = \sqrt{\nu}(X_t - \rho)$ with $\nu = 1 + a^{-1}$ and $\rho = \alpha$, where X is as in Case 3. An expression for the normalizing constant when ν is integer valued was derived in Nagahara (1996). By the transformation result above, the corresponding stochastic differential equation is

$$dZ_t = -\beta Z_t dt + \sqrt{2\beta(\nu - 1)^{-1}\{Z_t^2 + 2\rho\nu^{\frac{1}{2}} Z_t + (1 + \rho^2)\nu\}} dW_t. \quad (1.84)$$

For $\rho = 0$ the invariant distribution is the *t*-distribution with ν degrees of freedom.

The skew *t*-diffusion (1.84) has the eigenvalues $\lambda_n = n(\nu - n)(\nu - 1)^{-1}\beta$ for $n < \nu$. The four first eigenfunctions are

$$p_1(z) = z,$$
$$p_2(z) = z^2 - \frac{4\rho\nu^{\frac{1}{2}}}{\nu - 3} z - \frac{(1+\rho^2)\nu}{\nu - 2},$$
$$p_3(z) = z^3 - \frac{12\rho\nu^{\frac{1}{2}}}{\nu - 5} z^2 + \frac{24\rho^2\nu + 3(1+\rho^2)\nu(\nu - 5)}{(\nu - 5)(\nu - 4)} z + \frac{8\rho(1+\rho^2)\nu^{\frac{3}{2}}}{(\nu-5)(\nu-3)},$$

and

$$p_4(z) = z^4 - \frac{24\rho\nu^{\frac{1}{2}}}{\nu - 7}z^3 + \frac{144\rho^2\nu - 6(1 + \rho^2)\nu(\nu - 7)}{(\nu - 7)(\nu - 6)}z^2$$

$$+ \frac{8\rho(1 + \rho^2)\nu^{\frac{3}{2}}(\nu - 7) + 48\rho(1 + \rho^2)\nu^{\frac{3}{2}}(\nu - 6) - 192\rho^3\nu^{\frac{3}{2}}}{(\nu - 7)(\nu - 6)(\nu - 5)}z$$

$$+ \frac{3(1 + \rho^2)^2\nu(\nu - 7) - 72\rho^2(1 + \rho^2)\nu^2}{(\nu - 7)(\nu - 6)(\nu - 4)},$$

provided that $\nu > 4$. If $\nu > 2i$, the first i eigenfunctions are square integrable and thus satisfy (1.72). Hence (1.73) holds, and the eigenfunctions can be used to construct explicit martingale estimating functions.   □

Example 1.22 The model

$$dX_t = -\beta[X_t - (m + \gamma z)]dt + \sigma\sqrt{z^2 - (X_t - m)^2}dW_t, \qquad (1.85)$$

where $\beta > 0$ and $\gamma \in (-1, 1)$, has been proposed as a model for the random variation of the logarithm of an exchange rate in a target zone between realignments by De Jong, Drost, and Werker (2001) ($\gamma = 0$) and Larsen and Sørensen (2007). This is a diffusion on the interval $(m - z, m + z)$ with mean reversion around $m + \gamma z$. It is a *Jacobi diffusion* obtained by a location-scale transformation of the diffusion in Case 6 above. The parameter γ quantifies the asymmetry of the model. When $\beta(1 - \gamma) \geq \sigma^2$ and $\beta(1 + \gamma) \geq \sigma^2$, X is an ergodic diffusion, for which the stationary distribution is a Beta-distribution on $(m - z, m + z)$ with parameters $\kappa_1 = \beta(1 - \gamma)\sigma^{-2}$ and $\kappa_2 = \beta(1 + \gamma)\sigma^{-2}$. If the parameter restrictions are not satisfied, one or both of the boundaries can be hit in finite time, but if the boundaries are made instantaneously reflecting, a stationary process is obtained.

The eigenfunctions for the generator of the diffusion (1.85) are

$$\phi_i(x; \beta, \gamma, \sigma, m, z) = P_i^{(\kappa_1 - 1, \kappa_2 - 1)}((x - m)/z), \quad i = 1, 2, \dots$$

where $P_i^{(a,b)}(x)$ denotes the Jacobi polynomial of order i given by

$$P_i^{(a,b)}(x) = \sum_{j=0}^{i} 2^{-j}\binom{n + a}{n - j}\binom{a + b + n + j}{j}(x - 1)^j, \quad -1 < x < 1.$$

The eigenvalue of ϕ_i is $i(\beta + \frac{1}{2}\sigma^2(i - 1))$. Since (1.72) is obviously satisfied, (1.73) holds, so that the eigenfunctions can be used to construct explicit martingale estimating functions.   □

Explicit formulae for the *conditional moments* of a Pearson diffusion can be obtained from the eigenfunctions by means of (1.73). Specifically,

$$E(X_t^n \mid X_0 = x) = \sum_{k=0}^{n} \left(\sum_{\ell=0}^{n} q_{n,k,\ell} e^{-\lambda_\ell t} \right) x^k, \tag{1.86}$$

where $q_{n,k,n} = p_{n,k}$, $q_{n,n,\ell} = 0$ for $\ell \le n - 1$, and

$$q_{n,k,\ell} = - \sum_{j=k \vee \ell}^{n-1} p_{n,j} q_{j,k,\ell}$$

for $k, \ell = 0, \ldots, n - 1$ with λ_ℓ and $p_{n,j}$ given by (1.82) and (1.83). For details see Forman and Sørensen (2008).

Also the *moments* of the Pearson diffusions can, when they exist, be found explicitly by using the fact that the integral of the eigenfunctions with respect to the invariant probability measure is zero. We have seen above that $E(|X_t|^\kappa) < \infty$ if and only if $a < (\kappa - 1)^{-1}$. Thus if $a \le 0$ all moments exist, while for $a > 0$ only the moments satisfying that $\kappa < a^{-1} + 1$ exist. In particular, the expectation always exists. The moments of the invariant distribution can be found by the recursion

$$E(X_t^n) = a_n^{-1} \{ b_n \cdot E(X_t^{n-1}) + c_n \cdot E(X_t^{n-2}) \}, \quad n = 2, 3, \ldots, \tag{1.87}$$

where $a_n = n\{1 - (n-1)a\}\beta$, $b_n = n\{\alpha + (n-1)b\}\beta$, and $c_n = n(n-1)c\beta$. The initial conditions are given by $E(X_t^0) = 1$, and $E(X_t) = \alpha$. This can be found from the expressions for the eigenfunctions, but is more easily seen as follows. By Ito's formula

$$
\begin{aligned}
dX_t^n = & -\beta n X_t^{n-1}(X_t - \alpha)dt + \beta n(n-1)X_t^{n-2}(aX_t^2 + bX_t + c)dt \\
& + n X_t^{n-1} \sigma(X_t) dW_t,
\end{aligned}
$$

and if $E(X_t^{2n})$ is finite, i.e. if $a < (2n - 1)^{-1}$, the last term is a martingale with expectation zero.

Example 1.23 Equation (1.87) allows us to find the moments of the *skewed t-distribution*, in spite of the fact that the normalizing constant of the density is unknown. In particular, for the diffusion (1.84),

$$
\begin{aligned}
E(Z_t) &= 0, \\
E(Z_t^2) &= \frac{(1 + \rho^2)\nu}{\nu - 2}, \\
E(Z_t^3) &= \frac{4\rho(1 + \rho^2)\nu^{\frac{3}{2}}}{(\nu - 3)(\nu - 2)}, \\
E(Z_t^4) &= \frac{24\rho^2(1 + \rho^2)\nu^2 + 3(\nu - 3)(1 + \rho^2)^2\nu^2}{(\nu - 4)(\nu - 3)(\nu - 2)}.
\end{aligned}
$$

□

For a diffusion $T(X)$ obtained from a solution X to (1.81) by a twice differentiable and invertible transformation T, the eigenfunctions of the generator are $p_n\{T^{-1}(x)\}$, where p_n is an eigenfunction of the generator of X. The eigenvalues are the same as for the original eigenfunctions. Since the original eigenfunctions are polynomials, the eigenfunctions of $T(X)$ are of the form (1.77) with $\kappa = T^{-1}$. Hence *explicit optimal martingale estimating functions are also available for transformations of Pearson diffusions*, which is a very large and flexible class of diffusion processes. Their stochastic differential equations can, of course, be found by Ito's formula.

Example 1.24 For the Jacobi-diffusion (case 6) with $\alpha = -a = \frac{1}{2}$, i.e.

$$dX_t = -\beta(X_t - \tfrac{1}{2})dt + \sqrt{\beta X_t(1 - X_t)}dW_t$$

the invariant distribution is the uniform distribution on $(0,1)$ for all $\beta > 0$. For any strictly increasing and twice differentiable distribution function F, we therefore have a class of diffusions, given by $Y_t = F^{-1}(X_t)$, for which the invariant distribution has density $f = F'$. The diffusion Y_t solves

$$dY_t = -\beta\frac{(F(Y_t) - \frac{1}{2})f(Y_t)^2 + \frac{1}{2}F(Y_t)\{1 - F(Y_t)\}f'(Y_t)}{f(Y_t)^3}dt$$
$$+ \frac{\sqrt{\beta F(Y_t)\{1 - F(Y_t)\}}}{f(Y_t)}dW_t.$$

A particular example is the logistic distribution

$$F(x) = \frac{e^x}{1 + e^x}, \quad x \in \mathbb{R},$$

for which

$$dY_t = -\beta\left\{\sinh(Y_t) + 8\cosh^4(Y_t/2)\right\}dt + 2\sqrt{\beta}\cosh(Y_t/2)dW_t.$$

If the same transformation, $F^{-1}(y) = \log(y/(1-y))$, is applied to the general Jacobi diffusion (case 6), then we obtain

$$dY_t = -\beta\left\{1 - 2\alpha + (1 - \alpha)e^{Y_t} - \alpha e^{-Y_t} - 16a\cosh^4(Y_t/2)\right\}dt$$
$$+2\sqrt{-2a\beta}\cosh(Y_t/2)dW_t,$$

a diffusion for which the invariant distribution is the generalized logistic distribution with density

$$f(x) = \frac{e^{\kappa_1 x}}{(1 + e^x)^{\kappa_1 + \kappa_2}B(\kappa_1, \kappa_2)}, \quad x \in \mathbb{R},$$

where $\kappa_1 = -(1 - \alpha)/a$, $\kappa_2 = \alpha/a$ and B denotes the Beta-function. This distribution was introduced and studied in Barndorff–Nielsen, Kent, and Sørensen (1982).

☐

Example 1.25 Let again X be a general Jacobi-diffusion (case 6). If we apply the transformation $T(x) = \sin^{-1}(2x - 1)$ to X_t, we obtain the diffusion

$$dY_t = -\rho \frac{\sin(Y_t) - \varphi}{\cos(Y_t)} dt + \sqrt{-a\beta/2} dW_t,$$

where $\rho = \beta(1 + a/4)$ and $\varphi = (2\alpha - 1)/(1 + a/4)$. The state space is $(-\pi/2, \pi/2)$. Note that Y has dynamics that are very different from those of the Jacobi diffusion: the drift is highly non-linear and the diffusion coefficient is constant. This model was considered in Example 1.18.

☐

1.3.8 Implementation of martingale estimating functions

An R-package, where a number of methods for calculating estimators for diffusion models are implemented, has been developed by Stefano Iacus and is described in the book Iacus (2008), which outlines the underlying theory too. The R-package also contains implementations of methods for simulating solutions to stochastic differential equations. It is, however, useful to notice that for many martingales estimating functions the estimators, or asymptotically equivalent estimators, can be calculated by means of standard statistical software packages. Specifically, they can be calculated as weighted least squares estimators for non-linear regression models.

To see this, consider the weighted least squares estimator obtained by minimizing

$$C_n(\theta) = \tag{1.88}$$

$$\sum_{i=1}^{n} \left[f(X_{t_i}) - \pi_\Delta^\theta(f)(X_{t_{i-1}})\right]^T V_i^{-1} \left[f(X_{t_i}) - \pi_\Delta^\theta(f)(X_{t_{i-1}})\right],$$

with $f(x) = (f_1(x), \ldots, f_N(x))$ and

$$V_i = V_h(X_{t_{i-1}}; \tilde{\theta}_n), \tag{1.89}$$

where V_h is the $N \times N$-matrix given by (1.44), and $\tilde{\theta}_n$ is a consistent estimator of θ. As usual, π_Δ^θ denotes the transition operator (1.42). The consistent estimator can, for instance, be the non-weighted least squares estimator obtained by minimizing (1.88) with $V_i = I_N$, where I_N is the $N \times N$ identity matrix. The weighted least squares estimator obtained from (1.88) with V_i given by (1.89) solves the estimating equation

$$\sum_{i=1}^{n} B_h(X_{t_{i-1}}; \theta) V_h(X_{t_{i-1}}; \tilde{\theta}_n)^{-1} \left[f(X_{t_i}) - \pi_\Delta^\theta(f)(X_{t_{i-1}})\right] = 0 \tag{1.90}$$

with B_h given by (1.45). Therefore this estimator has the same asymptotic variance as the optimal G_n^*-estimator with h given by (1.40); see e.g. Jacod and Sørensen (2012). The estimating function (1.90) is similar in spirit to (1.39). The estimators obtained by minimizing (1.88) is the weighted least squares estimator for a regression model for the data $f(X_{t_i})$, $i = 1, \ldots, n$ with $X_{t_{i-1}}$, $i = 1, \ldots, n$, as explanatory regression variables, the non-linear regression function $\pi_\Delta^\theta(f)(X_{t_{i-1}})$, and the weight matrix V_i. In some particularly nice cases, the regression function is linear in the parameters, and the estimator is a linear regression estimator.

Example 1.26 Let X be the square root process (1.37), and suppose we have the observations $X_{i\Delta}$, $i = 0, \ldots, n$. Let us think of $(X_{i\Delta}, X_{i\Delta}^2)$, $i = 1, \ldots, n$, as data with explanatory regression variables $X_{(i-1)\Delta}$, $i = 1, \ldots, n$, and with the non-linear regression function

$$\pi_\Delta^\theta(f)(X_{(i-1)\Delta}) = \begin{pmatrix} F(X_{(i-1)\Delta}; \theta) \\ \phi(X_{(i-1)\Delta}; \theta) + F(X_{(i-1)\Delta}; \theta)^2 \end{pmatrix},$$

where F and ϕ are as in Example 1.9. Then we obtain a weighted least squares estimator for θ, by minimizing (1.88) with $f_1(x) = x$, $f_2(x) = x^2$, and

$$V_h(x; \theta) =$$

$$\begin{pmatrix} \phi(x; \theta) & \eta(x; \theta) + 2F(x; \theta)\phi(x; \theta) \\ \eta(x; \theta) + 2F(x; \theta)\phi(x; \theta) & \psi(x; \theta) + 4F(x; \theta)^2\phi(x; \theta) + 4F(x; \theta)\eta(x; \theta) \end{pmatrix},$$

where η and ψ are as in Example 1.9.

This estimator has the same efficiency as the estimator obtained from the optimal martingale estimating function of form (1.30) with $N = 2$ and

$$\begin{aligned} h_1(x, y; \theta) &= y - F(x; \theta) \\ h_2(x, y; \theta) &= y^2 - \phi(x; \theta) - F(x; \theta)^2. \end{aligned}$$

The optimal estimating function of this form is equivalent to the optimal estimating function in Example 1.9. For the square root process some simplification can be achieved by using the Gaussian approximation (1.38) in the definition of the matrix V_h.

□

Example 1.27 Consider the process (1.74), and suppose we have the observations $X_{i\Delta}$, $i = 0, \ldots, n$. Let us think of $\sin(X_{i\Delta})$, $i = 1, \ldots, n$, as data with explanatory regression variables $X_{(i-1)\Delta}$, $i = 1, \ldots, n$ and with the non-linear regression function $\pi_\Delta^\theta(\sin)(X_{(i-1)\Delta}) = e^{-(\theta+1/2)\Delta} \sin(X_{(i-1)\Delta})$. Again we can obtain a weighted least squares estimator for θ, by minimizing (1.88) with

$f(x) = \sin(x)$ and

$$V_h(x; \theta) = \tfrac{1}{2}(e^{2(\tilde{\theta}_n + 1)\Delta} - 1)/(\tilde{\theta}_n + 1) - (e^\Delta - 1)\sin^2(X_{(i-1)\Delta}),$$

where $\tilde{\theta}_n$ is a consistent estimator, for instance the simple estimator (1.76). Note that the non-linear regression is, in fact, a linear regression in the parameter $\xi = e^{-\theta\Delta}$. The regression estimator equals the estimator obtained from the estimating function

$$G_n^\bullet(\theta) = \sum_{i=1}^n \frac{\sin(X_{(i-1)\Delta})[\sin(X_{i\Delta}) - e^{-(\theta + \frac{1}{2})\Delta}\sin(X_{(i-1)\Delta})]}{\tfrac{1}{2}(e^{2(\tilde{\theta}_n + 1)\Delta} - 1)/(\tilde{\theta}_n + 1) - (e^\Delta - 1)\sin^2(X_{(i-1)\Delta})},$$

which has the same efficiency as the optimal estimator obtained in Example 1.20. If instead we minimize (1.88) with the approximation $V_h(x; \theta) = \cos^2(x)$, then we obtain the estimator $\tilde{\theta}_n$ from Example 1.20, and if we minimize (1.88) with the more crude approximation $V_h(x; \theta) = 1$, then we obtain the simple estimator (1.76) from Example 1.18.

$\square$

More generally, an estimator with the same efficiency as the optimal estimator from (1.30) with optimal weights (1.32) is obtained by minimizing the objective function

$$\sum_{i=1}^n h(X_{t_{i-1}}, X_{t_i}; \theta)^T V_i^{-1} h(X_{t_{i-1}}, X_{t_i}; \theta) \tag{1.91}$$

with V_i defined as in (1.89), but here with V_h given by (1.34). This estimator can be found by applying standard software for minimizing objective functions to (1.91).

Example 1.28 Let again X be the square root process (1.37), and consider the martingale estimating function of form (1.30) with $N = 2$ and h_1 and h_2 as in Example 1.7. In this case an optimal estimator is obtained by minimizing (1.91) with

$$V_h(x; \theta) = \begin{pmatrix} \phi(x; \theta) & \eta(x; \theta) \\ \eta(x; \theta) & \psi(x; \theta) \end{pmatrix},$$

where ϕ, η and ψ are as in Example 1.9. Here a considerable simplification can be obtained by the Gaussian approximation (1.38). With this approximation

$$V_h(x; \theta) = \begin{pmatrix} \phi(x; \theta) & 0 \\ 0 & 2\phi(x; \theta)^2 \end{pmatrix}.$$

$\square$

1.4 The likelihood function

The likelihood function for a discretely observed diffusion model, (1.24) is a product of transitions densities. Unfortunately, the transition density of a diffusion process is only rarely explicitly known, but several numerical approaches make likelihood inference feasible for diffusion models.

Pedersen (1995) proposed a method for obtaining an approximation to the likelihood function by rather extensive simulation. Pedersen's method was very considerably improved by Durham and Gallant (2002), whose method is computationally much more efficient. Poulsen (1999) obtained an approximation to the transition density by numerically solving a partial differential equation, whereas Aït–Sahalia (2002, 2008) proposed to approximate the transition density by means of expansions. A Gaussian approximation to the likelihood function obtained by local linearization of (1.11) was proposed by Ozaki (1985), while Forman and Sørensen (2008) proposed to use an approximation in terms of eigenfunctions of the generator of the diffusion. Bayesian estimators with the same asymptotic properties as the maximum likelihood estimator can be obtained by Markov chain Monte Carlo methods, see Elerian, Chib, and Shephard (2001), Eraker (2001), and Roberts and Stramer (2001). Finally, exact and computationally efficient likelihood-based estimation methods were presented by Beskos, Papaspiliopoulos, Roberts, and Fearnhead (2006). The latter approach is presented in Chapter 4. In the following we will outline the expansion approach of Aït–Sahalia (2002) for scalar diffusion models. The various other approaches to calculation of the likelihood function will not be considered further in this chapter.

Assume that the diffusion process (1.11) is one-dimensional and that the state space is either $(-\infty, \infty)$ or $(0, \infty)$, i.e. $r = \infty$ and ℓ is either $-\infty$ or 0. The coefficients b and σ are assumed to satisfy the following condition.

Condition 1.29
(i) The functions $b(x; \theta)$ and $\sigma(x; \theta)$ are infinitely often differentiable w.r.t. x and three times continuously differentiable w.r.t. θ for all $x \in (\ell, r)$ and $\theta \in \Theta$.

(ii-a) If $\ell = -\infty$, there exists a constant $c > 0$ such that $\sigma(x; \theta) > c$ for all $x \in (\ell, r)$ and all $\theta \in \Theta$.

(ii-b) If $\ell = 0$, then σ is non-degenerate on $(0, \infty)$ in the sense that for each $\xi > 0$ there exists a constant $c_\xi > 0$ such that $\sigma(x; \theta) \geq c_\xi$ for all $x \geq \xi$ and all $\theta \in \Theta$. Moreover, if $\lim_{x \to 0} \sigma(x; \theta) = 0$, then constants ξ_0, ω and ρ exist such that $\sigma(x; \theta) \geq \omega x^\rho$ for all $x \in (0, \xi_0)$ and all $\theta \in \Theta$.

The idea is to make an expansion of the transition density. However, the distribution of X_Δ given X_0 can be so far from a normal distribution that a convergent expansion with the normal density as the leading term is not possible.

This is possible for a diffusion with a constant diffusion coefficient. Therefore, the standard transformation

$$h(x; \theta) = \int_{x^*}^{x} \frac{1}{\sigma(u; \theta)} du,$$

where x^* is arbitrary, is applied to obtain the diffusion process

$$Y_t = h(X_t; \theta).$$

Since $\sigma > 0$, the transformation h is increasing, and by Ito's formula

$$dY_t = a(Y_t; \theta)dt + dW_t, \tag{1.92}$$

where

$$a(y; \theta) = \frac{b(h^{-1}(y; \theta); \theta)}{\sigma(h^{-1}(y; \theta); \theta)} - \tfrac{1}{2}\sigma'(h^{-1}(y; \theta); \theta)$$

with $\sigma'(x; \theta) = \partial_x \sigma(x; \theta)$. The state space of Y, (ℓ_Y, r_Y) could in principle depend on θ, but we assume that this is not the case. If only one of the boundaries ℓ_Y and r_Y is finite, it can always be arranged that the finite boundary equals zero by choosing x^* suitably. For instance if $r_Y = \infty$ and ℓ_Y is finite, then we can choose $x^* = \ell$ to obtain $\ell_Y = 0$. We will assume that ℓ_Y is either $-\infty$ or 0, and that r_Y is either 0 or ∞. It is further assumed that a satisfies the following condition (which can be translated into a condition on b and σ).

Condition 1.30
(i) For all $\theta \in \Theta$, the drift coefficient $a(y; \theta)$ and its derivatives w.r.t. y and θ have at most polynomial growth near the boundaries, and

$$\lim[a(y; \theta)^2 + \partial_y a(y; \theta)] > -\infty \quad \text{as } y \downarrow \ell_Y \quad \text{and} \quad y \uparrow r_Y.$$

(ii-a) If $\ell_Y = 0$, then there exist constants $\epsilon_0 > 0$, κ and α such that $a(y; \theta) \geq \kappa y^{-\alpha}$ for all $y \in (0, \epsilon_0)$ and all $\theta \in \Theta$, where either $\alpha > 1$ and $\kappa > 0$, or $\alpha = 1$ and $\kappa \geq 1$. If $\ell_Y = -\infty$, then there exists constants $E_0 > 0$ and $K > 0$ such that $a(y; \theta) \geq Ky$ for all $y \leq -E_0$ and all $\theta \in \Theta$.

(ii-b) If $r_Y = 0$, then there exist constants $\epsilon_0 > 0$, κ and α such that $a(y; \theta) \leq -\kappa|y|^{-\alpha}$ for all $y \in (-\epsilon_0, 0)$ and all $\theta \in \Theta$, where either $\alpha > 1$ and $\kappa > 0$, or $\alpha = 1$ and $\kappa \geq 1/2$. If $r_Y = \infty$, then there exist constants $E_0 > 0$ and $K > 0$ such that $a(y; \theta) \leq Ky$ for all $y \geq E_0$ and all $\theta \in \Theta$.

A real function f is said to be of polynomial growth near a boundary at ∞ or $-\infty$ if there exist constants $C > 0$, $K > 0$ and $p > 0$ such that $|f(x)| \leq C|x|^p$ for $x > K$ or $x < -K$. If the boundary is at zero, polynomial growth means that there exist constants $C > 0$, $\epsilon > 0$ and $p > 0$ such that $|f(x)| \leq C|x|^{-p}$ for $|x| \leq \epsilon$.

Under the assumptions imposed, a solution exists to (1.92) with a transition density that is sufficiently regular for likelihood inference. This is the content of the following proposition from Aït–Sahalia (2002).

Proposition 1.31 *Under the Conditions 1.29 and 1.30, the stochastic differential equation (1.92) has a unique weak solution for every initial distribution. The boundaries are unattainable. The solution Y has a transition density $p_Y(\Delta, y_0, y; \theta)$ that is continuously differentiable w.r.t. Δ, infinitely often differentiable w.r.t. $y \in (\ell_Y, r_Y)$, and three times continuously differentiable w.r.t. $\theta \in \Theta$.*

This result implies that the original stochastic differential equation (1.11) has a unique weak solution, and by the transformation theorem for density functions, it has a similarly regular transition density given by

$$p(\Delta, x_0, x; \theta) = p_Y(\Delta, h(x_0; \theta), h(x; \theta); \theta)/\sigma(x; \theta). \qquad (1.93)$$

Instead of expanding the transition density of Y, i.e. the conditional density function of Y_Δ given $Y_0 = y_0$, we expand the conditional density of the normalized increment

$$Z = \Delta^{-1/2}(Y_\Delta - y_0)$$

given $Y_0 = y_0$. This is because p_Y gets peaked around y_0 as Δ gets close to zero, whereas the distribution of Z is sufficiently close to the $N(0,1)$-distribution to make it the appropriate transformation of X_Δ to obtain a convergent expansion of the conditional density function with the standard normal density function as the leading term. Obviously,

$$p_Y(\Delta, y_0, y; \theta) = \Delta^{-1/2} p_Z(\Delta, \Delta^{-1/2}(y - y_0) \mid y_0; \theta), \qquad (1.94)$$

where $p_Z(\Delta, z \mid y_0; \theta)$ is the conditional density of Z given that $Y_0 = y_0$.

We can now obtain an approximation to the transition density of X, and hence an approximation to the likelihood function, by expanding the conditional density, p_Z, of Z given $Y_0 = y_0$ in terms of Hermite polynomials up to order J:

$$p_Z^J(\Delta, z \mid y_0; \theta) = \varphi(z) \sum_{j=0}^{J} \eta_j(\Delta, y_0; \theta) H_j(z), \qquad (1.95)$$

where φ denotes the density function of the standard normal distribution, and H_j is the jth Hermite polynomial, which is defined by

$$H_j(x) = (-1)^j e^{x^2/2} \frac{d^j}{dx^j} e^{-x^2/2}, \quad j = 0, 1, \dots.$$

The Hermite polynomials up to order 4 are

$$
\begin{aligned}
H_0(x) &= 1 \\
H_1(x) &= x \\
H_2(x) &= x^2 - 1 \\
H_3(x) &= x^3 - 3x \\
H_4(x) &= x^4 - 6x^2 + 3.
\end{aligned}
$$

The coefficients $\eta_j(\Delta, y_0; \theta)$ can be found by using that the Hermite polynomials are orthogonal in the space $L^2(\varphi)$:

$$\int_{-\infty}^{\infty} H_i(x)H_j(x)\varphi(x)dx = \begin{cases} 0 & \text{if } i \neq j \\ i! & \text{if } i = j. \end{cases}$$

Hence if

$$p_Z(\Delta, z \mid y_0; \theta) = \varphi(z) \sum_{j=0}^{\infty} \eta_j(\Delta, y_0; \theta)H_j(z),$$

it follows that

$$\begin{aligned} \int_{-\infty}^{\infty} H_i(z)p_Z(\Delta, z \mid y_0; \theta)dz &= \sum_{j=0}^{\infty} \eta_j(\Delta, y_0; \theta) \int_{-\infty}^{\infty} H_i(z)H_j(z)\varphi(z)dz \\ &= i!\,\eta_i(\Delta, y_0; \theta). \end{aligned}$$

By inserting the expansion (1.95) in (1.94) and (1.93), we obtain the following approximations to the transition densities p_Y and p

$$p_Y^J(\Delta, y_0, y; \theta) = \Delta^{-1/2}\varphi(\Delta^{-1/2}(y-y_0)) \sum_{j=0}^{J} \eta_j(\Delta, y_0; \theta)H_j(\Delta^{-1/2}(y-y_0))$$

(1.96)

and

$$p^J(\Delta, x_0, x; \theta) =$$

(1.97)

$$\frac{\varphi\left(\frac{h(x;\theta)-h(x_0;\theta)}{\sqrt{\Delta}}\right)}{\sqrt{\Delta}\,\sigma(x;\theta)} \sum_{j=0}^{J} \eta_j(\Delta, h(x_0;\theta); \theta)H_j\left(\frac{h(x;\theta)-h(x_0;\theta)}{\sqrt{\Delta}}\right).$$

Aït–Sahalia (2002) gave the following theorem about the convergence of the approximation p^J to the exact transition density p.

Theorem 1.32 *Under the Conditions 1.29 and 1.30, there exists $\bar{\Delta} > 0$ such that*

$$\lim_{J \to \infty} p^J(\Delta, x_0, x; \theta) = p(\Delta, x_0, x; \theta)$$

for all $\Delta \in (0, \bar{\Delta})$, $\theta \in \Theta$ and $(x_0, x) \in (\ell, r)^2$.

If $r_Y = \infty$ and $a(y; \theta) \leq 0$ near r_Y, and if $a(y; \theta) \geq 0$ near ℓ_Y (which is either 0 or $-\infty$), then $\bar{\Delta} = \infty$, see Proposition 2 in Aït–Sahalia (2002).

In order to use the expansions of the transition densities to calculate likelihood functions in practice, it is necessary to determine the coefficients $\eta_j(\Delta, y_0; \theta)$.

Note that by inserting (1.94) in the expression above for $\eta_i(\Delta, y_0; \theta)$ we find that

$$
\begin{aligned}
\eta_i(\Delta, y_0; \theta) &= \frac{1}{i!} \int_{-\infty}^{\infty} H_i(z) \Delta^{1/2} p_Y(\Delta, y_0, \Delta^{1/2} z + y_0; \theta) dz \\
&= \frac{1}{i!} \int_{-\infty}^{\infty} H_i(\Delta^{-1/2}(y - y_0)) p_Y(\Delta, y_0, y; \theta) dy \\
&= \frac{1}{i!} \mathrm{E}_\theta \left(H_i(\Delta^{-1/2}(Y_\Delta - y_0)) \,|\, Y_0 = y_0 \right).
\end{aligned}
$$

Thus the coefficients $\eta_i(\Delta, y_0; \theta)$, $i = 0, 1, \ldots$, are conditional moments of the process Y, and can therefore be found by simulation of Y or X. An approximation to $\eta_i(\Delta, y_0; \theta)$ can be obtained by applying the expansion (1.57) to the functions $(y - x)^i$, $i = 1, \ldots, J$. For instance, we find that

$$
\begin{aligned}
\eta_1(\Delta, y_0; \theta) &= \Delta^{1/2} a(y_0; \theta) + \tfrac{1}{2}\Delta^{3/2} \left(a(y_0; \theta)\partial_y a(y_0; \theta) + \tfrac{1}{2}\partial_y^2 a(y_0; \theta) \right) \\
&\qquad\qquad + O(\Delta^{5/2}) \\
\eta_2(\Delta, y_0; \theta) &= \Delta \left(a(y_0; \theta)^2 + \partial_y a(y_0; \theta) \right) + O(\Delta^2).
\end{aligned}
$$

By expanding the coefficients $\eta_i(\Delta, y_0; \theta)$ suitably and collecting terms of the same order in Δ, Aït–Sahalia (2002) found the following approximation to p_Y

$$
\tilde{p}_Y^K(\Delta, y_0, y; \theta) =
$$

$$
\Delta^{-1/2} \varphi\left(\frac{y - y_0}{\sqrt{\Delta}} \right) \exp\left(\int_{y_0}^{y} a(w, \theta) dw \right) \sum_{k=0}^{K} \frac{\Delta^k}{k!} c_k(y_0, y; \theta),
$$

where $c_0(y_0, y; \theta) = 1$, and

$$
c_k(y_0, y; \theta) =
$$

$$
k(y - y_0)^{-k} \int_{y_0}^{y} (w - y_0)^{k-1} \left[\lambda(w; \theta) c_{k-1}(y_0, w; \theta) + \tfrac{1}{2}\partial_w^2 c_{k-1}(y_0, w; \theta) \right] dw,
$$

for $k \geq 1$, where

$$
\lambda(w; \theta) = -\tfrac{1}{2} \left(a(w; \theta)^2 + \partial_w a(w; \theta) \right).
$$

1.5 Non-martingale estimating functions

1.5.1 Asymptotics

When the estimating function

$$
G_n(\theta) = \sum_{i=r}^{n} g(X_{(i-r+1)\Delta}, \ldots, X_{i\Delta}; \theta)
$$

is not a martingale under P_θ, further conditions on the diffusion process must be imposed to ensure the asymptotic normality in (1.5). A sufficient condition that (1.5) holds under P_{θ_0} with $V(\theta)$ given by (1.98) is that the diffusion process is stationary and geometrically α-mixing, that

$$
\begin{aligned}
V(\theta) \;=\; & Q_{\theta_0}\left(g(\theta)g(\theta)^T\right) \\[4pt]
& + \sum_{k=1}^{\infty} \left[E_{\theta_0}\left(g(X_\Delta,\ldots,X_{r\Delta})g(X_{(k+1)\Delta},\ldots,X_{(k+r)\Delta})^T \right) \right. \\[4pt]
& \left. \qquad + E_{\theta_0}\left(g(X_{(k+1)\Delta},\ldots,X_{(k+r)\Delta})g(X_\Delta,\ldots,X_{r\Delta})^T \right) \right],
\end{aligned}
$$
(1.98)

converges and is strictly positive definite, and that $Q_{\theta_0}(g_i(\theta)^{2+\epsilon}) < \infty$, $i = 1,\ldots,p$ for some $\epsilon > 0$, see e.g. Doukhan (1994). Here g_i is the ith coordinate of g, and Q_θ is the joint distribution of $X_\Delta,\ldots,X_{r\Delta}$ under P_θ. To define the concept of α-mixing, let $\mathcal{F}_t$ denote the σ-field generated by $\{X_s \mid s \le t\}$, and let $\mathcal{F}^t$ denote the σ-field generated by $\{X_s \mid s \ge t\}$. A stochastic process X is said to be α-*mixing* under P_{θ_0}, if

$$
\sup_{A \in \mathcal{F}_t, B \in \mathcal{F}^{t+u}} |P_{\theta_0}(A)P_{\theta_0}(B) - P_{\theta_0}(A \cap B)| \le \alpha(u)
$$

for all $t > 0$ and $u > 0$, where $\alpha(u) \to 0$ as $u \to \infty$. This means that X_t and X_{t+u} are almost independent, when u is large. If positive constants c_1 and c_2 exist such that

$$
\alpha(u) \le c_1 e^{-c_2 u},
$$

for all $u > 0$, then the process X is called geometrically α-mixing. For one-dimensional diffusions there are simple conditions for geometric α-mixing. If all non-zero eigenvalues of the generator (1.46) are larger than some $\lambda > 0$, then the diffusion is geometrically α-mixing with $c_2 = \lambda$. This is for instance the case if the spectrum of the generator is discrete. Ergodic diffusions with a linear drift $-\beta(x - \alpha)$, $\beta > 0$, for instance the Pearson diffusions, are geometrically α-mixing with $c_2 = \beta$; see Hansen, Scheinkman, and Touzi (1998).

Genon-Catalot, Jeantheau, and Larédo (2000) gave the following simple sufficient condition for the one-dimensional diffusion that solves (1.11) to be geometrically α-mixing, provided that it is ergodic with invariant probability density μ_θ.

Condition 1.33
(i) The function b is continuously differentiable with respect to x, and σ is twice continuously differentiable with respect to x, $\sigma(x;\theta) > 0$ for all $x \in (\ell,r)$, and a constant $K_\theta > 0$ exists such that $|b(x;\theta)| \le K_\theta(1+|x|)$ and $\sigma^2(x;\theta) \le K_\theta(1 + x^2)$ for all $x \in (\ell,r)$.

(ii) $\sigma(x;\theta)\mu_\theta(x) \to 0$ as $x \downarrow \ell$ and $x \uparrow r$.

(iii) $1/\gamma(x;\theta)$ *has a finite limit as* $x \downarrow \ell$ *and* $x \uparrow r$, *where* $\gamma(x;\theta) = \partial_x \sigma(x;\theta) - 2b(x;\theta)/\sigma(x;\theta)$.

Other conditions for geometric α-mixing were given by Veretennikov (1987), Hansen and Scheinkman (1995), and Kusuoka and Yoshida (2000).

For geometrically α-mixing diffusion processes and estimating functions G_n satisfying Condition 1.1, the existence of a $\bar{\theta}$-consistent and asymptotically normal G_n-estimator follows from Theorem 1.2, which also contains a result about eventual uniqueness of the estimator.

1.5.2 Explicit non-martingale estimating functions

Explicit martingale estimating functions are only available for the relatively small, but versatile, class of diffusions for which explicit eigenfunctions of the generator are available; see the Subsections 1.3.6 and 1.3.7. Explicit non-martingale estimating functions can be found for all diffusions, but cannot be expected to approximate the score functions as well as martingale estimating functions, and therefore usually give less efficient estimators. As usual we consider ergodic diffusion processes with invariant probability density μ_θ.

First we consider estimating functions of the form

$$G_n(\theta) = \sum_{i=1}^{n} h(X_{\Delta i};\theta), \qquad (1.99)$$

where h is a p-dimensional function. We assume that the diffusion is geometrically α-mixing, so that a central limit theorem holds (under an integrability condition), and that Condition 1.1 holds for $r = 1$ and $\bar{\theta} = \theta_0$. The latter condition simplifies considerably, because for estimating functions of the form (1.99), it does not involve the transition density, but only the invariant probability density μ_θ, which for one-dimensional ergodic diffusions is given explicitly by (1.15). In particular, (1.6) and (1.7) simplifies to

$$\mu_{\theta_0}(h(\theta_0)) = \int_{\ell}^{r} h(x;\theta_0)\mu_{\theta_0}(x)dx = 0 \qquad (1.100)$$

and

$$W = \mu_{\theta_0}(\partial_{\theta^T} h(\theta_0)) = \int_{\ell}^{r} \partial_{\theta^T} h(x;\theta_0)\mu_{\theta_0}(x)dx.$$

The condition for eventual uniqueness of the G_n-estimator (1.9) is here that θ_0 is the only root of $\mu_{\theta_0}(h(\theta))$.

Kessler (2000) proposed

$$h(x;\theta) = \partial_\theta \log \mu_\theta(x), \qquad (1.101)$$

which is the score function (the derivative of the log-likelihood function) if we pretend that the observations are an i.i.d. sample from the stationary distribution. If Δ is large, this might be a reasonable approximation. That (1.100) is satisfied for this specification of h follows under standard conditions that allow the interchange of differentiation and integration.

$$\int_\ell^r (\partial_\theta \log \mu_\theta(x)) \, \mu_\theta(x) dx = \int_\ell^r \partial_\theta \mu_\theta(x) dx = \partial_\theta \int_\ell^r \mu_\theta(x) dx = 0.$$

A modification of the simple estimating function (1.101) was shown by Kessler, Schick, and Wefelmeyer (2001) to be efficient in the sense of semiparametric models. The modified version of the estimating function was derived by Kessler and Sørensen (2005) in a completely different way.

Hansen and Scheinkman (1995) and Kessler (2000) proposed and studied the generally applicable specification

$$h_j(x; \theta) = A_\theta f_j(x; \theta), \tag{1.102}$$

where A_θ is the generator (1.46), and f_j, $j = 1, \dots, p$, are twice differentiable functions chosen such that Condition 1.1 holds. The estimating function with h given by (1.102) can easily be applied to multivariate diffusions, because an explicit expression for the invariant density μ_θ is not needed. The following lemma for one-dimensional diffusions shows that only weak conditions are needed to ensure that (1.100) holds for h_j given by (1.102).

Lemma 1.34 *Suppose* $f \in C^2((\ell, r))$, $A_\theta f \in L^1(\mu_\theta)$ *and*

$$\lim_{x \to r} f'(x)\sigma^2(x; \theta)\mu_\theta(x) = \lim_{x \to \ell} f'(x)\sigma^2(x; \theta)\mu_\theta(x). \tag{1.103}$$

Then

$$\int_\ell^r (A_\theta f)(x)\mu_\theta(x) dx = 0.$$

Proof: Note that by (1.15), the function $\nu(x; \theta) = \frac{1}{2}\sigma^2(x; \theta)\mu_\theta(x)$ satisfies that $\nu'(x; \theta) = b(x; \theta)\mu_\theta(x)$. In this proof all derivatives are with respect to x. It follows that

$$\int_\ell^r (A_\theta f)(x)\mu_\theta(x) dx$$

$$= \int_\ell^r \left(b(x; \theta)f'(x) + \tfrac{1}{2}\sigma^2(x; \theta)f''(x) \right) \mu_\theta(x) dx$$

$$= \int_\ell^r \left(f'(x)\nu'(x; \theta) + f''(x)\nu(x; \theta) \right) dx = \int_\ell^r (f'(x)\nu(x; \theta))' \, dx$$

$$= \lim_{x \to r} f'(x)\sigma^2(x; \theta)\mu_\theta(x) - \lim_{x \to \ell} f'(x)\sigma^2(x; \theta)\mu_\theta(x) = 0.$$

$\square$

Example 1.35 Consider the square-root process (1.37) with $\sigma = 1$. For $f_1(x) = x$ and $f_2(x) = x^2$, we see that

$$A_\theta f(x) = \begin{pmatrix} -\beta(x - \alpha) \\ -2\beta(x - \alpha)x + x \end{pmatrix},$$

which gives the simple estimators

$$\hat{\alpha}_n = \frac{1}{n} \sum_{i=1}^{n} X_{i\Delta}, \qquad \hat{\beta}_n = \frac{\dfrac{1}{n} \displaystyle\sum_{i=1}^{n} X_{i\Delta}}{2 \left(\dfrac{1}{n} \displaystyle\sum_{i=1}^{n} X_{i\Delta}^2 - \left(\dfrac{1}{n} \displaystyle\sum_{i=1}^{n} X_{i\Delta} \right)^2 \right)}.$$

The condition (1.103) is obviously satisfied because the invariant distribution is a normal distribution.

□

Conley, Hansen, Luttmer, and Scheinkman (1997) proposed a model-based choice of the f_js in (1.102): $f_j = \partial_{\theta_j} \log \mu_\theta(x)$, i.e. the i.i.d. score function used in (1.101). Thus they obtained an estimating function of the form (1.99) with

$$h(x; \theta) = A_\theta \partial_\theta \log \mu_\theta(x). \qquad (1.104)$$

H. Sørensen (2001) independently derived the same estimating function as an approximation to the score function for continuous-time observation of the diffusion process. Jacobsen (2001) showed that this estimating function is small Δ-optimal. This result was later rediscovered by Aït–Sahalia and Mykland (2008) who obtained a similar result for estimating functions given by (1.105).

An estimating function of the simple form (1.99) cannot be expected to yield as efficient estimators as an estimating function that depends on pairs of consecutive observations, and therefore can use the information contained in the transitions. Hansen and Scheinkman (1995) proposed non-martingale estimating functions of the form (1.12) with g given by

$$g_j(\Delta, x, y; \theta) = h_j(y) A_\theta f_j(x) - f_j(x) \hat{A}_\theta h_j(y), \qquad (1.105)$$

where the functions f_j and h_j satisfy weak regularity conditions ensuring that (1.6) holds for $\bar{\theta} = \theta_0$. The differential operator $\hat{A}_\theta$ is the generator of the time reversal of the observed diffusion X. For a multivariate diffusion it is given by

$$\hat{A}_\theta f(x) = \sum_{k=1}^{d} \hat{b}_k(x; \theta) \partial_{x_k} f(x) + \tfrac{1}{2} \sum_{k,\ell=1}^{d} C_{k\ell}(x; \theta) \partial_{x_k x_\ell}^2 f(x),$$

where $C = \sigma\sigma^T$ and

$$\hat{b}_k(x;\theta) = -b_k(x;\theta) + \frac{1}{\mu_\theta(x)} \sum_{\ell=1}^{d} \partial_{x_\ell} \left(\mu_\theta C_{k\ell} \right)(x;\theta).$$

For one-dimensional ergodic diffusions, $\hat{A}_\theta = A_\theta$. That $\hat{b} = b$ for a one-dimensional diffusion follows from (1.15). Obviously, the estimating function of the form (1.99) with $h_j(x;\theta) = A_\theta f_j(x)$ is a particular case of (1.105) with $h_j(y) = 1$.

1.5.3 Approximate martingale estimating functions

For martingale estimating functions of the form (1.30) and (1.40), we can always, as discussed in Subsection 1.3.3, obtain an explicit approximation to the optimal weight matrix by means of the expansion (1.47). For diffusion models where there is no explicit expression for the transition operator, it is tempting to go on and approximate the conditional moments $\pi_\Delta^\theta(f_j(\theta))(x)$ using (1.47), and thus, quite generally, obtain *explicit approximate martingale estimating functions*. Such estimators were the first type of estimators for discretely observed diffusion processes to be studied in the literature. They have been considered by Dorogovcev (1976), Prakasa Rao (1988), Florens-Zmirou (1989), Yoshida (1992), Chan, Karolyi, Longstaff, and Sanders (1992), Kloeden, Platen, Schurz, and Sørensen (1996), Kessler (1997), Kelly, Platen, and Sørensen (2004), and many others.

It is, however, important to note that there is a dangerous pitfall when using these simple approximate martingale estimating functions. They do not satisfy the condition that $Q_{\theta_0}(g(\theta_0)) = 0$, and hence the estimators are inconsistent. To illustrate the problem, consider an estimating function of the form (1.12) with

$$g(x,y;\theta) = a(x,\theta)[f(y) - f(x) - \Delta A_\theta f(x)], \qquad (1.106)$$

where A_θ is the generator (1.46), i.e., we have replaced $\pi_\Delta^\theta f(x)$ by a first order expansion. To simplify the exposition, we assume that θ, a and f are one-dimensional. We assume that the diffusion is geometrically α-mixing, that the other conditions mentioned above for the weak convergence result (1.5) hold, and that Condition 1.1 is satisfied. Then by Theorem 1.2, the estimator obtained using (1.106) converges to the solution, $\bar{\theta}$, of

$$Q_{\theta_0}(g(\bar{\theta})) = 0, \qquad (1.107)$$

where, as usual, θ_0 is the true parameter value. We assume that the solution is unique. Using the expansion (1.47), we find that

$$
\begin{aligned}
Q_{\theta_0}(g(\theta)) &= \mu_{\theta_0}\left(a(\theta)[\pi_\Delta^{\theta_0} f - f - \Delta A_\theta f]\right)\\
&= \Delta\mu_{\theta_0}\left(a(\theta)[A_{\theta_0} f - A_\theta f + \tfrac{1}{2}\Delta A_{\theta_0}^2 f]\right) + O(\Delta^3)\\[4pt]
&= (\theta_0 - \theta)\Delta\mu_{\theta_0}\left(a(\theta_0)\partial_\theta A_{\theta_0} f\right) + \tfrac{1}{2}\Delta^2 \mu_{\theta_0}\left(a(\theta_0) A_{\theta_0}^2 f\right)\\
&\qquad + O(\Delta|\theta - \theta_0|^2) + O(\Delta^2|\theta - \theta_0|) + O(\Delta^3).
\end{aligned}
$$

If we neglect all O-terms, we obtain that

$$
\bar\theta \doteq \theta_0 + \Delta\tfrac{1}{2}\mu_{\theta_0}\left(a(\theta_0) A_{\theta_0}^2 f\right)/\mu_{\theta_0}\left(a(\theta_0)\partial_\theta A_{\theta_0} f\right),
$$

which indicates that when Δ is small, the asymptotic bias is of order Δ. However, the bias can be huge when Δ is not sufficiently small, as the following example shows.

Example 1.36 Consider again a diffusion with linear drift,

$$
b(x;\theta) = -\beta(x - \alpha).
$$

In this case (1.106) with $f(x) = x$ gives the estimating function

$$
G_n(\theta) = \sum_{i=1}^n a(X_{\Delta(i-1)};\theta)[X_{\Delta i} - X_{\Delta(i-1)} + \beta\left(X_{\Delta(i-1)} - \alpha\right)\Delta],
$$

where a is 2-dimensional. For a diffusion with linear drift, we found in Example 1.11 that

$$
F(x;\alpha,\beta) = xe^{-\beta\Delta} + \alpha(1 - e^{-\beta\Delta}).
$$

Using this, we obtain that

$$
Q_{\theta_0}(g(\theta)) = c_1(e^{-\beta_0\Delta} - 1 + \beta\Delta) + c_2\beta(\alpha_0 - \alpha),
$$

where

$$
c_1 = \int_D a(x)x\mu_{\theta_0}(dx) - \mu_{\theta_0}(a)\alpha_0, \qquad c_2 = \mu_{\theta_0}(a)\Delta.
$$

Thus

$$
\bar\alpha = \alpha_0
$$

and

$$
\bar\beta = \frac{1 - e^{-\beta_0\Delta}}{\Delta} \le \frac{1}{\Delta}.
$$

We see that the estimator of α is consistent, while the estimator of β will tend to be small if Δ is large, whatever the true value β_0 is. We see that what determines how well $\hat\beta$ works is the magnitude of $\beta_0\Delta$, so it is not enough to know that Δ is small. Moreover, we cannot use $\hat\beta\Delta$ to evaluate whether there is a problem, because this quantity will always tend to be smaller than one. If $\beta_0 \Delta$ actually is small, then the bias is proportional to Δ as expected

$$\bar{\beta} = \beta_0 - \tfrac{1}{2}\Delta\beta_0^2 + O(\Delta^2).$$

We get an impression of how terribly misled we can be when estimating the parameter β by means of the dangerous estimating function given by (1.106) from a simulation study in Bibby and Sørensen (1995) for the square root process (1.37). The result is given in Table 1.1. For the weight function a, the approximately optimal weight function was used, cf. Example 1.11. For different values of Δ and the sample size, 500 independent datasets were simulated, and the estimators were calculated for each dataset. The expectation of the estimator $\hat{\beta}$ was determined as the average of the simulated estimators. The true parameter values were $\alpha_0 = 10$, $\beta_0 = 1$ and $\tau_0 = 1$, and the initial value was $x_0 = 10$. When Δ is large, the behaviour of the estimator is bizarre. □

Δ	# obs.	mean	Δ	# obs.	mean
0.5	200	0.81	1.5	200	0.52
	500	0.80		500	0.52
	1000	0.79		1000	0.52
1.0	200	0.65	2.0	200	0.43
	500	0.64		500	0.43
	1000	0.63		1000	0.43

Table 1.1 *Empirical mean of 500 estimates of the parameter β in the CIR model. The true parameter values are $\alpha_0 = 10$, $\beta_0 = 1$, and $\tau_0 = 1$.*

The asymptotic bias given by (1.107) is small when Δ is sufficiently small, and the results in the following section on high frequency asymptotics show that in this asymptotic scenario the approximate martingale estimating functions work well. However, how small Δ needs to be depends on the parameter values, and without prior knowledge about the parameters, it is safer to use an exact martingale estimating function, which gives consistent estimators at all sampling frequencies.

1.6 High-frequency asymptotics

An expression for the asymptotic variance of estimators was obtained in Theorem 1.5 using a low-frequency asymptotic scenario, where the time between observations is fixed. This expression is rather complicated and is not easy to use for comparing the efficiency of different estimators. Therefore the relative merits of estimators have often been investigated by simulation studies, and the general picture has been rather confusing. A much simpler and more manageable expression for the asymptotic variance of estimators can be obtained by considering the high frequency scenario,

$$n \to \infty, \qquad \Delta_n \to 0, \qquad n\Delta_n \to \infty. \qquad (1.108)$$

The assumption that $n\Delta_n \to \infty$ is needed to ensure that parameters in the drift coefficient can be consistently estimated.

For this type of asymptotics M. Sørensen (2008) obtained simple conditions for rate optimality and efficiency for ergodic diffusions, which allow identification of estimators that work well when the time between observations, Δ_n, is not too large. How small Δ_n needs to be for the high frequency scenario to be relevant, depends on the speed with which the diffusion moves. For financial data the speed of reversion is usually slow enough that this type of asymptotics works for daily, sometimes even weekly observations. A main result of the theory in this section is that under weak conditions optimal martingale estimating functions give rate optimal and efficient estimators.

It is also interesting that the high frequency asymptotics provide a very clear statement of the important fact that parameters in the diffusion coefficient can be estimated more exactly than drift parameters when the time between observations is small. A final advantage of high frequency asymptotics is that it also gives useful results about the approximate martingale estimating functions discussed in Subsection 1.5.3, in situations where they work.

To simplify the exposition, we restrict attention to a one-dimensional diffusion given by

$$dX_t = b(X_t; \alpha)dt + \sigma(X_t; \beta)dW_t, \qquad (1.109)$$

where $\theta = (\alpha, \beta) \in \Theta \subseteq \mathbb{R}^2$. The results below can be generalized to multivariate diffusions and parameters of higher dimension. We consider estimating functions of the general form (1.3), where the two-dimensional function $g = (g_1, g_2)$ for some $\kappa \geq 2$ and for all $\theta \in \Theta$ satisfies

$$E_\theta(g(\Delta_n, X_{\Delta_n i}, X_{\Delta_n(i-1)}; \theta) \,|\, X_{\Delta_n(i-1)}) \qquad (1.110)$$
$$= \Delta_n^\kappa R(\Delta_n, X_{\Delta_n(i-1)}; \theta).$$

Martingale estimating functions obviously satisfy (1.110) with $R = 0$, but for instance the approximate martingale estimating functions discussed at the end of the previous section satisfy (1.110) too. Here and later $R(\Delta, y, x; \theta)$ denotes a function such that $|R(\Delta, y, x; \theta)| \leq F(y, x; \theta)$, where F is of polynomial growth in y and x uniformly for θ in compact sets. This means that for any compact subset $K \subseteq \Theta$, there exist constants $C_1, C_2, C_3 > 0$ such that $\sup_{\theta \in K} |F(y, x; \theta)| \leq C_1(1 + |x|^{C_2} + |y|^{C_3})$ for all x and y in the state space of the diffusion.

The main results in this section are simple conditions on the estimating function that ensure rate optimality and efficiency of estimators. The condition for *rate optimality* is

Condition 1.37

$$\partial_y g_2(0, x, x; \theta) = 0 \qquad (1.111)$$

for all $x \in (\ell, r)$ and all $\theta \in \Theta$.

By $\partial_y g_2(0, x, x; \theta)$ we mean $\partial_y g_2(0, y, x; \theta)$ evaluated at $y = x$. This condition is called *the Jacobsen condition*, because it was first found in the theory of small Δ-optimal estimation developed in Jacobsen (2001), cf. (1.62) in Subsection 1.3.4.

The condition for *efficiency* is

Condition 1.38

$$\partial_y g_1(0, x, x; \theta) = \partial_\alpha b(x; \alpha)/\sigma^2(x; \beta) \qquad (1.112)$$

and

$$\partial_y^2 g_2(0, x, x; \theta) = \partial_\beta \sigma^2(x; \beta)/\sigma^4(x; \beta), \qquad (1.113)$$

for all $x \in (\ell, r)$ and all $\theta \in \Theta$.

Also (1.112) and (1.113) were found as conditions for small Δ-optimality in Jacobsen (2002), cf. (1.61) and (1.63). This is not surprising. The following theorem provides an interpretation of small Δ-optimality in terms of the classical statistical concepts rate optimality and efficiency. As usual, $\theta_0 = (\alpha_0, \beta_0)$ denotes the true parameter value.

Theorem 1.39 *Assume that the diffusion is ergodic, that $\theta_0 \in$ int Θ, and that the technical regularity Condition 1.40 given below holds. Denote the density function of the invariant probability measure by μ_θ. Suppose that $g(\Delta, y, x; \theta)$ satisfies Condition 1.37. Assume, moreover, that the following identifiability condition is satisfied*

$$\int_\ell^r [b(x, \alpha_0) - b(x, \alpha)] \partial_y g_1(0, x, x; \theta) \mu_{\theta_0}(x) dx \neq 0 \qquad when \ \alpha \neq \alpha_0,$$

$$\int_\ell^r [\sigma^2(x, \beta_0) - \sigma^2(x, \beta)] \partial_y^2 g_2(0, x, x; \theta) \mu_{\theta_0}(x) dx \neq 0 \qquad when \ \beta \neq \beta_0,$$

and that

$$S_1 = \int_\ell^r \partial_\alpha b(x; \alpha_0) \partial_y g_1(0, x, x; \theta_0) \mu_{\theta_0}(x) dx \neq 0,$$

$$S_2 = \frac{1}{2} \int_\ell^r \partial_\beta \sigma^2(x; \beta_0) \partial_y^2 g_2(0, x, x; \theta_0) \mu_{\theta_0}(x) dx \neq 0.$$

Then a consistent G_n–estimator $\hat{\theta}_n = (\hat{\alpha}_n, \hat{\beta}_n)$ exists and is unique in any compact subset of Θ containing θ_0 with probability approaching one as $n \to \infty$. If, moreover,

$$\partial_\alpha \partial_y^2 g_2(0, x, x; \theta) = 0, \tag{1.114}$$

then for a martingale estimating function, and for more general estimating functions if $n\Delta^{2(\kappa-1)} \to 0$,

$$\begin{pmatrix} \sqrt{n\Delta_n}(\hat{\alpha}_n - \alpha_0) \\ \sqrt{n}(\hat{\beta}_n - \beta_0) \end{pmatrix} \xrightarrow{\mathcal{D}} N_2\left(\begin{pmatrix} 0 \\ 0 \end{pmatrix}, \begin{pmatrix} \frac{W_1}{S_1^2} & 0 \\ 0 & \frac{W_2}{S_2^2} \end{pmatrix} \right) \tag{1.115}$$

where

$$W_1 = \int_\ell^r \sigma^2(x; \beta_0)[\partial_y g_1(0, x, x; \theta_0)]^2 \mu_{\theta_0}(x)dx$$

$$W_2 = \tfrac{1}{2}\int_\ell^r \sigma^4(x; \beta_0)[\partial_y^2 g_2(0, x, x; \theta_0)]^2 \mu_{\theta_0}(x)dx.$$

Note that the estimator of the diffusion coefficient parameter, β, converges faster than the estimator of the drift parameter, α, and that the two estimators are asymptotically independent. Gobet (2002) showed, under regularity conditions, that a discretely sampled diffusion model is locally asymptotically normal under high frequency asymptotics, and that the optimal rate of convergence for a drift parameter is $1/\sqrt{n\Delta_n}$, while it is $1/\sqrt{n}$ for a parameter in the diffusion coefficient. Thus under the conditions of Theorem 1.39 the estimators $\hat{\alpha}_n$ and $\hat{\beta}_n$ are rate optimal. More precisely, Condition 1.37 implies rate optimality. If this condition is not satisfied, the estimator of the diffusion coefficient parameter, β, does not use the information about the diffusion coefficient contained in the quadratic variation and therefore converges at the same relatively slow rate $1/\sqrt{n\Delta_n}$ as estimators of α, see M. Sørensen (2008).

Gobet gave the following expression for the Fisher information matrix

$$\mathcal{I} = \begin{pmatrix} W_1 & 0 \\ 0 & W_2 \end{pmatrix}, \tag{1.116}$$

where

$$W_1 = \int_\ell^r \frac{(\partial_\alpha b(x; \alpha_0))^2}{\sigma^2(x; \beta_0)} \mu_{\theta_0}(x)dx, \tag{1.117}$$

$$W_2 = \int_\ell^r \left[\frac{\partial_\beta \sigma^2(x; \beta_0)}{\sigma^2(x; \beta_0)}\right]^2 \mu_{\theta_0}(x)dx. \tag{1.118}$$

By comparing the covariance matrix in (1.115) to (1.116), we see that Condition 1.38 implies that $S_1 = W_1$ and $S_2 = W_2$, with W_1 and W_2 given by (1.117) and (1.118), and that hence the asymptotic covariance matrix of

$(\hat{\alpha}_n, \hat{\beta}_n)$ under Condition 1.38 equals the inverse of the Fisher information matrix (1.116). Thus Condition 1.38 ensures efficiency of $(\hat{\alpha}_n, \hat{\beta}_n)$. Under the conditions of Theorem 1.39 and Condition 1.38, we see that for a martingale estimating function, and more generally if $n\Delta^{2(\kappa-1)} \to 0$,

$$\begin{pmatrix} \sqrt{n\Delta_n}(\hat{\alpha}_n - \alpha_0) \\ \sqrt{n}(\hat{\beta}_n - \beta_0) \end{pmatrix} \xrightarrow{\mathcal{D}} N_2\left(\begin{pmatrix} 0 \\ 0 \end{pmatrix}, \mathcal{I}^{-1}\right). \tag{1.119}$$

Note that condition (1.114) is automatically satisfied under the efficiency Condition 1.38.

Proof of Theorem 1.39: Only a brief outline of the proof is given; for details see M. Sørensen (2008). Consider the normalized estimating function

$$G_n(\theta) = \frac{1}{n\Delta_n} \sum_{i=1}^n g(\Delta_n, X_{t_i^n}, X_{t_{i-1}^n}; \theta).$$

First the conditions of Theorem 1.58 must be checked. Using Lemma 9 in Genon-Catalot and Jacod (1993), it can be shown that $G_n(\theta_0) \to 0$ in P_{θ_0}-probability, and that $\partial_{\theta^T} G_n(\theta)$ under P_{θ_0} converges pointwise to a matrix, which for $\theta = \theta_0$ is upper triangular and has diagonal elements equal to S_1 and S_2, and thus is invertible. In order to prove that the convergence is uniform for θ in a compact set K, we show that the sequence

$$\zeta_n(\cdot) = \frac{1}{n\Delta_n} \sum_{i=1}^n g(\Delta_n, X_{t_i^n}, X_{t_{i-1}^n}, \cdot)$$

converges weakly to the limit $\gamma(\cdot, \theta_0)$ in the space, $C(K)$, of continuous functions on K with the supremum norm. Since the limit is non-random, this implies uniform convergence in probability for $\theta \in K$. We have proved pointwise convergence, so the weak convergence result follows because the family of distributions of $\zeta_n(\cdot)$ is tight. The tightness is shown by checking the conditions in Corollary 14.9 in Kallenberg (1997). Thus the conditions of Theorem 1.58 are satisfied, and we conclude the existence of a consistent and eventually unique G_n-estimator. The uniqueness on compact subsets follows from Theorem 1.59 because the identifiability condition in Theorem 1.39 implies (1.161).

The asymptotic normality of the estimators follows from Theorem 1.60 with

$$A_n = \begin{pmatrix} \sqrt{\Delta_n n} & 0 \\ 0 & \sqrt{n} \end{pmatrix}.$$

The weak convergence of $A_n G_n(\theta_0)$ follows from a central limit theorem for martingales, e.g. Corollary 3.1 in P. Hall and Heyde (1980). The uniform convergence of $A_n \partial_{\theta^T} G_n(\theta) A_n^{-1}$ was proved for three of the entries when the conditions of Theorem 1.58 were checked. The result for the last entry is proved in a similar way using (1.114). $\square$

The reader is reminded of the trivial fact that for any non-singular 2×2 matrix, M_n, the estimating functions $M_n G_n(\theta)$ and $G_n(\theta)$ have exactly the same roots and hence give the same estimator(s). We call them *versions* of the same estimating function. The matrix M_n may depend on Δ_n. The point is that a version must exist which satisfies the conditions (1.111) – (1.113), but not all versions of an estimating function satisfy these conditions.

It follows from results in Jacobsen (2002) that to obtain a rate optimal and efficient estimator from an estimating function of the form (1.41), we need that $N \geq 2$ and that the matrix

$$D(x) = \begin{pmatrix} \partial_x f_1(x; \theta) & \partial_x^2 f_1(x; \theta) \\ \partial_x f_2(x; \theta) & \partial_x^2 f_2(x; \theta) \end{pmatrix}$$

is invertible for μ_θ-almost all x. Under these conditions, M. Sørensen (2008) showed that Godambe–Heyde optimal martingale estimating functions give rate optimal and efficient estimators. For a d-dimensional diffusion, Jacobsen (2002) gave the conditions $N \geq d(d+3)/2$, and that the $N \times (d+d^2)$-matrix $D(x) = \left(\partial_x f(x; \theta) \, \partial_x^2 f(x; \theta) \right)$ has full rank $d(d+3)/2$, which are needed to ensure the existence of a rate optimal and efficient estimator from an estimating function of the form (1.41).

We conclude this section by an example, but first we state technical conditions under which the results in this section hold. The assumptions about polynomial growth are far too strong, but simplify the proofs. These conditions can most likely be weakened considerably.

Condition 1.40 *The diffusion is ergodic with invariant probability density μ_θ, and the following conditions hold for all $\theta \in \Theta$:*

(1) $\int_\ell^r x^k \mu_\theta(x) dx < \infty$ *for all* $k \in \mathbb{N}$.

(2) $\sup_t E_\theta(|X_t|^k) < \infty$ *for all* $k \in \mathbb{N}$.

(3) $b, \sigma \in C_{p,4,1}((\ell, r) \times \Theta)$.

(4) *There exists a constant C_θ such that for all $x, y \in (\ell, r)$*

$$|b(x; \alpha) - b(y; \alpha)| + |\sigma(x; \beta) - \sigma(y; \beta)| \leq C_\theta |x - y|$$

(5) $g(\Delta, y, x; \theta) \in C_{p,2,6,2}(\mathbb{R}_+ \times (\ell, r)^2 \times \Theta)$ *and has an expansion in powers of Δ:*

$$g(\Delta, y, x; \theta) =$$
$$g(0, y, x; \theta) + \Delta g^{(1)}(y, x; \theta) + \tfrac{1}{2}\Delta^2 g^{(2)}(y, x; \theta) + \Delta^3 R(\Delta, y, x; \theta),$$

where

$$g(0, y, x; \theta) \in C_{p,6,2}((\ell, r)^2 \times \Theta),$$
$$g^{(1)}(y, x; \theta) \in C_{p,4,2}((\ell, r)^2 \times \Theta),$$
$$g^{(2)}(y, x; \theta) \in C_{p,2,2}((\ell, r)^2 \times \Theta).$$

We define $C_{p,k_1,k_2,k_3}(\mathbb{R}_+ \times (\ell, r)^2 \times \Theta)$ as the class of real functions $f(t, y, x; \theta)$ satisfying that

(i) $f(t, y, x; \theta)$ is k_1 times continuously differentiable with respect to t, k_2 times continuously differentiable with respect to y, and k_3 times continuously differentiable with respect to α and with respect to β.

(ii) f and all partial derivatives $\partial_t^{i_1} \partial_y^{i_2} \partial_\alpha^{i_3} \partial_\beta^{i_4} f$, $i_j = 1, \ldots k_j$, $j = 1, 2$, $i_3 + i_4 \le k_3$, are of polynomial growth in x and y uniformly for θ in a compact set (for fixed t).

The classes $C_{p,k_1,k_2}((\ell, r) \times \Theta)$ and $C_{p,k_1,k_2}((\ell, r)^2 \times \Theta)$ are defined similarly for functions $f(y; \theta)$ and $f(y, x; \theta)$, respectively.

Example 1.41 We can now interpret the findings in Example 1.14 as follows. The general quadratic martingale estimating function (1.64) gives rate optimal estimators in the high frequency asymptotics considered in this section. Moreover, the estimators are efficient in three particular cases: the optimal estimating function given in Example 1.9 and the approximations (1.28) and (1.51).

Kessler (1997) considered an approximation to the Gaussian quasi-likelihood presented in Subsection 1.3.2, where the conditional mean F and the conditional variance Φ are approximated as follows. The conditional mean is replaced by the expansion

$$r_k(\Delta, x; \theta) = \sum_{i=0}^{k} \frac{\Delta^i}{i!} A_\theta^i f(x) = x + \Delta \sum_{i=0}^{k-1} \frac{\Delta^i}{(i+1)!} A_\theta^i b(x; \alpha),$$

where $f(x) = x$, cf. (1.47). For fixed x, y and θ the function $(y - r_k(\Delta, x; \theta))^2$ is a polynomial in Δ of order $2k$. Define $g_{x,\theta}^j(y)$, $j = 0, 1, \cdots, k$ by

$$(y - r_k(\Delta, x; \theta))^2 = \sum_{j=0}^{k} \Delta^j g_{x,\theta}^j(y) + O(\Delta^{k+1}).$$

For instance, for $k = 2$

$$(y - r_2(\Delta, x; \theta))^2 =$$
$$(y - x)^2 - 2(y - x)b(x; \alpha)\Delta + \left[(y - x)A_\theta b(x; \alpha) + b(x; \alpha)^2\right] \Delta^2 + O(\Delta^3),$$

from which we can see the expressions for $g_{x,\theta}^j(y)$, $j = 0, 1, 2$. The conditional variance can be approximated by

$$\Gamma_k(\Lambda, x; \theta) = \sum_{j=0}^{k} \Lambda^j \sum_{r=0}^{k-j} \frac{\Delta^r}{r!} A_0^r g_{x,0}^j(x).$$

In particular,

$$\Gamma_2(\Delta, x; \theta) = \Delta\sigma^2(x; \beta) + \tfrac{1}{2}\Delta^2 \left[A_\theta \sigma^2(x; \beta) - \sigma^2(x; \beta)\partial_x b(x; \alpha) \right].$$

By inserting these approximations in (1.28), we obtain the approximate martingale estimating function

$$H_n^{(k)}(\theta) = \sum_{i=1}^{n} \frac{\partial_\theta r_k(\Delta_i, X_{t_{i-1}}; \theta)}{\Gamma_{k+1}(\Delta_i, X_{t_{i-1}}; \theta)} [X_{t_i} - r_k(\Delta_i, X_{t_{i-1}}; \theta)] \qquad (1.120)$$

$$+ \sum_{i=1}^{n} \frac{\partial_\theta \Gamma_{k+1}(\Delta_i, X_{t_{i-1}}; \theta)}{2\Gamma_{k+1}(\Delta_i, X_{t_{i-1}}; \theta)^2} [(X_{t_i} - r_k(\Delta_i, X_{t_{i-1}}; \theta))^2 - \Gamma_{k+1}(\Delta_i, X_{t_{i-1}}; \theta)].$$

Kessler (1997) (essentially) showed that for ergodic diffusions satisfying Condition 1.40 (1) – (4), the estimator obtained from $H_n^{(k)}(\theta)$ satisfies (1.119) provided that $n\Delta^{2k+1} \to 0$.

$\square$

1.7 High-frequency asymptotics in a fixed time-interval

We will now briefly consider a more extreme type of high-frequence asymptotics, where the observation times are restricted to a bounded interval, which, without loss of generality, we can take to be $[0, 1]$. Suppose that the d-dimensional diffusion X which solves (1.109) has been observed at the time points $t_i = i/n, i = 0, \dots, n$. Note that in this section W in equation (1.109) is a d-dimensional standard Wiener process, and σ is a $d \times d$-matrix. We assume that the matrix $C(x; \beta) = \sigma(x; \beta)\sigma(x; \beta)^T$ is invertible for all x in the state space, D, of X. Because the observation times are bounded, the drift parameter, α, cannot be consistently estimated as $n \to \infty$, so in the following we consider estimation of β only, and concentrate on the following Gaussian quasi-likelihood function:

$$Q_n(\beta) = \qquad\qquad\qquad\qquad\qquad\qquad\qquad\qquad\qquad\qquad\qquad (1.121)$$

$$\sum_{i=1}^{n} \left[\log \det C(X_{t_{i-1}}; \beta) + n(X_{t_i} - X_{t_{i-1}})^T C(X_{t_{i-1}}; \beta)^{-1}(X_{t_i} - X_{t_{i-1}}) \right].$$

This is an approximation to a multivariate version of the Gaussian quasi-likelihood in Subsection 1.3.2 with $b = 0$, where the conditional mean $F(x; \theta)$ is approximated by x, and the conditional covariance matrix Φ is approximated by $n^{-1}C$. An estimator is obtained by minimizing $Q_n(\beta)$. This estimator can also be obtained from the approximate martingale estimating function which we get by differentiating $Q_n(\beta)$ with respect to β. The drift may be known, but in general we allow it to depend on an unknown parameter α. We assume that $\theta = (\alpha, \beta) \in A \times B = \Theta$, and we denote the true parameter value by $\theta_0 = (\alpha_0, \beta_0)$.

Genon-Catalot and Jacod (1993) showed the following theorem under the assumption that $\beta \in B$, where B is a compact subset of $\mathbb{R}^q$, which ensures that a $\hat{\beta}_n \in B$ that minimizes $Q_n(\beta)$ always exists. The results in the theorem hold for any $\hat{\beta}_n$ that minimizes $Q_n(\beta)$.

Theorem 1.42 *Assume that Condition 1.43 given below holds. Then the estimator $\hat{\beta}_n$ is consistent, and provided that $\beta_0 \in$ int B,*

$$\sqrt{n}(\hat{\beta}_n - \beta_0) \xrightarrow{\mathcal{D}} Z,$$

where the distribution of Z is a normal variance mixture with characteristic function

$$s \mapsto E_{\theta_0} \left(\exp \left(-\tfrac{1}{2} s^T W(\beta_0)^{-1} s \right) \right)$$

with $W(\beta)$ given by (1.122). Conditional on $W(\beta_0)$, the asymptotic distribution of $\sqrt{n}(\hat{\beta}_n - \beta_0)$ is a centered q-dimensional normal distribution with covariate matrix $W(\beta_0)^{-1}$.

We will not prove Theorem 1.42 here. Note, however, that to do so we need the full generality of the Theorems 1.58, 1.59 and 1.60, where the matrix $W(\theta)$ (equal to $W_0(\theta)$ in Theorem 1.60) is random. Only if the matrix $B(x; \beta)$ defined below does not depend on x, is $W(\beta)$ non-random, in which case the limit distribution is simply the centered q-dimensional normal distribution with covariate matrix $W(\beta_0)^{-1}$. A simple example of a non-random $W(\beta)$ is when β is one-dimensional and a $d \times d$-matrix $F(x)$ exists such that $C(x; \beta) = \beta F(x)$. So for instance for the Ornstein-Uhlenbeck process and the square-root diffusion (1.37), $W(\beta)$ is non-random, and the limit distribution is normal.

Condition 1.43 *The stochastic differential equation (1.109) has a non-exploding, unique strong solution for $t \in [0, 1]$, and the following conditions hold for all $\theta = (\alpha, \beta) \in \Theta$:*

(1) $b(x; \alpha)$ is a continuous function of x, and the partial derivatives $\partial_x^2 \sigma(x; \beta)$, $\partial_x \partial_\beta \sigma(x; \beta)$, $\partial_\beta^2 \sigma(x; \beta)$ exist and are continuous functions of $(x, \beta) \in D \times B$.

(2) With P_θ-probability one it holds that for all $\beta_1 \neq \beta$, the functions $t \mapsto C(X_t; \beta_1)$ and $t \mapsto C(X_t; \beta)$ are not equal.

(3) The random $q \times q$- matrix

$$W(\beta) = \int_0^1 B(X_t; \beta)dt, \qquad (1.122)$$

where the ijth entry of $B(x; \beta)$ is given by

$$B(x; \beta)_{ij} = \tfrac{1}{2} \operatorname{tr} \left(\partial_{\beta_i} C(x; \beta) C(x; \beta)^{-1} \partial_{\beta_j} C(x; \beta) C(x; \beta)^{-1} \right),$$

is invertible P_θ-almost surely.

The Condition 1.43 (2) can be difficult to check because it depends on the path of the process X. It is implied by the stronger condition that for all $\beta_1 \neq \beta$, $C(x; \beta_1) \neq C(x; \beta)$ for almost all $x \in D$.

Gobet (2001) showed, under regularity conditions, that for the high-frequency asymptotics in a fixed time-interval considered in this section, the diffusion model is locally asymptotically mixed normal (LAMN) with rate $\sqrt{n}$ and conditional variance given by $W(\beta)$; see e.g. Le Cam and Yang (2000) for the definition of LAMN. Therefore the estimator discussed above is efficient in the sense of Jeganathan (1982, 1983).

Example 1.44 Consider the one-dimensional model given by

$$dX_t = -(X_t - \alpha)dt + \sqrt{\beta + X_t^2}dW_t,$$

where $\alpha > 0$ and $\beta > 0$. In this case $c(x; \beta) = \beta + x^2$, so

$$W(\beta) = \int_0^1 \frac{X_t^4}{2(\beta + X_t^2)^2}dt,$$

which is random.

□

1.8 Small-diffusion asymptotics

Under the high-frequency asymptotics with bounded observation times considered in the previous section, drift parameters could not be consistently estimated. Here we combine the high-frequency asymptotics with small-diffusion asymptotics to show that if the diffusion coefficient is small, we can find accurate estimators of drift parameters even when we have only observations in a bounded time-interval, which we again take to be $[0, 1]$.

We consider observations that the time points $t_i = i/n$, $i = 1, \ldots, n$, of a d-dimensional diffusion process that solves the stochastic differential equation

$$dX_t = b(X_t, \alpha)dt + \varepsilon\sigma(X_t, \beta)dW_t, \quad X_0 = x_0, \qquad (1.123)$$

with $\varepsilon > 0$ and $(\alpha, \beta) \in A \times B$, where $A \subseteq \mathbb{R}^{q_1}$ and $B \subseteq \mathbb{R}^{q_2}$ are convex, compact subsets. It is assumed that ϵ is known, while the parameter $\theta = (\alpha, \beta) \in \Theta = A \times B$ must be estimated. In (1.123) W is a d-dimensional standard Wiener process, and σ is a $d \times d$-matrix. We assume that the matrix $C(x; \beta) = \sigma(x; \beta)\sigma(x; \beta)^T$ is invertible for all x in the state space, D, of X.

In this section the asymptotic scenario is that $n \to \infty$ and $\varepsilon \to 0$ with a suitable balance between the rate of convergence of the two. Small diffusion asymptotics, where $\varepsilon \to 0$, has been widely studied and has proved fruitful in applied problems, see e.g. Freidlin and Wentzell (1998). Applications to contingent claim pricing and other financial problems can be found in Takahashi and Yoshida (2004) and Uchida and Yoshida (2004a), and applications to filtering problems in Picard (1986, 1991). The estimation problem outlined above was studied by Genon-Catalot (1990), M. Sørensen and Uchida (2003), and Gloter and Sørensen (2009). Here we follow Gloter and Sørensen (2009), which generalize results in the other papers, and consider the asymptotic scenario:

$$n \to \infty \qquad \varepsilon_n \to 0 \qquad \liminf_{n \to \infty} \varepsilon_n n^\rho > 0 \qquad (1.124)$$

for some $\rho > 0$. When ρ is large, ϵ can go faster to zero than when ρ is relatively small. The value of ρ depends on the quasi-likelihood, as we shall see below.

The solution to (1.123) for $\epsilon = 0$ plays a crucial role in the theory. It is obviously non-random. More generally, we define the flow $\xi_t(x, \alpha)$ as the solution to the equation

$$\partial_t \xi_t(x, \alpha) = b(\xi_t(x, \alpha), \alpha), \qquad \xi_0(x, \alpha) = x, \qquad (1.125)$$

for all $x \in D$. The solution to (1.123) for $\epsilon = 0$ is given by $\xi_t(x_0, \alpha)$. A related function of central importance is

$$\tilde{\delta}_n(x, \alpha) = \xi_{1/n}(x, \alpha) - x. \qquad (1.126)$$

When ε is small, $\tilde{\delta}_n(X_{t_{i-1}}, \alpha) + X_{t_{i-1}}$ approximates the conditional expectation of X_{t_i} given $X_{t_{i-1}}$, and can be used to define a Gaussian quasi-likelihood. However, equation (1.125) does not generally have an explicit solution, so $\xi_t(x, \alpha)$ is usually not explicitly available. Therefore we replace it by an approximation $\delta(x, \alpha)$ that satisfies Condition 1.46 (5) given below. Using this approximation, we define a Gaussian quasi-log-likelihood by

$$U_{\varepsilon, n}(\theta) = \sum_{k=1}^n \left\{ \log \det C_{k-1}(\beta) + \varepsilon^{-2} n P_k(\alpha)^T C_{k-1}(\beta)^{-1} P_k(\alpha) \right\},$$

$$(1.127)$$

where

$$P_k(\alpha) = X_{k/n} - X_{(k-1)/n} - \delta_n(X_{(k-1)/n}, \alpha)$$
$$C_k(\beta) = \sigma(X_{k/n}, \beta)\sigma(X_{k/n}, \beta)^T.$$

This is the log-likelihood function that would have been obtained if the conditional distribution of X_{t_i} given $X_{t_{i-1}}$ were a normal distribution with mean $\delta_n(X_{t_{i-1}}, \alpha) + X_{t_{i-1}}$ and covariance matrix $(t_i - t_{i-1})\varepsilon^2 C_{k-1}(\beta)$.

When ξ is explicitly available, a natural choice is $\delta_n(x, \alpha) = \tilde{\delta}_n(x, \alpha)$. Otherwise, simple useful approximations to $\tilde{\delta}_n(x, \alpha)$ are given by

$$\delta_n^k(x, \alpha) = \sum_{j=1}^{k} \frac{n^{-j}}{j!} \left(\mathcal{L}_\alpha\right)^{j-1} (b(\cdot, \alpha))(x),$$

$k = 1, 2 \ldots$, where the operator $\mathcal{L}_\alpha$ is defined by

$$\mathcal{L}_\alpha(f)(x) = \sum_{i=1}^{d} b_i(x, \alpha)\partial_{x_i} f(x).$$

By $(\mathcal{L}_\alpha)^j$ we denote j-fold application of the operator $\mathcal{L}_\alpha$. The approximation δ_n^k satisfies Conditions 1.46 (5)–(6), when $k - 1/2 \geq \rho$. The first two approximations are

$$\delta_n^1(x, \alpha) = n^{-1}b(x, \alpha),$$

for which the quasi-likelihood studied in M. Sørensen and Uchida (2003) is obtained, and

$$\delta_n^2(x, \alpha) = n^{-1}b(x, \alpha) + \tfrac{1}{2}n^{-2} \sum_{i=1}^{d} b_i(x, \alpha)\partial_{x_i} b(x, \alpha).$$

Since the parameter space Θ is compact, a $\hat{\theta}_{\varepsilon,n} = (\hat{\alpha}_{\varepsilon,n}, \hat{\beta}_{\varepsilon,n})$ that minimizes the Gaussian quasi- log-likelihood $U_{\varepsilon,n}(\theta)$ always exists. The results in the following theorem hold for any $\hat{\theta}_{\varepsilon,n}$ that minimizes $U_{\varepsilon,n}(\theta)$. As usual, $\theta_0 = (\alpha_0, \beta_0)$ denotes the true parameter value.

Theorem 1.45 *Assume that Condition 1.46 given below holds, that $\theta_0 \in$ int Θ, and that the matrix*

$$I(\theta_0) = \begin{pmatrix} I_1(\theta_0) & 0 \\ 0 & I_2(\theta_0) \end{pmatrix}$$

is invertible, where the ijth entries of the $q_1 \times q_1$ matrix I_1 and of the $q_2 \times q_2$ matrix I_2 are given by

$$I_1^{i,j}(\theta_0) =$$
$$\int_0^1 \partial_{\alpha_i} b(\xi_s(x_0, \alpha_0), \alpha_0)^T C^{-1}(\xi_s(x_0, \alpha_0), \beta_0)\partial_{\alpha_j} b(\xi_s(x_0, \alpha_0), \alpha_0)ds$$

and

$$I_\sigma^{i,j}(\theta_0) = \tfrac{1}{2}\int_0^1 \mathrm{tr}\left[(\partial_{\beta_i} C)C^{-1}(\partial_{\beta_j} C)C^{-1}(\xi_s(x_0, \alpha_0), \beta_0)\right] ds.$$

Then, under the asymptotic scenario (1.124), the estimator $\hat{\theta}_{\varepsilon,n}$ is consistent, and

$$\begin{pmatrix} \varepsilon^{-1}(\hat{\alpha}_{\varepsilon,n} - \alpha_0) \\ \sqrt{n}(\hat{\beta}_{\varepsilon,n} - \beta_0) \end{pmatrix} \xrightarrow{\mathcal{D}} N\left(0, I(\theta_0)^{-1}\right).$$

We do not prove the theorem here, but a similar result for the estimating function obtained by differentiation of $U_{\varepsilon,n}(\theta)$ with respect to θ can be proved using the asymptotic results in Section 1.10. Note that the estimators of the drift and diffusion coefficient parameters are asymptotically independent. The two parameters are not estimated at the same rate. For the approximation δ_n^1 the conditions below are satisfied if ε^{-1} converges at a rate smaller than or equal to $\sqrt{n}$, so in this case the rate of convergence of $\hat{\alpha}_{\varepsilon,n}$ is slower than or equal to that of $\hat{\beta}_{\varepsilon,n}$. For the approximations δ_n^k, $k \geq 2$, the rate of convergence of $\hat{\alpha}_{\varepsilon,n}$ can be slower than or faster than that of $\hat{\beta}_{\varepsilon,n}$, dependent on how fast ε goes to zero.

The matrix I_1 equals the Fisher information matrix when the data is a continuous sample path in $[0, 1]$ and $\varepsilon \to 0$, cf. Kutoyants (1994), so $\hat{\alpha}_{\varepsilon,n}$ is efficient. Probably $\hat{\beta}_{\varepsilon,n}$ is efficient too, but this cannot be seen in this simple way and has not yet been proved.

We now give the technical conditions that imply Theorem 1.45.

Condition 1.46 *The following holds for all $\varepsilon > 0$:*

(1) The stochastic differential equation (1.123) has a unique strong solution for $t \in [0, 1]$ for all $\theta = (\alpha, \beta) \in \Theta$.

(2) $b(x; \alpha)$ is a smooth (i.e. C^{∞}) function of (x, α), and a constant c exists such that for all $x, y \in D$ and all $\alpha_1, \alpha_2 \in A$:

$$|b(x; \alpha_1) - b(y; \alpha_2)| \leq c(|x - y| + |\alpha_1 - \alpha_2|).$$

(3) $\sigma(x; \beta)$ is continuous, and there exists an open convex subset $\mathcal{U} \subseteq D$ such that $\xi_t(x_0, \alpha_0) \in \mathcal{U}$ for all $t \in [0, 1]$, and $\sigma(x; \beta)$ is smooth on $\mathcal{U} \times B$.

(4) If $\alpha \neq \alpha_0$, then the two functions $t \mapsto b(\xi_t(x_0, \alpha_0); \alpha)$ and $t \mapsto b(\xi_t(x_0, \alpha_0); \alpha_0)$ are not equal. If $\beta \neq \beta_0$, then the two functions $t \mapsto C(\xi_t(x_0, \alpha_0); \beta)$ and $t \mapsto C(\xi_t(x_0, \alpha_0); \beta_0)$ are not equal.

(5) The function $\delta_n(x; \alpha)$ is smooth, and for any compact subset $K \subseteq D$, a constant $c(K)$ exists such that

$$\sup_{x \in K, \alpha \in A} \left| \delta_n(x; \alpha) - \tilde{\delta}_n(x; \alpha) \right| \leq c(K)\varepsilon n^{-3/2}.$$

Similar bounds hold for the first two derivatives of δ_n w.r.t. α.

(6) *For any compact subset $K \subseteq D \times A$, there exists a constant $c(K)$, independent of n, such that*

$$|n\delta_n(x;\alpha_1) - n\delta_n(x;\alpha_2)| \leq c(K)|\alpha_1 - \alpha_2|$$

for all $(x,\alpha_1),(x,\alpha_2) \in K$ and for all $n \in \mathbb{N}$. The same holds for derivatives of any order w.r.t. α of $n\delta_n$.

It can be shown that $\delta_n(x,\alpha) = \tilde{\delta}_n(x,\alpha)$ satisfies Condition 1.46 (6), under Condition 1.46 (2). This choice of δ_n trivially satisfies Condition 1.46 (5).

Example 1.47 Consider the two dimensional diffusion $X = (Y, R)$ given by

$$
\begin{aligned}
dY_t &= (R_t + \mu_1)dt + \varepsilon\kappa_1 dW_t^1 \\
dR_t &= -\mu_2(R_t - m)dt + \varepsilon\kappa_2\sqrt{R_t}\left(\rho dW_t^1 + \sqrt{1-\rho^2}dW_t^2\right),
\end{aligned}
$$

where $(Y_0, R_0) = (y_0, r_0)$ with $r_0 > 0$. This model was used in finance by Longstaff and Schwartz (1995). In their mode, the second component represents the short term interest rate, while Y is the logarithm of the price of some asset. The second component is the square-root diffusion. The parameters are $\theta = (\alpha, \beta)$, where $\alpha = (\mu_1, \mu_2, m)$ and $\beta = (\kappa_1^2, \kappa_2^2, \rho)$. The parameter ρ allows correlation between the innovation terms of the two coordinates. The diffusion process (Y, R) satisfies Condition 1.46 (1) – (3), and (4) is holds if $r_0 \neq m_0$. The equation (1.125) is linear and has the solution

$$
\xi_t(y, r, \mu_1, \mu_2, m) = \begin{pmatrix} y + (\mu_1 + m)t + \mu_2^{-1}(r - m)(1 - e^{-\mu_2 t}) \\ m + (r - m)e^{-\mu_2 t} \end{pmatrix}.
$$

Therefore we can choose $\delta_n(x,\alpha) = \tilde{\delta}_n(x,\alpha)$, which satisfies Condition 1.46 (5) – (6). The matric $I(\theta_0)$ is invertible when $r_0 \neq m_0$ and is given by

$$
I_1(\theta) = (1-\rho^2)^{-1} \begin{pmatrix} \kappa_1^{-2} & 0 & 0 \\ 0 & \frac{-m(\mu_2 + \log(q)) + (m - r_0)(e^{-\mu_2} - 1))}{\kappa_2^2 \mu_2} & \frac{-\mu_1 + \log(q)}{\kappa_2^2} \\ 0 & \frac{-\mu_1 + \log(q)}{\kappa_2^2} & -\frac{\mu_2 \log(q)}{m\kappa_2^2} \end{pmatrix},
$$

where $q = r_0/(r_0 + m(e^{\mu_2} - 1))$, and

$$
I_2(\theta) = \begin{pmatrix} 2\kappa_1^4 & 2\rho^2\kappa_1^2\kappa_2^2 & \rho(1 - \rho^2)\kappa_1^2 \\ \rho^2\kappa_1^2\kappa_2^2 & 2\kappa_2^4 & \rho(1 - \rho^2)\kappa_2^2 \\ \rho(1 - \rho^2)\kappa_1^2 & \rho(1 - \rho^2)\kappa_2^2 & (1 - \rho^2)^2 \end{pmatrix}.
$$

Note that the asymptotic variance of the estimators of the drift parameter goes to zero, as the correlation parameter ρ goes to one.

□

Several papers have studied other aspects of small diffusion asymptotics for estimators of parameters in diffusion models. First estimation of the parameter α based on a continuously observed sample path of the diffusion process was considered by Kutoyants (1994). Semiparametric estimation for the same type of data was studied later by Kutoyants (1998) and Iacus and Kutoyants (2001). Information criteria were investigated by Uchida and Yoshida (2004b). Uchida (2004, 2008) studied approximations to martingale estimating functions for discretely sampled diffusions under small diffusion asymptotics. Martingale estimating functions were studied by M. Sørensen (2000b) under an extreme type of small diffusion asymptotics where n is fixed.

1.9 Non-Markovian models

In this section we consider estimating functions that can be used when the observed process is not a Markov process. In this situation, it is usually not easy to find a tractable martingale estimating function. For instance, a simple estimating function of the form (1.41) is not a martingale. To obtain a martingale, the conditional expectation given $X_{(i-1)\Delta}$ in (1.41) must be replaced by the conditional expectation given all previous observations, which can only very rarely be found explicitly, and which it is rather hopeless to find by simulation. Instead we will consider a generalization of the martingale estimating functions, called the prediction-based estimating functions, which can be interpreted as approximations to martingale estimating functions.

To clarify our thoughts, we will consider a concrete model type. Let the D-dimensional process X be the stationary solution to the stochastic differential equation

$$dX_t = b(X_t; \theta)dt + \sigma(X_t; \theta)dW_t, \tag{1.128}$$

where b is D-dimensional, σ is a $D \times D$-matrix, and W a D-dimensional standard Wiener process. As usual the parameter θ varies in a subset Θ of $\mathbb{R}^p$. However, we do not observe X directly. What we observe is

$$Y_i = k(X_{t_i}) + Z_i, \quad i = 1, \dots, n, \tag{1.129}$$

or

$$Y_i = \int_{t_{i-1}}^{t_i} k(X_s)ds + Z_i, \quad i = 1, \dots, n, \tag{1.130}$$

where k maps $\mathbb{R}^D$ into $\mathbb{R}^d$ $(d < D)$, and $\{Z_i\}$ is a sequence of independent identically distributed d-dimensional measurement errors with mean zero. We assume that the measurement errors are independent of the process X. Obviously, the discrete time process $\{Y_i\}$ is not a Markov-process.

1.9.1 Prediction-based estimating functions

In the following we will outline the method of prediction-based estimating functions introduced in M. Sørensen (2000a). Assume that $f_j, j = 1, \ldots, N$, are functions that map $\mathbb{R}^{s+1} \times \Theta$ into $\mathbb{R}$ such that $E_\theta(f_j(Y_{s+1}, \ldots, Y_1; \theta)^2) < \infty$ for all $\theta \in \Theta$. Let $\mathcal{P}^\theta_{i-1,j}$ be a closed linear subset of the L_2-space, L^θ_{i-1}, of all functions of $Y_1, \ldots, Y_{i-1}$ with finite variance under P_θ. The set $\mathcal{P}^\theta_{i-1,j}$ can be interpreted as a set of predictors of $f_j(Y_i, \ldots, Y_{i-s}; \theta)$ based on $Y_1, \ldots, Y_{i-1}$. A prediction-based estimating function has the form

$$G_n(\theta) = \sum_{i=s+1}^{n} \sum_{j=1}^{N} \Pi_j^{(i-1)}(\theta) \left[f_j(Y_i, \ldots, Y_{i-s}; \theta) - \breve{\pi}_j^{(i-1)}(\theta) \right],$$

where $\Pi_j^{(i-1)}(\theta)$ is a p-dimensional vector, the coordinates of which belong to $\mathcal{P}^\theta_{i-1,j}$, and $\breve{\pi}_j^{(i-1)}(\theta)$ is the minimum mean square error predictor in $\mathcal{P}^\theta_{i-1,j}$ of $f_j(Y_i, \ldots, Y_{i-s}; \theta)$ under P_θ. When $s = 0$ and $\mathcal{P}^\theta_{i-1,j}$ is the set of all functions of $Y_1, \ldots, Y_{i-1}$ with finite variance, then $\breve{\pi}_j^{(i-1)}(\theta)$ is the conditional expectation under P_θ of $f_j(Y_i; \theta)$ given $Y_1, \ldots, Y_{i-1}$, so in this case we obtain a martingale estimating function. Thus for a Markov process, a martingale estimating function of the form (1.41) is a particular case of a prediction-based estimating function.

The minimum mean square error predictor in $\mathcal{P}^\theta_{i-1,j}$ of $f_j(Y_i, \ldots, Y_{i-s}; \theta)$ is the projection in L^θ_{i-1} of $f_j(Y_i, \ldots, Y_{i-s}; \theta)$ onto the subspace $\mathcal{P}_{i-1,j}$. Therefore $\breve{\pi}_j^{(i-1)}(\theta)$ satisfies the normal equation

$$E_\theta \left(\pi_j^{(i-1)} \left[f_j(Y_i, \ldots, Y_{i-s}; \theta) - \breve{\pi}_j^{(i-1)}(\theta) \right] \right) = 0 \qquad (1.131)$$

for all $\pi_j^{(i-1)} \in \mathcal{P}^\theta_{i-1,j}$. This implies that a prediction-based estimating function satisfies that

$$E_\theta(G_n(\theta)) = 0. \qquad (1.132)$$

We can interpret the minimum mean square error predictor as an approximation to the conditional expectation of $f_j(Y_i, \ldots, Y_{i-s}; \theta)$ given $X_1, \ldots, X_{i-1}$, which is the projection of $f_j(Y_i, \ldots, Y_{i-s}; \theta)$ onto the subspace of all functions of $X_1, \ldots, X_{i-1}$ with finite variance.

To obtain estimators that can relatively easily be calculated in practice, we will from now on restrict attention to predictor sets, $\mathcal{P}^\theta_{i-1,j}$, that are finite dimensional. Let $h_{jk}, j = 1, \ldots, N, k = 0, \ldots, q_j$ be functions from $\mathbb{R}^r$ into $\mathbb{R}$ ($r \geq s$), and define (for $i \geq r + 1$) random variables by

$$Z_{jk}^{(i-1)} = h_{jk}(Y_{i-1}, Y_{i-2}, \ldots, Y_{i-r}).$$

We assume that $E_\theta((Z_{jk}^{(i-1)})^2) < \infty$ for all $\theta \in \Theta$, and let $\mathcal{P}_{i-1,j}$ denote the subspace spanned by $Z_{j0}^{(i-1)}, \ldots, Z_{jq_j}^{(i-1)}$. We set $h_{j0} = 1$ and make the natural assumption that the functions $h_{j0}, \ldots, h_{jq_j}$ are linearly independent. We write the elements of $\mathcal{P}_{i-1,j}$ in the form $a^T Z_j^{(i-1)}$, where $a^T = (a_0, \ldots, a_{q_j})$ and

$$Z_j^{(i-1)} = \left(Z_{j0}^{(i-1)}, \ldots, Z_{jq_j}^{(i-1)} \right)^T$$

are $(q_j + 1)$-dimensional vectors. With this specification of the predictors, the estimating function can only include terms with $i \geq r + 1$:

$$G_n(\theta) = \sum_{i=r+1}^{n} \sum_{j=1}^{N} \Pi_j^{(i-1)}(\theta) \left[f_j(Y_i, \ldots, Y_{i-s}; \theta) - \breve{\pi}_j^{(i-1)}(\theta) \right]. \quad (1.133)$$

It is well-known that the minimum mean square error predictor, $\breve{\pi}_j^{(i-1)}(\theta)$, is found by solving the normal equations (1.131). Define $C_j(\theta)$ as the covariance matrix of $(Z_{j1}^{(r)}, \ldots, Z_{jq_j}^{(r)})^T$ under P_θ, and $b_j(\theta)$ as the vector for which the ith coordinate is

$$b_j(\theta)_i = \text{Cov}_\theta(Z_{ji}^{(r)}, f_j(Y_{r+1}, \ldots, Y_{r+1-s}; \theta)), \quad (1.134)$$

$i = 1, \ldots, q_j$. Then we have

$$\breve{\pi}_j^{(i-1)}(\theta) = \breve{a}_j(\theta)^T Z_j^{(i-1)}, \quad (1.135)$$

where $\breve{a}_j(\theta)^T = (\breve{a}_{j0}(\theta), \breve{a}_{j*}(\theta)^T)$ with

$$\breve{a}_{j*}(\theta) = C_j(\theta)^{-1} b_j(\theta) \quad (1.136)$$

and

$$\breve{a}_{j0}(\theta) = E_\theta(f_j(Y_{s+1}, \ldots, Y_1; \theta)) - \sum_{k=1}^{q_j} \breve{a}_{jk}(\theta) E_\theta \left(Z_{jk}^{(r)} \right). \quad (1.137)$$

That $C_j(\theta)$ is invertible follows from the assumption that the functions h_{jk} are linearly independent. If $f_j(Y_i, \ldots, Y_{i-s}; \theta)$ has mean zero under P_θ for all $\theta \in \Theta$, we need not include a constant in the space of predictors, i.e. we need only the space spanned by $Z_{j1}^{(i-1)}, \ldots, Z_{jq_j}^{(i-1)}$.

Example 1.48 An important particular case when $d = 1$ is $f_j(y) = y^j$, $j = 1, \ldots, N$. For each $i = r + 1, \ldots, n$ and $j = 1, \ldots, N$, we let $\{Z_{jk}^{(i-1)} \mid k = 0, \ldots, q_j\}$ be a subset of $\{Y_{i-\ell}^{\kappa} \mid \ell = 1, \ldots, r, \kappa = 0, \ldots, j\}$, where $Z_{j0}^{(i-1)}$ is always equal to 1. Here we need to assume that $E_\theta(Y_i^{2N}) < \infty$ for all $\theta \in \Theta$. To find $\breve{\pi}_j^{(i-1)}(\theta)$, $j = 1, \ldots, N$, by means of (1.136) and (1.137), we must calculate moments of the form

$$E_\theta(Y_1^{\kappa} Y_k^j), \quad 0 \leq \kappa \leq j \leq N, \quad k = 1, \ldots, r. \quad (1.138)$$

To avoid the matrix inversion in (1.136), the vector of coefficients $\breve{a}_j$ can be found by means of the N-dimensional Durbin-Levinson algorithm applied to the process $\{(Y_i, Y_i^2, \ldots, Y_i^N)\}_{i \in \mathbb{N}}$, see Brockwell and Davis (1991), provided that the predictor spaces consist of lagged values of this N-dimensional process. Suppose the diffusion process X is exponentially ρ-mixing, see Doukhan (1994) for a definition. This is for instance the case for a Pearson diffusion (see Subsection 1.3.7) or for a one-dimensional diffusion that satisfies Condition 1.33. Then the observed process Y inherits this property, which implies that constants $K > 0$ and $\lambda > 0$ exist such that $|\text{Cov}_\theta(Y_1^j, Y_k^j)| \leq K e^{-\lambda k}$. Therefore a small value of r can usually be used.

In many situations it is reasonable to choose $N = 2$ with the following simple predictor sets where $q_1 = r$ and $q_2 = 2r$. The predictor sets are generated by $Z_{j0}^{(i-1)} = 1$, $Z_{jk}^{(i-1)} = Y_{i-k}$, $k = 1, \ldots, r$, $j = 1, 2$ and $Z_{2k}^{(i-1)} = Y_{i+r-k}^2$, $k = r + 1, \ldots, 2r$. In this case the minimum mean square error predictor of Y_i can be found using the Durbin-Levinson algorithm for real processes, while the predictor of Y_i^2 can be found by applying the two-dimensional Durbin-Levinson algorithm to the process (Y_i, Y_i^2). Including predictors in the form of lagged terms $Y_{i-k}Y_{i-k-l}$ for a number of lags l's might also be of relevance.

We illustrate the use of the Durbin-Levinson algorithm in the simplest possible case, where $N = 1$, $f(x) = x$, $Z_0^{(i-1)} = 1$, $Z_k^{(i-1)} = Y_{i-k}$, $k = 1, \ldots, r$. We suppress the superfluous j in the notation. Let $K_\ell(\theta)$ denote the covariance between Y_1 and $Y_{\ell+1}$ under P_θ, and define $\phi_{1,1}(\theta) = K_1(\theta)/K_0(\theta)$ and $v_0(\theta) = K_0(\theta)$. Then the Durbin-Levinson algorithm works as follows

$$\phi_{\ell,\ell}(\theta) = \left(K_\ell(\theta) - \sum_{k=1}^{\ell-1} \phi_{\ell-1,k}(\theta) K_{\ell-k}(\theta) \right) v_{\ell-1}(\theta)^{-1},$$

$$\begin{pmatrix} \phi_{\ell,1}(\theta) \\ \vdots \\ \phi_{\ell,\ell-1}(\theta)) \end{pmatrix} = \begin{pmatrix} \phi_{\ell-1,1}(\theta) \\ \vdots \\ \phi_{\ell-1,\ell-1}(\theta)) \end{pmatrix} - \phi_{\ell,\ell}(\theta) \begin{pmatrix} \phi_{\ell-1,\ell-1}(\theta) \\ \vdots \\ \phi_{\ell-1,1}(\theta)) \end{pmatrix}$$

and

$$v_\ell(\theta) = v_{\ell-1}(\theta) \left(1 - \phi_{\ell,\ell}(\theta)^2 \right).$$

The algorithm is run for $\ell = 2, \ldots, r$. Then

$$\breve{a}_*(\theta) = (\phi_{r,1}(\theta), \ldots, \phi_{r,r}(\theta)),$$

while $\breve{a}_0$ can be found from (1.137), which here simplifies to

$$\breve{a}_0(\theta) = E_\theta(Y_1) \left(1 - \sum_{k=1}^{r} \phi_{r,k}(\theta) \right).$$

The quantity $v_r(\theta)$ is the prediction error $E_\theta\left((Y_i - \breve{\pi}^{(i-1)})^2\right)$. Note that if we want to include a further lagged value of Y in the predictor, we just iterate the algorithm once more.

□

We will now find the optimal prediction-based estimating function of the form (1.133) in the sense explained in Section 1.11. First we express the estimating function in a more compact way. The ℓth coordinate of the vector $\Pi_j^{(i-1)}(\theta)$ can be written as

$$\pi_{\ell,j}^{(i-1)}(\theta) = \sum_{k=0}^{q_j} a_{\ell jk}(\theta)Z_{jk}^{(i-1)}, \quad \ell = 1, \ldots, p.$$

With this notation, (1.133) can be written in the form

$$G_n(\theta) = A(\theta) \sum_{i=r+1}^{n} H^{(i)}(\theta), \tag{1.139}$$

where

$$A(\theta) = \begin{pmatrix} a_{110}(\theta) & \cdots & a_{11q_1}(\theta) & \cdots\cdots & a_{1N0}(\theta) & \cdots & a_{1Nq_N}(\theta) \\ \vdots & & \vdots & & \vdots & & \vdots \\ a_{p10}(\theta) & \cdots & a_{p1q_1}(\theta) & \cdots\cdots & a_{pN0}(\theta) & \cdots & a_{pNq_N}(\theta) \end{pmatrix},$$

and

$$H^{(i)}(\theta) = Z^{(i-1)}\left(F(Y_i, \ldots, Y_{i-s}; \theta) - \breve{\pi}^{(i-1)}(\theta)\right), \tag{1.140}$$

with $F = (f_1, \ldots, f_N)^T$, $\breve{\pi}^{(i-1)}(\theta) = (\breve{\pi}_1^{(i-1)}(\theta), \ldots, \breve{\pi}_N^{(i-1)}(\theta))^T$, and

$$Z^{(i-1)} = \begin{pmatrix} Z_1^{(i-1)} & 0_{q_1+1} & \cdots & 0_{q_1+1} \\ 0_{q_2+1} & Z_2^{(i-1)} & \cdots & 0_{q_2+1} \\ \vdots & \vdots & & \vdots \\ 0_{q_N+1} & 0_{q_N+1} & \cdots & Z_N^{(i-1)} \end{pmatrix}. \tag{1.141}$$

Here 0_{q_j+1} denotes the $(q_j + 1)$-dimensional zero-vector. When we have chosen the functions f_j and the predictor spaces, the quantities $H^{(i)}(\theta)$ are completely determined, whereas we are free to choose the matrix $A(\theta)$ in an optimal way, i.e. such that the asymptotic variance of the estimators is minimized.

We will find an explicit expression for the optimal weight matrix, $A^*(\theta)$, under the following condition, in which we need one further definition:

$$\breve{a}(\theta) = (\breve{a}_{10}(\theta), \ldots, \breve{a}_{1q_1}(\theta), \ldots, \breve{a}_{N0}(\theta), \ldots \breve{a}_{Nq_N}(\theta))^T, \tag{1.142}$$

where the quantities $\breve{a}_{jk}$ define the minimum mean square error predictors, cf. (1.135).

Condition 1.49

(1) *The function $F(y_1,\ldots,y_{s+1};\theta)$ and the coordinates of $\breve{a}(\theta)$ are continuously differentiable functions of θ.*

(2) $p \leq \bar{p} = N + q_1 + \cdots + q_N$.

(3) *The $\bar{p} \times p$-matrix $\partial_{\theta^T}\breve{a}(\theta)$ has rank p.*

(4) *The functions $1, f_1, \ldots, f_N$ are linearly independent (for fixed θ) on the support of the conditional distribution of $(Y_i,\ldots,Y_{i-s})$ given $(Y_{i-1},\ldots,Y_{i-r})$.*

(5) *The $\bar{p} \times p$-matrix*

$$U(\theta)^T = E_\theta\left(Z^{(i-1)}\partial_{\theta^T}F(Y_i,\ldots,Y_{i-s};\theta)\right) \tag{1.143}$$

exists.

If we denote the optimal prediction-based estimating function by $G_n^*(\theta)$, then

$$E_\theta\left(G_n(\theta)G_n^*(\theta)^T\right) = (n-r)A(\theta)\bar{M}_n(\theta)A_n^*(\theta)^T,$$

where

$$\bar{M}_n(\theta) = E_\theta\left(H^{(r+1)}(\theta)H^{(r+1)}(\theta)^T\right) \tag{1.144}$$

$$+ \sum_{k=1}^{n-r-1}\frac{(n-r-k)}{(n-r)}\left\{E_\theta\left(H^{(r+1)}(\theta)H^{(r+1+k)}(\theta)^T\right)\right.$$

$$\left. + E_\theta\left(H^{(r+1+k)}(\theta)H^{(r+1)}(\theta)^T\right)\right\},$$

which is the covariance matrix of $\sum_{i=r+1}^n H^{(i)}(\theta)/\sqrt{n-r}$. The sensitivity function (1.167) is given by

$$S_{G_n}(\theta) = (n-r)A(\theta)\left[U(\theta)^T - D(\theta)\partial_{\theta^T}\breve{a}(\theta)\right],$$

where $D(\theta)$ is the $\bar{p}\times\bar{p}$-matrix

$$D(\theta) = E_\theta\left(Z^{(i-1)}(Z^{(i-1)})^T\right). \tag{1.145}$$

It follows from Theorem 1.61 that $A_n^*(\theta)$ is optimal if $E_\theta\left(G_n(\theta)G_n^*(\theta)^T\right) = S_{G_n}(\theta)$. Under Condition 1.49 (4) the matrix $\bar{M}_n(\theta)$ is invertible, see **M. Sørensen** (2000a), so it follows that

$$A_n^*(\theta) = (U(\theta) - \partial_\theta\breve{a}(\theta)^T D(\theta))\bar{M}_n(\theta)^{-1}, \tag{1.146}$$

and that the estimating function

$$G_n^*(\theta) = A_n^*(\theta) \sum_{i=r+1}^{n} Z^{(i-1)} \left(F(Y_i, \ldots, Y_{i-s}; \theta) - \breve{\pi}^{(i-1)}(\theta) \right), \quad (1.147)$$

is Godambe optimal. When the function F does not depend on θ, the expression for $A_n^*(\theta)$ simplifies slightly as in this case $U(\theta) = 0$.

Example 1.50 Consider again the type of prediction-based estimating function discussed in Example 1.48. In order to calculate (1.144), we need mixed moments of the form

$$E_\theta[Y_{t_1}^{k_1} Y_{t_2}^{k_2} Y_{t_3}^{k_3} Y_{t_4}^{k_4}], \quad (1.148)$$

for $t_1 \leq t_2 \leq t_3 \leq t_4$ and $k_1 + k_2 + k_3 + k_4 \leq 4N$, where k_i, $i = 1, \ldots, 4$ are non-negative integers.

$\square$

1.9.2 Asymptotics

Consider a prediction-based estimating function of the form (1.139) and (1.140), where the function F does not depend on the parameter θ. Then the corresponding estimator is consistent and asymptotically normal under the following condition, where θ_0 denotes the true parameter value. Asymptotic theory for more general prediction-based estimating functions can be found in M. Sørensen (2011).

Condition 1.51
(1) *The diffusion process X is stationary and geometrically α-mixing.*

(2) *There exists a $\delta > 0$ such that*

$$E_{\theta_0} \left(\left| Z_{jk}^{(r)} f_j(Y_{r+1}, \ldots, Y_{r+1-s}; \theta_0) \right|^{2+\delta} \right) < \infty$$

and

$$E_{\theta_0} \left(\left| Z_{jk}^{(r)} Z_{j\ell}^{(r)} \right|^{2+\delta} \right) < \infty,$$

for $j = 1, \ldots, N$, $k, \ell = 0, \ldots q_j$.

(3) *The components of $A(\theta)$ and $\breve{a}(\theta)$, given by (1.142) are continuously differentiable functions of θ.*

(4) *The matrix* $W = -A(\theta_0)D(\theta_0)\partial_{\theta^T}\breve{a}(\theta_0)$ *has full rank* p. *The matrix* $D(\theta)$ *is given by* (1.145).

(5)

$$A(\theta)\left(E_{\theta_0}\left(Z^{(r)}F(Y_{r+1},\ldots,Y_{r+1-s})\right) - D(\theta_0)\partial_{\theta^T}\breve{a}(\theta))\right) \neq 0$$

for all $\theta \neq \theta_0$.

Condition 1.51 (1) and (2) ensures that the central limit theorem (1.5) holds and that $\bar{M}_n(\theta_0) \to M(\theta_0)$, where

$$M(\theta) = E_\theta\left(H^{(r+1)}(\theta)H^{(r+1)}(\theta)^T\right)$$
$$+ \sum_{k=1}^{\infty}\left\{E_\theta\left(H^{(r+1)}(\theta)H^{(r+1+k)}(\theta)^T\right)\right.$$
$$\left. + E_\theta\left(H^{(r+1+k)}(\theta)H^{(r+1)}(\theta)^T\right)\right\}.$$

The asymptotic covariance matrix in (1.5) is $V(\theta) = A(\theta)M(\theta)A(\theta)^T$. The concept of geometric α-mixing was explained in Subsection 1.5.1, where also conditions for geometric α-mixing were discussed. It is not difficult to see that if the basic diffusion process X is geometrically α-mixing, then the observed process Y inherits this property. We only need to check Condition 1.1 with $\bar{\theta} = \theta_0$ to obtain asymptotic results for prediction-based estimators. The condition (1.6) is satisfied because of (1.132). It is easy to see that Condition 1.51 (3) and (4) implies that $\theta \mapsto g(y_1,\ldots,y_{r+1})$ is continuously differentiable and that g as well as $\partial_{\theta^T}g$ are locally dominated integrable under P_{θ_0}. Finally, for a prediction-based estimating function, the condition (1.9) is identical to Condition 1.51 (5). Therefore it follows from Theorem 1.2 that a consistent G_n–estimator $\hat{\theta}_n$ exists and is the unique G_n–estimator on any bounded subset of Θ containing θ_0 with probability approaching one as $n \to \infty$. The estimator satisfies that

$$\sqrt{n}(\hat{\theta}_n - \theta_0) \xrightarrow{\mathcal{D}} N_p\left(0, W^{-1}A(\theta_0)M(\theta_0)A(\theta_0)^T W^{T^{-1}}\right)$$

as $n \to \infty$. For an optimal estimating function, this simplifies to

$$\sqrt{n}(\hat{\theta}_n - \theta_0) \xrightarrow{\mathcal{D}} N_p\left(0, W^{-1}\right).$$

1.9.3 Measurement errors

Suppose a one-dimensional diffusion has been observed with measurement errors so that the data are

$$Y_i = X_{t_i} + Z_i, \quad i = 1,\ldots,n,$$

where X solves (1.11), and the measurement errors Z_i are independent and identically distributed and independent of X. Since the observed process (Y_i) is not a Markov process, it is usually not possible to find a feasible martingale estimating function. Instead we can use a prediction-based estimating function of the type considered Example 1.48. To find the minimum mean square error predictor, we must find mixed moments of the form (1.138). By the binomial formula,

$$
\begin{aligned}
E_\theta(Y_1^{k_1} Y_\ell^{k_\ell}) &= E_\theta\left((X_{t_1} + Z_1)^{k_1}(X_{t_\ell} + Z_\ell)^{k_2}\right) \\
&= \sum_{i_1=0}^{k_1} \sum_{i_2=0}^{k_2} \binom{k_1}{i_1}\binom{k_2}{i_2} E_\theta(X_{t_1}^{i_1} X_{t_\ell}^{i_2}) E_\theta(Z_1^{k_1-i_1}) E_\theta(Z_\ell^{k_2-i_2}).
\end{aligned}
$$

Note that the distribution of the measurement error Z_i can depend on components of the unknown parameter θ. We need to find the mixed moments $E_\theta(X_{t_1}^{i_1} X_{t_2}^{i_2})$, $(t_1 < t_2)$. If expressions for the moments and conditional moments of X_t are available, these mixed moments can be found explicitly. As an example, consider the Pearson diffusions discussed in Subsection 1.3.7, for which the conditional moments are given by (1.86). Thus

$$
\begin{aligned}
E_\theta(X_{t_1}^{i_1} X_{t_2}^{i_2}) &= E_\theta(X_{t_1}^{i_1} E_\theta(X_{t_2}^{i_2} | X_{t_1})) \qquad\qquad\qquad (1.149) \\
&= \sum_{k=0}^{i_2}\left(\sum_{\ell=0}^{i_2} q_{i_2,k,\ell} e^{-\lambda_\ell(t_2-t_1)}\right) E_\theta(X_{t_1}^{i_1+k}),
\end{aligned}
$$

where $E_\theta(X_{t_1}^{i_1+k})$ can be found by (1.87), provided, of course, that it exists. For stationary, ergodic one-dimensional diffusions, the polynomial moments can usually be found because we have an explicit expression for the marginal density functions, at least up to a multiplicative constant, cf. (1.15). In order to find the optimal prediction-based estimating functions of the form considered in Example 1.48, we must find the mixed moments of the form (1.148), which can be calculated in a similar way.

1.9.4 Integrated diffusions and hypoelliptic stochastic differential equations

Sometimes a diffusion process, X, cannot be observed directly, but data of the form

$$
Y_i = \frac{1}{\Delta}\int_{(i-1)\Delta}^{i\Delta} X_s\, ds, \quad i = 1,\dots,n, \qquad\qquad (1.150)
$$

are available for some fixed Δ. Such observations might be obtained when the process X is observed after passage through an electronic filter. Another example is provided by ice-core records. The isotope ratio $^{18}O/^{16}O$ in the ice is a proxy for paleo-temperatures. The average isotope ratio is measured in pieces of ice, each of which represent a time interval. The variation of the paleo-temperature can be modelled by a stochastic differential equation, and hence the ice-core data can be modelled as an integrated diffusion process, see P. D. Ditlevsen, Ditlevsen, and Andersen (2002). Estimation based on this type of data was considered by Gloter (2000, 2006), Bollerslev and Wooldridge (1992), S. Ditlevsen and Sørensen (2004), and Baltazar-Larios and Sørensen (2010). Non-parametric inference was studied in Comte, Genon-Catalot, and Rozenholc (2009).

The model for data of the type (1.150) is a particular case of (1.128) with

$$d \begin{pmatrix} X_{1,t} \\ X_{2,t} \end{pmatrix} = \begin{pmatrix} b(X_{1,t}; \theta) \\ X_{1,t} \end{pmatrix} dt + \begin{pmatrix} \sigma(X_{1,t}; \theta) \\ 0 \end{pmatrix} dW_t,$$

with $X_{2,0} = 0$, where W and the two components are one-dimensional, and only the second coordinate, $X_{2,t}$, is observed. The second coordinate is not stationary, but if the first coordinate is a stationary process, then the observed increments $Y_i = (X_{2,i\Delta} - X_{2,(i-1)\Delta})/\Delta$ form a stationary sequence. A stochastic differential equation of the form (1.151) is called *hypoelliptic*. Hypoelliptic stochastic differential equations are, for instance, used to model molecular dynamics, see e.g. Pokern, Stuart, and Wiberg (2009). The unobserved component, $X_{1,t}$, can more generally be multivariate and have coefficients that depend on the observed component $X_{2,t}$ too. The observed smooth component can also be multivariate. The drift is typically minus the derivative of a potential. A simple example is the stochastic harmonic oscillator

$$\begin{aligned} dX_{1,t} &= -(\beta_1 X_{1,t} + \beta_2 X_{2,t}) \, dt + \gamma dW_t \\ dX_{2,t} &= X_{1,t} \, dt, \end{aligned}$$

$\beta_1, \beta_2, \gamma > 0$. Observations of the form (1.150), or more generally discrete time observations of the smooth components of a hypoelliptic stochastic differential equation, do not form a Markov process, so usually a feasible martingale estimating function is not available, but prediction-based estimating functions can be used instead. For instance, the stochastic harmonic oscillator above is a Gaussian process. Therefore all the mixed moments needed in the optimal prediction-based estimating function of the form considered in Example 1.48 can be found explicitly.

In the following we will again denote the basic diffusion by X (rather than X_1), and assume that the data are given by (1.150). Suppose that $4N$'th moment of X_t is finite. The moments (1.138) and (1.148) can be calculated by

$$E\left[Y_1^{k_1} Y_{t_1}^{k_2} Y_{t_2}^{k_3} Y_{t_3}^{k_4} \right] =$$

$$\frac{\int_A E[X_{v_1} \cdots X_{v_{k_1}} X_{u_1} \cdots X_{u_{k_2}} X_{s_1} \cdots X_{s_{k_3}} X_{r_1} \cdots X_{r_{k_4}}] \, dt}{\Delta^{k_1 + k_2 + k_3 + k_4}}$$

where $1 \le t_1 \le t_2 \le t_3$, $A = [0, \Delta]^{k_1} \times [(t_1 - 1)\Delta, t_1 \Delta]^{k_2} \times [(t_2 -$

1)Δ, $t_2\Delta]^{k_3} \times [(t_3-1)\Delta$, $t_3\Delta]^{k_4}$, and $dt = dr_{k_4} \cdots dr_1\, ds_{k_3} \cdots ds_1\, du_{k_2} \cdots du_1\, dv_{k_1} \cdots dv_1$. The domain of integration can be reduced considerably by symmetry arguments, but the point is that we need to calculate mixed moments of the type $E(X_{t_1}^{\kappa_1} \cdots X_{t_k}^{\kappa_k})$, where $t_1 < \cdots < t_k$. For the Pearson diffusions discussed in Subsection 1.3.7, these mixed moments can be calculated by a simple iterative formula obtained from (1.86) and (1.87), as explained in the previous subsection. Moreover, for the Pearson diffusions, $E(X_{t_1}^{\kappa_1} \cdots X_{t_k}^{\kappa_k})$ depends on $t_1, \dots, t_k$ through sums and products of exponential functions, cf. (1.86) and (1.149). Therefore the integral above can be explicitly calculated, and thus explicit optimal estimating functions of the type considered in Example 1.48 are available for observations of integrated Pearson diffusions.

Example 1.52 Consider observation of an integrated square root process (1.37) and a prediction-based estimating function with $f_1(x) = x$ and $f_2(x) = x^2$ with predictors given by $\pi_1^{(i-1)} = a_{1,0} + a_{1,1}Y_{i-1}$ and $\pi_2^{(i-1)} = a_{2,0}$. Then the minimum mean square error predictors are

$$\check{\pi}_1^{(i-1)}(Y_{i-1}; \theta) = \mu(1 - \check{a}(\beta)) + \check{a}(\beta)Y_{i-1},$$
$$\check{\pi}_2^{(i-1)}(\theta) = \alpha^2 + \alpha\tau^2\beta^{-3}\Delta^{-2}(e^{-\beta\Delta} - 1 + \beta\Delta)$$

with

$$\check{a}(\beta) = \frac{(1 - e^{-\beta\Delta})^2}{2(\beta\Delta - 1 + e^{-\beta\Delta})}.$$

The optimal prediction-based estimating function is

$$\sum_{i=1}^{n} \begin{pmatrix} 1 \\ Y_{i-1} \\ 0 \end{pmatrix} [Y_i - \check{\pi}_1^{(i-1)}(Y_{i-1}; \theta)] + \sum_{i=1}^{n} \begin{pmatrix} 0 \\ 0 \\ 1 \end{pmatrix} [Y_i^2 - \check{\pi}_2^{(i-1)}(\theta)],$$

from which we obtain the estimators

$$\hat{\alpha} = \frac{1}{n}\sum_{i=1}^{n}Y_i + \frac{\check{a}(\hat{\beta})Y_n - Y_1}{(n-1)(1 - \check{a}(\hat{\beta}))}$$

$$\sum_{i=2}^{n}Y_{i-1}Y_i = \hat{\alpha}(1 - \check{a}(\hat{\beta}))\sum_{i=2}^{n}Y_{i-1} + \check{a}(\hat{\beta})\sum_{i=2}^{n}Y_{i-1}^2$$

$$\hat{\tau}^2 = \frac{\hat{\beta}^3\Delta^2\sum_{i=2}^{n}(Y_i^2 - \hat{\alpha}^2)}{(n-1)\hat{\alpha}(e^{-\hat{\beta}\Delta} - 1 + \hat{\beta}\Delta)}.$$

The estimators are explicit apart from $\hat{\beta}$, which can easily be found numerically by solving a non-linear equation in one variable. For details, see S. Ditlevsen and Sørensen (2004).

□

1.9.5 Sums of diffusions

An autocorrelation function of the form

$$\rho(t) = \phi_1 \exp(-\beta_1 t) + \ldots + \phi_D \exp(-\beta_D t), \tag{1.151}$$

where $\sum_{i=1}^{D} \phi_i = 1$ and $\phi_i, \beta_i > 0$, is found in many observed time series. Examples are financial time series, see Barndorff–Nielsen and Shephard (2001), and turbulence, see Barndorff–Nielsen, Jensen, and Sørensen (1990) and Bibby et al. (2005).

A simple model with autocorrelation function of the form (1.151) is the sum of diffusions

$$Y_t = X_{1,t} + \cdots + X_{D,t}, \tag{1.152}$$

where the D diffusions

$$dX_{i,t} = -\beta_i(X_{i,t} - \alpha_i) + \sigma_i(X_{i,t})dW_{i,t}, \quad i = 1, \ldots, D,$$

are independent. In this case

$$\phi_i = \frac{\mathrm{Var}(X_{i,t})}{\mathrm{Var}(X_{1,t}) + \cdots + \mathrm{Var}(X_{D,t})}.$$

Sums of diffusions of this type with a pre-specified marginal distribution of Y were considered by Bibby and Sørensen (2003) and Bibby et al. (2005), while Forman and Sørensen (2008) studied sums of Pearson diffusions. The same type of autocorrelation function is obtained for sums of independent Ornstein-Uhlenbeck processes driven by Lévy processes. This class of models was introduced and studied in Barndorff–Nielsen, Jensen, and Sørensen (1998).

Example 1.53 *Sum of square root processes.* If $\sigma_i^2(x) = 2\beta_i bx$ and $\alpha_i = \kappa_i b$ for some $b > 0$, then the stationary distribution of Y_t is a gamma-distribution with shape parameter $\kappa_1 + \cdots + \kappa_D$ and scale parameter b. The weights in the autocorrelation function are $\phi_i = \kappa_i / (\kappa_1 + \cdots + \kappa_D)$.

$\square$

For sums of the Pearson diffusions presented in Subsection 1.3.7, we have explicit formulae that allow calculation of (1.138) and (1.148), provided these mixed moments exist. Thus for sums of Pearson diffusions we have explicit optimal prediction-based estimating functions of the type considered in Example 1.48. By the multinomial formula,

$$E(Y_{t_1}^{\kappa} Y_{t_2}^{\nu}) =$$

$$\sum \sum \binom{\kappa}{\kappa_1, \ldots, \kappa_D} \binom{\nu}{\nu_1, \ldots, \nu_D} E(X_{1,t_1}^{\kappa_1} X_{1,t_2}^{\nu_1}) \ldots E(X_{D,t_1}^{\kappa_D} X_{D,t_2}^{\nu_D})$$

where

$$\binom{\kappa}{\kappa_1, \ldots, \kappa_D} = \frac{\kappa!}{\kappa_1! \cdots \kappa_D!}$$

is the multinomial coefficient, and where the first sum is over $0 \leq \kappa_1, \ldots, \kappa_D$ such that $\kappa_1 + \ldots \kappa_D = \kappa$, and the second sum is analogous for the ν_is. Higher order mixed moments of the form (1.148) can be found by a similar formula with four sums and four multinomial coefficients. Such formulae may appear daunting, but are easy to program. For a Pearson diffusion, mixed moments of the form $E(X_{t_1}^{\kappa_1} \cdots X_{t_k}^{\kappa_k})$ can be calculated by a simple iterative formula obtained from (1.86) and (1.87), as explained in Subsection 1.9.3.

Example 1.54 *Sum of two skew t-diffusions.* If $\alpha_i = 0$ and

$$\sigma_i^2(x) = 2\beta_i(\nu_i - 1)^{-1}\{x^2 + 2\rho\sqrt{\nu_i}x + (1 + \rho^2)\nu_i\}, \quad i = 1, 2,$$

with $\nu_i > 3$, then the stationary distribution of $X_{i,t}$ is a skew t-distribution, and in (1.151) the weights are given by $\phi_i = \nu_i(\nu_i - 2)^{-1}/\{\nu_1(\nu_1 - 2)^{-1} + \nu_2(\nu_2 - 2)^{-1}\}$. To simplify the exposition we assume that the sampling times are $t_i = \Delta i$, and that the correlation parameters β_1, β_2, ϕ_1, and ϕ_2 are known or have been estimated in advance, for instance by fitting (1.151) with $D = 2$ to the empirical autocorrelation function. We denote the observations by Y_i. We will find the optimal estimating function in the simple case where predictions of Y_i^2 are made based on predictors of the form $\pi^{(i-1)} = a_0 + a_1 Y_{i-1}$. The estimating equations take the form

$$\sum_{i=2}^{n} \left[\begin{array}{c} Y_i^2 - \sigma^2 - \zeta Y_{i-1} \\ Y_{i-1}Y_i^2 - \sigma^2 Y_{i-1} - \zeta Y_{i-1}^2 \end{array} \right] = 0, \tag{1.153}$$

with

$$\sigma^2 = \text{Var}(Y_i) = (1 + \rho^2)\left\{ \frac{\nu_1}{\nu_1 - 2} + \frac{\nu_2}{\nu_2 - 2} \right\},$$

$$\zeta = \frac{\text{Cov}(Y_{i-1}, Y_i^2)}{\text{Var}(Y_i)} = 4\rho\left\{ \frac{\sqrt{\nu_1}}{\nu_1 - 3}\phi_1 e^{-\beta_1\Delta} + \frac{\sqrt{\nu_2}}{\nu_2 - 3}\phi_2 e^{-\beta_2\Delta} \right\}.$$

Solving equation (1.153) for ζ and σ^2 we get

$$\hat{\zeta} = \frac{\frac{1}{n-1}\sum_{i=2}^{n} Y_{i-1}Y_i^2 - (\frac{1}{n-1}\sum_{i=2}^{n} Y_{i-1})(\frac{1}{n-1}\sum_{i=2}^{n} Y_i^2)}{\frac{1}{n-1}\sum_{i=2}^{n} Y_{i-1}^2 - (\frac{1}{n-1}\sum_{i=2}^{n} Y_{i-1})^2},$$

$$\hat{\sigma}^2 = \frac{1}{n-1}\sum_{i=2}^{n} Y_i^2 + \hat{\zeta}\frac{1}{n-1}\sum_{i=2}^{n} Y_{i-1}.$$

In order to estimate ρ we restate ζ as

$$\zeta = \sqrt{32(1+\rho^2)} \cdot \rho \cdot \left\{ \frac{\sqrt{9(1+\rho^2)-\phi_1\sigma^2}}{3(1+\rho^2)-\phi_1\sigma^2} \phi_1 e^{-\beta_1\Delta} \right.$$

$$\left. + \frac{\sqrt{9(1+\rho^2)-\phi_2\sigma^2}}{3(1+\rho^2)-\phi_2\sigma^2} \phi_2 e^{-\beta_2\Delta} \right\}$$

and insert $\hat{\sigma}^2$ for σ^2. Thus, we get a one-dimensional estimating equation, $\zeta(\beta, \phi, \hat{\sigma}^2, \rho) = \hat{\zeta}$, which can be solved numerically. Finally, by inverting $\phi_i = \frac{1+\rho^2}{\sigma^2} \frac{\nu_i}{\nu_i - 2}$, we find the estimates $\hat{\nu}_i = \frac{2\phi_i\hat{\sigma}^2}{\phi_i\hat{\sigma}^2 - (1+\hat{\rho}^2)}$, $i = 1, 2$.

$\square$

A more complex model is obtained if the observations are integrals of the process Y given by (1.152). In this case the data are

$$Z_i = \frac{1}{\Delta} \int_{(i-1)\Delta}^{i\Delta} Y_s \, ds = \frac{1}{\Delta} \left(\int_{(i-1)\Delta}^{i\Delta} X_{1,s} ds + \cdots + \int_{(i-1)\Delta}^{i\Delta} X_{D,s} ds \right), \quad (1.154)$$

$i = 1, \ldots, n$. Also here the moments of form (1.138) and (1.148), and hence optimal prediction-based estimating functions, can be found explicitly for Pearson diffusions. This is because each of the observations Z_i is a sum of processes of the type considered in Subsection 1.9.4. In order to calculate the moment $E(Z_{t_1}^{k_1} Z_{t_2}^{k_2} Z_{t_3}^{k_3} Z_{t_4}^{k_4})$, first apply the multinomial formula as above to express it in terms of moments of the form $E(Y_{j,t_1}^{\ell_1} Y_{j,t_2}^{\ell_2} Y_{j,t_3}^{\ell_3} Y_{j,t_4}^{\ell_4})$, where

$$Y_{j,i} = \frac{1}{\Delta} \int_{(i-1)\Delta}^{i\Delta} X_{j,s} \, ds.$$

Now proceed as in Subsection 1.9.4.

1.9.6 Stochastic volatility models

A stochastic volatility model is a generalization of the Black-Scholes model for the logarithm of an asset price $dX_t = (\kappa + \beta\sigma^2)dt + \sigma dW_t$, that takes into account the empirical finding that the volatility σ^2 varies randomly over time, see e.g. Ghysels, Harvey, and Renault (1996) or Shephard (1996). The model is given by

$$dX_t = (\kappa + \beta v_t)dt + \sqrt{v_t}dW_t, \quad (1.155)$$

where the volatility, v_t, is a stochastic process that cannot be observed directly. If the data are observations at the time points Δi, $i = 0, 1, 2, \ldots, n$, then the returns $Y_i = X_{i\Delta} - X_{(i-1)\Delta}$ can be written in the form

$$Y_i = \kappa\Delta + \beta S_i + \sqrt{S_i}A_i, \quad (1.156)$$

where

$$S_i = \int_{(i-1)\Delta}^{i\Delta} v_t dt, \qquad (1.157)$$

and where the A_i's are independent, standard normal distributed random variables. Prediction-based estimating functions for stochastic volatility models were considered in detail in M. Sørensen (2000a).

Here we consider the case where the volatility process v is a sum of independent Pearson diffusions with state-space $(0, \infty)$ (the cases 2, 4 and 5). Barndorff–Nielsen and Shephard (2001) demonstrated that an autocorrelation function of the type (1.151) fits empirical autocorrelation functions of volatility well, while an autocorrelation function like that of a single Pearson diffusion is too simple to obtain a good fit. Stochastic volatility models where the volatility process is a sum of independent square root processes were considered by Bollerslev and Zhou (2002) and Bibby and Sørensen (2003). We assume that v and W are independent, so that the sequences $\{A_i\}$ and $\{S_i\}$ are independent. By the multinomial formula we find that

$$\mathrm{E}\left(Y_1^{k_1} Y_{t_1}^{k_2} Y_{t_2}^{k_3} Y_{t_3}^{k_4}\right) =$$
$$\sum K_{k_{11},\dots,k_{43}} \mathrm{E}\left(S_1^{k_{12}+k_{13}/2} S_{t_1}^{k_{22}+k_{23}/2} S_{t_2}^{k_{32}+k_{33}/2} S_{t_3}^{k_{42}+k_{43}/2}\right)$$
$$\cdot \mathrm{E}\left(A_1^{k_{13}}\right) \mathrm{E}\left(A_{t_1}^{k_{23}}\right) \mathrm{E}\left(A_{t_2}^{k_{33}}\right) \mathrm{E}\left(A_{t_3}^{k_{43}}\right),$$

where the sum is over all non-negative integers k_{ij}, $i = 1, 2, 3, 4$, $j = 1, 2, 3$ such that $k_{i1} + k_{i2} + k_{i3} = k_i$ $(i = 1, 2, 3, 4)$, and where

$$K_{k_{11},\dots,k_{43}} =$$
$$\binom{k_1}{k_{11}, k_{12}, k_{13}} \binom{k_2}{k_{21}, k_{22}, k_{23}} \binom{k_3}{k_{31}, k_{32}, k_{33}} \binom{k_4}{k_{41}, k_{42}, k_{43}} (\kappa\Delta)^{k_{\cdot1}} \beta^{k_{\cdot2}}$$

with $k_{\cdot j} = k_{1j} + k_{2j} + k_{3j} + k_{4j}$. The moments $E(A_i^{k_{i3}})$ are the well-known moments of the standard normal distribution. When k_{i3} is odd, these moments are zero. Thus we only need to calculate the mixed moments of the form $E(S_1^{\ell_1} S_{t_1}^{\ell_2} S_{t_2}^{\ell_3} S_{t_3}^{\ell_4})$, where $\ell_1, \dots, \ell_4$ are integers. When the volatility process is a sum of independent Pearson diffusions, S_i of the same form as Z_i in (1.154) (apart from the factor $1/\Delta$), so we can proceed as in the previous subsection to calculate the necessary mixed moments. Thus also for the stochastic volatility models defined in terms of Pearson diffusions, we can explicitly find optimal estimating functions based on prediction of powers of returns, cf. Example 1.48.

1.9.7 Compartment models

Diffusion compartment models are D-dimensional diffusions with linear drift,

$$dX_t = [B(\theta)X_t - b(\theta)]\, dt + \sigma(X_t;\theta)dW_t, \qquad (1.158)$$

where only a subset of the coordinates are observed. Here $B(\theta)$ is a $D \times D$-matrix, $b(\theta)$ is a D-dimensional vector, $\sigma(x;\theta)$ is a $D \times D$-matrix, and W a D-dimensional standard Wiener process. Compartment models are used to model the dynamics of the flow of a substance between different parts (compartments) of, for instance, an ecosystem or the body of a human being or an animal. The process X_t is the concentration in the compartments, and flow from a given compartment into other compartments is proportional to the concentration in the given compartment, but modified by the random perturbation given by the diffusion term. The vector $b(\theta)$ represents input to or output from the system, for instance infusion or degradation of the substance. The complication is that only a subset of the compartments can be observed, for instance the first compartment, in which case the data are $Y_i = X_{1,t_i}$.

Example 1.55 The two-compartment model given by

$$B = \begin{pmatrix} -\beta_1 & \beta_2 \\ \beta_1 & -(\beta_1 + \beta_2) \end{pmatrix}, \quad b = \begin{pmatrix} 0 \\ 0 \end{pmatrix}, \quad \sigma = \begin{pmatrix} \tau_1 & 0 \\ 0 & \tau_2 \end{pmatrix},$$

where all parameters are positive, was used by Bibby (1995) to model how a radioactive tracer moved between the water and the biosphere in a certain ecosystem. Samples could only be taken from the water, the first compartment, so $Y_i = X_{1,t_i}$. Likelihood inference, which is feasible because the model is Gaussian, was studied by Bibby (1995). All mixed moments of the form (1.138) and (1.148) can be calculated explicitly, again because the model is Gaussian. Therefore also explicit optimal prediction-based estimating functions of the type considered in Example 1.48 are available to estimate the parameters and were studied by Düring (2002).

□

Example 1.56 A non-Gaussian diffusion compartment model is obtained by the specification $\sigma(x,\theta) = \mathrm{diag}(\tau_1\sqrt{x_1},\dots,\tau_D\sqrt{x_D})$. This multivariate version of the square root process was studied by Düring (2002), who used methods in Down, Meyn, and Tweedie (1995) to show that the D-dimensional process is geometrically α-mixing and established the asymptotic normality of prediction-based estimators of the type considered in Example 1.48. As in the previous example, only the first compartment is observed, i.e. the data are $Y_i = X_{1,t_i}$. For the multivariate square root model, the mixed moments (1.138) and (1.148) must be calculated numerically.

□

1.10 General asymptotic results for estimating functions

In this section we review some general asymptotic results for estimators obtained from estimating functions for stochastic process models. Proofs can be found in Jacod and Sørensen (2012).

Suppose as a statistical model for the data $X_1, X_2, \ldots, X_n$ that they are observations from a stochastic process. In this section and in the following section we allow any stochastic process, in particular the process does not have to be a diffusion-type process or a Markov process. The corresponding probability measures (P_θ) are indexed by a p-dimensional parameter $\theta \in \Theta$. An estimating function is a function, $G_n(\theta; X_1, X_2, \ldots, X_n)$, of the parameter and the observations with values in $\mathbb{R}^p$. Usually we suppress the dependence on the observations in the notation and write $G_n(\theta)$. We get an estimator by solving the equation (1.1) and call such an estimator a G_n-*estimator*. It should be noted that n might indicate more than just the sample size: the distribution of the data $X_1, X_2, \ldots, X_n$ might depend on n. For instance, the data might be observations of a diffusion process at time points $i\Delta_n$, $i = 1, \ldots, n$, where Δ_n decreases as n increases; see Sections 1.6 and 1.7. Another example is that the diffusion coefficient might depend on n; see Section 1.8.

We will not necessarily assume that the data are observations from one of the probability measures $(P_\theta)_{\theta \in \Theta}$. We will more generally denote the *true probability measure* by P. If the statistical model contains the true model, in the sense that there exists a $\theta_0 \in \Theta$ such that $P = P_{\theta_0}$, then we call θ_0 the *true parameter value*.

A priori, there might be more than one solution or no solution at all to the estimating equation (1.1), so conditions are needed to ensure that a unique solution exists when n is sufficiently large. Moreover, we need to be careful when formally defining our estimator. In the following definition, δ denotes a "special" point, which we take to be outside Θ and $\Theta_\delta = \Theta \cup \{\delta\}$.

Definition 1.57 a) *The domain of G_n-estimators (for a given n) is the set A_n of all observations $x = (x_1, \ldots, x_n)$ for which $G_n(\theta) = 0$ for at least one value $\theta \in \Theta$.*

b) *A G_n-estimator, $\hat{\theta}_n(x)$, is any function of the data with values in Θ_δ, such that for P–almost all observations we have either $\hat{\theta}_n(x) \in \Theta$ and $G_n(\hat{\theta}_n(x), x) = 0$ if $x \in A_n$, or $\hat{\theta}_n(x) = \delta$ if $x \notin A_n$.*

We usually suppress the dependence on the observations in the notation and write $\hat{\theta}_n$.

The following theorem gives conditions which ensure that, for n large enough, the estimating equation (1.1) has a solution that converges to a particular parameter value $\bar{\theta}$. When the statistical model contains the true model, the estimating function should preferably be chosen such that $\bar{\theta} = \theta_0$. To facilitate the following discussion, we will refer to an estimator that converges to $\bar{\theta}$ in probability as a $\bar{\theta}$–consistent estimator, meaning that it is a (weakly) consistent estimator of $\bar{\theta}$. We assume that $G_n(\theta)$ is differentiable with respect to θ and denote by $\partial_{\theta^T} G_n(\theta)$ the $p \times p$-matrix, where the ijth entry is $\partial_{\theta_j} G_n(\theta)_i$.

Theorem 1.58 *Suppose the existence of a parameter value $\bar{\theta} \in$ int Θ (the interior of Θ), a connected neighbourhood M of $\bar{\theta}$, and a (possibly random) function W on M taking its values in the set of $p \times p$ matrices, such that the following holds:*

(i) $G_n(\bar{\theta}) \xrightarrow{P} 0$ (convergence in probability, w.r.t. the true measure P) as $n \to \infty$.

(ii) $G_n(\theta)$ is continuously differentiable on M for all n, and

$$\sup_{\theta \in M} \| \partial_{\theta^T} G_n(\theta) - W(\theta) \| \xrightarrow{P} 0. \tag{1.159}$$

(iii) The matrix $W(\bar{\theta})$ is non-singular with P–probability one.

Then a sequence $(\hat{\theta}_n)$ of G_n-estimators exists which is $\bar{\theta}$-consistent. Moreover this sequence is eventually unique, *that is if $(\hat{\theta}'_n)$ is any other $\bar{\theta}$–consistent sequence of G_n–estimators, then $P(\hat{\theta}_n \neq \hat{\theta}'_n) \to 0$ as $n \to \infty$.*

Note that the condition (1.159) implies the existence of a subsequence $\{n_k\}$ such that $\partial_{\theta^T} G_{n_k}(\theta)$ converges uniformly to $W(\theta)$ on M with probability one. Hence W is a continuous function of θ (up to a null set), and it follows from elementary calculus that outside some P–null set there exists a unique continuously differentiable function G satisfying $\partial_{\theta^T} G(\theta) = W(\theta)$ for all $\theta \in M$ and $G(\bar{\theta}) = 0$. When M is a bounded set, (1.159) implies that

$$\sup_{\theta \in M} |G_n(\theta) - G(\theta)| \xrightarrow{P} 0. \tag{1.160}$$

This observation casts light on the result of Theorem 1.58. Since $G_n(\theta)$ can be made arbitrarily close to $G(\theta)$ by choosing n large enough, and since $G(\theta)$ has a root at $\bar{\theta}$, it is intuitively clear that $G_n(\theta)$ must have a root near $\bar{\theta}$ when n is sufficiently large.

If we impose an identifiability condition, we can give a stronger result on any sequence of G_n–estimators. By $\bar{B}_\epsilon(\theta)$ we denote the closed ball with radius ϵ centered at θ.

Theorem 1.59 *Assume (1.160) for some subset M of Θ containing $\bar{\theta}$, and* *that*

$$P\left(\inf_{M\setminus\bar{B}_\epsilon(\bar{\theta})} |G(\theta)| > 0\right) = 1 \qquad (1.161)$$

for all $\epsilon > 0$. Then for any sequence $(\hat{\theta}_n)$ of G_n–estimators

$$P(\hat{\theta}_n \in M\setminus\bar{B}_\epsilon(\bar{\theta})) \to 0 \qquad (1.162)$$

as $n \to \infty$ for every $\epsilon > 0$

If $M = \Theta$, we see that any sequence $(\hat{\theta}_n)$ of G_n–estimators is $\bar{\theta}$–consistent. If the conditions of Theorem 1.59 hold for any compact subset M of Θ, then a sequence $(\hat{\theta}_n)$ of G_n–estimators is $\bar{\theta}$–consistent or converges to the boundary of Θ.

Finally, we give a result on the asymptotic distribution of a sequence $(\hat{\theta}_n)$ of $\bar{\theta}$–consistent G_n–estimators.

Theorem 1.60 *Assume the estimating function G_n satisfies the conditions of Theorem 1.58 and that there is a sequence of invertible matrices A_n such that each entry of A_n^{-1} tends to zero,*

$$\begin{pmatrix} A_n G_n(\bar{\theta}) \\ A_n \partial_{\theta^T} G_n(\bar{\theta}) A_n^{-1} \end{pmatrix} \xrightarrow{\mathcal{D}} \begin{pmatrix} Z \\ W_0(\bar{\theta}) \end{pmatrix}, \qquad (1.163)$$

and there exists a connected neighbourhood M of $\bar{\theta}$ such that

$$\sup_{\theta\in M} \| A_n \partial_{\theta^T} G_n(\theta) A_n^{-1} - W_0(\theta) \| \xrightarrow{P} 0. \qquad (1.164)$$

Here Z is a non-degenerate random variable, and W_0 is a random function taking values in the set of $p \times p$-matrices satisfying that $W_0(\bar{\theta})$ is invertible. Under these conditions, we have for any $\bar{\theta}$–consistent sequence $(\hat{\theta}_n)$ of G_n–estimators that

$$A_n(\hat{\theta}_n - \bar{\theta}) \xrightarrow{\mathcal{D}} -W_0(\bar{\theta})^{-1} Z. \qquad (1.165)$$

When Z is normal distributed with expectation zero and covariance matrix V, and when Z is independent of $W_0(\bar{\theta})$, then the limit distribution is the normal variance-mixture with characteristic function

$$s \mapsto E\left(\exp\left(-\tfrac{1}{2} s^T W_0(\bar{\theta})^{-1} V W_0(\bar{\theta})^{T^{-1}} s\right)\right). \qquad (1.166)$$

If, moreover, $W_0(\bar{\theta})$ is non-random, then the limit distribution is a normal distribution with expectation zero and covariance matrix $W_0(\bar{\theta})^{-1} V W_0(\bar{\theta})^{T^{-1}}$.

In the often occurring situation, where $W_0(\bar{\theta})$ is non-random, joint convergence of $A_n \partial_{\theta^T} G_n(\bar{\theta}) A_n^{-1}$ and $A_n G_n(\bar{\theta})$ is not necessary — marginal convergence of $A_n G_n(\bar{\theta})$ is enough.

1.11 Optimal estimating functions: General theory

The modern theory of optimal estimating functions dates back to the papers by Godambe (1960) and Durbin (1960); however, the basic idea was in a sense already used in Fisher (1935). The theory was extended to stochastic processes by Godambe (1985), Godambe and Heyde (1987), Heyde (1988), and several others; see the references in Heyde (1997). Important particular instances are likelihood inference, the quasi-likelihood of Wedderburn (1974) and the generalized estimating equations developed by Liang and Zeger (1986) to deal with problems of longitudinal data analysis, see also Prentice (1988) and Li (1997). A modern review of the theory of optimal estimating functions can be found in Heyde (1997). The theory is very closely related to the theory of the generalized method of moments developed independently in parallel in the econometrics literature, where the foundation was laid by Hansen (1982), who followed Sagan (1958) by using selection matrices. Important extensions to the theory were made by Hansen (1985), Chamberlain (1987), Newey and West (1987), and Newey (1990); see also the discussion and references in A. R. Hall (2005). Particular attention is given to the time series setting in Hansen (1985, 1993), West (2001), and Kuersteiner (2002). A discussion of links between the econometrics and statistics literature can be found in Hansen (2001). In the following we present the theory as it was developed in the statistics literature by Godambe and Heyde.

The general setup is as in the previous section. We will only consider *unbiased* estimating functions, i.e., estimating functions satisfying that $E_\theta(G_n(\theta)) = 0$ for all $\theta \in \Theta$. This natural requirement is also called Fisher consistency. It often implies condition (i) of Theorem 1.58 for $\bar\theta = \theta_0$, which is an essential part of the condition for existence of a consistent estimator. Suppose we have a class $\mathcal{G}_n$ of unbiased estimating functions. How do we choose the best member in $\mathcal{G}_n$? And in what sense are some estimating functions better than others? These are the main problems in the theory of estimating functions.

To simplify the discussion, let us first assume that $p = 1$. The quantity

$$S_{G_n}(\theta) = E_\theta(\partial_\theta G_n(\theta)) \tag{1.167}$$

is called the *sensitivity* function for G_n. As in the previous section, it is assumed that $G_n(\theta)$ is differentiable with respect to θ. A large absolute value of the sensitivity implies that the equation $G_n(\theta) = 0$ tends to have a solution near the true parameter value, where the expectation of $G_n(\theta)$ is equal to zero. Thus a good estimating function is one with a large absolute value of the sensitivity.

Ideally, we would base the statistical inference on the likelihood function $L_n(\theta)$, and hence use the score function $U_n(\theta) = \partial_\theta \log L_n(\theta)$ as our estimating function. This usually yields an efficient estimator. However, when $L_n(\theta)$ is not available or is difficult to calculate, we might prefer to use an estimating function that is easier to obtain and is in some sense close to the score function. Suppose that both $U_n(\theta)$ and $G_n(\theta)$ have finite variance. Then it can be proven under usual regularity conditions that

$$S_{G_n}(\theta) = -\mathrm{Cov}_\theta(G_n(\theta), U_n(\theta)).$$

Thus we can find an estimating function $G_n(\theta)$ that maximizes the absolute value of the correlation between $G_n(\theta)$ and $U_n(\theta)$ by finding one that maximizes the quantity

$$K_{G_n}(\theta) = S_{G_n}(\theta)^2 / \mathrm{Var}_\theta(G_n(\theta)) = S_{G_n}(\theta)^2 / E_\theta(G_n(\theta)^2), \qquad (1.168)$$

which is known as the *Godambe information*. This makes intuitive sense: the ratio $K_{G_n}(\theta)$ is large when the sensitivity is large and when the variance of $G_n(\theta)$ is small. The Godambe information is a natural generalization of the Fisher information. Indeed, $K_{U_n}(\theta)$ is the Fisher information. For a discussion of information quantities in a stochastic process setting, see Barndorff–Nielsen and Sørensen (1991, 1994). In a short while, we shall see that the Godambe information has a large sample interpretation too. An estimating function $G_n^* \in \mathcal{G}_n$ is called *Godambe optimal* in $\mathcal{G}_n$ if

$$K_{G_n^*}(\theta) \geq K_{G_n}(\theta) \qquad (1.169)$$

for all $\theta \in \Theta$ and for all $G_n \in \mathcal{G}_n$.

When the parameter θ is multivariate ($p > 1$), the sensitivity function is the $p \times p$-matrix

$$S_{G_n}(\theta) = E_\theta(\partial_{\theta^T} G_n(\theta)), \qquad (1.170)$$

and the Godambe information is the $p \times p$-matrix

$$K_{G_n}(\theta) = S_{G_n}(\theta)^T E_\theta \left(G_n(\theta) G_n(\theta)^T\right)^{-1} S_{G_n}(\theta), \qquad (1.171)$$

An optimal estimating function G_n^* can be defined by (1.169) with the inequality referring to the partial ordering of the set of positive semi-definite $p \times p$-matrices. Whether a Godambe optimal estimating function exists and whether it is unique depends on the class $\mathcal{G}_n$. In any case, it is only unique up to multiplication by a regular matrix that might depend on θ. Specifically, if $G_n^*(\theta)$ satisfies (1.169), then so does $M_\theta G_n^*(\theta)$ where M_θ is an invertible deterministic $p \times p$-matrix. Fortunately, the two estimating functions give rise to the same estimator(s), and we refer to them as *versions* of the same estimating function. For theoretical purposes a standardized version of the estimating functions is useful. The *standardized version* of $G_n(\theta)$ is given by

$$G_n^{(s)}(\theta) = -S_{G_n}(\theta)^T E_\theta \left(G_n(\theta) G_n(\theta)^T\right)^{-1} G_n(\theta).$$

The rationale behind this standardization is that $G_n^{(s)}(\theta)$ satisfies the *second*

Bartlett-identity

$$E_\theta \left(G_n^{(s)}(\theta) G_n^{(s)}(\theta)^T \right) = -E_\theta (\partial_{\theta^T} G_n^{(s)}(\theta)), \qquad (1.172)$$

an identity usually satisfied by the score function. The standardized estimating function $G_n^{(s)}(\theta)$ is therefore more directly comparable to the score function. Note that when the second Bartlett identity is satisfied, the Godambe information is equal to minus one times the sensitivity matrix.

An Godambe optimal estimating function is close to the score function U_n in an L_2-sense. Suppose G_n^* is Godambe optimal in $\mathcal{G}_n$. Then the standardized version $G_n^{*(s)}(\theta)$ satisfies the inequality

$$E_\theta \left((G_n^{(s)}(\theta) - U_n(\theta))^T (G_n^{(s)}(\theta) - U_n(\theta)) \right)$$
$$\geq E_\theta \left((G_n^{*(s)}(\theta) - U_n(\theta))^T (G_n^{*(s)}(\theta) - U_n(\theta)) \right)$$

for all $\theta \in \Theta$ and for all $G_n \in \mathcal{G}_n$, see Heyde (1988). In fact, if $\mathcal{G}_n$ is a closed subspace of the L_2-space of all square integrable functions of the data, then the optimal estimating function is the orthogonal projection of the score function onto $\mathcal{G}_n$. For further discussion of this Hilbert space approach to estimating functions, see McLeish and Small (1988). The interpretation of an optimal estimating function as an approximation to the score function is important. By choosing a sequence of classes $\mathcal{G}_n$ that, as $n \to \infty$, converges to a subspace containing the score function U_n, a sequence of estimators that is asymptotically fully efficient can be constructed.

The following result by Heyde (1988) can often be used to find the optimal estimating function.

Theorem 1.61 *If $G_n^* \in \mathcal{G}_n$ satisfies the equation*

$$S_{G_n}(\theta)^{-1} E_\theta \left(G_n(\theta) G_n^*(\theta)^T \right) = S_{G_n^*}(\theta)^{-1} E_\theta \left(G_n^*(\theta) G_n^*(\theta)^T \right) \qquad (1.173)$$

for all $\theta \in \Theta$ and for all $G_n \in \mathcal{G}_n$, then it is Godambe optimal in $\mathcal{G}_n$. When $\mathcal{G}_n$ is closed under addition, any Godambe optimal estimating function G_n^* satisfies (1.173).

The condition (1.173) can often be verified by showing that $E_\theta (G_n(\theta) G_n^*(\theta)^T) = -E_\theta (\partial_{\theta^T} G_n(\theta))$ for all $\theta \in \Theta$ and for all $G_n \in \mathcal{G}_n$. In such situations, G_n^* satisfies the *second Bartlett-identity*, (1.172), so that

$$K_{G_n^*}(\theta) = E_\theta \left(G_n^*(\theta) G_n^*(\theta)^T \right).$$

Example 1.62 Suppose we have a number of functions $h_{ij}(x_1, \ldots, x_i; \theta), j = 1, \ldots, N, i = 1, \ldots n$ satisfying that

$$E_\theta (h_{ij}(X_1, \ldots, X_i; \theta)) = 0.$$

Such functions define relationships (dependent on θ) between an observation X_i and the previous observations $X_1, \ldots, X_{i-1}$ (or some of them) that are on average equal to zero. It is natural to use such relationships to estimate θ by solving the equations $\sum_{i=1}^{n} h_{ij}(X_1, \ldots, X_i; \theta) = 0$. In order to estimate θ it is necessary that $N \geq p$, but if $N > p$ we have too many equations. The theory of optimal estimating functions tells us how to combine the N relations in an optimal way.

Let h_i denote the N-dimensional vector $(h_{i1}, \ldots, h_{iN})^T$, and define an N-dimensional estimating function by $H_n(\theta) = \sum_{i=1}^{n} h_i(X_1, \ldots, X_i; \theta)$. First we consider the class of p-dimensional estimating functions of the form

$$G_n(\theta) = A_n(\theta) H_n(\theta),$$

where $A_n(\theta)$ is a non-random $p \times N$-matrix that is differentiable with respect to θ. By $A_n^*(\theta)$ we denote the optimal choice of $A_n(\theta)$. It is not difficult to see that

$$S_{G_n}(\theta) = A_n(\theta) S_{H_n}(\theta)$$

and

$$E_\theta \left(G_n(\theta) G_n^*(\theta)^T \right) = A_n(\theta) E_\theta \left(H_n(\theta) H_n(\theta)^T \right) A_n^*(\theta)^T,$$

where $S_{H_n}(\theta) = E_\theta(\partial_{\theta^T} H_n(\theta))$. If we choose

$$A_n^*(\theta) = -S_{H_n}(\theta)^T E_\theta \left(H_n(\theta) H_n(\theta)^T \right)^{-1},$$

then (1.173) is satisfied for all $G_n \in \mathcal{G}_n$, so that $G_n^*(\theta) = A_n^*(\theta) H_n(\theta)$ is Godambe optimal.

Sometimes there are good reasons to use functions h_{ij} satisfying that

$$E_\theta(h_{ij}(X_1, \ldots, X_i; \theta) h_{i'j'}(X_1, \ldots, X_{i'}; \theta)) = 0 \qquad (1.174)$$

for all $j, j' = 1, \ldots, N$ when $i \neq i'$. For such functions the random variables $h_{ij}(X_1, \ldots, X_i; \theta)$, $i = 1, 2, \ldots$ are uncorrelated, and in this sense the "new" random variation of $h_{ij}(X_1, \ldots, X_i; \theta)$ depends only on the innovation in the ith observation. This is for instance the case for martingale estimating functions, see (1.181). In this situation it is natural to consider the larger class of estimating functions given by

$$G_n(\theta) = \sum_{i=1}^{n} a_i(\theta) h_i(X_1, \ldots, X_i; \theta),$$

where $a_i(\theta)$, $i = 1, \ldots n$, are $p \times N$ matrices that do not depend on the data and are differentiable with respect to θ. Here

$$S_{G_n}(\theta) = \sum_{i=1}^{n} a_i(\theta) E_\theta(\partial_{\theta^T} h_i(X_1, \ldots, X_i; \theta))$$

and

$$E_\theta \left(G_n(\theta) G_n^*(\theta)^T \right) =$$

$$\sum_{i=1}^{n} a_i(\theta) E_\theta \left(h_i(X_1, \ldots, X_i; \theta) h_i(X_1, \ldots, X_i; \theta)^T \right) a_i^*(\theta)^T,$$

where $a_i^*(\theta)$ denotes the optimal choice of $a_i(\theta)$. We see that with

$$a_i^*(\theta) =$$

$$-E_\theta(\partial_{\theta^T} h_i(X_1, \ldots, X_i; \theta))^T \left(E_\theta \left(h_i(X_1, \ldots, X_i; \theta) h_i(X_1, \ldots, X_i; \theta)^T \right) \right)^{-1}$$

the condition (1.173) is satisfied.   □

1.11.1 Martingale estimating functions

More can be said about martingale estimating functions, i.e. estimating functions G_n satisfying that

$$E_\theta(G_n(\theta)|\mathcal{F}_{n-1}) = G_{n-1}(\theta), \quad n = 1, 2, \ldots,$$

where $\mathcal{F}_{n-1}$ is the σ-field generated by the observations $X_1, \ldots, X_{n-1}$ ($G_0 = 0$ and $\mathcal{F}_0$ is the trivial σ-field). In other words, the stochastic process $\{G_n(\theta) : n = 1, 2, \ldots\}$ is a martingale under the model given by the parameter value θ. Since the score function is usually a martingale (see e.g. Barndorff–Nielsen and Sørensen (1994)), it is natural to approximate it by families of martingale estimating functions.

The well-developed martingale limit theory allows a straightforward discussion of the asymptotic theory, and motivates an optimality criterion that is particular to martingale estimating functions. Suppose the estimating function $G_n(\theta)$ satisfies the conditions of the central limit theorem for martingales and let $\hat{\theta}_n$ be a solution of the equation $G_n(\theta) = 0$. Under the regularity conditions of the previous section, it can be proved that

$$\langle G(\theta) \rangle_n^{-\frac{1}{2}} \bar{G}_n(\theta)(\hat{\theta}_n - \theta_0) \xrightarrow{\mathcal{D}} N(0, I_p). \tag{1.175}$$

Here $\langle G(\theta) \rangle_n$ is the *quadratic characteristic* of $G_n(\theta)$ defined by

$$\langle G(\theta) \rangle_n = \sum_{i=1}^{n} E_\theta \left((G_i(\theta) - G_{i-1}(\theta))(G_i(\theta) - G_{i-1}(\theta))^T | \mathcal{F}_{i-1} \right),$$

and $\partial_{\theta^T} G_n(\theta)$ has been replaced by its compensator

$$\bar{G}_n(\theta) = \sum_{i=1}^{n} E_\theta \left(\partial_{\theta^T} G_i(\theta) - \partial_{\theta^T} G_{i-1}(\theta) | \mathcal{F}_{i-1} \right),$$

using the extra assumption that $\bar{G}_n(\theta)^{-1}\partial_{\theta^T} G_n(\theta) \xrightarrow{P_\theta} I_p$. Details can be found in Heyde (1988). We see that the inverse of the data-dependent matrix

$$I_{G_n}(\theta) = \bar{G}_n(\theta)^T \langle G(\theta) \rangle_n^{-1} \bar{G}_n(\theta) \tag{1.176}$$

estimates the co-variance matrix of the asymptotic distribution of the estimator $\hat{\theta}_n$. Therefore $I_{G_n}(\theta)$ can be interpreted as an information matrix, called the *Heyde information*. It generalizes the incremental expected information of the likelihood theory for stochastic processes; see Barndorff–Nielsen and Sørensen (1994). Since $\bar{G}_n(\theta)$ estimates the sensitivity function, and $\langle G(\theta) \rangle_n$ estimates the variance of the asymptotic distribution of $G_n(\theta)$, the Heyde information has a heuristic interpretation similar to that of the Godambe information. In fact,

$$E_\theta\left(\bar{G}_n(\theta)\right) = S_{G_n}(\theta) \quad \text{and} \quad E_\theta\left(\langle G(\theta) \rangle_n\right) = E_\theta\left(G_n(\theta)G_n(\theta)^T\right).$$

We can thus think of the Heyde information as an estimated version of the Godambe information.

Let $\mathcal{G}_n$ be a class of martingale estimating functions with finite variance. We say that a martingale estimating function G_n^* is *Heyde optimal* in $\mathcal{G}_n$ if

$$I_{G_n^*}(\theta) \geq I_{G_n}(\theta) \tag{1.177}$$

P_θ-almost surely for all $\theta \in \Theta$ and for all $G_n \in \mathcal{G}_n$.

The following useful result from Heyde (1988) is similar to Theorem 1.61. In order to formulate it, we need the concept of the *quadratic co-characteristic* of two martingales, G and $\tilde{G}$, both of which are assumed to have finite variance:

$$\langle G, \tilde{G} \rangle_n = \sum_{i=1}^n E\left((G_i - G_{i-1})(\tilde{G}_i - \tilde{G}_{i-1})^T | \mathcal{F}_{i-1}\right). \tag{1.178}$$

Theorem 1.63 *If $G_n^* \in \mathcal{G}_n$ satisfies that*

$$\bar{G}_n(\theta)^{-1} \langle G(\theta), G^*(\theta) \rangle_n = \bar{G}_n^*(\theta)^{-1} \langle G^*(\theta) \rangle_n \tag{1.179}$$

for all $\theta \in \Theta$ and all $G_n \in \mathcal{G}_n$, then it is Heyde optimal in $\mathcal{G}_n$. When $\mathcal{G}_n$ is closed under addition, any Heyde optimal estimating function G_n^* satisfies (1.179). Moreover, if $\bar{G}_n^*(\theta)^{-1} \langle G^*(\theta) \rangle_n$ is non-random, then G_n^* is also Godambe optimal in $\mathcal{G}_n$.

Since in many situations condition (1.179) can be verified by showing that $\langle G(\theta), G^*(\theta) \rangle_n = -\bar{G}_n(\theta)$ for all $G_n \in \mathcal{G}_n$, it is in practice often the case that Heyde optimality implies Godambe optimality.

Example 1.64 Let us discuss an often occurring type of martingale estimating functions. To simplify the exposition we assume that the observed process is

Markovian. For Markov processes it is natural to base martingale estimating functions on functions $h_{ij}(y, x; \theta)$, $j = 1, \ldots, N$, $i = 1, \ldots, n$ satisfying that

$$E_\theta(h_{ij}(X_i, X_{i-1}; \theta)|\mathcal{F}_{i-1}) = 0. \tag{1.180}$$

As in Example 1.62, such functions define relationships (dependent on θ) between consecutive observation X_i and X_{i-1} that are, on average, equal to zero and can be used to estimate θ. We consider the class of p-dimensional estimating functions of the form

$$G_n(\theta) = \sum_{i=1}^{n} a_i(X_{i-1}; \theta) h_i(X_i, X_{i-1}; \theta), \tag{1.181}$$

where h_i denotes the N-dimensional vector $(h_{i1}, \ldots, h_{iN})^T$, and $a_i(x; \theta)$ is a function from $\mathbb{R} \times \Theta$ into the set of $p \times N$-matrices that are differentiable with respect to θ. It follows from (1.180) that $G_n(\theta)$ is a p-dimensional unbiased martingale estimating function.

We will now find the matrices a_i that combine the N functions h_{ij} in an optimal way. Let $\mathcal{G}_n$ be the class of martingale estimating functions of the form (1.181) that have finite variance. Then

$$\bar{G}_n(\theta) = \sum_{i=1}^{n} a_i(X_{i-1}; \theta) E_\theta(\partial_{\theta^T} h_i(X_i, X_{i-1}; \theta)|\mathcal{F}_{i-1})$$

and

$$\langle G(\theta), G^*(\theta)\rangle_n = \sum_{i=1}^{n} a_i(X_{i-1}; \theta) V_{h_i}(X_{i-1}; \theta) a_i^*(X_{i-1}; \theta)^T,$$

where

$$G_n^*(\theta) = \sum_{i=1}^{n} a_i^*(X_{i-1}; \theta) h_i(X_i, X_{i-1}; \theta), \tag{1.182}$$

and

$$V_{h_i}(X_{i-1}; \theta) = E_\theta\left(h_i(X_i, X_{i-1}; \theta) h_i(X_i, X_{i-1}; \theta)^T|\mathcal{F}_{i-1}\right)$$

is the conditional covariance matrix of the random vector $h_i(X_i, X_{i-1}; \theta)$ given $\mathcal{F}_{i-1}$. If we assume that $V_{h_i}(X_{i-1}; \theta)$ is invertible and define

$$a_i^*(X_{i-1}; \theta) = -E_\theta(\partial_{\theta^T} h_i(X_i, X_{i-1}; \theta)|\mathcal{F}_{i-1})^T V_{h_i}(X_{i-1}; \theta)^{-1}, \tag{1.183}$$

then the condition (1.179) is satisfied. Hence by Theorem 1.63 the estimating function $G_n^*(\theta)$ with a_i^* given by (1.183) is Heyde optimal — provided, of course, that it has finite variance. Since $\bar{G}_n^*(\theta)^{-1}\langle G^*(\theta)\rangle_n = -I_p$ is nonrandom, the estimating function $G_n^*(\theta)$ is also Godambe optimal. If a_i^* were defined without the minus, $G_n^*(\theta)$ would obviously also be optimal. The reason for the minus will be clear in the following.

We shall now see, in exactly what sense the optimal estimating function (1.182) approximates the score function. The following result was first given by Kessler (1996). Let $p_i(y; \theta|x)$ denote the conditional density of X_i given that $X_{i-1} = x$. Then the likelihood function for θ based on the data $(X_1, \ldots, X_n)$ is

$$L_n(\theta) = \prod_{i=1}^{n} p_i(X_i; \theta|X_{i-1})$$

(with p_1 denoting the unconditional density of X_1). If we assume that all p_is are differentiable with respect to θ, the score function is

$$U_n(\theta) = \sum_{i=1}^{n} \partial_\theta \log p_i(X_i; \theta|X_{i-1}). \tag{1.184}$$

Let us fix i, x_{i-1} and θ and consider the L_2-space $\mathcal{K}_i(x_{i-1}, \theta)$ of functions $f : \mathbb{R} \mapsto \mathbb{R}$ for which $\int f(y)^2 p_i(y; \theta|x_{i-1}) dy < \infty$. We equip $\mathcal{K}_i(x_{i-1}, \theta)$ with the usual inner product

$$\langle f, g \rangle = \int f(y)g(y)p_i(y; \theta|x_{i-1}) dy,$$

and let $\mathcal{H}_i(x_{i-1}, \theta)$ denote the N-dimensional subspace of $\mathcal{K}_i(x_{i-1}, \theta)$ spanned by the functions $y \mapsto h_{ij}(y, x_{i-1}; \theta)$, $j = 1, \ldots, N$. That the functions are linearly independent in $\mathcal{K}_i(x_{i-1}, \theta)$ follows from the earlier assumption that the covariance matrix $V_{h_i}(x_{i-1}; \theta)$ is regular.

Now, assume that $\partial_{\theta_j} \log p_i(y|x_{i-1}; \theta) \in \mathcal{K}_i(x_{i-1}, \theta)$ for $j = 1, \ldots, p$, denote by g_{ij}^* the orthogonal projection with respect to $\langle \cdot, \cdot \rangle$ of $\partial_{\theta_j} \log p_i$ onto $\mathcal{H}_i(x_{i-1}, \theta)$, and define a p-dimensional function by $g_i^* = (g_{i1}^*, \ldots, g_{ip}^*)^T$. Then (under weak regularity conditions)

$$g_i^*(x_{i-1}, x; \theta) = a_i^*(x_{i-1}; \theta)h_i(x_{i-1}, x; \theta), \tag{1.185}$$

where a_i^* is the matrix defined by (1.183). To see this, note that g^* must have the form (1.185) with a_i^* satisfying the normal equations

$$\langle \partial_{\theta_j} \log p_i - g_j^*, \ h_{ik} \rangle = 0,$$

$j = 1, \ldots, p$ and $k = 1, \ldots, N$. These equations can be expressed in the form $B_i = a_i^* V_{h_i}$, where B_i is the $p \times p$-matrix whose (j, k)th element is $\langle \partial_{\theta_j} \log p_i, h_{ik} \rangle$. The main regularity condition needed to prove (1.185) is that we can interchange differentiation and integration so that

$$\int \partial_{\theta_j} \left[h_{ik}(y, x_{i-1}; \theta)p(y, x_{i-1}; \theta) \right] dy =$$

$$\partial_{\theta_j} \int h_{ik}(y, x_{i-1}; \theta)p(x_{i-1}, y; \theta) dy = 0,$$

from which it follows that

$$B_i = - \int \partial_{\theta^T} h_i(y, x_{i-1}; \theta) p(x_{i-1}, y; \theta) dy.$$

Thus a_i^* is given by (1.183).

$\square$

Acknowledgements

I am grateful to Susanne Ditlevsen, Anders Christian Jensen, Nina Munkholt and Per Mykland for their careful readings of the manuscript that have improved the presentation and reduced the number of misprints considerably. The research was supported by the Danish Center for Accounting and Finance funded by the Danish Social Science Research Council, by the Center for Research in Econometric Analysis of Time Series funded by the Danish National Research Foundation, and by the Excellence Programme of the University of Copenhagen.

References

Aït–Sahalia, Y. (2002). Maximum likelihood estimation of discretely sampled diffusions: a closed-form approximation approach. *Econometrica, 70,* 223–262.

Aït–Sahalia, Y. (2008). Closed-form likelihood expansions for multivariate diffusions. *Ann. Statist., 36,* 906–937.

Aït–Sahalia, Y., & Mykland, P. (2003). The effects of random and discrete sampling when estimating continuous-time diffusions. *Econometrica, 71,* 483–549.

Aït–Sahalia, Y., & Mykland, P. A. (2004). Estimators of diffusions with randomly spaced discrete observations: a general theory. *Ann. Statist., 32,* 2186–2222.

Aït–Sahalia, Y., & Mykland, P. A. (2008). An analysis of Hansen–Scheinkman moment estimators for discretely and randomly sampled diffusions. *Journal of Econometrics, 144,* 1–26.

Baltazar-Larios, F., & Sørensen, M. (2010). Maximum likelihood estimation for integrated diffusion processes. In C. Chiarella & A. Novikov (Eds.), *Contemporary quantitative finance: Essays in honour of Eckhard Platen* (pp. 407–423). Springer.

Barndorff–Nielsen, O. E., Jensen, J. L., & Sørensen, M. (1990). Parametric modelling of turbulence. *Phil. Trans. R. Soc. Lond. A, 332,* 439–455.

Barndorff–Nielsen, O. E., Jensen, J. L., & Sørensen, M. (1998). Some stationary processes in discrete and continuous time. *Advances in Applied Probability, 30,* 989–1007.

Barndorff–Nielsen, O. E., Kent, J., & Sørensen, M. (1982). Normal variance-mean mixtures and z-distributions. *International Statistical Review, 50,* 145–159.

Barndorff–Nielsen, O. E., & Shephard, N. (2001). Non-Gaussian Ornstein–Uhlenbeck-based models and some of their uses in financial econometrics (with discussion). *Journal of the Royal Statistical Society* **B,** *63,* 167–241.

Barndorff–Nielsen, O. E., & Sørensen, M. (1991). Information quantities in non-classical settings. *Computational Statistics and Data Analysis, 12,* 143–158.

Barndorff–Nielsen, O. E., & Sørensen, M. (1994). A review of some aspects of asymptotic likelihood theory for stochastic processes. *International Statistical Review*, *62*, 133–165.

Beskos, A., Papaspiliopoulos, O., & Roberts, G. O. (2006). Retrospective exact simulation of diffusion sample paths with applications. *Bernoulli*, *12*, 1077–1098.

Beskos, A., Papaspiliopoulos, O., Roberts, G. O., & Fearnhead, P. (2006). Exact and computationally efficient likelihood-based estimation for discretely observed diffusion processes. *J. Roy. Statist. Soc. B*, *68*, 333–382.

Beskos, A., & Roberts, G. O. (2005). Exact simulation of diffusions. *Ann. Appl. Prob.*, *15*, 2422–2444.

Bibby, B. M. (1995). *Inference for diffusion processes with particular emphasis on compartmental diffusion processes*. Unpublished doctoral dissertation, University of Aarhus.

Bibby, B. M., Skovgaard, I. M., & Sørensen, M. (2005). Diffusion-type models with given marginals and autocorrelation function. *Bernoulli*, *11*, 191–220.

Bibby, B. M., & Sørensen, M. (1995). Martingale estimation functions for discretely observed diffusion processes. *Bernoulli*, *1*, 17–39.

Bibby, B. M., & Sørensen, M. (1996). On estimation for discretely observed diffusions: a review. *Theory of Stochastic Processes*, *2*, 49–56.

Bibby, B. M., & Sørensen, M. (2003). Hyperbolic processes in finance. In S. Rachev (Ed.), *Handbook of heavy tailed distributions in finance* (pp. 211–248). Elsevier Science.

Billingsley, P. (1961). The Lindeberg–Lévy theorem for martingales. *Proc. Amer. Math. Soc.*, *12*, 788–792.

Bollerslev, T., & Wooldridge, J. (1992). Quasi-maximum likelihood estimators and inference in dynamic models with time-varying covariances. *Econometric Review*, *11*, 143–172.

Bollerslev, T., & Zhou, H. (2002). Estimating stochastic volatility diffusion using conditional moments of integrated volatility. *Journal of Econometrics*, *109*, 33–65.

Brockwell, P. J., & Davis, R. A. (1991). *Time series: Theory and methods*. New York: Springer-Verlag.

Campbell, J. Y., Lo, A. W., & MacKinlay, A. C. (1997). *The econometrics of financial markets*. Princeton: Princeton University Press.

Chamberlain, G. (1987). Asymptotic efficiency in estimation with conditional moment restrictions. *Journal of Econometrics*, *34*, 305–34.

Chan, K. C., Karolyi, G. A., Longstaff, F. A., & Sanders, A. B. (1992). An empirical comparison of alternative models of the short-term interest rate. *Journal of Finance*, *47*, 1209–1227.

Christensen, B. J., & Sørensen, M. (2008). *Optimal inference in dynamic moels with conditional moment restrictions* (Working Paper). CREATES, Center for Research in Econometric Analysis of Time Series.

Clement, E. (1997). Estimation of diffusion processes by simulated moment methods. *Scand. J. Statist.*, *24*, 353–369.

Comte, F., Genon-Catalot, V., & Rozenholc, Y. (2009). Nonparametric adaptive estimation for integrated diffusions. *Stochastic processes and their applications*, *119*, 811–834.

Conley, T. G., Hansen, L. P., Luttmer, E. G. J., & Scheinkman, J. A. (1997). Short-term interest rates as subordinated diffusions. *Review of Financial Studies*, *10*, 525–577.

Dacunha-Castelle, D., & Florens-Zmirou, D. (1986). Estimation of the coefficients of a diffusion from discrete observations. *Stochastics*, *19*, 263–284.

De Jong, F., Drost, F. C., & Werker, B. J. M. (2001). A jump-diffusion model for exchange rates in a target zone. *Statistica Neerlandica*, *55*, 270–300.

Ditlevsen, P. D., Ditlevsen, S., & Andersen, K. K. (2002). The fast climate fluctuations during the stadial and interstadial climate states. *Annals of Glaciology*, *35*, 457–462.

Ditlevsen, S., & Sørensen, M. (2004). Inference for observations of integrated diffusion processes. *Scand. J. Statist.*, *31*, 417–429.

Dorogovcev, A. J. (1976). The consistency of an estimate of a parameter of a stochastic differential equation. *Theor. Probability and Math. Statist.*, *10*, 73–82.

Doukhan, P. (1994). *Mixing, properties and examples.* New York: Springer. (Lecture Notes in Statistics 85)

Down, D., Meyn, S., & Tweedie, R. (1995). Exponential and uniform ergodicity of Markov processes. *Annals of Probability*, *23*, 1671–1691.

Duffie, D., & Singleton, K. (1993). Simulated moments estimation of Markov models of asset prices. *Econometrica*, *61*, 929–952.

Durbin, J. (1960). Estimation of parameters in time-series regression models. *J. Roy. Statist. Soc. B*, *22*, 139–153.

Durham, G. B., & Gallant, A. R. (2002). Numerical techniques for maximum likelihood estimation of continuous-time diffusion processes. *J. Business & Econom. Statist.*, *20*, 297–338.

Düring, M. (2002). *Den prediktions-baserede estimationsfunktion for diffusions puljemodeller.* Unpublished master's thesis, University of Copenhagen. (In Danish)

Elerian, O., Chib, S., & Shephard, N. (2001). Likelihood inference for discretely observed non-linear diffusions. *Econometrica*, *69*, 959–993.

Eraker, B. (2001). Mcmc analysis of diffusion models with application to finance. *J. Business & Econom. Statist.*, *19*, 177–191.

Fisher, R. A. (1935). The logic of inductive inference. *J. Roy. Statist. Soc.*, *98*, 39–54.

Florens-Zmirou, D. (1989). Approximate discrete-time schemes for statistics of diffusion processes. *Statistics*, *20*, 547–557.

Forman, J. L., & Sørensen, M. (2008). The Pearson diffusions: A class of statistically tractable diffusion processes. *Scand. J. Statist.*, *35*, 438–465.

Freidlin, M. I., & Wentzell, A. D. (1998). *Random pertubations of dynamical systems, 2nd edition*. New York: Springer.

Friedman, A. (1975). *Stochastic differential equations and applications, volume 1*. New York: Academic Press.

Genon-Catalot, V. (1990). Maximum contrast estimation for diffusion processes from discrete observations. *Statistics*, *21*, 99–116.

Genon-Catalot, V., & Jacod, J. (1993). On the estimation of the diffusion coefficient for multi-dimensional diffusion processes. *Ann. Inst. Henri Poincaré, Probabilités et Statistiques*, *29*, 119–151.

Genon-Catalot, V., Jeantheau, T., & Larédo, C. (2000). Stochastic volatility models as hidden Markov models and statistical applications. *Bernoulli*, *6*, 1051–1079.

Ghysels, E., Harvey, A. C., & Renault, E. (1996). Stochastic volatility. In C. Rao & G. Maddala (Eds.), *Statistical methods in finance* (pp. 119–191). Amsterdam: North-Holland.

Gloter, A. (2000). Parameter estimation for a discrete sampling of an integrated Ornstein–Uhlenbeck process. *Statistics*, *35*, 225–243.

Gloter, A. (2006). Parameter estimation for a discretely observed integrated diffusion process. *Scand. J. Statist.*, *33*, 83–104.

Gloter, A., & Sørensen, M. (2009). Estimation for stochastic differential equations with a small diffusion coefficient. *Stoch. Proc. Appl.*, *119*, 679–699.

Gobet, E. (2001). Local asymptotic mixed normality property for elliptic diffusion: a malliavin calculus approach. *Bernoulli*, *7*, 899–912.

Gobet, E. (2002). Lan property for ergodic diffusions with discrete observations. *Ann. Inst. Henri Poincaré, Probabilités et Statistiques*, *38*, 711–737.

Godambe, V. P. (1960). An optimum property of regular maximum likelihood estimation. *Ann. Math. Stat.*, *31*, 1208–1212.

Godambe, V. P. (1985). The foundations of finite sample estimation in stochastic processes. *Biometrika*, *72*, 419–428.

Godambe, V. P., & Heyde, C. C. (1987). Quasi likelihood and optimal estimation. *International Statistical Review*, *55*, 231–244.

Gourieroux, C., & Jasiak, J. (2006). Multivariate Jacobi process and with application to smooth transitions. *Journal of Econometrics*, *131*, 475–505.

Gradshteyn, I. S., & Ryzhik, I. M. (1965). *Table of integrals, series, and products, 4th edition*. New-York: Academic Press.

Hall, A. R. (2005). *Generalized method of moments*. New York: Oxford University Press.

Hall, P., & Heyde, C. C. (1980). *Martingale limit theory and its applications*. New York: Academic Press.

Hansen, L. P. (1982). Large sample properties of generalized method of moments estimators. *Econometrica, 50*, 1029–1054.

Hansen, L. P. (1985). A method for calculating bounds on the asymptotic covariance matrices of generalized method of moments estimators. *Journal of Econometrics, 30*, 203–238.

Hansen, L. P. (1993). Semiparametric efficiency bounds for linear time-series models. In P. C. B. Phillips (Ed.), *Models, methods and applications of econometrics: Essays in honor of A. R. Bergstrom* (pp. 253–271). Cambridge, MA: Blackwell.

Hansen, L. P. (2001). Method of moments. In *International encyclopedia of the social and behavior sciences*. New York: Elsevier.

Hansen, L. P., & Scheinkman, J. A. (1995). Back to the future: generating moment implications for continuous-time Markov processes. *Econometrica, 63*, 767–804.

Hansen, L. P., Scheinkman, J. A., & Touzi, N. (1998). Spectral methods for identifying scalar diffusions. *Journal of Econometrics, 86*, 1–32.

Heyde, C. C. (1988). Fixed sample and asymptotic optimality for classes of estimating functions. *Contemporary Mathematics, 80*, 241–247.

Heyde, C. C. (1997). *Quasi-likelihood and its application*. New York: Springer-Verlag.

Hildebrandt, E. H. (1931). Systems of polynomials connected with the charlier expansions and the Pearson differential and difference equations. *Ann. Math. Statist., 2*, 379–439.

Iacus, S. M. (2008). *Simulation and inference for stochastic differential equations*. New York: Springer.

Iacus, S. M., & Kutoyants, Y. (2001). Semiparametric hypotheses testing for dynamical systems with small noise. *Math. Methods Statist., 10*, 105–120.

Jacobsen, M. (2001). Discretely observed diffusions; classes of estimating functions and small δ-optimality. *Scand. J. Statist., 28*, 123–150.

Jacobsen, M. (2002). Optimality and small δ-optimality of martingale estimating functions. *Bernoulli, 8*, 643–668.

Jacod, J., & Sørensen, M. (2012). *Aspects of asymptotic statistical theory for stochastic processes*. (Preprint). Department of Mathematical Sciences, University of Copenhagen.

Jeganathan, P. (1982). On the asymptotic theory of estimation when the limit of the log-likelihood ratios is mixed normal. *Sankhyā A, 44*, 173–212.

Jeganathan, P. (1983). Some asymptotic properties of risk functions when the limit of the experiment is mixed normal. *Sankhyā A*, *45*, 66–87.

Kallenberg, O. (1997). *Foundations of modern probability*. New York: Springer-Verlag.

Kelly, L., Platen, E., & Sørensen, M. (2004). Estimation for discretely observed diffusions using transform functions. *J. Appl. Prob.*, *41*, 99–118.

Kessler, M. (1996). *Estimation paramétrique des coefficients d'une diffusion ergodique à partir d'observations discrètes*. Unpublished doctoral dissertation, Laboratoire de Probabilités, Université Paris VI.

Kessler, M. (1997). Estimation of an ergodic diffusion from discrete observations. *Scand. J. Statist.*, *24*, 211–229.

Kessler, M. (2000). Simple and explicit estimating functions for a discretely observed diffusion process. *Scand. J. Statist.*, *27*, 65–82.

Kessler, M., & Paredes, S. (2002). Computational aspects related to martingale estimating functions for a discretely observed diffusion. *Scand. J. Statist.*, *29*, 425–440.

Kessler, M., Schick, A., & Wefelmeyer, W. (2001). The information in the marginal law of a Markov chain. *Bernoulli*, *7*, 243–266.

Kessler, M., & Sørensen, M. (1999). Estimating equations based on eigenfunctions for a discretely observed diffusion process. *Bernoulli*, *5*, 299–314.

Kessler, M., & Sørensen, M. (2005). On time-reversibility and estimating functions for Markov processes. *Statistical Inference for Stochastic Processes*, *8*, 95–107.

Kimball, B. F. (1946). Sufficient statistical estimation functions for the parameters of the distribution of maximum values. *Ann. Math. Statist.*, *17*, 299–309.

Kloeden, P. E., & Platen, E. (1999). *Numerical solution of stochastic differential equations*. New York: 3rd revised printing. Springer-Verlag.

Kloeden, P. E., Platen, E., Schurz, H., & Sørensen, M. (1996). On effects of discretization on estimates of drift parameters for diffusions processes. *J. Appl. Prob.*, *33*, 1061–1076.

Kuersteiner, G. (2002). Efficient instrumental variables estimation for autoregressive models with conditional heteroskedasticity. *Econometric Theory*, *18*, 547–583.

Kusuoka, S., & Yoshida, N. (2000). Malliavin calculus, geometric mixing, and expansion of diffusion functionals. *Probability Theory and Related Fields*, *116*, 457–484.

Kutoyants, Y. (1994). *Identification of dynamical systems with small noise*. Kluwer.

Kutoyants, Y. (1998). Semiparametric estimation for dynamical systems with small noise. *Math. Methods Statist.*, *7*, 457–465.

Larsen, K. S., & Sørensen, M. (2007). A diffusion model for exchange rates in a target zone. *Mathematical Finance*, *17*, 285–306.

Le Cam, L., & Yang, G. L. (2000). *Asymptotics in statistical: Some basic concepts. 2nd edition.* New York: Springer.

Li, B. (1997). On the consistency of generalized estimating equations. In I. V. Basawa, V. P. Godambe, & R. L. Taylor (Eds.), *Selected proceedings of the symposium on estimating functions* (pp. 115–136). Hayward: Institute of Mathematical Statistics. (IMS Lecture Notes – Monograph Series, Vol. 32)

Liang, K.-Y., & Zeger, S. L. (1986). Longitudinal data analysis using generalized linear model. *Biometrika, 73,* 13–22.

Longstaff, F., & Schwartz, E. (1995). A simple approach to valuing risky fixed and floating rate debt. *J. Finance, 1,* 789–819.

McLeish, D. L., & Small, C. G. (1988). *The theory and applications of statistical inference functions.* New York: Springer-Verlag. (Lecture Notes in Statistics 44)

Nagahara, Y. (1996). Non-Gaussian distribution for stock returns and related stochastic differential equation. *Financial Engineering and the Japanese Markets, 3,* 121–149.

Newey, W. K. (1990). Efficient instrumental variables estimation of nonlinear models. *Econometrica, 58,* 809–837.

Newey, W. K., & West, K. D. (1987). Hypothesis testing with efficient method of moments estimation. *International Economic Review, 28,* 777–787.

Overbeck, L., & Rydén, T. (1997). Estimation in the Cox–Ingersoll–Ross model. *Econometric Theory, 13,* 430–461.

Ozaki, T. (1985). Non-linear time series models and dynamical systems. In E. J. Hannan, P. R. Krishnaiah, & M. M. Rao (Eds.), *Handbook of statistics, vol. 5* (pp. 25–83). Elsevier Science Publishers.

Pearson, K. (1895). Contributions to the mathematical theory of evolution ii. Skew variation in homogeneous material. *Philosophical Transactions of the Royal Society of London. A, 186,* 343–414.

Pedersen, A. R. (1994). *Quasi-likelihood inference for discretely observed diffusion processes* (Research Report No. No. 295). Department of Theoretical Statistics, University of Aarhus.

Pedersen, A. R. (1995). A new approach to maximum likelihood estimation for stochastic differential equations based on discrete observations. *Scand. J. Statist., 22,* 55–71.

Picard, J. (1986). Nonlinear filtering of one-dimensional diffusions in the case of a high signal-to-noise ratio. *SIAM J. Appl. Math., 46,* 1098–1125.

Picard, J. (1991). Efficiency of the extended kalman filter for nonlinear system with small noise. *SIAM J. Appl. Math., 51,* 843–885.

Pokern, Y., Stuart, A. M., & Wiberg, P. (2009). Parameter estimation for partially observed hypoelliptic diffusions. *J. Roy. Statist. Soc. B, 71,* 49–73.

Poulsen, R. (1999). *Approximate maximum likelihood estimation of discretely observed diffusion processes* (Working Paper No. 29). Centre for Analytical Finance, Aarhus.

Prakasa Rao, B. L. S. (1988). Statistical inference from sampled data for stochastic processes. *Contemporary Mathematics, 80,* 249–284.

Prentice, R. L. (1988). Correlated binary regression with covariates specific to each binary observation. *Biometrics, 44,* 1033–1048.

Roberts, G. O., & Stramer, O. (2001). On inference for partially observed nonlinear diffusion models using Metropolis–Hastings algorithms. *Biometrika, 88,* 603–621.

Romanovsky, V. (1924). Generalization of some types of the frequency curves of Professor Pearson. *Biometrika, 16,* 106–117.

Sagan, J. D. (1958). The estimation of economic relationsships using instrumental variables. *Econometrica, 26,* 393–415.

Shephard, N. (1996). Statistical aspects of ARCH and stochastic volatility. In D. Cox, D. Hinkley, & O. Barndorff–Nielsen (Eds.), *Time series models in econometrics, finance and other fields* (pp. 1–67). London: Chapman & Hall.

Skorokhod, A. V. (1989). *Asymptotic methods in the theory of stochastic differential equations.* Providence, Rhode Island: American Mathematical Society.

Sørensen, H. (2001). Discretely observed diffusions: Approximation of the continuous-time score function. *Scand. J. Statist., 28,* 113–121.

Sørensen, M. (2000a). Prediction-based estimating functions. *Econometrics Journal, 3,* 123–147.

Sørensen, M. (2000b). *Small dispersion asymptotics for diffusion martingale estimating functions* (Preprint No. 2). Department of Statistics and Operation Research, University of Copenhagen.

Sørensen, M. (2008). *Efficient estimation for ergodic diffusions sampled at high frequency* (Preprint). Department of Mathematical Sciences, University of Copenhagen.

Sørensen, M. (2011). Prediction-based estimating functions: review and new developments. *Brazilian Journal of Probability and Statistics, 25,* 362–391.

Sørensen, M., & Uchida, M. (2003). Small-diffusion asymptotics for discretely sampled stochastic differential equations. *Bernoulli, 9,* 1051–1069.

Takahashi, A., & Yoshida, N. (2004). An asymptotic expansion scheme for optimal investment problems. *Statistical Inference for Stochastic Processes, 7,* 153–188.

Uchida, M. (2004). Estimation for discretely observed small diffusions based on approximate martingale estimating functions. *Scand. J. Statist., 31,* 553–566.

Uchida, M. (2008). Approximate martingale estimating functions for stochastic differential equations with small noises. *Stoch. Proc. Appl.*, *118*, 1706–1721.

Uchida, M., & Yoshida, N. (2004a). Asymptotic expansion for small diffusions applied to option pricing. *Stat. Inference Stoch. Process.*, *7*, 189–223.

Uchida, M., & Yoshida, N. (2004b). Information criteria for small diffusions via the theory of Malliavin–Watanabe. *Stat. Inference Stoch. Process.*, *7*, 35–67.

Veretennikov, A. Y. (1987). Bounds for the mixing rate in the theory of stochastic equations. *Theory of Probability and its Applications*, *32*, 273–281.

Wedderburn, R. W. M. (1974). Quasi-likelihood functions, generalized linear models, and the Gauss–Newton method. *Biometrika*, *61*, 439–447.

West, K. D. (2001). On optimal instrumental variables estimation of stationary time series models. *International Economic Review*, *42*, 1043–1050.

Wong, E. (1964). The construction of a class of stationary Markoff processes. In R. Bellman (Ed.), *Stochastic processes in mathematical physics and engineering* (pp. 264–276). American Mathematical Society, Rhode Island.

Yoshida, N. (1992). Estimation for diffusion processes from discrete observations. *Journal of Multivariate Analysis*, *41*, 220–242.

CHAPTER 2

The econometrics
of high-frequency data

Per A. Mykland and Lan Zhang

Department of Statistics, University of Chicago
5734 University Avenue, Chicago, IL 60637, USA
and
Department of Finance, University of Illinois at Chicago
601 S Morgan Street, Chicago, IL 60607-7124, USA

2.1 Introduction

2.1.1 Overview

This is a course on estimation in high-frequency data. It is intended for an audience that includes people interested in finance, econometrics, statistics, probability and financial engineering.

There has in recent years been a vast increase in the amount of high-frequency data available. There has also been an explosion in the literature on the subject. In this course, we start from scratch, introducing the probabilistic model for such data, and then turn to the estimation question in this model. We shall be focused on the (for this area) emblematic problem of estimating volatility. Similar techniques to those we present can be applied to estimating leverage effects, realized regressions, semivariances, doing analyses of variance, detecting jumps, measuring liquidity by measuring the size of the microstructure noise, and many other objects of interest.

The applications are mainly in finance, ranging from risk management to options hedging (see Section 2.2.6 below), execution of transactions, portfolio optimization (Fleming, Kirby, and Ostdiek (2001; 2003)), and forecasting. The latter literature has been particularly active, with contributions including

Andersen and Bollerslev (1998), Andersen, Bollerslev, Diebold, and Labys (2001, 2003), Andersen, Bollerslev, and Meddahi (2005), Dacorogna, Gençay, Müller, Olsen, and Pictet (2001), and Meddahi (2001). Methodologies based on high-frequency data can also be found in neural science (see, for example, Valdés-Sosa, Bornot-Sánchez, Melie-García, Lage-Castellanos, and Canales-Rodriguez (2007)) and climatology (see Ditlevsen, Ditlevsen and Andersen (2002) and Ditlevsen and Sørensen (2004) on Greenlandic ice cores).

The purpose of this article, however, is not so much to focus on the applications as on the probabilistic setting and the estimation methods. The theory was started, on the probabilistic side, by Jacod (1994) and Jacod and Protter (1998), and on the econometric side by Foster and Nelson (1996) and Comte and Renault (1998). The econometrics of integrated volatility was pioneered in Andersen et al. (2001, 2003), Barndorff–Nielsen and Shephard (2002, 2004b) and Dacorogna et al. (2001). The authors of this article started to work in the area through Zhang (2001), Zhang, Mykland, and Aït-Sahalia (2005), and Mykland and Zhang (2006). For further references, see Section 2.5.5.

Parametric estimation for discrete observations in a fixed time interval is also an active field. This problem has been studied by Genon-Catalot and Jacod (1994), Genon-Catalot, Jeantheau, and Larédo (1999; 2000), Gloter (2000), Gloter and Jacod (2001b, 2001a), Barndorff–Nielsen and Shephard (2001), Bibby, Jacobsen, and Sørensen (2009), Elerian, Siddhartha, and Shephard (2001), Jacobsen (2001), Sørensen (2001), and Hoffmann (2002). This is, of course, only a small sample of the literature available. Also, these references only concern the type of asymptotics considered in this paper, where the sampling interval is $[0, T]$. There is also substantial literature on the case where $T \to \infty$ (see Section 1.2 of Mykland (2010b) for some of the main references in this area).

This article is meant to be a moderately self-contained course on the basics of this material. The introduction assumes some degree of statistics/econometric literacy, but at a lower level than the standard probability text. Some of the material is on the research front and not published elsewhere. This is not meant as a full review of the area. Readers with a good probabilistic background can skip most of Section 2.2, and occasional other sections.

The text also mostly overlooks (except Sections 2.3.5 and 2.6.3) the questions that arise in connection with multidimensional processes. For further literature in this area, one should consult Barndorff–Nielsen and Shephard (2004a), Hayashi and Yoshida (2005) and Zhang (2011).

2.1.2 high-frequency data

Recent years have seen an explosion in the amount of financial high frequency data. These are the records of transactions and quotes for stocks, bonds, currencies, options, and other financial instruments.

A main source of such data is the *Trades and Quotes (TAQ)* database, which covers the stocks traded on the New York Stock Exchange (NYSE). For example, here is an excerpt of the transactions for Monday, April 4, 2005, for the pharmaceutical company Merck (MRK):

```
symbol date       time      price   size
MRK 20050405      9:41:37 32.69    100
MRK 20050405      9:41:42 32.68    100
MRK 20050405      9:41:43 32.69    300
MRK 20050405      9:41:44 32.68    1000
MRK 20050405      9:41:48 32.69    2900
MRK 20050405      9:41:48 32.68    200
MRK 20050405      9:41:48 32.68    200
MRK 20050405      9:41:51 32.68    4200
MRK 20050405      9:41:52 32.69    1000
MRK 20050405      9:41:53 32.68    300
MRK 20050405      9:41:57 32.69    200
MRK 20050405      9:42:03 32.67    2500
MRK 20050405      9:42:04 32.69    100
MRK 20050405      9:42:05 32.69    300
MRK 20050405      9:42:15 32.68    3500
MRK 20050405      9:42:17 32.69    800
MRK 20050405      9:42:17 32.68    500
MRK 20050405      9:42:17 32.68    300
MRK 20050405      9:42:17 32.68    100
MRK 20050405      9:42:20 32.69    6400
MRK 20050405      9:42:21 32.69    200
MRK 20050405      9:42:23 32.69    3000
MRK 20050405      9:42:27 32.70    8300
MRK 20050405      9:42:29 32.70    5000
MRK 20050405      9:42:29 32.70    1000
MRK 20050405      9:42:30 32.70    1100
```

"Size" here refers to the number of stocks that changed hands in the given transaction. This is often also called "volume."

There are 6302 transactions recorded for Merck for this day. On the same day, Microsoft (MSFT) had 80982 transactions. These are massive amounts of data, and they keep growing. Four years later, on April 3, 2009, there were 74637

Merck transactions, and 211577 Microsoft transactions. What can we do with such data? This course is about how to approach this question.

2.1.3 A first model for financial data: The GBM

Finance theory suggests the following description of prices, that they must be so-called *semimartingales*. We defer a discussion of the general concept until later (see also Delbaen and Schachermayer (1995)), and go instead to the most commonly used such semimartingale: the *Geometric Brownian Motion (GBM)*. This is a model where the stock price movement is additive on the log scale, as follows.

Set

$$X_t = \log S_t = \text{ the logarithm of the stock price } S_t \text{ at time } t. \qquad (2.1)$$

The GBM model is now that

$$X_t = X_0 + \mu t + \sigma W_t, \qquad (2.2)$$

where μ and σ are constants, and W_t is a *Brownian Motion (BM)*, a concept we now define. The "time zero" is an arbitrary reference time.

Definition 2.1 *The process $(W_t)_{0 \leq t \leq T}$ is a Brownian motion provided*
(1) $W_0 = 0$;
(2) $t \to W_t$ is a continuous function of t;
(3) W has independent increments: if $t > s > u > v$, then $W_t - W_s$ is independent of $W_u - W_v$;
(4) for $t > s$, $W_t - W_s$ is normal with mean zero and variance $t - s$ ($N(0,t\text{-}s)$).

2.1.4 Estimation in the GBM model

It is instructive to consider estimation in this model. We take time $t = 0$ to be the beginning of the trading day, and time $t = T$ to be the end of the day.

Let's assume that there are n observations of the process (transactions). We suppose for right now that the transactions are spaced equally in time, so that an observation is had every $\Delta t_n = T/n$ units of time. This assumption is quite unrealistic, but it provides a straightforward development which can then be modified later.

The observations (log transaction prices) are therefore $X_{t_{n,i}}$, where $t_{n,i} = i\Delta t_n$. If we take differences, we get observations

$$\Delta X_{t_{n,i+1}} = X_{t_{n,i+1}} - X_{t_{n,i}}, \ i = 0, ..., n - 1.$$

The $\Delta X_{t_{n,i+1}}$ are independent and identically distributed (iid) with law $N(\mu\Delta t_n, \sigma^2\Delta t_n)$. The natural estimators are:

$$\hat{\mu}_n = \frac{1}{n\Delta t_n}\sum_{i=0}^{n-1}\Delta X_{t_{n,i+1}} = (X_T - X_0)/T \text{ both MLE and UMVU;}$$

and

$$\hat{\sigma}^2_{n,MLE} = \frac{1}{n\Delta t_n}\sum_{i=0}^{n-1}(\Delta X_{t_{n,i+1}} - \overline{\Delta X}_{t_n})^2 \text{ MLE; or} \qquad (2.3)$$

$$\hat{\sigma}^2_{n,UMVU} = \frac{1}{(n-1)\Delta t_n}\sum_{i=0}^{n-1}(\Delta X_{t_{n,i+1}} - \overline{\Delta X}_{t_n})^2 \text{ UMVU.}$$

Here, MLE is the maximum likelihood estimator, and UMVU is the uniformly minimum variance unbiased estimator (see Lehmann (1983) or Rice (2006)). Also, $\overline{\Delta X}_{t_n} = \frac{1}{n}\sum_{i=0}^{n-1}\Delta X_{t_{n,i+1}} = \hat{\mu}_n\Delta t_n$.

The estimators (2.3) clarify some basics. First of all, μ *cannot be consistently estimated* for fixed length T of time interval. In fact, the $\hat{\mu}_n$ does not depend on n, but only on T and the value of the process at the beginning and end of the time period. This is reassuring from a common sense perspective. If we could estimate μ for actual stock prices, we would know much more about the stock market than we really do, and in the event that μ changes over time, benefit from a better allocation between stocks and bonds. – Of course, if $T \to \infty$, then μ *can* be estimated consistently. Specifically, $(X_T - X_0)/T \xrightarrow{p} \mu$ as $T \to \infty$. This is because $\text{Var}((X_T - X_0)/T) = \sigma^2/T \to 0$.

It is perhaps more surprising that σ^2 *can* be estimated consistently for fixed T, as $n \to \infty$. In other words, $\hat{\sigma}^2_n \xrightarrow{p} \sigma^2$ as $n \to \infty$. Set $U_{n,i} = \Delta X_{t_{n,i}}/(\sigma\Delta t_n^{1/2})$. Then the $U_{n,i}$ are iid with distribution $N((\mu/\sigma)\Delta t_n^{1/2}, 1)$. Set $\bar{U}_{n,\cdot} = n^{-1}\sum_{i=0}^{n-1}U_{n,i}$. It follows from considerations for normal random variables (Cochran (1934)) that

$$\sum_{i=0}^{n-1}(U_{n,i} - \bar{U}_{n,\cdot})^2$$

is χ^2 distributed with $n-1$ degrees of freedom (and independent of $\bar{U}_{n,\cdot}$). Hence, for the UMVU estimator,

$$\hat{\sigma}^2_n = \sigma^2\Delta t_n \frac{1}{(n-1)\Delta t_n}\sum_{i=0}^{n-1}(U_{n,i} - \bar{U}_{n,\cdot})^2$$

$$\overset{\mathcal{L}}{=} \sigma^2\frac{\chi^2_{n-1}}{n-1}.$$

It follows that

$$E(\hat{\sigma}_n^2) = \sigma^2 \text{ and } \text{Var}(\hat{\sigma}_n^2) = \frac{2\sigma^4}{n-1},$$

since $E\chi_m^2 = m$ and $\text{Var}(\chi_m^2) = 2m$. Hence $\hat{\sigma}_n^2$ is consistent for σ^2: $\hat{\sigma}_n^2 \to \sigma^2$ in probability as $n \to \infty$.

Similarly, since χ_{n-1}^2 is the sum of $n-1$ iid χ_1^2 random variables, by the central limit theorem we have the following convergence in law:

$$\frac{\chi_{n-1}^2 - E\chi_{n-1}^2}{\sqrt{\text{Var}(\chi_{n-1}^2)}} = \frac{\chi_{n-1}^2 - (n-1)}{\sqrt{2(n-1)}} \xrightarrow{\mathcal{L}} N(0,1),$$

and so

$$n^{1/2}(\hat{\sigma}_n^2 - \sigma^2) \sim (n-1)^{1/2}(\hat{\sigma}_n^2 - \sigma^2)$$

$$\overset{\mathcal{L}}{=} \sqrt{2}\sigma^2 \frac{\chi_{n-1}^2 - (n-1)}{\sqrt{2(n-1)}}$$

$$\xrightarrow{\mathcal{L}} \sigma^2 N(0,2) = N(0, 2\sigma^4).$$

This provides an asymptotic distribution which permits the setting of intervals. For example, $\sigma^2 = \hat{\sigma}_n^2 \pm 1.96 \times \sqrt{2}\hat{\sigma}_n^2$ would be an asymptotic 95 % confidence interval for σ^2.

Since $\hat{\sigma}_{n,MLE}^2 = \frac{n-1}{n}\hat{\sigma}_{n,UMVU}^2$, the same asymptotics apply to the MLE.

2.1.5 Behavior of non-centered estimators

The above discussion of $\hat{\sigma}_{n,UMVU}^2$ and $\hat{\sigma}_{n,MLE}^2$ is exactly the same as in the classical case of estimating variance on the basis of iid observations. More unusually, for high-frequency data, the mean is often not removed in estimation. The reason is as follows. Set

$$\hat{\sigma}_{n,nocenter}^2 = \frac{1}{n\Delta t_n} \sum_{i=0}^{n-1} (\Delta X_{t_n, i+1})^2.$$

Now note that for the MLE version of $\hat{\sigma}_n$,

$$\hat{\sigma}^2_{n,MLE} = \frac{1}{n\Delta t_n}\sum_{i=0}^{n-1}(\Delta X_{t_n,i+1} - \overline{\Delta X}_{t_n})^2$$

$$= \frac{1}{n\Delta t_n}\left(\sum_{i=0}^{n-1}(\Delta X_{t_n,i+1})^2 - n(\overline{\Delta X}_{t_n})^2\right)$$

$$= \hat{\sigma}^2_{n,nocenter} - \Delta t_n \hat{\mu}^2_n$$

$$= \hat{\sigma}^2_{n,nocenter} - \frac{T}{n}\hat{\mu}^2_n.$$

Since $\hat{\mu}^2_n$ does not depend on n, it follows that

$$n^{1/2}\left(\hat{\sigma}^2_{n,MLE} - \hat{\sigma}^2_{n,nocenter}\right) \xrightarrow{p} 0.$$

Hence, $\hat{\sigma}^2_{n,nocenter}$ is consistent and has the same asymptotic distribution as $\hat{\sigma}^2_{n,UMVU}$ and $\hat{\sigma}^2_{n,MLE}$. It can therefore also be used to estimate variance. This is quite common for high-frequency data.

2.1.6 GBM and the Black–Scholes–Merton formula

The GBM model is closely tied in to other parts of finance. In particular, following the work of Black and Scholes (1973), Merton (1973), Harrison and Kreps (1979), and Harrison and Pliska (1981), precise option prices can be calculated in this model. See also Duffie (1996), Neftci (2000), Øksendal (2003), or Shreve (2004) for book sized introductions to the theory.

In the case of the *call option*, the price is as follows. A *European call option* on stock S_t with *maturity (expiration) time* T and *strike price* K is the option to buy one unit of stock at price K at time T. It is easy to see that the value of this option at time T is $(S_T - K)^+$, where $x^+ = x$ if $x \geq 0$, and $x^+ = 0$ otherwise.

If we make the assumption that S_t is a GBM, which is to say that it follows (2.1) – (2.2), and also the assumption that the short term interest rate r is constant (in time), then the price at time t, $0 \leq t \leq T$ of this option must be

$$\text{price} = C(S_t, \sigma^2(T-t), r(T-t)),$$

where

$$C(S,\Xi,R) = S\Phi(d_1(S,\Xi,R)) - K\exp(-R)\Phi(d_2(S,\Xi,R)), \text{ where}$$
$$d_{1,2}(S,\Xi,R) = (\log(S/K) + R \pm \Xi/2)/\sqrt{\Xi} \ (+ \text{ in } d_1 \text{ and } - \text{ in } d_2) \text{ and}$$
$$(2.4)$$

$$\Phi(x) = P(N(0,1) \leq x), \text{ the standard normal cdf.}$$

This is the Black–Scholes–Merton formula.

We shall see later on how high-frequency estimates can be used in this formula. For the moment, note that the price only depends on quantities that are either observed (the interest rate r) or (perhaps) nearly so (the volatility σ^2). It does not depend on μ. Unfortunately, the assumption of constant r and σ^2 is unrealistic, as we shall discuss in the following.

The GBM model is also heavily used in portfolio optimization

2.1.7 Our problem to be solved: Inadequacies in the GBM model

We here give a laundry list of questions that arise and have to be dealt with.

The volatility depends on t

It is empirically the case that σ^2 depends on t. We shall talk about the *instantaneous volatility* σ_t^2. This concept will be defined carefully in Section 2.2.

Non-normal returns

Returns are usually assumed to be non-normal. This behavior can be explained through random volatility and/or jumps.

- *The volatility is random; leverage effect.* Non-normality can be achieved in a continuous model by letting σ_t^2 have random evolution. It is also usually assumed that σ_t^2 can be correlated with the (log) stock price. This is often referred to as the *leverage effect*. More about this in Section 2.2.

- *Jumps.* The GBM model assumes that the log stock price X_t is continuous as a function of t. The evolution of the stock price, however, is often thought to have a jump component. The treatment of jumps is largely not covered in this article, though there is some discussion in Section 2.6.4, which also gives some references.

Jumps and random volatility are often confounded, since any martingale can be embedded in a Brownian motion (Dambis (1965), Dubins and Schwartz (1965), see also Mykland (1995) for a review and further discussion). The difficulty in distinguishing these two sources of non-normality is also studied by Bibby, Skovgaard, and Sørensen (2005).

Microstructure noise

An important feature of actual transaction prices is the existence of *microstructure noise*. Transaction prices, as actually observed, are typically best modeled on the form $Y_t = \log S_t =$ the logarithm of the stock price S_t at time t, where for transaction at time t_i,

$$Y_{t_i} = X_{t_i} + \text{noise},$$

and X_t is a semimartingale. This is often called the *hidden semimartingale model*. This issue is an important part of our narrative, and is further discussed in Section 2.5, see also Section 2.6.4.

Unequally spaced observations

In the above, we assumed that the transaction times t_i are equally spaced. A quick glance at the data snippet in Section 2.1.2 reveals that this is typically not the case. This leads to questions that will be addressed as we go along.

2.1.8 A note on probability theory, and other supporting material

We will extensively use probability theory in these notes. To avoid making a long introduction on stochastic processes, we will define concepts as we need them, but not always in the greatest depth. We will also omit other concepts and many basic proofs. As a compromise between the rigorous and the intuitive, we follow the following convention: the notes will (except when the opposite is clearly stated) use mathematical terms as they are defined in Jacod and Shiryaev (2003). Thus, in case of doubt, this work can be consulted.

Other recommended reference books on stochastic process theory are Karatzas and Shreve (1991), Øksendal (2003), Protter (2004), and Shreve (2004). For an introduction to measure theoretic probability, one can consult Billingsley (1995). Mardia, Kent, and Bibby (1979) provides a handy reference on normal distribution theory.

2.2 A more general model: time varying drift and volatility

2.2.1 Stochastic integrals, Itô processes

We here make some basic definitions. We consider a process X_t, where the time variable $t \in [0, T]$. We mainly develop the univariate case here.

Information sets, σ-fields, filtrations

Information is usually described with so-called σ-fields. The setup is as follows. Our basic space is $(\Omega, \mathcal{F})$, where Ω is the set of all possible outcomes ω, and $\mathcal{F}$ is the collection of subsets $A \subseteq \Omega$ that will eventually be decidable (it will be observed whether they occured or not). All random variables are thought to be a function of the basic outcome $\omega \in \Omega$.

We assume that $\mathcal{F}$ is a so-called σ-field. In general,

Definition 2.2 *A collection $\mathcal{A}$ of subsets of Ω is a σ-field if*

(i) $\emptyset$, $\Omega \in \mathcal{A}$;
(ii) if $A \in \mathcal{A}$, then $A^c = \Omega - A \in \mathcal{A}$; and
(iii) if $A_n, n = 1, 2, \ldots$ are all in $\mathcal{A}$, then $\cup_{n=1}^{\infty} A_n \in \mathcal{A}$.

If one thinks of $\mathcal{A}$ as a collection of decidable sets, then the interpretation of this definition is as follows:

(i) $\emptyset$, Ω are decidable ($\emptyset$ didn't occur, Ω did);
(ii) if A is decidable, so is the complement A^c (if A occurs, then A^c does not occur, and vice versa);
(iii) if all the A_n are decidable, then so is the event $\cup_{n=1}^{\infty} A_n$ (the union occurs if and only if at least one of the A_i occurs).

A random variable X is called $\mathcal{A}$-*measurable* if the value of X can be decided on the basis of the information in $\mathcal{A}$. Formally, the requirement is that for all x, the set $\{X \leq x\} = \{\omega \in \Omega : X(\omega) \leq x\}$ be decidable $(\in \mathcal{A})$.

The evolution of knowledge in our system is described by the *filtration* (or sequence of σ-fields) $\mathcal{F}_t$, $0 \leq t \leq T$. Here $\mathcal{F}_t$ is the knowledge available at time t. Since increasing time makes more sets decidable, the family $(\mathcal{F}_t)$ is taken to satisfy that if $s \leq t$, then $\mathcal{F}_s \subseteq \mathcal{F}_t$.

Most processes will be taken to be *adapted* to $(\mathcal{F}_t)$: (X_t) is adapted to $(\mathcal{F}_t)$ if for all $t \in [0, T]$, X_t is $\mathcal{F}_t$-measurable. A vector process is adapted if each component is adapted.

We define the filtration $(\mathcal{F}_t^X)$ *generated* by the process (X_t) as the smallest filtration to which X_t is adapted. By this we mean that for any filtration $\mathcal{F}_t'$ to which (X_t) is adapted, $\mathcal{F}_t^X \subseteq \mathcal{F}_t'$ for all t. (Proving the existence of such a filtration is left as an exercise for the reader).

Wiener processes

A Wiener process is Brownian motion relative to a filtration. Specifically,

Definition 2.3 *The process* $(W_t)_{0 \le t \le T}$ *is an* $(\mathcal{F}_t)$*-Wiener process if it is adpted to* $(\mathcal{F}_t)$ *and*
(1) $W_0 = 0$*;*
(2) $t \to W_t$ *is a continuous function of t;*
(3) W has independent increments relative to the filtration $(\mathcal{F}_t)$*: if* $t > s$*, then* $W_t - W_s$ *is independent of* $\mathcal{F}_s$*;*
(4) for $t > s$*,* $W_t - W_s$ *is normal with mean zero and variance* $t - s$ *(N(0,t-s)).*

Note that a Brownian motion (W_t) is an $(\mathcal{F}_t^W)$-Wiener process.

Predictable processes

For defining stochastic integrals, we need the concept of *predictable process.* "Predictable" here means that one can forecast the value over infinitesimal time intervals. The most basic example would be a "simple process." This is given by considering break points $0 = s_0 = t_0 \le s_1 < t_1 \le s_2 < t_2 < ... \le s_n < t_n \le T$, and random variables $H^{(i)}$, observable (measurable) with respect to $\mathcal{F}_{s_i}$.

$$ H_t = \begin{cases} H^{(0)} \text{ if } t = 0 \\ H^{(i)} \text{ if } s_i < t \le t_i \end{cases} \qquad (2.5) $$

In this case, at any time t (the beginning time $t = 0$ is treated separately), the value of H_t is known *before* time t.

Definition 2.4 *More generally, a process* H_t *is predictable if it can be written as a limit of simple functions* $H_t^{(n)}$*. This means that* $H_t^{(n)}(\omega) \to H_t(\omega)$ *as* $n \to \infty$*, for all* $(t, \omega) \in [0, T] \times \Omega$*.*

All adapted continuous processes are predictable. More generally, this is also true for adapted processes that are left continuous (*càg*, for *continue à gauche*). (Proposition I.2.6 (p. 17) in Jacod and Shiryaev (2003)).

Stochastic integrals

We here consider the meaning of the expression

$$ \int_0^T H_t dX_t. \qquad (2.6) $$

The ingredients are the integrand H_t, which is assumed to be predictable, and the integrator X_t, which will generally be a semi-martingale (to be defined below in Section 2.2.3).

The expression (2.6) is defined for simple process integrands as

$$ \sum_i H^{(i)}(X_{t_i} - X_{s_i}). \qquad (2.7) $$

For predictable integrands H_t that are bounded and limits of simple processes $H_t^{(n)}$, the integral (2.6) is the limit in probability of $\int_0^T H_t^{(n)} dX_t$. This limit is well defined, i.e., independent of the sequence $H_t^{(n)}$.

If X_t is a Wiener process, the integral can be defined for any predictable process H_t satisfying

$$\int_0^T H_t^2 dt < \infty.$$

It will always be the case that the integrator X_t is right continuous with left limits (*càdlàg*, for *continue à droite, limites à gauche*).

The integral process

$$\int_0^t H_s dX_s = \int_0^T H_s I\{s \le t\} dX_s$$

can also be taken to be *càdlàg*. If (X_t) is continuous, the integral is then automatically continuous.

Itô processes

We now come to our main model, the Itô process. X_t is an Itô process relative to filtration $(\mathcal{F}_t)$ provided (X_t) is $(\mathcal{F}_t)$ adapted; and if there is an $(\mathcal{F}_t)$-Wiener process (W_t), and $(\mathcal{F}_t)$-adapted processes (μ_t) and (σ_t), with

$$\int_0^T |\mu_t| dt < \infty, \text{ and}$$

$$\int_0^T \sigma_t^2 dt < \infty$$

so that

$$X_t = X_0 + \int_0^t \mu_s ds + \int_0^t \sigma_s dW_s. \tag{2.8}$$

The process is often written in differential form:

$$dX_t = \mu_t dt + \sigma_t dW_t. \tag{2.9}$$

We note that the Itô process property is preserved under stochastic integration. If H_t is bounded and predictable, then

$$\int_0^t H_s dX_s = \int_0^t H_s \mu_s dt + \int_0^t H_s \sigma_s dW_s.$$

It is clear from this formula that predictable processes H_t can be used for integration w.r.t. X_t provided

$$\int_0^T |H_t \mu_t| dt < \infty \text{ and}$$

$$\int_0^T (H_t \sigma_t)^2 dt < \infty.$$

2.2.2 *Two interpretations of the stochastic integral*

One can use the stochastic integral in two different ways: as a model, or as a description of trading profit and loss (P/L).

Stochastic integral as trading profit or loss (P/L)

Suppose that X_t is the value of a security. Let H_t be the number of this stock that is held at time t. In the case of a simple process (2.5), this means that we hold $H^{(i)}$ units of X from time s_i to time t_i. The trading profit and loss (P/L) is then given by the stochastic integral (2.7). In this description, it is quite clear that $H^{(i)}$ must be known at time s_i, otherwise we would base the portfolio on future information. More generally, for predictable H_t, we similarly avoid using future information.

Stochastic integral as model

This is a different genesis of the stochastic integral model. One simply uses (2.8) as a model, in the hope that this is a sufficiently general framework to capture most relevant processes. The advantage of using predictable integrands comes from the simplicity of connecting the model with trading gains.

For simple μ_t and σ_t^2, the integral

$$\sum_i \mu^{(i)} (t_i - s_i) + \sum_i \sigma^{(i)} (W_{t_i} - W_{s_i})$$

is simply a sum of conditionally normal random variables, with mean $\mu^{(i)}(t_i - s_i)$ and variance $(\sigma^{(i)})^2 (t_i - s_i)$. The sum need not be normal, since μ and σ^2 can be random.

It is worth noting that in this model, $\int_0^T \mu_t dt$ is the sum of instantaneous means (drift), and $\int_0^T \sigma_t^2 dt$ is the sum of intstantaneous variances. To make the latter statement precise, note that in the model (2.8), one can show the following: Let $\text{Var}(\cdot|\mathcal{F}_t)$ be the conditional variance given the information at time t. If X_t is an Itô process, and if $0 = t_{n,0} < t_{n,i} < ... < t_{n,n} = T$, then

$$\sum_i \mathrm{Var}(X_{t_{n,i+1}} - X_{t_{n,i}}|\mathcal{F}_{t_{n,i}}) \xrightarrow{p} \int_0^T \sigma_t^2 dt \qquad (2.10)$$

when

$$\max_i |t_{n,i+1} - t_{n,i}| \to 0.$$

If the μ_t and σ_t^2 processes are nonrandom, then X_t is a Gaussian process, and X_T is normal with mean $X_0 + \int_0^T \mu_t dt$ and variance $\int_0^T \sigma_t^2 dt$.

The Heston model

A popular model for volatility is due to Heston (1993). In this model, the process X_t is given by

$$dX_t = \left(\mu - \frac{\sigma_t^2}{2}\right) dt + \sigma_t dW_t$$
$$d\sigma_t^2 = \kappa(\alpha - \sigma_t^2)dt + \gamma\sigma_t dZ_t \ , \ \text{with}$$
$$Z_t = \rho W_t + (1 - \rho^2)^{1/2} B_t$$

where (W_t) and (B_t) are two independent Wiener processes, $\kappa > 0$, and $|\rho| \leq 1$. To assure that σ_t^2 does not hit zero, one must also require (Feller (1951)) that $2\kappa\alpha \geq \gamma^2$.

2.2.3 Semimartingales

Conditional expectations

Denote by $E(\cdot|\mathcal{F}_t)$ the conditional expectation given the information available at time t. Formally, this concept is defined as follows:

Theorem 2.5 *Let $\mathcal{A}$ be a σ-field, and let X be a random variable so that $E|X| < \infty$. There is a $\mathcal{A}$-measurable random variable Z so that for all $A \in \mathcal{A}$,*

$$EZI_A = EXI_A,$$

where I_A is the indicator function of A. Z is unique "almost surely," which means that if Z_1 and Z_2 satisfy the two criteria above, then $P(Z_1 = Z_2) = 1$.

We thus define

$$E(X|\mathcal{A}) = Z$$

where Z is given in the theorem. The conditional expectation is well defined "almost surely."

For further details and proof of the theorem, see Section 34 (pp. 445–455) of Billingsley (1995).

This way of defining conditional expectation is a little counterintuitive if un-familiar. In particular, the conditional expectation is a random variable. The heuristic is as follows. Suppose that Y is a random variable, and that $\mathcal{A}$ carries the information in Y. Introductory textbooks often introduce conditional expectation as a non-random quantity $E(X|Y = y)$. To make the connection, set

$$f(y) = E(X|Y = y).$$

The conditional expectation we have just defined then satisfies

$$E(X|\mathcal{A}) = f(Y). \tag{2.11}$$

The expression in (2.11) is often written $E(X|Y)$.

Properties of conditional expectations

- Linearity: for constant c_1, c_2:

$$E(c_1 X_1 + c_2 X_2 \mid \mathcal{A}) = c_1 E(X_1 \mid \mathcal{A}) + c_2 E(X_2 \mid \mathcal{A})$$

- Conditional constants: if Z is $\mathcal{A}$-measurable, then

$$E(ZX|\mathcal{A}) = ZE(X|\mathcal{A})$$

- Law of iterated expectations (iterated conditioning, Tower property): if $\mathcal{A}' \subseteq \mathcal{A}$, then

$$E[E(X|\mathcal{A})|\mathcal{A}'] = E(X|\mathcal{A}')$$

- Independence: if X is independent of $\mathcal{A}$:

$$E(X|\mathcal{A}) = E(X)$$

- Jensen's inequality: if $g : x \to g(x)$ is convex:

$$E(g(X)|\mathcal{A}) \geq g(E(X|\mathcal{A}))$$

Note: g is convex if $g(ax + (1 - a)y) \leq ag(x) + (1 - a)g(y)$ for $0 \leq a \leq 1$.

For example: $g(x) = e^x$, $g(x) = (x - K)^+$. Or g'' exists and is continuous, and $g''(x) \geq 0$.

Martingales

An $(\mathcal{F}_t)$ adapted process M_t is called a *martingale* if $E|M_t| < \infty$, and if, for all $s < t$,

$$E(M_t|\mathcal{F}_s) = M_s.$$

This is a central concept in our narrative. A martingale is also known as a *fair game*, for the following reason. In a gambling situation, if M_s is the amount of money the gambler has at time s, then the gambler's expected wealth at time $t > s$ is also M_s. (The concept of martingale applies equally to discrete and continuous time axis).

Example 2.6 *A Wiener process is a martingale. To see this, for* $t > s$, *since* $W_t - W_s$ *is N(0,t-s) given* $\mathcal{F}_s$, *we get that*

$$
\begin{aligned}
E(W_t|\mathcal{F}_s) &= E(W_t - W_s|\mathcal{F}_s) + W_s \\
&= E(W_t - W_s) + W_s \text{ by independence} \\
&= W_s.
\end{aligned}
$$

A useful fact about martingales is the *representation by final value*: M_t is a martingale for $0 \leq t \leq T$ if and only if one can write (with $E|X| < \infty$)

$$M_t = E(X|\mathcal{F}_t) \text{ for all } t \in [0, T]$$

(only if by definition ($X = M_T$), if by Tower property). Note that for $T = \infty$ (which we do not consider here), this property may not hold. (For a full discussion, see Chapter 1.3.B (pp. 17–19) of Karatzas and Shreve (1991)).

Example 2.7 *If* H_t *is a bounded predictable process, then for any martingale* X_t,

$$M_t = \int_0^t H_s dX_s$$

is a martingale. To see this, consider first a simple process (2.5), for which $H_s = H^{(i)}$ *when* $s_i < s \leq t_i$. *For given* t, *if* $s_i > t$, *by the properties of conditional expectations,*

$$
\begin{aligned}
E\left(H^{(i)}(X_{t_i} - X_{s_i})|\mathcal{F}_t\right) &= E\left(E(H^{(i)}(X_{t_i} - X_{s_i})|\mathcal{F}_{s_i})|\mathcal{F}_t\right) \\
&= E\left(H^{(i)}E(X_{t_i} - X_{s_i}|\mathcal{F}_{s_i})|\mathcal{F}_t\right) \\
&= 0,
\end{aligned}
$$

and similarly, if $s_i \leq t \leq t_i$, *then*

$$E\left(H^{(i)}(X_{t_i} - X_{s_i})|\mathcal{F}_t\right) = H^{(i)}(X_t - X_{s_i})$$

so that

$$
\begin{aligned}
E(M_T|\mathcal{F}_t) &= E\left(\sum_i H^{(i)}(X_{t_i} - X_{s_i})|\mathcal{F}_t\right) \\
&= \sum_{i:t_i<t} H^{(i)}(X_{t_i} - X_{s_i}) + I\{t_i \leq t \leq s_i\}H^{(i)}(X_t - X_{s_i}) \\
&= M_t.
\end{aligned}
$$

The result follows for general bounded predictable integrands by taking limits and using uniform integrability. (For definition and results on uniform integrability, see Billingsley (1995).)

Thus, any bounded trading strategy H in an asset M which is a martingale results in a martingale profit and loss (P/L).

Stopping times and local martingales

The concept of local martingale is perhaps best understood by considering the following integral with respect to a Wiener process (see also Duffie (1996)):

$$X_t = \int_0^t \frac{1}{\sqrt{T-s}} dW_s$$

Note that for $0 \le t < T$, X_t is a zero mean Gaussian process with independent increments. We shall show below (in Section 2.2.4) that the integral has variance

$$\begin{aligned} \mathrm{Var}(X_t) &= \int_0^t \frac{1}{T-s} ds \\ &= \int_{T-t}^T \frac{1}{u} du \\ &= \log \frac{T}{T-t}. \end{aligned} \tag{2.12}$$

Since the dispersion of X_t goes to infinity as we approach T, X_t is not defined at T. However, one can *stop* the process at a convenient time, as follows: Set, for $A > 0$,

$$\tau = \inf\{t \ge 0 : X_t = A\}. \tag{2.13}$$

One can show that $P(\tau < T) = 1$. Define the modified integral by

$$\begin{aligned} Y_t &= \int_0^t \frac{1}{\sqrt{T-s}} I\{s \le \tau\} dW_s \\ &= X_{\tau \wedge t}, \end{aligned} \tag{2.14}$$

where

$$s \wedge t = \min(s, t).$$

The process (2.14) has the following trading interpretation. Suppose that W_t is the value of a security at time t (the value can be negative, but that is possible for many securities, such as futures contracts). We also take the short term interest rate to be zero. The process X_t comes about as the value of a portfolio which holds $1/\sqrt{T-t}$ units of this security at time t. The process Y_t is obtained by holding this portfolio until such time that $X_t = A$, and then liquidating the portfolio.

In other words, we have displayed a trading strategy which starts with wealth $Y_0 = 0$ at time $t = 0$, and ends with wealth $Y_T = A > 0$ at time $t = T$. In trading terms, this is an arbitrage. In mathematical terms, this is a stochastic integral w.r.t. a martingale which is no longer a martingale.

We note that from (2.12), the conditions for the existence of the integral (2.14) are satisfied.

For trading, the lesson we can learn from this is that some condition has to be imposed to make sure that a trading strategy in a martingale cannot result in arbitrage profit. The most popular approach to this is to require that the traders wealth at any time cannot go below some fixed amount $-K$. This is the so-called credit constraint. (So strategies are required to satisfy that the integral never goes below $-K$). This does not quite guarantee that the integral w.r.t. a martingale is a martingale, but it does prevent arbitrage profit. The technical result is that the integral is a *super-martingale* (see the next section).

For the purpose of characterizing the stochastic integral, we need the concept of a *local martingale*. For this, we first need to define:

Definition 2.8 *A stopping time is a random variable τ satisfying $\{\tau \le t\} \in \mathcal{F}_t$, for all t.*

The requirement in this definition is that we must be able to know at time t whether τ occurred or not. The time (2.13) given above is a stopping time. On the other hand, the variable $\tau = \inf\{t : W_t = \max_{0 \le s \le T} W_s\}$ is not a stopping time. Otherwise, we would have a nice investment strategy.

Definition 2.9 *A process M_t is a local martingale for $0 \le t \le T$ provided there is a sequence of stopping times τ_n so that*
(i) $M_{\tau_n \wedge t}$ is a martingale for each n; and
(ii) $P(\tau_n \to T) = 1$ as $n \to \infty$.

The basic result for stochastic integrals is now that the integral with respect to a local martingale is a local martingale, cf. result I.4.34(b) (p. 47) in Jacod and Shiryaev (2003).

Semimartingales

X_t is a semimartingale if it can be written

$$X_t = X_0 + M_t + A_t, 0 \le t \le T,$$

where X_0 is $\mathcal{F}_0$-measurable, M_t is a local martingale, and A_t is a process of finite variation, i.e.,

$$\sup \sum_i |A_{t_{i+1}} - A_{t_i}| < \infty,$$

where the supremum is over all grids $0 = t_0 < t_1 < ... < t_n = T$, and all n.

In particular, an Itô process is a semimartingale, with

$$M_t = \int_0^t \sigma_s dW_s \text{ and}$$

$$A_t = \int_0^t \mu_s ds.$$

A *supermartingale* is semimartingale for which A_t is nonincreasing. A *submartingale* is a semimartingale for which A_t is nondecreasing.

2.2.4 Quadratic variation of a semimartingale

Definitions

We start with some notation. A grid of observation times is given by

$$\mathcal{G} = \{t_0, t_1, ..., t_n\},$$

where we suppose that

$$0 = t_0 < t_1 < ... < t_n = T.$$

Set

$$\Delta(\mathcal{G}) = \max_{1 \le i \le n} (t_i - t_{i-1}).$$

For any process X, we define its *quadratic variation* relative to grid $\mathcal{G}$ by

$$[X, X]_t^{\mathcal{G}} = \sum_{t_{i+1} \le t} (X_{t_{i+1}} - X_{t_i})^2. \tag{2.15}$$

We note that the quadratic variation is path-dependent. One can more generally define the quadratic covariation

$$[X, Y]_t^{\mathcal{G}} = \sum_{t_{i+1} \le t} (X_{t_{i+1}} - X_{t_i})(Y_{t_{i+1}} - Y_{t_i}).$$

An important theorem of stochastic calculus now says that

Theorem 2.10 *For any semimartingale, there is a process $[X, Y]_t$ so that*

$$[X, Y]_t^{\mathcal{G}} \xrightarrow{P} [X, Y]_t \text{ for all } t \in [0, T], \text{ as } \Delta(\mathcal{G}) \to 0.$$

The limit is independent of the sequence of grids $\mathcal{G}$.

The result follows from Theorem I.4.47 (p. 52) in Jacod and Shiryaev (2003). The t_i can even be stopping times.

For an Itô process,

$$[X, X]_t = \int_0^t \sigma_s^2 ds. \qquad (2.16)$$

(Cf Thm I.4.52 (p. 55) and I.4.40(d) (p. 48) of Jacod and Shiryaev (2003)). In particular, for a Wiener process W, $[W, W]_t = \int_0^t 1 ds = t$.

The process $[X, X]_t$ is usually referred to as the quadratic variation of the semimartingale (X_t). This is an important concept, as seen in Section 2.2.2. The theorem asserts that this quantity can be estimated consistently from data.

Properties

Important properties are as follows:

(1) Bilinearity: $[X, Y]_t$ is linear in each of X and Y: so for example, $[aX + bZ, Y]_t = a[X, Y]_t + b[Z, Y]_t$.

(2) If (W_t) and (B_t) are two independent Wiener processes, then

$$[W, B]_t = 0.$$

Example 2.11 *For the Heston model in Section 2.2.2, one gets from first principles that*

$$[W, Z]_t = \rho[W, W]_t + (1 - \rho^2)^{1/2}[W, B]_t$$
$$= \rho t,$$

since $[W, W]_t = t$ *and* $[W, B]_t = 0$.

(3) For stochastic integrals over Itô processes X_t and Y_t,

$$U_t = \int_0^t H_s dX_s \text{ and } V_t = \int_0^t K_s dY_s,$$

then

$$[U, V]_t = \int_0^t H_s K_s d[X, Y]_s.$$

This is often written on "differential form" as

$$d[U, V]_t = H_t K_t d[X, Y]_t.$$

by invoking the same results that led to (2.16). For a rigorous statement, see Property I.4.54 (p.55) of Jacod and Shiryaev (2003).

(4) For any Itô process X, $[X, t] = 0$.

Example 2.12 *(Leverage effect in the Heston model).*

$$d[X, \sigma^2] = \gamma \sigma_t^2 d[W, Z]_t$$
$$= \gamma \sigma^2 \rho dt.$$

(5) Invariance under discounting by the short term interest rate. Discounting is important in finance theory. The typical discount rate is the risk free short term interest rate r_t. Recall that $S_t = \exp\{X_t\}$. The discounted stock price is then given by

$$S_t^* = \exp\{-\int_0^t r_s ds\} S_t.$$

The corresponding process on the log scale is $X_t^* = X_t - \int_0^t r_s ds$, so that if X_t is given by (2.9), then

$$dX_t^* = (\mu_t - r_t)dt + \sigma_t dW_t.$$

The quadratic variation of X_t^* is therefore the same as for X_t.

It should be emphasized that while this result remains true for certain other types of discounting (such as those incorporating cost-of-carry), it is not true for many other relevant types of discounting. For example, if one discounts by the zero coupon bond Λ_t maturing at time T, the discounted log price becomes $X_t^* = X_t - \log \Lambda_t$. Since the zero coupon bond will itself have volatility, we get

$$[X^*, X^*]_t = [X, X]_t + [\log \Lambda, \log \Lambda]_t - 2[X, \log \Lambda]_t.$$

Variance and quadratic variation

Quadratic variation has a representation in terms of variance. The main result concerns martingales. For $E(X^2) < \infty$, define the conditional variance by

$$\text{Var}(X|\mathcal{A}) = E((X - E(X|\mathcal{A}))^2|\mathcal{A}) = E(X^2|\mathcal{A}) - E(X|\mathcal{A})^2$$

and similarly $\text{Cov}(X, Y|\mathcal{A}) = E((X - E(X|\mathcal{A}))(Y - E(Y|\mathcal{A})|\mathcal{A})$.

Theorem 2.13 *Let M_t be a martingale, and assume that $E[M, M]_T < \infty$. Then, for all $s < t$,*

$$\text{Var}(M_t|\mathcal{F}_s) = E((M_t - M_s)^2|\mathcal{F}_s) = E([M, M]_t - [M, M]_s|\mathcal{F}_s). \quad (2.17)$$

This theorem is the beginning of something important: the left-hand side of (2.17) relates to the central limit theorem, while the right-hand side only concerns the law of large numbers. We shall see this effect in more detail in the sequel.

A quick argument for (2.17) is as follows. Let $\mathcal{G} = \{t_0, t_1, ..., t_n\}$, and let $t_* = \max\{u \in \mathcal{G} : u \leq t\}$, and similarly for s_*. Suppose for simplicity that $s, t \in \mathcal{G}$. Then, for $s_* \leq t_i < t_j$,

$$\begin{aligned} &E((M_{t_{i+1}} - M_{t_i})(M_{t_{j+1}} - M_{t_j})|\mathcal{F}_{t_j}) \\ &= (M_{t_{i+1}} - M_{t_i})E((M_{t_{j+1}} - M_{t_j})|\mathcal{F}_{t_j}) \\ &= 0, \end{aligned}$$

so that by the Tower rule (since $\mathcal{F}_{s_*} \subseteq \mathcal{F}_{t_j}$)

$$
\begin{aligned}
\text{Cov}(M_{t_{i+1}} &- M_{t_i}, M_{t_{j+1}} - M_{t_j}|\mathcal{F}_{s_*}) \\
&= E((M_{t_{i+1}} - M_{t_i})(M_{t_{j+1}} - M_{t_j})|\mathcal{F}_{s_*}) \\
&= 0.
\end{aligned}
$$

It follows that, for $s < t$,

$$
\begin{aligned}
\text{Var}(M_{t_*} - M_{s_*}|\mathcal{F}_{s_*}) &= \sum_{s_* \le t_i < t_*} \text{Var}(M_{t_{i+1}} - M_{t_i}|\mathcal{F}_{s_*}) \\
&= \sum_{s_* \le t_i < t_*} E((M_{t_{i+1}} - M_{t_i})^2|\mathcal{F}_{s_*}) \\
&= E(\sum_{s_* \le t_i < t_*} (M_{t_{i+1}} - M_{t_i})^2|\mathcal{F}_{s_*}) \\
&= E([M, M]_{t_*}^{\mathcal{G}} - [M, M]_{s_*}^{\mathcal{G}}|\mathcal{F}_{s_*}).
\end{aligned}
$$

The result as $\Delta(\mathcal{G}) \to 0$ then follows by uniform integrability (Theorem 25.12 (p. 338) in Billingsley (1995)).

On the basis of this, one can now show for an Itô process that

$$
\lim_{h \downarrow 0} \frac{1}{h} \text{Cov}(X_{t+h} - X_t, Y_{t+h} - Y_t|\mathcal{F}_t) = \frac{d}{dt}[X, Y]_t.
$$

A similar result holds in the integrated sense, cf. formula (2.10). The reason this works is that the dt terms are of smaller order than the martingale terms.

Sometimes instantaneous correlation is important. We define

$$
\text{cor}(X, Y)_t = \lim_{h \downarrow 0} \text{cor}(X_{t+h} - X_t, Y_{t+h} - Y_t|\mathcal{F}_t),
$$

and note that

$$
\text{cor}(X, Y)_t = \frac{d[X, Y]_t/dt}{\sqrt{(d[X, X]_t/dt)(d[Y, Y]_t/dt)}}.
$$

We emphasize that these results only hold for Itô processes. For general semi-martingales, one needs to involve the concept of predictable quadratic variation, cf. Section 2.2.4.

To see the importance of the instantaneous correlation, note that in the Heston model,

$$
\text{cor}(X, \sigma^2)_t = \rho.
$$

In general, if $dX_t = \sigma_t dW_t + dt\text{-term}$, and $dY_t = \gamma_t dB_t + dt\text{-term}$, where W_t and B_t are two Wiener processes, then

$$
\text{cor}(X, Y)_t = \text{sgn}(\sigma_t \gamma_t)\text{cor}(W, B)_t. \tag{2.18}
$$

Lévy's Theorem

An important result is now the following:

Theorem 2.14 *Suppose that M_t is a continuous $(\mathcal{F}_t)$-local martingale, $M_0 = 0$, so that $[M, M]_t = t$. Then M_t is an $(\mathcal{F}_t)$-Wiener process.*

(Cf. Thm II.4.4 (p. 102) in Jacod and Shiryaev (2003)). More generally, from properties of normal random variables, the same result follows in the vector case: If $M_t = (M_t^{(1)}, ..., M_t^{(p)})$ is a continuous $(\mathcal{F}_t)$-martingale, $M_0 = 0$, so that $[M^{(i)}, M^{(j)}]_t = \delta_{ij}t$, then M_t is a vector Wiener process. (δ_{ij} is the Kronecker delta: $\delta_{ij} = 1$ for $i = j$, and $= 0$ otherwise.)

Predictable quadratic variation

One can often see the symbol $\langle X, Y \rangle_t$. This can be called the predictable quadratic variation. Under regularity conditions, it is defined as the limit of $\sum_{t_i \leq t} \text{Cov}(X_{t_{i+1}} - X_{t_i}, Y_{t_{i+1}} - Y_{t_i} | \mathcal{F}_{t_i})$ as $\Delta(\mathcal{G}) \to 0$.

For Itô processes, $\langle X, Y \rangle_t = [X, Y]_t$. For general semimartingales this equality does not hold. Also, except for Itô processes, $\langle X, Y \rangle_t$ cannot generally be estimated consistently from data without further assumptions. For example, if N_t is a Poisson process with intensity λ, then $M_t = N_t - \lambda t$ is a martingale. In this case, $[M, M]_t = N_t$ (observable), while $\langle M, M \rangle_t = \lambda t$ (cannot be estimated in finite time). For further discussion of such discontinuous processes, see the references mentioned in Section 2.1.8, and also, in the context of survival analysis, Andersen, Borgan, Gill, and Keiding (1992).

For continuous semimartingales, The symbol $\langle X, Y \rangle_t$ is commonly used in the literature in lieu of $[X, Y]_t$ (including in our papers).

2.2.5 Itô's Formula for Itô processes

Main theorem

Theorem 2.15 *Suppose that f is a twice continuously differentiable function, and that X_t is an Itô process. Then*

$$df(X_t) = f'(X_t)dX_t + \frac{1}{2}f''(X_t)d[X, X]_t. \qquad (2.19)$$

Similarly, in the multivariate case, for $X_t = (X_t^{(1)}, ..., X_t^{(p)})$,

$$df(X_t) = \sum_{i=1}^{p} \frac{\partial f}{\partial x^{(i)}}(X_t)dX_t^{(i)} + \frac{1}{2}\sum_{i,j=1}^{p} \frac{\partial^2 f}{\partial x^{(i)}\partial x^{(j)}}(X_t)d[X^{(i)}, X^{(j)}]_t.$$

(Reference: Theorem I.4.57 in Jacod and Shiryaev (2003).)

We emphasize that (2.19) is the same as saying that

$$f(X_t) = f(X_0) + \int_0^t f'(X_s)dX_s + \frac{1}{2}\int_0^t f''(X_s)d[X,X]_s.$$

If we write out $dX_t = \mu_t dt + \sigma_t dW_t$ and $d[X,X]_t = \sigma_t^2 dt$, then equation (2.19) becomes

$$df(X_t) = f'(X_t)(\mu_t dt + \sigma_t dW_t) + \frac{1}{2}f''(X_t)\sigma_t^2 dt$$

$$= (f'(X_t)\mu_t + \frac{1}{2}f''(X_t)\sigma_t^2)dt + f'(X_t)\sigma_t dW_t.$$

We note, in particular, that if X_t is an Itô process, then so is $f(X_t)$.

Example of Itô's Formula: Stochastic equation for a stock price

We have so far discussed the model for a stock on the log scale, as $dX_t = \mu_t dt + \sigma_t dW_t$. The price is given as $S_t = \exp(X_t)$. Using Itô's formula, with $f(x) = \exp(x)$, we get

$$dS_t = S_t(\mu_t + \frac{1}{2}\sigma_t^2)dt + S_t\sigma_t dW_t. \qquad (2.20)$$

Example of Itô's Formula: Proof of Lévy's Theorem (Section 2.2.4)

Take $f(x) = e^{ihx}$, and go on from there. Left to the reader.

Example of Itô's Formula: Genesis of the leverage effect

We here see a case where quadratic covariation between a process and its volatility can arise from basic economic principles. The following is the origin of the use of the term "leverage effect" to describe such covariation. We emphasize that this kind of covariation can arise from many considerations, and will later use the term leverage effect to describe the phenomenon broadly.

Suppose that the log value of a firm is Z_t, given as a GBM,

$$dZ_t = \nu dt + \gamma dW_t.$$

For simplicity, suppose that the interest rate is zero, and that the firm has borrowed C dollars (or euros, yuan, ...). If there are M shares in the company, the value of one share is therefore

$$S_t = (\exp(Z_t) - C)/M.$$

On the log scale, therefore, by Itô's Formula,

$$dX_t = d\log(S_t)$$
$$= \frac{1}{S_t}dS_t - \frac{1}{2}\frac{1}{S_t^2}d[S,S]_t$$
$$= \frac{M}{\exp(Z_t) - C}dS_t - \frac{1}{2}\left(\frac{M}{\exp(Z_t) - C}\right)^2 d[S,S]_t.$$

Since, in the same way as for (2.20)

$$dS_t = \frac{1}{M}d\exp(Z_t)$$
$$= \frac{1}{M}\exp(Z_t)[(\nu + \frac{1}{2}\gamma^2)dt + \gamma dW_t].$$

Hence, if we set

$$U_t = \frac{\exp(Z_t)}{\exp(Z_t) - C},$$

$$dX_t = U_t[(\nu + \frac{1}{2}\gamma^2)dt + \gamma dW_t] - \frac{1}{2}U_t^2\gamma^2 dt$$
$$= (\nu U_t + \frac{1}{2}\gamma^2(U_t - U_t^2))dt + U_t\gamma dW_t.$$

In other words,

$$dX_t = \mu_t dt + \sigma_t dW_t$$

where

$$\mu_t = \nu U_t + \frac{1}{2}\gamma^2(U_t - U_t^2) \text{ and}$$
$$\sigma_t = U_t\gamma.$$

In this case, the log stock price and the volatility are, indeed, correlated. When the stock price goes down, the volatility goes up (and the volatility will go to infinity if the value of the firm approaches the borrowed amount C, since in this case $U_t \to \infty$. In terms of quadratic variation, the leverage effect is given as

$$d[X, \sigma^2]_t = U_t\gamma^3 d[W, U^2]_t$$
$$= 2U_t^2\gamma^3 d[W, U]_t \text{ since } dU_t^2 = 2U_t dU_t + d[U, U]_t$$
$$= -2U_t^4\gamma^4 C\exp(-Z_t)dt.$$

The last transition follows, since by taking $f(x) = (1 - C\exp(-x))^{-1}$

$$dU_t = df(Z_t)$$
$$= f'(Z_t)dZ_t + dt\text{-terms}$$

so that

$$d[W, U]_t = f'(Z_t)d[W, Z]_t$$
$$= f'(Z_t)\gamma dt$$
$$= -U_t^2 C \exp(-Z_t)\gamma dt,$$

since $f'(x) = -f(x)^2 C \exp(-x)$.

A perhaps more intuitive result is obtained from (2.18), by observing that $\text{sgn}(d[X, \sigma^2]_t/dt) = -1$: on the correlation scale, the leverage effect is

$$\text{cor}(X, \sigma^2)_t = -1.$$

2.2.6 Non-parametric hedging of options

Suppose we can set the following prediction intervals at time $t = 0$:

$$R^+ \geq \int_0^T r_u du \geq R^- \text{ and } \Xi^+ \geq \int_0^T \sigma_u^2 du \geq \Xi^- \qquad (2.21)$$

Is there any sense that we can hedge an option based on this interval?

We shall see that for a European call there is a strategy, beginning with wealth $C(S_0, \Xi^+, R^+)$, which will be solvent for the option payoff so long as the intervals in (2.21) are realized.

First note that by direct differentiation in (2.4), one obtains the two (!!!) Black–Scholes–Merton differential equations

$$\frac{1}{2}C_{SS}S^2 = C_\Xi \text{ and } -C_R = C - C_S S \qquad (2.22)$$

(recall that $C(S, \Xi, R) = S\Phi(d_1) - K \exp(-R)\Phi(d_2)$ and $d_{1,2} = (\log(S/K) + R \pm \Xi/2)/\sqrt{\Xi}$).

In analogy with Section 2.1.6, consider the financial instrument with price at time t:

$$V_t = C(S_t, \Xi_t, R_t),$$

where

$$R_t = R^+ - \int_0^t r_u du \text{ and } \Xi_t = \Xi^+ - \int_0^t \sigma_u^2 du.$$

We shall see that the instrument V_t can be self financed by holding, at each time t,

$C_S(S_t, \Xi_t, R_t)$ units of stock, in other words $S_t C_S(S_t, \Xi_t, R_t)$ \$ of stock, and $V_t - S_t C_S(S_t, \Xi_t, R_t) = -C_R(S_t, \Xi_t, R_t)$ \$ in bonds. $\qquad (2.23)$

where the equality follows from the first equation in (2.22). Note first that, from Itô's formula,

$$dV_t = dC(S_t, \Xi_t, R_t)$$

$$= C_S dS_t + C_R dR_t + C_\Xi d\Xi_t + \frac{1}{2} C_{SS} d[S, S]_t$$

$$= C_S dS_t - C_R r_t dt - C_\Xi \sigma_t^2 dt + \frac{1}{2} C_{SS} S_t^2 \sigma_t^2 dt$$

$$= C_S dS_t - C_R r_t dt \qquad (2.24)$$

because of the second equation in (2.22).

From equation (2.24), we see that holding C_S units of stock, and $-C_R$ \$ of bonds at all times t does indeed produce a P/L $V_t - V_0$, so that starting with V_0 \$ yields V_t \$ at time t.

From the second equation in (2.23), we also see that V_t \$ is exactly the amount needed to maintain these positions in stock and bond. Thus, V_t has a self financing strategy.

Estimated volatility can come into this problem in two ways:

(1) In real time, to set the hedging coefficients: under discrete observation, use

$$\hat{\Xi}_t = \Xi^+ - \text{estimate of integrated volatility from 0 to } t.$$

(2) As an element of a forecasting procedure, to set intervals of the form (2.21).

For further literature on this approach, consult Mykland (2000, 2003a, 2003b, 2005, 2010b). The latter paper discusses, among other things, the use of this method for setting reserve requirements based on an exit strategy in the event of model failure.

For other ways of using realized volatility and similar estimators in options trading, we refer to Zhang (2001), Hayashi and Mykland (2005), Mykland and Zhang (2008), and Zhang (2009).

2.3 Behavior of estimators: Variance

2.3.1 The emblematic problem: Estimation of volatility

In this section, we develop the tools to show convergence in high frequency data. As an example throughout, we consider the problem of estimation of volatility. (In the absence of microstructure.) This classical problem is that of estimating $\int_0^t \sigma_s^2 ds$. The standard estimator, *Realized Volatility (RV)*, is simply $[X, X]_t^{\mathcal{G}}$ in (2.15). The estimator is consistent as $\Delta(\mathcal{G}) \to 0$, from the very definition of quadratic variation.

This raises the question of what other properties one can associate with this estimator. For example, does the asymptotic normality continue to hold? This is a rather complex matter, as we shall see.

There is also the question of what to do in the presence of microstructure, to which we return in Section 2.5.

2.3.2 A temporary martingale assumption

For now consider the case where

$$X_t = X_0 + \int_0^t \sigma_s dW_s, \tag{2.25}$$

i.e., X_t is a local martingale. We shall see in Section 2.4.4 that drift terms can easily be incorporated into the analysis.

We shall also, for now, assume that σ_t is bounded, i.e., there is a nonrandom σ_+ so that

$$\sigma_t^2 \le \sigma_+^2 \text{ for all } t. \tag{2.26}$$

This makes X_t a martingale. We shall see in Section 2.4.5 how to remove this assumption.

2.3.3 The error process

On a grid $\mathcal{G} = \{t_0, t_1, ..., t_n\}$, we get from Itô's formula that

$$(X_{t_{i+1}} - X_{t_i})^2 = 2 \int_{t_i}^{t_{i+1}} (X_s - X_{t_i}) dX_s + \int_{t_i}^{t_{i+1}} \sigma_s^2 ds.$$

If we set

$$t_* = \max\{t_i \in \mathcal{G} : t_i \le t\}, \tag{2.27}$$

the same equation will hold with (t_*, t) replacing (t_i, t_{i+1}). Hence

$$M_t = \sum_{t_{i+1} \le t} (X_{t_{i+1}} - X_{t_i})^2 + (X_t - X_{t_*})^2 - \int_0^t \sigma_s^2 ds$$

is a local martingale of the form

$$M_t = 2 \sum_{t_{i+1} \le t} \int_{t_i}^{t_{i+1}} (X_s - X_{t_i}) dX_s + 2 \int_{t_*}^t (X_s - X_{t_*}) dX_s.$$

On differential form $dM_t = 2(X_t - X_{t_*}) dX_t$. We shall study the behavior of martingales such as M_t.

Of course, we only observe $[X, X]_t^{\mathcal{G}} = \sum_{t_{i+1} \le t} (X_{t_{i+1}} - X_{t_i})^2$, but we shall see next that the same results apply to this quantity. ($(X_t - X_{t_*})^2$ is negligible.)

2.3.4 Stochastic order symbols

We also make use of the following notation:

Definition 2.16 *(stochastic order symbols) Let Z_n be a sequence of random variables. We say that $Z_n = o_p(1)$ if $Z_n \to 0$ in probability, and that $Z_n = o_p(u_n)$ if $Z_n/u_n = o_p(1)$. Similarly, we say that $Z_n = O_p(1)$ – "bounded in probability" – if for all $\epsilon > 0$, there is an M so that $\sup_n P(|Z_n| > M) \leq \epsilon$. There is a theorem to the effect that this is the same as saying that for every subsequence n_k, there is a further subsequence n_{k_l} so that $Z_{n_{k_l}}$ converges in law. (See Theorem 29.3 (p. 380) in Billingsley (1995)). Finally, $Z_n = O_p(u_n)$ if $Z_n/u_n = O_p(1)$.*

For further discusion of this notation, see the Appendix A in Pollard (1984). (This book is out of print, but can at the time of writing be downloaded from http://www.stat.yale.edu/~pollard/).

To see an illustration of the usage: under (2.26), we have that

$$E(X_t - X_{t_*})^2 = E([X,X]_t - [X,X]_{t_*})$$
$$= E \int_{t_*}^t \sigma_s^2 ds$$
$$\leq E(t - t_*)\sigma_+^2$$
$$\leq E\Delta(\mathcal{G})\sigma_+^2$$

so that $(X_t - X_{t_*})^2 = O_p(E\Delta(\mathcal{G}))$, by Chebychev's inequality.

2.3.5 Quadratic variation of the error process: Approximation by quarticity

An important result

To find the variance of our estimate, we start by computing the quadratic variation

$$[M,M]_t = 4 \sum_{t_{i+1} \leq t} \int_{t_i}^{t_{i+1}} (X_s - X_{t_i})^2 d[X,X]_s$$
$$+ 4 \int_{t_*}^t (X_s - X_{t_*})^2 d[X,X]_s. \tag{2.28}$$

It is important here that we mean $[M,M]_t$, and not $[M,M]_t^{\mathcal{G}}$.

A nice result, originally due to Barndorff–Nielsen and Shephard (2002), concerns the estimation of this variation. Define the quarticity by

$$[X,X,X,X]_t^{\mathcal{G}} = \sum_{t_{i+1} \leq t} (X_{t_{i+1}} - X_{t_i})^4 + (X_t - X_{t_*})^4.$$

Use Itô's formula to see that (where M_t is the error process from Section 2.3.3)

$$d(X_t - X_{t_i})^4 = 4(X_t - X_{t_i})^3 dX_t + 6(X_t - X_{t_i})^2 d[X, X]_t$$
$$= 4(X_t - X_{t_i})^3 dX_t + \frac{6}{4} d[M, M]_t,$$

since $d[M, M]_t = 4(X_t - X_{t_i})^2 d[X, X]_t$. It follows that if we set

$$M_t^{(2)} = \sum_{t_{i+1} \leq t} \int_{t_i}^{t_{i+1}} (X_s - X_{t_i})^3 dX_s + \int_{t_*}^{t} (X_s - X_{t_*})^3 dX_s$$

we obtain

$$[X, X, X, X]_t^{\mathcal{G}} = \frac{3}{2}[M, M]_t + 4M_t^{(2)}.$$

It turns out that the $M_t^{(2)}$ term is of order $o_p(n^{-1})$, so that $(2/3)n[X, X, X, X]_t^{\mathcal{G}}$ is a consistent estimate of the quadratic variation (2.28):

Proposition 2.17 *Assume (2.26). Consider a sequence of grids $\mathcal{G}_n = \{0 = t_{n,0} < ... < t_{n,n} = T\}$. Suppose that, as $n \to \infty$, $\Delta(\mathcal{G}_n) = o_p(1)$, and*

$$\sum_{i=0}^{n-1} (t_{n,i+1} - t_{n,i})^3 = O_p(n^{-2}). \tag{2.29}$$

Then

$$\sup_{0 \leq t \leq T} | [M, M]_t - \frac{2}{3}[X, X, X, X]_t^{\mathcal{G}_n} | = o_p(n^{-1}) \text{ as } n \to \infty.$$

Note that in the following, we typically suppress the double subscript on the times:

$$t_i \text{ means } t_{n,i}.$$

The conditions on the times – why they are reasonable

Example 2.18 *We first provide a simple example to emphasize that Proposition 2.17 does the right thing. Assume for simplicity that the observation times are equidistant: $t_i = t_{n,i} = iT/n$, and that the volatility is constant: $\sigma_t \equiv \sigma$. It is then easy to see that the conditions, including (2.29), are satisfied. On the other hand, $[X, X, X, X]_t^{\mathcal{G}}$ has the distribution of $(T/n)^2 \sigma^4 \sum_{i=1}^{n} U_i^4$, where the U_i are iid standard normal. Hence, $n\frac{2}{3}[X, X, X, X]_t^{\mathcal{G}} \xrightarrow{P} \frac{2}{3}T^2\sigma^4 \times E(N(0, 1)^4) = 2T^2\sigma^4$. It then follows from Proposition 2.17 that $n[M, M]_t^{\mathcal{G}} \xrightarrow{P} 2T^2\sigma^4$.*

Example 2.19 *To see more generally why (2.29) is a natural condition, consider a couple of cases for the spacings.*

(i) The spacings are sufficiently regular to satisfy

$$\Delta(\mathcal{G}) = \max_i(t_{i+1} - t_i) = O_p(n^{-1}).$$

Then

$$\sum_{i=0}^{n}(t_{i+1} - t_i)^3 \le \sum_{i=0}^{n}(t_{i+1} - t_i)\left(\max_i(t_{i+1} - t_i)\right)^2$$
$$= T \times O_p(n^{-2}).$$

(ii) On the other hand, suppose that the sampling times follow a Poisson process with parameter λ (still with $t_0 = 0$). Denote by N the number of sampling points in the interval $[0, T]$, i.e., $N = \inf\{i : t_i > T\}$. If one conditions on N, say, $N = n$, the conditional distribution of the points $t_i, i = 1, ..., n - 1$, behave like the order statistics of $n - 1$ uniformly distributed random variables (see, for example, Chapter 2.3 in Ross (1996)). In other words, $t_i = TU_{(i)}$ (for $0 < i < n$), where $U_{(i)}$ is the i'th order statistic of $U_1, ..., U_{n-1}$, which are iid $U[0,1]$. Without any asymptotic impact, now also impose $t_n = T$ (to formally match the rest of our theory).

Now define $U_{(0)} = 0$ and $U_{(n)} = 1$. With these definitions, note that for $i = 1, ..., n$, $U_{(i)} - U_{(i-1)}$ are identically distributed with the same distribution as $U_{(1)}$, which has density $(n - 1)(1 - x)^{n-2}$. (See, for example, Exercise 3.67 (p. 110) in Rice (2006).) The expression in (2.29) becomes

$$\sum_{i=0}^{n-1}(t_{i+1} - t_i)^3 = T^3 \sum_{i=1}^{n}(U_{(i)} - U_{(i-1)})^3$$
$$= T^3 n E U_{(1)}^3 (1 + o_p(1))$$

by the law of large numbers. Since $EU_{(1)}^3 = \frac{6}{(n+1)n(n-1)} = O(n^{-3})$, (2.29) follows.

Application to refresh times

We here briefly consider the case of multidimensional processes of the form $(X_t^{(1)}, ..., X_t^{(p)})$. It will often be the case that the observation occurs at asynchronous times. In other words, process $(X_t^{(r)})$ is observed at times $\mathcal{G}_n^{(r)} = \{0 \le t_{n,0}^{(r)} < t_{n,1}^{(r)} < ... < t_{n,n_r}^{(r)} \le T\}$, and the grids $\mathcal{G}_n^{(r)}$ are not the same. Note that in this case, there is latitude in what meaning to assign to the symbol n. It is an index that goes to infinity with each n_r, for example $n = n_1 + ... + n_p$. One would normally require that n_r/n is bounded away from zero.

A popular way of dealing with this problem is to use *refresh times*, as follows.

Set $u_{n,0} = 0$, and then define recursively for $i > 0$

$$u_{n,i} = \max_{r=1,\ldots,p} \min\{t \in \mathcal{G}_n^{(r)} : t > u_{n,i-1}\}.$$

The $u_{n,i}$ is called the i'th refresh time, and is the time when all the p processes have undergone an update of observation. Successful uses of refresh times can be found in Barndorff–Nielsen, Hansen, Lunde, and Shephard (2011) and Zhang (2011).

Now note that if the conditions (on times) in Proposition 2.17 are satisfied for each grid $\mathcal{G}_n^{(r)}$, the conditions are also satisfied for the grid of refresh times. This is because each $\Delta u_{n,i+1}$ must be matched or dominated by a spacing in one of each grid $\mathcal{G}_n^{(r)}$. Specifically, for each i, define $j_{r,i} = \max\{j : t_{n,j}^{(r)} \leq u_{n,i}\}$ and note that $j_{r,i} + 1 = \min\{j : t_{n,j}^{(r)} > u_{n,i}\}$. Hence, there is an r_i so that

$$u_{n,i+1} = \max_r t_{n,j_{r,i}+1}^{(r)} = t_{n,j_{r_i,i}+1}^{(r_i)}$$

and so

$$u_{n,i+1} - u_{n,i} \leq t_{n,j_{r_i,i}+1}^{(r_i)} - t_{n,j_{r_i,i}}^{(r_i)} \leq \max_r \left(t_{n,j_{r,i}+1}^{(r)} - t_{n,j_{r,i}}^{(r)}\right).$$

In particular, for (2.29),

$$\sum_i (u_{n,i+1} - u_{n,i})^3 \leq \sum_i \max_r \left(t_{n,j_{r,i}+1}^{(r)} - t_{n,j_{r,i}}^{(r)}\right)^3$$

$$\leq \sum_i \sum_r \left(t_{n,j_{r,i}+1}^{(r)} - t_{n,j_{r,i}}^{(r)}\right)^3$$

$$\leq \sum_{r=1}^p \sum_i \left(t_{n,i+1}^{(r)} - t_{n,i}^{(r)}\right)^3,$$

and similarly for the condition $\Delta(\mathcal{G}_n) = o_p(1)$.

The theory in this article is therefore amenable to developments involving refresh times. This issue is not further pursued here, though we return to asynchronous times in Section 2.6.3.

2.3.6 Moment inequalities, and proof of Proposition 2.17

L^p Norms, moment inequalities, and the Burkholder–Davis–Gundy inequality

For $1 \leq p < \infty$, define the L^p-norm:

$$||X||_p = (E|X|^p)^{\frac{1}{p}},$$

The Minkowski and Hölder inequalities say that

$$||X + Y||_p \leq ||X||_p + ||Y||_p$$

$$||XY||_1 \leq ||X||_p ||Y||_q \text{ for } \frac{1}{p} + \frac{1}{q} = 1.$$

Example 2.20 *A special case of the Hölder inequalitiy is* $||X||_1 \leq ||X||_p$ *(take* $Y = 1$*). In particular, under (2.29), for for* $1 \leq v \leq 3$:

$$\left(\frac{1}{n}\sum_{i=0}^{n-1}(t_{i+1} - t_i)^v\right)^{\frac{1}{v}} \leq \left(\frac{1}{n}\sum_{i=0}^{n-1}(t_{i+1} - t_i)^3\right)^{\frac{1}{3}}$$

$$= \left(\frac{1}{n} \times O_p(n^{-2})\right)^{\frac{1}{3}} = \left(O_p(n^{-3})\right)^{\frac{1}{3}} = O_p(n^{-1}),$$

so that

$$\sum_{i=0}^{n}(t_{i+1} - t_i)^v = O_p(n^{1-v}). \tag{2.30}$$

To show Proposition 2.17, we need the Burkholder–Davis–Gundy inequality (see Section 3 of Ch. VII of Dellacherie and Meyer (1982), or p. 193 and 222 in Protter (2004)), as follows. For $1 \leq p < \infty$, there are universal constants c_p and C_p so that for all continuous martingales N_t,

$$c_p||[N, N]_T||_{p/2}^{1/2} \leq || \sup_{0 \leq t \leq T} |N_t| \,||_p \leq C_p||[N, N]_T||_{p/2}^{1/2}.$$

Note, in particular, that for $1 < p < \infty$,

$$C_p^2 = q^p \left(\frac{p(p-1)}{2}\right)$$

where q is given by $p^{-1} + q^{-1} = 1$.

Proof of Proposition 2.17

From applying Itô's Formula to $(X_t - X_{t_i})^8$:

$$[M^{(2)}, M^{(2)}]_t = \sum_{t_{i+1} \leq t} \int_{t_i}^{t_{i+1}} (X_s - X_{t_i})^6 d[X, X]_s$$

$$+ \int_{t_*}^{t} (X_s - X_{t_*})^6 d[X, X]_s$$

$$= \frac{1}{28}[X; 8]_t^{\mathcal{G}} + \text{martingale term}$$

where $[X; 8]_t^{\mathcal{G}} = \sum_{t_{i+1} \leq t}(X_{t_{i+1}} - X_{t_i})^8 + (X_t - X_{t_*})^8$ is the *ochticity*.

Note that for stopping time $\tau \le T$, $[X; 8]_\tau^{\mathcal{G}} = \sum_i (X_{t_{i+1} \wedge \tau} - X_{t_i \wedge \tau})^8$. Hence, by the Burkholder–Davis–Gundy inequality (with $p = 8$)

$$
\begin{aligned}
E[M^{(2)}, M^{(2)}]_\tau &= \frac{1}{28} E[X; 8]_\tau^{\mathcal{G}} \\
&\le \frac{1}{28} C_8^8 E \sum_i ([X, X]_{t_{i+1} \wedge \tau} - [X, X]_{t_i \wedge \tau})^4 \\
&\le \frac{1}{28} C_8^8 \sigma_+^8 E \sum_i (t_{i+1} \wedge \tau - t_i \wedge \tau)^4.
\end{aligned}
$$

Let $\epsilon > 0$, and set

$$
\tau_n = \inf\{\, t \in [0, T] : n^2 \sum_i (t_{i+1} \wedge t - t_i \wedge t)^4 > \epsilon \,\}.
$$

Then

$$
E[M^{(2)}, M^{(2)}]_{\tau_n} \le n^{-2} \frac{1}{28} C_8^8 \sigma_+^8 \epsilon. \tag{2.31}
$$

By assumption, $n^2 \sum_i (t_{i+1} \wedge t - t_i \wedge t)^4 \le \Delta(\mathcal{G}) n^2 \sum_i (t_{i+1} - t_i)^3 \overset{P}{\to} 0$, and hence

$$
P(\tau_n \ne T) \to 0 \text{ as } n \to \infty. \tag{2.32}
$$

Hence, for any $\delta > 0$,

$$
P(n \sup_{0 \le t \le T} |M_t^{(2)}| > \delta) \tag{2.33}
$$

$$
\begin{aligned}
&\le \quad P(n \sup_{0 \le t \le \tau_n} |M_t^{(2)}| > \delta) + P(\tau_n \ne T) \\
&\le \quad \frac{1}{\delta^2} E\left(n \sup_{0 \le t \le \tau_n} |M_t^{(2)}|\right)^2 + P(\tau_n \ne T) \text{ (Chebychev)} \\
&\le \quad \frac{1}{\delta^2} C_2^2 n^2 E[M^{(2)}, M^{(2)}]_{\tau_n} + P(\tau_n \ne T) \text{ (Burkholder–Davis–Gundy)} \\
&\le \quad \frac{1}{\delta^2} C_2^2 \frac{1}{28} C_8^8 \sigma_+^8 \epsilon + P(\tau_n \ne T) \text{ (from (2.31))} \\
&\to \quad \frac{1}{\delta^2} C_2^2 \frac{1}{28} C_8^8 \sigma_+^8 \epsilon \text{ as } n \to \infty \text{ (from (2.32))}.
\end{aligned}
$$

Hence Proposition 2.17 has been shown.

2.3.7 Quadratic variation of the error process: When observation times are independent of the process

Main approximation

We here assume that the observation times are independent of the process X. The basic insight for the following computation is that over small intervals,

$(X_t - X_{t_*})^2 \approx [X, X]_t - [X, X]_{t_*}$. To the extent that this approximation is valid, it follows from (2.28) that

$$
\begin{aligned}
[M, M]_t &\approx 4 \sum_{t_{i+1} \leq t} \int_{t_i}^{t_{i+1}} ([X, X]_s - \lfloor X, X \rfloor_{t_i}) d\lfloor X, X \rfloor_s \\
&\quad + 4 \int_{t_*}^{t} ([X, X]_s - [X, X]_{t_*}) d[X, X]_s \\
&= 2 \sum_{t_{i+1} \leq t} ([X, X]_{t_{i+1}} - [X, X]_{t_i})^2 + 2([X, X]_t - [X, X]_{t_*})^2.
\end{aligned}
$$

We shall use this device several times in the following, and will this first time do it rigorously.

Proposition 2.21 *Assume (2.26), and that σ_t^2 is continuous in mean square:*

$$
\sup_{0 \leq t - s \leq \delta} E(\sigma_t^2 - \sigma_s^2)^2 \to 0 \text{ as } \delta \to \infty.
$$

Also suppose that the grids $\mathcal{G}_n$ are nonrandom, or independent of the process X_t. Also suppose that, as $n \to \infty$, $\Delta(\mathcal{G}) = o_p(n^{-1/2})$, and assume (2.29). Then

$$
\begin{aligned}
[M, M]_t &= 2 \sum_{t_{i+1} \leq t} ([X, X]_{t_{i+1}} - [X, X]_{t_i})^2 + 2([X, X]_t - [X, X]_{t_*})^2 \\
&\quad + o_p(n^{-1}).
\end{aligned}
\tag{2.34}
$$

If σ_t is continuous, it is continuous in mean square (because of (2.26)). More generally, σ_t can, for example, also have Poisson jumps.

In the rest of this section, we shall write all expectations implicitly as conditional on the times.

To show Proposition 2.21, we need some notation and a lemma, as follows:

Lemma 2.22 *Let $t_* = \max\{t_i \in \mathcal{G} : t_i \leq t\}$ (as in (2.27)). Let N_t be an Itô process martingale, for which (for $a, b > 0$), for all t,*

$$
\frac{d}{dt} E[N, N]_t \leq a(t - t_*)^b.
$$

Let H_t be a predictable process, satisfying $|H_t| \leq H_+$ for some constant H_+. Set

$$
R_v(\mathcal{G}) = \left(\sum_{i=0}^{n-1} (t_{i+1} - t_i)^v \right).
$$

Then

$$\|\sum_{t_{i+1}\leq t}\int_{t_i}^{t_{i+1}}(N_s-N_{t_i})H_s ds+\int_{t_*}^{t}(N_s-N_{t_*})H_s ds\|_1$$

$$\leq\left(H_+^2\frac{a}{b+3}R_{b+3}(\mathcal{G})\right)^{1/2}$$

$$+R_{(b+3)/2}(\mathcal{G})\frac{2}{b+3}\left(\frac{a}{b+1}\right)^{1/2}\sup_{0\leq t-s\leq\Delta(\mathcal{G})}\|H_s-H_t\|_2.$$

$$(2.35)$$

Proof of Proposition 2.21. Set $N_t=M_t$ and $H_t=\sigma_t^2$. Then

$$d[M,M]_t=4(X_t-X_{t_i})^2 d[X,X]_t$$
$$=4([X,X]_t-[X,X]_{t_i})d[X,X]_t+4((X_t-X_{t_i})^2$$
$$-([X,X]_t-[X,X]_{t_i}))d[X,X]_t$$
$$=4([X,X]_t-[X,X]_{t_i})d[X,X]_t+2(N_t-N_{t_i})\sigma_t^2 dt.$$

Thus, the approximation error in (2.34) is exactly of the form of the left-hand side in (2.35). We note that

$$Ed[N,N]_t=4E(X_t-X_{t_i})^2 d[X,X]_t$$
$$=4E(X_t-X_{t_i})^2\sigma_+^2 dt$$
$$=4(t-t_i)\sigma_+^4 dt$$

hence the conditions of Lemma 2.22 are satisfied with $a=4\sigma_+^4$ and $b=1$. The result follows from (2.30). □

Proof of Lemma 2.22 (Technical material, can be omitted).

Decompose the original problem as follows:

$$\int_{t_i}^{t_{i+1}}(N_s-N_{t_i})H_s ds$$

$$=\int_{t_i}^{t_{i+1}}(N_s-N_{t_i})H_{t_i}ds+\int_{t_i}^{t_{i+1}}(N_s-N_{t_i})(H_s-H_{t_i})ds.$$

For the first term, from Itô's formula, $d(t_{i+1}-s)(N_s-N_{t_i})=-(N_s-N_{t_i})ds+(t_{i+1}-s)dN_s$, so that

$$\int_{t_i}^{t_{i+1}}(N_s-N_{t_i})H_{t_i}ds=H_{t_i}\int_{t_i}^{t_{i+1}}(t_{i+1}-s)dN_s$$

hence

$$\sum_{t_{i+1} \le t} \int_{t_i}^{t_{i+1}} (N_s - N_{t_i}) H_s ds$$

$$= \sum_{t_{i+1} \le t} H_{t_i} \int_{t_i}^{t_{i+1}} (t_{i+1} - t) dN_s$$

$$+ \sum_{t_{i+1} \le t} \int_{t_i}^{t_{i+1}} (N_s - N_{t_i})(H_s - H_{t_i}) ds. \tag{2.36}$$

The first term is the end point of a martingale. For each increment,

$$E \left(\int_{t_i}^{t_{i+1}} (N_s - N_{t_i}) H_{t_i} ds \right)^2 = E \left(H_{t_i} \int_{t_i}^{t_{i+1}} (t_{i+1} - s) dN_s \right)^2$$

$$\le H_+^2 E \left(\int_{t_i}^{t_{i+1}} (t_{i+1} - s) dN_s \right)^2$$

$$= H_+^2 E \left(\int_{t_i}^{t_{i+1}} (t_{i+1} - s)^2 d[N, N]_s \right)$$

$$= H_+^2 \int_{t_i}^{t_{i+1}} (t_{i+1} - s)^2 dE[N, N]_s$$

$$= H_+^2 \int_{t_i}^{t_{i+1}} (t_{i+1} - s)^2 \frac{d}{ds} E[N, N]_s ds$$

$$= H_+^2 \int_{t_i}^{t_{i+1}} (t_{i+1} - s)^2 a(s - t_i)^b ds$$

$$= H_+^2 \frac{a}{b+3} (t_{i+1} - t_i)^{b+3}$$

and so, by the uncorrelatedness of martingale increments,

$$E \left(\sum_{t_{i+1} \le t} H_{t_i} \int_{t_i}^{t_{i+1}} (t_{i+1} - t) dN_s \right)^2 \le H_+^2 \frac{a}{b+3} \left(\sum_{t_{i+1} \le t} (t_{i+1} - t_i)^3 \right)$$

$$\le H_+^2 \frac{a}{b+3} R_{b+3}(\mathcal{G}). \tag{2.37}$$

On the other hand, for the second term in (2.36),

$$
\begin{aligned}
\|(N_s - N_{t_i})(H_s - H_{t_i})\|_1 & \\
&\leq \|N_s - N_{t_i}\|_2 \|H_s - H_{t_i}\|_2 \\
&\leq \left(E(N_s - N_{t_i})^2\right)^{1/2} \|H_s - H_{t_i}\|_2 \\
&= \left(E([N,N]_s - [N,N]_{t_i})\right)^{1/2} \|H_s - H_{t_i}\|_q \\
&= \left(\int_{t_i}^{s} \frac{d}{du} E[N,N]_u du\right)^{1/2} \|H_s - H_{t_i}\|_2 \\
&\leq \left(\int_{t_i}^{s} a(u - t_i)^b du\right)^{1/2} \|H_s - H_{t_i}\|_2 \\
&= \left(\frac{a}{b+1}(s - t_i)^{b+1}\right)^{1/2} \|H_s - H_{t_i}\|_2 \\
&= (s - t_i)^{(b+1)/2} \left(\frac{a}{b+1}(s - t_i)^{b+1}\right)^{1/2} \|H_s - H_{t_i}\|_2,
\end{aligned}
$$

and from this

$$
\begin{aligned}
\left\| \int_{t_i}^{t_{i+1}} (N_s - N_{t_i})(H_s - H_{t_i}) ds \right\|_1 & \\
&\leq \int_{t_i}^{t_{i+1}} \|(N_s - N_{t_i})(H_s - H_{t_i})\|_1 ds \\
&\leq \int_{t_i}^{t_{i+1}} (s - t_i)^{(b+1)/2} ds \left(\frac{a}{b+1}\right)^{1/2} \sup_{t_i \leq s \leq t_{i+1}} \|H_s - H_{t_i}\|_2 \\
&= (t_{i+1} - t_i)^{(b+3)/2} \frac{2}{b+3} \left(\frac{a}{b+1}\right)^{1/2} \sup_{t_i \leq s \leq t_{i+1}} \|H_s - H_{t_i}\|_2.
\end{aligned}
$$

Hence, finally, for the second term in (2.36),

$$
\begin{aligned}
\left\| \sum_{t_{i+1} \leq t} \int_{t_i}^{t_{i+1}} (N_s - N_{t_i})(H_s - H_{t_i}) dt \right\|_1 & \\
&\leq \left(\sum_{t \leq t_{i+1}} (t_{i+1} - t_i)^{(b+3)/2}\right) \frac{2}{b+3} \left(\frac{a}{b+1}\right)^{1/2} \sup_{0 \leq t - s \leq \Delta(\mathcal{G})} \|H_s - H_t\|_2 \\
&= R_{(b+3)/2}(\mathcal{G}) \frac{2}{b+3} \left(\frac{a}{b+1}\right)^{1/2} \sup_{0 \leq t - s \leq \Delta(\mathcal{G})} \|H_s - H_t\|_2. \qquad (2.38)
\end{aligned}
$$

Hence, for the overall sum (2.36), from (2.37) and (2.38) and

$$\| \sum_{t_{i+1}\leq t} \int_{t_i}^{t_{i+1}} (N_s - N_{t_i}) H_s ds \|_1$$

$$\leq \| \sum_{t_{i+1}\leq t} H_{t_i} \int_{t_i}^{t_{i+1}} (t_{i+1} - t) dN_s \|_1$$

$$+ \| \sum_{t_{i+1}\leq t} \int_{t_i}^{t_{i+1}} (N_s - N_{t_i})(H_s - H_{t_i}) ds \|_1$$

$$\leq \| \sum_{t_{i+1}\leq t} H_{t_i} \int_{t_i}^{t_{i+1}} (t_{i+1} - t) dN_s \|_2$$

$$+ \| \sum_{t_{i+1}\leq t} \int_{t_i}^{t_{i+1}} (N_s - N_{t_i})(H_s - H_{t_i}) ds \|_1$$

$$\leq \left(H_+^2 \frac{a}{b+3} R_{b+3}(\mathcal{G}) \right)^{1/2}$$

$$+ R_{(b+3)/2}(\mathcal{G}) \frac{2}{b+3} \left(\frac{a}{b+1} \right)^{1/2} \sup_{0\leq t-s\leq\Delta(\mathcal{G})} \|H_s - H_t\|_2.$$

The part from t_* to t can be included similarly, showing the result. $\square$

Quadratic variation of the error process, and quadratic variation of time

To give the final form to this quadratic variation, define the "Asymptotic Quadratic Variation of Time" (AQVT), given by

$$H_t = \lim_{n\to\infty} \frac{n}{T} \sum_{t_{n,j+1}\leq t} (t_{n,j+1} - t_{n,j})^2, \qquad (2.39)$$

provided that the limit exists. From Example 2.19, we know that dividing by n is the right order. We now get

Proposition 2.23 *Assume the conditions of Proposition 2.21, and that the AQVT exists. Then*

$$n[M, M]_t \xrightarrow{p} 2T \int_0^t \sigma_s^4 dH_s.$$

The proof is a straight exercise in analysis. The heuristic for the result is as follows. From (2.34),

$$[M,M]_t = 2 \sum_{t_{i+1}\le t} ([X,X]_{t_{i+1}} - [X,X]_{t_i})^2 + 2([X,X]_t - [X,X]_{t_*})^2$$

$$+ o_p(n^{-1})$$

$$= 2 \sum_{t_{i+1}\le t} (\int_{t_i}^{t_{i+1}} \sigma_s^2 ds)^2 + 2(\int_{t_*}^t \sigma_s^2 ds)^2 + o_p(n^{-1})$$

$$= 2 \sum_{t_{i+1}\le t} ((t_{i+1}-t_i)\sigma_{t_i}^2)^2 + 2((t-t_*)\sigma_{t_*}^2)^2 + o_p(n^{-1})$$

$$= 2\frac{T}{n} \int_0^t \sigma_s^4 dH_s + o_p(n^{-1}).$$

Example 2.24 *We here give a couple of examples of the AQVT:*
(i) When the times are equidistant: $t_{i+1} - t_i = T/n$, *then*

$$H_t \approx \frac{n}{T} \sum_{t_{n,j+1}\le t} \left(\frac{T}{n}\right)^2$$

$$= \frac{T}{n} \#\{t_{i+1} \le t\}$$

$$= T \times \text{ fraction of } t_{i+1} \text{ in } [0,t]$$

$$\approx T \times \frac{t}{T} = t.$$

(ii) When the times follow a Poisson process with parameter λ, *we proceed as in case (ii) in Example 2.19. We condition on the number of sampling points* n, *and get* $t_i = TU_{(i)}$ *(for* $0 < i < n$), *where* $U_{(i)}$ *is the i'th order statistic of* $U_1, ..., U_n$, *which are iid U[0,1]. Hence (again taking* $U_{(0)} = 0$ *and* $U_{(n)} = 1$)

$$H_t \approx \frac{n}{T} \sum_{t_{n,j+1}\le t} (t_{i+1}-t_i)^2$$

$$= T^2 \frac{n}{T} \sum_{t_{n,j+1}\le t} (U_{(i)} - U_{(i-1)})^2$$

$$= T^2 \frac{n}{T} \sum_{t_{n,j+1}\le t} EU_{(1)}^2 (1 + o_p(1))$$

$$= T^2 \frac{n}{T} \#\{t_{i+1} \le t\} EU_{(1)}^2 (1 + o_p(1))$$

$$= Tn^2 \frac{t}{T} EU_{(1)}^2 (1 + o_p(1))$$

$$= 2t(1 + o_p(1))$$

by the law of large numbers, since the spacings have identical distribution, and since $EU_{(1)}^2 = 2/n(n+1)$. Hence $H_t = 2t$.

The quadratic variation of time in the general case

We now go back to considering the times as possibly dependent with the process X. Note that by using the Burkholder–Davis–Gundy Inequality conditionally, we obtain that

$$c_4^4 E(([X,X]_{t_{i+1}} - [X,X]_{t_i})^2 \mid \mathcal{F}_{t_i})$$
$$\leq \quad E((X_{t_{i+1}} - X_{t_i})^4 \mid \mathcal{F}_{t_i}) \quad \leq \quad C_4^4 E(([X,X]_{t_{i+1}} - [X,X]_{t_i})^2 \mid \mathcal{F}_{t_i}),$$

where c_4 and C_4 are as in Section 2.3.6. In the typical law of large numbers setting, $[X,X,X,X]_t - \sum_i E((X_{t_{i+1}} - X_{t_i})^4 \mid \mathcal{F}_{t_i})$ is a martingale which is of lower order than $[X,X,X,X]_t$ itself, and the same goes for

$$\sum_i \left[([X,X]_{t_{i+1}} - [X,X]_{t_i})^2 - E(([X,X]_{t_{i+1}} - [X,X]_{t_i})^2 \mid \mathcal{F}_{t_i}) \right].$$

By the argument in Proposition 2.23, therefore, it follows that under suitable regularity conditions, if $n[X,X,X,X]_t \xrightarrow{p} U_t$ as $n \to \infty$, and if the AQVT H_t is absolutely continuous in t, then U_t is also absolutely continuous, and

$$c_4^4 2T \sigma_t^4 H_t' \leq U_t' \leq C_4^4 2T \sigma_t^4 H_t'.$$

This is of some theoretic interest in that it establishes the magnitude of the limit of $n[X,X,X,X]_t$. However, it should be noted that $C_4^4 = 2^{18}/3^6 \approx 359.6$, so the bounds are of little practical interest.

A slightly closer analysis of this particular case uses the Bartlett type identities for martingales to write

$$E((X_{t_{i+1}} - X_{t_i})^4 \mid \mathcal{F}_{t_i})$$
$$= \quad -3E(([X,X]_{t_{i+1}} - [X,X]_{t_i})^2 \mid \mathcal{F}_{t_i})$$
$$\quad + 6E((X_{t_{i+1}} - X_{t_i})^2([X,X]_{t_{i+1}} - [X,X]_{t_i}) \mid \mathcal{F}_{t_i})$$
$$\leq \quad -3E(([X,X]_{t_{i+1}} - [X,X]_{t_i})^2 \mid \mathcal{F}_{t_i})$$
$$\quad + 6E((X_{t_{i+1}} - X_{t_i})^4 \mid \mathcal{F}_{t_i})^{1/2} E(([X,X]_{t_{i+1}} - [X,X]_{t_i})^2 \mid \mathcal{F}_{t_i})^{1/2}.$$

Solving this quadratic inequality yields that we can take $c_4^4 = (3 - \sqrt{6})^2 \approx 0.3$ and $C_4^4 = (3 + \sqrt{6})^2 \approx 29.7$.

2.3.8 Quadratic variation, variance, and asymptotic normality

We shall later see that $n^{1/2}([X,X]_t^{\mathcal{G}} - [X,X]_t)$ is approximately normal. In the simplest case, where the times are independent of the process, the normal distribution has mean zero and variance $n[M,M]_t \approx 2T \int_0^t \sigma_s^4 dH_s$. From standard central limit considerations, this is unsurprising when the σ_t process is nonrandom, or more generally independent of the W_t process. (In the latter case, one simply conditions on the σ_t process).

What is surprising, and requires more concepts, is that the normality result also holds when σ_t process has dependence with the W_t process. For this we shall need new concepts, to be introduced in Section 2.4.

2.4 Asymptotic normality

2.4.1 Stable convergence

In order to define convergence in law, we need to deal with the following issue. Suppose $\hat{\theta}_n$ is an estimator of θ, say, $\hat{\theta}_n = [X, X]_T^{\mathcal{G}_n}$ and $\theta = [X, X]_T = \int_0^T \sigma_t^2 dt$. As suggested in Section 2.3.7, the variance of $Z_n = n^{1/2}(\hat{\theta}_n - \theta)$ converges to $2T \int_0^t \sigma_s^4 dH_s$. We shall now go on to show the following convergence in law:

$$n^{1/2}(\hat{\theta}_n - \theta) \xrightarrow{\mathcal{L}} U \times \left(2T \int_0^T \sigma_s^4 dH_s \right)^{1/2},$$

where U is a standard normal random variable, independent of the σ_t^2 process. In order to show this, we need to be able to bring along prelimiting information into the limit: U only exists in the limit, while as argued in Section 2.3.5, the asymptotic variance $2T \int_0^T \sigma_s^4 dH_s$ can be estimated consistently, and so is a limit in probability of a prelimiting quantity.

To operationalize the concept in our setting, we need the filtration $(\mathcal{F}_t)$ to which all relevant processes (X_t, σ_t, etc) are adapted. We shall assume Z_n (the quantity that is converging in law) to be measurable with respect to a σ-field χ, $\mathcal{F}_T \subseteq \chi$. The reason for this is that it is often convenient to exclude microstructure noise from the filtration $\mathcal{F}_t$. Hence, for example, the TSRV (in Section 2.5 below) is not $\mathcal{F}_T$-measurable.

Definition 2.25 *Let Z_n be a sequence of χ-measurable random variables, $\mathcal{F}_T \subseteq \chi$. We say that Z_n converges $\mathcal{F}_T$-stably in law to Z as $n \to \infty$ if Z is measurable with respect to an extension of χ so that for all $A \in \mathcal{F}_T$ and for all bounded continuous g, $EI_A g(Z_n) \to EI_A g(Z)$ as $n \to \infty$.*

The definition means, up to regularity conditions, that Z_n converges jointly in law with all $\mathcal{F}_T$ measurable random variables. This intuition will be important in the following. For further discussion of stable convergence, see Rényi (1963), Aldous and Eagleson (1978), Chapter 3 (p. 56) of Hall and Heyde (1980), Rootzén (1980), and Section 2 (pp. 169–170) of Jacod and Protter (1998). We now move to the main result.

2.4.2 Asymptotic normality

We shall be concerned with a sequence of martingales M_t^n, $0 \leq t \leq T$, $n = 1, 2, ...$, and how it converges to a limit M_t. We consider here only continuous martingales, which are thought of as random variables taking values in the set $\mathbb{C}$ of continuous functions $[0, T] \to \mathbb{R}$.

To define weak, and stable, convergence, we need a concept of continuity. We say that g is a continuous function $\mathbb{C} \to \mathbb{R}$ if:

$$\sup_{0 \leq t \leq T} |x_n(t) - x(t)| \to 0 \text{ implies } g(x_n) \to g(x).$$

We note that if $(M_t^n) \xrightarrow{\mathcal{L}} (M_t)$ in this process sense, then, for example, $M_T^n \xrightarrow{\mathcal{L}} M_T$ as a random variable. This is because the function $x \to g(x) = x(T)$ is continuous. The reason for going via process convergence is (1) sometimes this is really the result one needs, and (2) since our theory is about continuous processes converging to a continuous process, one does not need asymptotic negligibility conditions à la Lindeberg (these kinds of conditions are in place in the usual CLT precisely to avoid jumps in the asymptotic process). For a related development based on discrete time predictable quadratic variations, and Lindeberg conditions, see Theorem IX.7.28 (p. 590-591) of Jacod and Shiryaev (2003).

In order to show results about continuous martingales, we shall use the following assumption

Condition 2.26 *There are Brownian motions $W_t^{(1)}, ..., W_t^{(p)}$ (for some p) that generate $(\mathcal{F}_t)$.*

It is also possible to proceed with assumptions under which there are jumps in some processes, but for simplicity, we omit any discussion of that here.

Under Condition 2.26, it follows from Lemma 2.1 (p. 270) in Jacod and Protter (1998) that stable convergence in law of a local martingale M^n to a process M is equivalent to (straight) convergence in law of the process $(W^{(1)}, ..., W^{(p)}, M^n)$ to the process $(W^{(1)}, ..., W^{(p)}, M)$. This result does not extend to all processes and spaces, cf. the discussion in the cited paper.

Another main fact about stable convergence is that limits and quadratic variation can be interchanged:

Proposition 2.27 *(Interchangeability of limits and quadratic variation). Assume that M^n is a sequence of continuous local martingales which converges stably to a process M. Then $(M^n, [M^n, M^n])$ converges stably to $(M, [M, M])$.*

For proof, we refer to Corollary VI.6.30 (p. 385) in Jacod and Shiryaev (2003), which also covers the case of bounded jumps. More generally, consult ibid., Chapter VI.6.

We now state the main central limit theorem (CLT).

Theorem 2.28 *Assume Condition 2.26. Let (M_t^n) be a sequence of continuous local martingales on $[0, T]$, each adapted to $(\mathcal{F}_t)$, with $M_0^n = 0$. Suppose that there is an $(\mathcal{F}_t)$ adapted process f_t so that*

$$[M^n, M^n]_t \xrightarrow{p} \int_0^t f_s^2 ds \text{ for each } t \in [0, T]. \tag{2.40}$$

Also suppose that, for each $i = 1, .., p$,

$$[M^n, W^{(i)}]_t \xrightarrow{p} 0 \text{ for each } t \in [0, T]. \tag{2.41}$$

There is then an extension $(\mathcal{F}_t')$ of $(\mathcal{F}_t)$, and an $(\mathcal{F}_t')$-martingale M_t so that (M_t^n) converges stably to (M_t). Furthermore, there is a Brownian motion (W_t') so that $(W_t^{(1)}, ..., W_t^{(p)}, W_t')$ is an $(\mathcal{F}_t')$-Wiener process, and so that

$$M_t = \int_0^t f_s dW_s'. \tag{2.42}$$

It is worthwhile to understand the proof of this result, and hence we give it here. The proof follows more or less *verbatim* that of Theorem B.4 in Zhang (2001) (pp. 65–67). The latter is slightly more general.

Proof of Theorem 2.28. Since $[M^n, M^n]_t$ is a non-decreasing process and has non-decreasing continuous limit, the convergence (2.40) is also in law in $\mathbb{D}(\mathbb{R})$ by Theorem VI.3.37 (p. 354) in Jacod and Shiryaev (2003). Thus, in their terminology (ibid., Definition VI.3.25, p. 351), $[M^n, M^n]_t$ is C-tight. From this fact, ibid., Theorem VI.4.13 (p. 358) yields that the sequence M^n is tight.

From this tightness, it follows that for any subsequence M^{n_k}, we can find a further subsequence $M^{n_{k_l}}$ which converges in law (as a process) to a limit M, jointly with $W^{(1)}, ..., W^{(p)}$; in other words, $(W^{(1)}, ..., W^{(p)}, M^{n_{k_l}})$ converges in law to $(W^{(1)}, ..., W^{(p)}, M)$. This M is a local martingale by ibid., Proposition IX.1.17 (p. 526), using the continuity of M_t^n. Using Proposition 2.27 above, $(M^{n_{k_l}}, [M^{n_{k_l}}, M^{n_{k_l}}])$ converge jointly in law (and jointly with the $W^{(i)}$'s) to $(M, [M, M])$. From (2.40) this means that $[M, M]_t = \int_0^t f_s^2 ds$. The continuity of $[M, M]_t$ assures that M_t is continuous. By the same reasoning, from (2.41), $[M, W^{(i)}] \equiv 0$ for each $i = 1, .., p$. Now let $W_t' = \int_0^t f_s^{-1/2} dM_s$ (if f_t is zero on a set of Lebesgue measure greater than zero, follow the alternative construction in Volume III of Gikhman and Skorohod (1969)). By Property (3) in Section 2.2.4 (or refer directly to Property I.4.54 (p.55) of Jacod and Shiryaev (2003)), $[W', W']_t = t$, while $[W', W^{(i)}] \equiv 0$. By the multivariate version of Lévy's Theorem (Section 2.2.4, or refer directly to Theorem II.4.4 (p. 102) of Jacod and Shiryaev (2003)), it therefore follows that $(W_t^{(1)}, ..., W_t^{(p)}, W_t')$ is a Wiener process. The equality (2.42) follows by construction of W_t'. Hence the Theorem is shown for subsequence $M^{n_{k_l}}$. Since the subsequence M^{n_k} was arbitrary, Theorem 2.28 follows (cf. the Corollary on p. 337 of Billingsley (1995)). $\square$

2.4.3 Application to realized volatility

Independent times

We now turn our attention to the simplest application: the estimator from Section 2.3. Consider the normalized (by $\sqrt{n}$) error process

$$M_t^n = 2n^{1/2} \sum_{t_{i+1} \leq t} \int_{t_i}^{t_{i+1}} (X_s - X_{t_i}) dX_s + 2n^{1/2} \int_{t_*}^{t} (X_s - X_{t_*}) dX_s.$$

$$(2.43)$$

From Section 2.3.7, we have that Condition (2.40) of Theorem 2.28 is satisfied, with

$$f_t^2 = 2T\sigma_t^4 H_t'.$$

It now remains to check Condition (2.41). Note that

$$d[M^n, W^{(i)}]_t = 2n^{1/2}(X_t - X_{t_*})d[X, W^{(i)}]_t.$$

We can now apply Lemma 2.22 with $N_t = X_t$ and $H_t = (d/dt)[X, W^{(i)}]_t$. From the Cauchy-Schwarz inequality (in this case known as the Kunita-Watanabe inequality)

$$|[X, W^{(i)}]_{t+h} - [X, W^{(i)}]_t|$$
$$\leq \sqrt{[X, X]_{t+h} - [X, X]_t} \sqrt{[W^{(i)}, W^{(i)}]_{t+h} - [W^{(i)}, W^{(i)}]_t}$$
$$\leq \sqrt{\sigma_+^2 h} \sqrt{h} = \sigma_+ h$$

(recall that the quadratic variation is a limit of sums of squares), so we can take $H_+ = \sigma_+$. On the other hand, $(d/dt)E[N, N]_t \leq \sigma_+^2 = a(t - t_*)^b$ with $a = \sigma_+^2$ and $b = 0$.

Thus, from Lemma 2.22,

$$||[M^n, W^{(i)}]_t||_1$$

$$= 2n^{1/2}||\sum_{t_{i+1}\leq t}\int_{t_i}^{t_{i+1}}(N_s - N_{t_i})H_s ds + \int_{t_*}^t (N_s - N_{t_*})H_s ds||_1$$

$$\leq 2n^{1/2}\left(H_+^2 \frac{a}{b+3}R_{b+3}(\mathcal{G})\right)^{1/2}$$

$$+ R_{(b+3)/2}(\mathcal{G})\frac{2}{b+3}\left(\frac{a}{b+1}\right)^{1/2}\sup_{0\leq t-s\leq\Delta(\mathcal{G})}||H_s - H_t||_2$$

$$= O_p(n^{1/2}R_3(\mathcal{G})^{1/2}) + O_p(n^{1/2}R_{3/2}(\mathcal{G})\sup_{0\leq t-s\leq\Delta(\mathcal{G})}||H_s - H_t||)$$

$$= o_p(1)$$

under the conditions of Proposition 2.21, since $R_v(\mathcal{G}) = O_p(n^{1-v})$ from (2.30), and since $\sup_{0\leq t-s\leq\Delta(\mathcal{G})}||H_s - H_t|| = o_p(1)$ (The latter fact is somewhat complex. One shows that one can take $W^{(1)} = W$ by a use of Lévy's theorem, and the result follows).

We have therefore shown:

Theorem 2.29 *Assume Condition 2.26, as well as the conditions of Proposition 2.21, and also that the AQVT $H(t)$ exists and is absolutely continuous. Let M_t^n be given by (2.43). Then (M_t^n) converges stably in law to M_t, given by*

$$M_t = \sqrt{2T}\int_0^t \sigma_s^2\sqrt{H_s'}dW_s'.$$

As a special case:

Corollary 2.30 *Under the conditions of the above theorem, for fixed t,*

$$\sqrt{n}\left([X,X]_t^{\mathcal{G}_n} - [X,X]_t\right) \xrightarrow{\mathcal{L}} U \times \left(2T\int_0^t \sigma_s^4 dH_s\right)^{1/2}, \qquad (2.44)$$

where U is a standard normal random variable independent of $\mathcal{F}_T$.

Similar techniques can now be used on other common estimators, such as the TSRV. We refer to Section 2.5.

In the context of equidistant times, this result goes back to Jacod (1994), Jacod and Protter (1998), and Barndorff–Nielsen and Shephard (2002). We emphasize that the method of proof in Jacod and Protter (1998) is quite different from the one used here, and gives rise to weaker conditions. The reason for our different treatment is that we have found the current framework more conducive to generalization to other observation time structures and other estimators. In the long run, it is an open question which general framework is the most useful.

Endogenous times

The assumption of independent sampling times is not necessary for a limit result, though a weakening of conditions will change the result. To see what happens, we follow the development in Li, Mykland, Renault, Zhang, and Zheng (2009), and define the *tricicity* by $[X, X, X]_t^{\mathcal{G}} = \sum_{t_{i+1} \leq t} (X_{t_{i+1}} - X_{t_i})^3 + (X_t - X_{t_*})^3$, and assume that

$$n[X, X, X, X]_t^{\mathcal{G}_n} \xrightarrow{p} U_t \text{ and } n^{1/2}[X, X, X]_t^{\mathcal{G}_n} \xrightarrow{p} V_t. \qquad (2.45)$$

By the reasoning in Section 2.3.7, n and $n^{1/2}$ are the right rates for $[X, X, X, X]^{\mathcal{G}}$ and $[X, X, X]^{\mathcal{G}}$, respectively. Hence U_t and V_t will exist under reasonable regularity conditions. Also, from Section 2.3.7, if the AQVT exists and is absolutely continuous, then so are U_t and V_t. We shall use

$$U_t = \int_0^t u_s ds \text{ and } V_t = \int_0^t v_s ds. \qquad (2.46)$$

Triticity is handled in much the same way as quarticity. In analogy to the development in Section 2.3.5, observe that

$$d(X_t - X_{t_i})^3 = 3(X_t - X_{t_i})^2 dX_t + 3(X_t - X_{t_i}) d[X, X]_t$$
$$= 3(X_t - X_{t_i})^2 dX_t + \frac{3}{2} d[M, X]_t,$$

since $d[M, M]_t = 4(X_t - X_{t_i})^2 d[X, X]_t$. It follows that if we set

$$M_t^{(3/2)} = \sum_{t_{i+1} \leq t} \int_{t_i}^{t_{i+1}} (X_s - X_{t_i})^3 dX_s + \int_{t_*}^t (X_s - X_{t_*})^3 dX_s$$

we get

$$[X, X, X]_t^{\mathcal{G}} = \frac{3}{2}[M, X]_t + 3M_t^{(3/2)}.$$

In analogy with Proposition 2.17, we hence obtain:

Proposition 2.31 *Assume the conditions of Proposition 2.17. Then*

$$\sup_{0 \leq t \leq T} | [M, X]_t - \frac{2}{3}[X, X, X]_t^{\mathcal{G}} | = o_p(n^{-1/2}) \text{ as } n \to \infty.$$

It follows that unless $V_t \equiv 0$, the condition (2.41) is Theorem 2.28 will not hold. To solve this problem, normalize as in (2.43), and define an auxiliary martingale

$$\tilde{M}_t^n = M_t^n - \int_0^t g_s dX_s,$$

where g is to be determined. We now see that

$$[\tilde{M}^n, X]_t = [M^n, X]_t - \int_0^t g_s d[X, X]_s$$

$$\overset{p}{\to} \int_0^t (\frac{2}{3}v_s - g_s\sigma_s^2)ds \text{ and}$$

$$[\tilde{M}^n, \tilde{M}^n] = [M^n, M^n] + \int_0^t g_s^2 d[X, X]_s - 2\int_0^t g_s d[M^n, X]$$

$$\overset{p}{\to} \int_0^t (\frac{2}{3}u_s + g_s^2\sigma_s^2 - 2\frac{2}{3}g_s v_s)ds.$$

Hence, if we choose $g_t = 2v_t/3\sigma_t^2$, we obtain that $[\tilde{M}^n, X]_t \overset{p}{\to} 0$ and $[\tilde{M}^n, \tilde{M}^n]$ $\overset{p}{\to} \int_0^t (u_s - v_s\sigma_s^{-2})ds$.

By going through the same type of arguments as above, we obtain:

Theorem 2.32 *Assume Condition 2.26, as well as the conditions of Proposition 2.17. Also assume that (2.45) holds for each $t \in [0, T]$, and that the absolute continuity (2.46) holds. Then (M_t^n) converges stably in law to M_t, given by*

$$M_t = \frac{2}{3}\int_0^t \frac{v_s}{\sigma_s^2}dX_s + \int_0^t \left(\frac{2}{3}u_s - \frac{4}{9}\frac{v_s^2}{\sigma_s^2}\right)^{1/2} dW_s',$$

where W' is independent of $W^{(1)}, ..., W^{(p)}$.

Again as a special case:

Corollary 2.33 *Under the conditions of the above theorem, for fixed t,*

$$\sqrt{n}\left([X, X]_t^{\mathcal{G}_n} - [X, X]_t\right) \overset{\mathcal{L}}{\to} \frac{2}{3}\int_0^t \frac{v_s}{\sigma_s^2}dX_s + U \times \int_0^t (\frac{2}{3}u_s - \frac{4}{9}\frac{v_s^2}{\sigma_s^2})ds,$$

where U is a standard normal random variable independent of $\mathcal{F}_T$.

It is clear from this that the assumption of independent sampling times implies that $v_t \equiv 0$.

A similar result was shown in Li et al. (2009), where implications of this result are discussed further.

2.4.4 Statistical risk neutral measures

We have so far ignored the drift μ_t. We shall here provide a trick to reinstate the drift in any analysis, without too much additional work. It will turn out that stable convergence is a key element in the discussion. Before we go there, we need to introduce the concept of absolute continuity.

We refer to a probability where there is no drift as a "statistical" risk neutral measure. This is in analogy to the use of equivalent measures in asset pricing. See, in particular, Ross (1976), Harrison and Kreps (1979), Harrison and Pliska (1981), Delbaen and Schachermayer (1995), and Duffie (1996).

Absolute continuity

We shall in the following think about having two different probabilities on the same observables. For example, P can correspond to the system

$$dX_t = \sigma_t dW_t, X_0 = x_0, \tag{2.47}$$

while Q can correspond to the system

$$dX_t = \mu_t dt + \sigma_t dW_t^Q, X_0 = x_0. \tag{2.48}$$

In this case, W_t is a Wiener process under P, and W_t^Q is a Wiener process under Q. Note that since we are modeling the process X_t, this process is the observable quantity whose distribution we seek. Hence, the process X_t does not change from P to Q, but its distribution changes. If we equate (2.47) and (2.48), we get

$$\mu_t dt + \sigma_t dW_t^Q = \sigma_t dW_t,$$

or

$$\frac{\mu_t}{\sigma_t} dt + dW_t^Q = dW_t.$$

As we discussed in the constant μ and σ case, when carrying out inference for observations in a fixed time interval $[0, T]$, the process μ_t cannot be consistently estimated. A precise statement to this effect (Girsanov's Theorem) is given below.

The fact that μ cannot be observed means that one cannot fully distinguish between P and Q, even with infinite data. This concept is captured in the following definition:

Definition 2.34 *For a given σ-field $\mathcal{A}$, two probabilities P and Q are mutually absolutely continuous (or equivalent) if, for all $A \in \mathcal{A}$, $P(A) = 0 <=> Q(A) = 0$. More generally, Q is absolutely continuous with respect to P if, for all $A \in \mathcal{A}$, $P(A) = 0 => Q(A) = 0$.*

We shall see that P and Q from (2.47) and (2.48) are, indeed, mutually absolutely continuous.

The Radon–Nikodym Theorem, and the likelihood ratio

Theorem 2.35 *(Radon–Nikodym) Suppose that Q is absolutely continuous under P on a σ-field $\mathcal{A}$. Then there is a random variable ($\mathcal{A}$ measurable) dQ/dP so that for all $A \in \mathcal{A}$,*

$$Q(A) = E_P \left(\frac{dQ}{dP} I_A \right).$$

For proof and a more general theorem, see Theorem 32.2 (p. 422) in Billingsley (1995).

The quantity dQ/dP is usually called either the Radon–Nikodym derivative or the likelihood ratio. It is easy to see that dQ/dP is unique "almost surely" (in the same way as the conditional expectation).

Example 2.36 *The simplest case of a Radon–Nikodym derivative is where X_1, $X_2, ..., X_n$ are iid, with two possible distributions P and Q. Suppose that X_i has density f_P and f_Q under P and Q, respectively. Then*

$$\frac{dQ}{dP} = \frac{f_Q(X_1)f_Q(X_2)...f_Q(X_n)}{f_P(X_1)f_P(X_2)...f_P(X_n)}.$$

Likelihood ratios are of great importance in statistical inference generally.

Properties of likelihood ratios

- $P(\frac{dQ}{dP} \geq 0) = 1$

- If Q is equivalent to P: $P(\frac{dQ}{dP} > 0) = 1$

- $E_P \left(\frac{dQ}{dP} \right) = 1$

- For all $\mathcal{A}$-measurable Y: $E_Q (Y) = E_P \left(Y \frac{dQ}{dP} \right)$

- If Q is equivalent to P: $\frac{dP}{dQ} = \left(\frac{dQ}{dP} \right)^{-1}$

Girsanov's Theorem

We now get to the relationship between P and Q in systems (2.47) and (2.48). To give the generality, we consider the vector process case (where μ is a vector, and σ is a matrix). The superscript "T" here stands for "transpose."

Theorem 2.37 (*Girsanov*). *Subject to regularity conditions, P and Q are mutually absolutely continuous, and*

$$\frac{dP}{dQ} = \exp\left\{ -\int_0^T \sigma_t^{-1} \mu_t dW_t^Q - \frac{1}{2} \int_0^T \mu_t^T (\sigma_t \sigma_t^T)^{-1} \mu_t dt \right\},$$

The regularity conditons are satisfied if $\sigma_- \leq \sigma_t \leq \sigma_+$, and $|\mu_t| \leq \mu_+$, but they also cover much more general situations. For a more general statement, see, for example, Section 5.5 of Karatzas and Shreve (1991).

How to get rid of μ: Interface with stable convergence

The idea is borrowed from asset pricing theory. We think that the true distribution is Q, but we prefer to work with P because then calculations are much simpler.

Our plan is the following: carry out the analysis under P, and adjust results back to Q using the likelihood ratio (Radon–Nikodym derivative) dP/dQ. Specifically, suppose that θ is a quantity to be estimated (such as $\int_0^T \sigma_t^2 dt$, $\int_0^T \sigma_t^4 dt$, or the leverage effect). An estimator $\hat{\theta}_n$ is then found with the help of P, and an asymptotic result is established whereby, say,

$$n^{1/2}(\hat{\theta}_n - \theta) \xrightarrow{\mathcal{L}} N(b, a^2) \text{ stably}$$

under P. It then follows directly from the measure theoretic equivalence that $n^{1/2}(\hat{\theta}_n - \theta)$ also converges in law under Q. *In particular, consistency and rate of convergence are unaffected by the change of measure.* We emphasize that this is due to the finite (fixed) time horizon T.

The asymptotic law may be different under P and Q. While the normal distribution remains, the distributions of b and a^2 (if random) may change.

The technical result is as follows.

Proposition 2.38 *Suppose that Z_n is a sequence of random variables which converges stably to $N(b, a^2)$ under P. By this we mean that $N(b, a^2) = b + aN(0,1)$, where $N(0,1)$ is a standard normal variable independent of $\mathcal{F}_T$, also a and b are $\mathcal{F}_T$ measurable. Then Z_n converges stably in law to $b + aN(0,1)$ under Q, where $N(0,1)$ remains independent of $\mathcal{F}_T$ under Q.*

Proof of Proposition. $E_Q I_A g(Z_n) = E_P \frac{dQ}{dP} I_A g(Z_n) \to E_P \frac{dQ}{dP} I_A g(Z) = E_Q I_A g(Z)$ by uniform integrability of $\frac{dQ}{dP} I_A g(Z_n)$.   □

Proposition 2.38 substantially simplifies calculations and results. In fact, the same strategy will be helpful for the localization results that come next in the paper. It will turn out that the relationship between the localized and continuous process can also be characterized by absolute continuity and likelihood ratios.

Remark 2.39 *It should be noted that after adjusting back from P to Q, the process μ_t may show up in expressions for asymptotic distributions. For instances of this, see Sections 2.5 and 4.3 of Mykland and Zhang (2009). One should always keep in mind that drift most likely is present, and may affect inference.*

Remark 2.40 *As noted, our device is comparable to the use of equivalent martingale measures in options pricing theory (Ross (1976), Harrison and Kreps (1979), Harrison and Pliska (1981), see also Duffie (1996)) in that it affords a convenient probability distribution with which to make computations. In our econometric case, one can always take the drift to be zero, while in the options pricing case, this can only be done for discounted securities prices. In both cases, however, the computational purpose is to get rid of a nuisance "dt term."*

The idea of combining stable convergence with measure change appears to go back to Rootzén (1980).

2.4.5 Unbounded σ_t

We have so far assumed that $\sigma_t^2 \leq \sigma_+^2$. With the help of stable convergence, it is also easy to weaken this assumption. One can similarly handle restrictions on μ_t, and on σ_t^2 being bounded away from zero.

The much weaker requirement is that σ_t be *locally bounded*. This is to say that there is a sequence of stopping times τ_m and of constants $\sigma_{m,+}$ so that

$$P(\tau_m < T) \to 0 \text{ as } m \to \infty \text{ and}$$
$$\sigma_t^2 \leq \sigma_{m,+}^2 \text{ for } 0 \leq t \leq \tau_m.$$

For example, this is automatically satisfied if σ_t is a continuous process.

As an illustration of how to incorporate such local boundedness in existing results, take Corollary 2.30. If we replace the condition $\sigma_t^2 \leq \sigma_+^2$ by local boundedness, the corollary continues to hold (for fixed m) with $\sigma_{\tau_n \wedge t}$ replacing σ_t. On the other hand we note that $[X, X]^{\mathcal{G}_n}$ is the same for $\sigma_{\tau_n \wedge t}$ and σ_t on the set $\{\tau_n = T\}$. Thus, the corollary tells us that for any set $A \in \mathcal{F}_T$, and for any bounded continuous function g,

$$EI_{A \cap \{\tau_m = T\}} g\left(\sqrt{n}\left([X, X]_t^{\mathcal{G}_n} - [X, X]_t\right)\right)$$
$$\to EI_{A \cap \{\tau_m = T\}} g\left(U \times \left(2T \int_0^t \sigma_s^4 dH_s\right)^{1/2}\right)$$

as $n \to \infty$ (and for fixed m), where U has the same meaning as in the corollary.

Hence,

$$\left| EI_A g\left(\sqrt{n}\left([X,X]_t^{\mathcal{G}_n} - [X,X]_t\right)\right) - EI_A g\left(U \times \left(2T\int_0^t \sigma_s^4 dH_s\right)^{1/2}\right)\right|$$

$$\leq \left| EI_{A\cap\{\tau_m=T\}} g\left(\sqrt{n}\left([X,X]_t^{\mathcal{G}_n} - [X,X]_t\right)\right)\right.$$

$$\left. - EI_{A\cap\{\tau_m=T\}} g\left(U \times \left(2T\int_0^t \sigma_s^4 dH_s\right)^{1/2}\right)\right|$$

$$+ 2\max|g(x)|P(\tau_m \neq T)$$

$$\rightarrow 2\max|g(x)|P(\tau_m \neq T)$$

as $n \rightarrow \infty$. By choosing m large, the right-hand side of this expression can be made as small as we wish. Hence, the left-hand side actually converges to zero. We have shown:

Corollary 2.41 *Theorem 2.29, Corollary 2.30, and Theorem 2.32 all remain true if the condition $\sigma_t^2 \leq \sigma_+^2$ is replaced by a requirement that σ_t^2 be locally bounded.*

2.5 Microstructure

2.5.1 The problem

The basic problem is that the semimartingale X_t is actually contaminated by noise. One observes

$$Y_{t_i} = X_{t_i} + \epsilon_i. \tag{2.49}$$

We do not right now take a position on the structure of the ϵ_is.

The reason for going to this structure is that the convergence (consistency) predicted by Theorem 2.10 manifestly does not hold. To see this, in addition to $\mathcal{G}$, we also use subgrids of the form $\mathcal{H}_k = \{t_k, t_{K+k}, t_{2K+k}, ...\}$. This gives rise to the *average realized volatility (ARV)*

$$ARV(Y, \mathcal{G}, K) = \frac{1}{K}\sum_{k=1}^{K}[Y,Y]^{\mathcal{H}_k}.$$

Note that $ARV(Y, \mathcal{G}, 1) = [Y,Y]^{\mathcal{G}}$ in obvious notation. If one believes Theorem 2.10, then the $ARV(Y, \mathcal{G}, K)$ should be close for small K. In fact, the convergence in the theorem should be visible as K decreases to 1. Figure 2.1 looks at the $ARV(Y, \mathcal{G}, K)$ for Alcoa Aluminun (AA) for January 4, 2001. As can be seen in the figure, the actual data behaves quite differently from what the theory predicts. It follows that the semimartingale assumption does not hold, and we have to move to a model like (2.49).

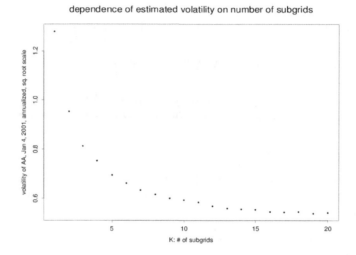

dependence of estimated volatility on number of subgrids

Figure 2.1 *RV as One Samples More Frequently. The plot gives* $ARV(Y, \mathcal{G}, K)$ *for* $K = 1, ..., 20$ *for Alcoa Aluminum for the transactions on January 4, 2001. It is clear that consistency does not hold for the quadratic variation. The semimartingale model, therefore, does not hold.*

2.5.2 An initial approach: Sparse sampling

Plots of the type given in Figures 2.1 and 2.2 were first considered by T. G. Andersen, Bollerslev, Diebold, and Labys (2000) and called *signature plots*. The authors concluded that the most correct values for the volatility were the lower ones on the left-hand side of the plot, based mainly on the stabilization of the curve in this region. On the basis of this, the authors recommended to estimate volatility using $[Y, Y]^{\mathcal{H}}$, where $\mathcal{H}$ is a sparsely sampled subgrid of $\mathcal{G}$. In this early literature, the standard approach was to subsample about every five minutes.

The philosophy behind this approach is that the size of the noise ϵ is very small, and if there are not too many sampling points, the effect of noise will be limited. While true, this uses the data inefficiently, and we shall see that better methods can be found. The basic subsampling scheme does, however, provide some guidance on how to proceed to more complex schemes. For this reason, we shall analyze its properties.

The model used for most analysis is that ϵ_i is independent of X, and iid. One can still, however, proceed under weaker conditions. For example, if the ϵ_i have serial dependence, a similar analysis will go through.

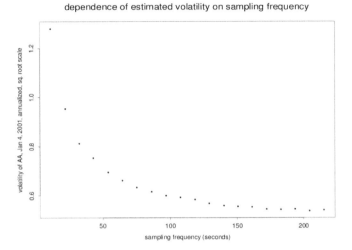

Figure 2.2 *RV as One Samples More Frequently. This is the same figure as Figure 2.1, but the x axis is the average number of observations between each transaction for each* $ARV(Y, \mathcal{G}, K)$. *There is one transaction about each 50 seconds in this particular data.*

The basic decomposition is

$$[Y, Y]^{\mathcal{H}} = [X, X]^{\mathcal{H}} + [\epsilon, \epsilon]^{\mathcal{H}} + 2[X, \epsilon]^{\mathcal{H}},$$

where the cross term is usually (but not always) ignorable. Thus, if the ϵ's are independent of X, and $E(\epsilon) = 0$, we get

$$E([Y, Y]^{\mathcal{H}} | X \text{ process}) = [X, X]^{\mathcal{H}} + E[\epsilon, \epsilon]^{\mathcal{H}}.$$

If the ϵ are identically distributed, then

$$E[\epsilon, \epsilon]^{\mathcal{H}} = n_{sparse} E(\epsilon_K - \epsilon_0)^2,$$

where $n_{sparse} = $ (number of points in $\mathcal{H}$) $- 1$. Smaller n_{sparse} gives smaller bias, but bigger variance.

At this point, if you would like to follow this line of development, please consult the discussion in Section 2 in Zhang et al. (2005). This shows that there is an optimal subsampling frequency, given by equation (31) (p. 1399) in the paper. A similar analysis for $ARV(Y, \mathcal{G}, K)$ *is carried out in Sections 3.1–3.3 of the paper.*

2.5.3 Two scales realized volatility (TSRV)

To get a consistent estimator, we go to the *two scales realized volatility (TSRV)*. The TRSV is defined as follows.

$$\widehat{[X,X]}_T^{(\text{tsrv})} = a_n ARV(Y, \mathcal{G}, K) - b_n ARV(Y, \mathcal{G}, J) \qquad (2.50)$$

where we shall shortly fix a_n and b_n. It will turn out to be meaningful to use

$$b_n = a_n \times \frac{\bar{n}_K}{\bar{n}_J},$$

where $\bar{n}_K = (n - K + 1)/K$. For asymptotic purposes, we can take $a_n = 1$, but more generally will assume that $a_n \to 1$ as $n \to \infty$. Choices with good small sample properties are given in Section 4.2 in Zhang et al. (2005), and equation (4.22) in Aït-Sahalia, Mykland, and Zhang (2011).

This estimator is discussed in Section 4 in Zhang et al. (2005), though only in the case where $J = 1$. In the more general case, J is not necessarily 1, but $J << K$.

One can prove under weak assumptions, that

$$\sum_{i=0}^{n-J} (X_{t_{i+j}} - X_{t_i})(\epsilon_{t_{i+j}} - \epsilon_{t_i}) = O_p(J^{-1/2}).$$

This is important because it gives rise to the sum of squares decomposition

$$ARV(Y, \mathcal{G}, J) = ARV(X, \mathcal{G}, J) + ARV(\epsilon, \mathcal{G}, J) + O_p(J^{-1/2}).$$

Thus, if we look at linear combinations of the form (2.50), one obtains, for $a_n = 1$,

$$\widehat{[X,X]}_T^{(\text{tsrv})}$$

$$= \underbrace{ARV(X, \mathcal{G}, K) - \frac{\bar{n}_K}{\bar{n}_J} ARV(X, \mathcal{G}, J)}_{\text{signal term}}$$

$$\underbrace{+ ARV(\epsilon, \mathcal{G}, K) - \frac{\bar{n}_K}{\bar{n}_J} ARV(\epsilon, \mathcal{G}, J)}_{\text{noise term}} + O_p(K^{-1/2}), \quad (2.51)$$

so long as

$$1 \leq J \leq K \quad \text{and} \quad K = o(n).$$

The noise term behaves as follows:

$$[\epsilon, \epsilon]_T^{(J)} = \frac{1}{J} \sum_{i=0}^{n} c_i^{(J)} \epsilon_{t_i}^2 - \frac{2}{J} \sum_{i=0}^{n-J} \epsilon_{t_i} \epsilon_{t_{i+J}},$$

where $c_i^{(J)} = 2$ for $J \leq i \leq n - J$, and $= 1$ for other i. By construction

$$\sum_i c_i^{(J)} = 2J\bar{n}_J,$$

so that

$$\text{noise term} \approx -\frac{2}{K} \sum_{i=0}^{n-K} \epsilon_{t_i}\epsilon_{t_{i+K}} + \frac{\bar{n}_K}{\bar{n}_J}\frac{2}{J} \sum_{i=0}^{n-J} \epsilon_{t_i}\epsilon_{t_{i+J}} \qquad (2.52)$$

so that (1) the ϵ^2 terms have been removed, and (2) the estimator is unbiased if J is chosen to be bigger than the range of dependence of the ϵ's.

THE CASE FOR TWO SCALES ESTIMATOR
100 days 2001

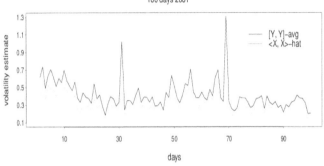

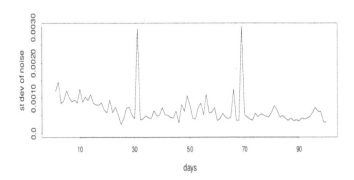

Figure 2.3 *ARV$(Y, \mathcal{G}, K)$ and the two scales estimator for Alcoa Aluminum for the first 100 trading days of 2001. Square root, annualized scale. Also estimated size of the microstructure noise. One can see from the plot that the microstructure has a substantially bigger impact on the ARV than on the TSRV.*

ERROR PROPORTIONAL TO VOLATILITY?

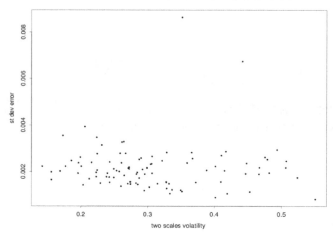

Figure 2.4 *Data as in Figure 2.3. Here size of microstructure noise plotted vs. TSRV. The figure suggests that the size of the microstructure is largely unaffected by the volatility.*

2.5.4 Asymptotics for the TSRV

If the noise is assumed to be independent of the X process, one can deal separately with the signal and noise terms in (2.51). The signal term is analyzed with the kind of technique developed in Sections 2.3–2.4 above. The precise derivation is given in Section 3.4 (pp. 1400–1401) and Appendix A.3 (pp. 1410–1411) of Zhang et al. (2005). Meanwhile, the noise term (2.52) is a U-statistic which can be handled with methods from discrete process limit theory, based either on martingales or mixing, using the limit theory in Hall and Heyde (1980) (see, in particular, ibid., Theorem 3.2 (pp. 58–59)). For concrete implementation, see Zhang et al. (2005) for the case of iid noise. By combining the two sources of error in (2.51), a rate of convergence of the TSRV to the true integrated volatility is $O_p(n^{-1/6})$, and the limit is, again, of mixed normal type.

2.5.5 The emerging literature on estimation of volatility under microstructure

This estimation problem has by now become somewhat of an industry. The following approaches are now in the process of becoming available:

- *Extensions of the two scales approach.* Zhang (2006) studies a multi scale realized volatility (MSRV), and obtains that the estimator of integrated volatility converges at the rate of $O_p(n^{-1/4})$. This rate is optimal, as it also comes up in the case where σ_t is constant and the noise is normal, which is a parametric problem. Also, the conditions on the noise structure in the TSRV has been weakened. Aït-Sahalia et al. (2011) studies noise that is internally dependent but independent of the signal, and Li and Mykland (2007) discuss a formulation where the noise can depend on the process X.

- *An approach based on autocovariances.* Often called the "kernel" approach. This work has been pioneered by Barndorff–Nielsen, Hansen, Lunde, and Shephard (2008). We refer to this paper for further discussion.

- *Preaveraging.* The idea here is to try to reduce the noise by averaging observations before computing volatilities. The two main papers for the moment are Jacod, Li, Mykland, Podolskij, and Vetter (2009) and Podolskij and Vetter (2009).

- *Likelihood based methods.* There are two approaches under development. On the one hand, Xiu (2010) uses the likelihood function for constant σ and normal noise as a quasi-likelihood to estimate $[X, X]$. On the other hand, Mykland and Zhang (2009) show that in sufficiently small neighboorhoods of observations, one can act as if σ_t really is constant. We return to a discussion of this in Section 2.6.

To first order, all of these estimators do similar things. The main difference between them seems to be the handling of end effects. This topic will, no doubt, be the subject of future research.

2.5.6 A wider look at subsampling and averaging

We have seen above that subsampling and averaging can help with several problems.

1. It is a first order remedy for microstructure.

2. If an estimator $\widehat{[X, X]}$ is based on the noise being an independent series, one can ameliorate the effects of actual dependence by subsampling every J'th observation, and then average across the J resulting estimators. We have explicitly described this for the TSRV, but the same *rationale* can be used to subsample and then average any of the estimators mentioned in Section 2.5.5.

3. It should be noted that subsampling and averaging is robust to noise that depends on the latent process X, cf. Delattre and Jacod (1997) and Li and Mykland (2007).

4. A further use of subsampling is that it seems in some instances to regularize time. This is further discussed in Section 2.7.2.

2.6 Methods based on contiguity

We have seen in Section 2.4.4 that measure changes can be a powerful tool in high-frequency data problems. We here pursue this matter further, by considering measure changes that are asymptotically absolutely continuous. This is closely related to the concept of *contiguity*, which is discussed further in Section 2.6.1. This first section mainly abstracts the results in Mykland and Zhang (2009), which should be consulted for details and proofs. The later sections are new material.

2.6.1 Block discretization

In addition to the grid $\mathcal{G}_n = \{0 = t_{n,0} < t_{n,1} < ... < t_{n,n} = T\}$, which we again take to be independent of the underlying process to be observed, we consider a subgrid

$$\mathcal{H}_n = \{0 = \tau_{n,0} < \tau_{n,1} < ... < \tau_{n,K_n} = T\} \subseteq \mathcal{G}_n.$$

We shall now define a new measure, *on the observations* $X_{t_{n,j}}$ *only*, for which the volatility is constant on each of the K_n blocks $(\tau_{n,i-1}, \tau_{n,i}]$. Specifically, consider the approximate measure, called Q_n, satisfying $X_0 = x_0$ and

$$\text{for each } i = 1, ..., K_n : \Delta X_{t_{j+1}} = \sigma_{\tau_{n,i-1}} \Delta W^Q_{t_{j+1}} \text{ for } t_{n,j+1} \in (\tau_{n,i-1}, \tau_{n,i}].$$
$$(2.53)$$

Formally, we define the approximation Q_n recursively (with $\Delta t_{n,j} = t_{n,j} - t_{n,j-1}$).

Definition 2.42 *(Block approximation). Define the probability Q_n recursively by:*
(i) U_0 has same distribution under Q_n as under P;
(ii) The conditional Q_n-distribution of $U^{(1)}_{t_{n,j+1}}$ given $U_0, ..., U_{t_{n,j}}$ is given by (2.53), where $\Delta W^Q_{t_{j+1}}$ is conditionally normal $N(0, \Delta t_{n,j+1})$, and
(iii) The conditional P^*_n-distribution of $U^{(2)}_{t_{n,j+1}}$ given $U_0, ..., U_{t_{n,j}}, U^{(1)}_{t_{n,j+1}}$ is the same as under P.

Note that we often drop the subscript "n" on $\Delta t_{n,j}$, and write Δt_j.

Denote by $M_{n,i} = \#\{t_{n,j} \in (\tau_{n,i-1}, \tau_{n,i}]\}$. We shall require that $\max_i M_{n,i} = O(1)$ as $n \to \infty$, from which it follows that K_n is of exact order $O(n)$. To measure the extent to which we hold the volatility constant, we define the following

"Asymptotic Decoupling Delay" (ADD) by

$$K(t) = \lim_{n \to \infty} \sum_i \sum_{t_{n,i} \in (\tau_{n,i-1}, \tau_{n,i}] \cap [0,t]} (t_{n,j} - \tau_{n,i-1}), \qquad (2.54)$$

provided the limit exists. In the case of equidistant observations and equally sized blocks of M observations, the ADD takes the form $K(t) = \frac{1}{2}(M-1)t$.

It is shown in Theorems 1 and 3 in Mykland and Zhang (2009) that, subject to regularity conditions, P and Q_n are mutually absolutely continuous on the σ-field $\mathcal{X}_{n,n}$ generated by $U_{t_{n,j}}$, $j = 0, ..., n$. Furthermore, let $(dP/dQ_n)(U_{t_0}, ..., U_{t_{n,j}}, ..., U_{t_{n,n}})$ be the likelihood ratio (Radon–Nikodym derivative) on $\mathcal{X}_{n,n}$. Then,

$$\frac{dP}{dQ_n}(U_{t_0}, ..., U_{t_{n,j}}, ..., U_{t_{n,n}}) \overset{\mathcal{L}}{\to} \exp\{\Gamma^{1/2}N(0,1) - \frac{1}{2}\Gamma\} \qquad (2.55)$$

stably in law, under Q_n, as $n \to \infty$. The asymptotic variance is given by $\Gamma = \Gamma_0 + \Gamma_1$, where

$$\Gamma_0 = \frac{3}{8} \int_0^T \left(\frac{1}{\sigma_t^2} \frac{d}{dt} [\sigma^2, W]_t \right)^2 dt = \frac{3}{2} \int_0^T \left(\frac{1}{\sigma_t} \frac{d}{dt} [\sigma, W]_t \right)^2 dt, \text{ and}$$

$$\Gamma_1 = \frac{1}{2} \int_0^T \frac{1}{\sigma_t^4} \left(\frac{d}{dt} [\sigma^2, \sigma^2]_t \right) dK(t) = 2 \int_0^T \frac{1}{\sigma_t^2} \left(\frac{d}{dt} [\sigma, \sigma]_t \right) dK(t).$$
$$(2.56)$$

Hence, Γ_0 is related to the leverage effect, while Γ_1 is related to the volatility of volatility.

The important consequence of this result is that that P and the approximation Q_n are *contiguous* in the sense of Hájek and Sidak (1967) (Chapter VI), LeCam (1986), LeCam and Yang (2000), and Jacod and Shiryaev (2003) (Chapter IV). This is to say that for a sequence A_n of sets, $P(A_n) \to 0$ if and only if $Q_n(A_n) \to 0$. This follows from (2.55) since dP/dQ_n is uniformly integrable under Q_n (since the sequence dP/dQ_n is nonnegative, the limit also integrates to one under Q_n). In it follows that consistency and orders of convergence are maintained from one measure to the other. In particular, a martingale Z^n is $O_p(1)$ under P if and only if it has the same order of convergence under Q_n. Hence consistency and order of convergence are maintained from one measure to the other. The result in Mykland and Zhang (2009) also covers the multivariate case. Also consult ibid., Section 3.3, for connections to Hermite polynomials. Finally note that contiguity also holds if the sequence $\frac{dP}{dQ_n}$ is tight, which requires even weaker conditions. (For example, K need only exist through subsequences, which is assured under weak conditions by Helly's Theorem, (see, for example, p. 336 in Billingsley (1995)).

The good news: This means that one can construct estimators *as if σ_t^2 is locally constant*. If the resulting estimator $\hat{\theta}_n$ is such that $n^\alpha(\hat{\theta}_n - \theta) = O_p(1)$ under

Q_n (where local constancy is satisfied), then $n^\alpha(\hat{\theta}_n - \theta) = O_p(1)$ also under P. In other words, the change from P to Q_n has much the same simplifying function as the measure change in Section 2.4.4.

Remark 2.43 *The (potentially) bad news: This change of measure is not completely innocuous. A sequence Z^n of martingales may not have exactly the same limit distribution under P and Q_n. The reason is that the Q_n martingale part of $\log dP/dQ_n$ may have nonzero asymptotic covariation with Z^n. This is the same phenomenon which occurs (in a different context) in Section 2.4.3. An adjustment then has to be carried out along the lines of "LeCam's Third Lemma" (Hájek and Sidak (1967), Chapter VI.1.4., p. 208). We refer to Section 2.4 and 3.4 of Mykland and Zhang (2009) for the methodology for adjusting the limit distribution. Ibid., Section 4.3, provides an example where such an adjustment actually has to be made. Other parts of Section 4 of the paper provide examples of how the methodology can be used, and where adjustment is not necessary.*

The approximation above depends on the following Itô process structure on σ:

$$d\sigma_t = \tilde{\sigma}_t dt + f_t dW_t + g_t dB_t, \qquad (2.57)$$

where B is a Brownian motion independent of W. (It is an open question what happens in, say, the long range dependent case). We also require $\inf_{0\le t\le T} \sigma^2 > 0$. Note that in the representation (2.57), (2.56) becomes

$$\Gamma_0 = \frac{3}{2}\int_0^T \left(\frac{f_t}{\sigma_t}\right)^2 dt \text{ and } \Gamma_1 = 2\int_0^T \left(\frac{f_t^2 + g_t^2}{\sigma_t^2}\right) dt.$$

Example 2.44 *In the case of a Heston model (Section 2.2.2), we obtain that*

$$\Gamma_0 = \frac{3}{8}(\rho\gamma)^2 \int_0^T \sigma_t^{-2} dt \text{ and } \Gamma_1 = \frac{1}{4}\gamma^2(M-1)\int_0^T \sigma_t^{-2} dt.$$

Remark 2.45 *(One step discretization). Let P_n^* be the measure Q_n which arises when the block length is $M = 1$. Observe that even with this one-step discretization, dP/dP_n^* does not necessarily converge to unity. In this case, $\Gamma_1 = 0$, but Γ_0 does not vanish when there is leverage effect.*

2.6.2 Moving windows

The paper so far has considered chopping n data up into non-overlapping windows of size M each. We here show by example that the methodology can be adapted to the moving window case. We consider the estimation of $\theta = \int_0^T |\sigma_t|^p dt$, as in Section 4.1 of Mykland and Zhang (2009). It should be noted that the moving window is close to the concept of a moving kernel,

METHODS BASED ON CONTIGUITY

and this may be a promising avenue of further investigation. See, in particular, Linton (2007).

We use block length M, and equidistant times with spacing $\Delta t_n = T/n$. Also, we use for simplicity

$$\tilde{\sigma}^2_{\tau_{n,i}} = \frac{1}{\Delta t_n M_n} \sum_{t_{n,j} \in (\tau_{n,i}, \tau_{n,i+1}]} (\Delta X_{t_{n,j+1}})^2,$$

as estimator of $\sigma^2_{\tau_{n,i}}$. The moving window estimate of θ is now

$$\tilde{\theta}_n^{MW} = (\Delta t) \sum_{i=0}^{n-M} \widetilde{|\sigma_{t_{n,i}}|^r}.$$

It is easy to see that

$$\tilde{\theta}_n^{MW} = \frac{1}{M} \sum_{m=1}^{M} \tilde{\theta}_{n,m} + O_p(n^{-1}),$$

where $\tilde{\theta}_{n,m}$ is the non-overlapping block estimator, with block number one starting at $t_{n,m}$. In view of this representation, it is once again clear from sufficiency considerations that the moving window estimator will have an asymptotic variance which is smaller (or, at least, no larger) than the estimator based on non-overlapping blocks. We now carry out the precise asymptotic analysis.

To analyze this estimator, let $\mathcal{M} > M$, and let

$$A_n = \{i = 0, ..., n - M : [t_{n,i}, t_{n,i+M}] \subseteq [k\mathcal{M}, (k+1)\mathcal{M}] \text{ for some } k\},$$

with $B_n = \{0, ..., n - M\} - A_n$. Write

$$n^{1/2}(\tilde{\theta}_n^{MW} - \theta)$$

$$= n^{1/2}\Delta t \sum_k \sum_{i:[t_{n,i}, t_{n,i+M}] \subseteq [k\mathcal{M}/n, (k+1)\mathcal{M}/n]} (\widetilde{|\sigma_{t_{n,i}}|^r} - |\sigma_{t_{k\mathcal{M}}}|^r)$$

$$+ n^{1/2}\Delta t \sum_{i \in B_n} (\widetilde{|\sigma_{t_{n,i}}|^r} - |\sigma_{t_{n,i}}|^r) + O_p(n^{-1/2}). \quad (2.58)$$

Now apply our methodology from Section 2.6.1, *with block size $\mathcal{M}$*, to the first term in (2.58). Under this block approximation, the inner sum in the first term is based on conditionally i.i.d. observations, in fact, for $[t_{n,i}, t_{n,i+M}] \subseteq [k\mathcal{M}/n, (k+1)\mathcal{M}/n]$, $\tilde{\sigma}^2_{t_{n,i}} = \sigma^2_{k\mathcal{M}/n} S_i$, in law, where

$$S_i = M^{-1} \sum_{j=i}^{i+M-1} U_j^2, \quad U_0, U_1, U_2, ... \text{ iid standard normal.} \quad (2.59)$$

As in Section 4.1 of Mykland and Zhang (2009), there is no adjustment (*à la* Remark 2.43) due to covariation with the asymptotic likelihood ratios, and so the first term in (2.58) converges stably to a mixed normal with random variance as the limit of $n\Delta t_n^2 \sum_k |\sigma|_{kM/n}^r \mathrm{Var}\left(c_{M,r}^{-1} \sum_{i=0}^{M-M} S_i^{r/2}\right)$, which is

$$Tc_{M,r}^{-2} \frac{1}{M} \mathrm{Var}\left(\sum_{i=0}^{M-M} S_i^{r/2}\right) \int_0^T |\sigma|_t^r dt. \tag{2.60}$$

Similarly, one can apply the same technique to the second term in (2.58), but now with the k'th block ($k \geq 2$) starting at $kM - M$. This analysis yields that the second term is also asymptotically mixed normal, but with a variance that is of order $o_p(1)$ as $M \to \infty$. (In other words, once again, first send n to infinity, and then, afterwards, do the same to M). This yields that, overall, and in the sense of stable convergence,

$$n^{1/2}(\tilde{\theta}_n^{MW} - \theta) \xrightarrow{\mathcal{L}} N(0,1) \times \left(c_{M,r}^{-2} \alpha_{M,r} T \int_0^T |\sigma|_t^r dt\right)^{1/2},$$

where, from (2.60), $\alpha_{M,r} = \lim_{M\to\infty} \mathrm{Var}\left(\sum_{i=0}^{M-M} S_i^{r/2}\right)/M$, i.e.,

$$\alpha_{M,r} = \mathrm{Var}(S_0^{r/2}) + 2 \sum_{i=1}^{M-1} \mathrm{Cov}(S_0^{r/2}, S_i^{r/2}),$$

where the S_i are given in (2.59).

2.6.3 Multivariate and asynchronous data

The results discussed in Section 2.6.1 also apply to vector processes (see Mykland and Zhang (2009) for details). Also, for purposes of analysis, asynchronous data does not pose any conceptual difficulty when applying the results. One includes all observation times when computing the likelihood ratios in the contiguity theorems. It does not matter that some components of the vector are not observed at all these times. In a sense, they are just treated as missing data. Just as in the case of irregular times for scalar processes, this does not necessarily mean that it is straightforward to write down sensible estimators.

For example, consider a bivariate process $(X_t^{(1)}, X_t^{(2)})$. If the process $(X_t^{(r)})$ is observed at times:

$$\mathcal{G}_n^{(r)} = \{0 \leq t_{n,0}^{(r)} < t_{n,1}^{(r)} < \ldots < t_{n,n_r}^{(r)} \leq T\}, \tag{2.61}$$

one would normally use the grid $\mathcal{G}_n = \mathcal{G}_n^{(1)} \cup \mathcal{G}_n^{(2)} \cup \{0, T\}$ to compute the likelihood ratio dP/dQ_n.

To focus the mind with an example, consider the estimation of covariation under asynchronous data. It is shown in Mykland (2010a) that the Hayashi-Yoshida estimator (Hayashi and Yoshida (2005)) can be seen as a non-parametric maximum likelihood estimator (MLE). We shall here see that blocking induces an additional class of local likelihood based MLEs. The difference between the former and the latter depends on the continuity assumptions made on the volatility process, and is a little like the difference between the Kaplan-Meier (Kaplan and Meier (1958)) and Nelson-Aalen (Nelson (1969), Aalen (1976, 1978)) estimators in survival analysis. (Note that the variance estimate for the Haysahi-Yoshida estimator from Section 5.3 of Mykland (2010a) obviously also remains valid in the setting of this paper).

For simplicity, work with a bivariate process, and let the grid $\mathcal{G}_n$ be given by (2.61). For now, let the block dividers τ be any subset of $\mathcal{G}_n$. Under the approximate measure Q_n, note that for

$$\tau_{n,i-1} \leq t^{(1)}_{n,j-1} < t^{(1)}_{n,j} \leq \tau_{n,i} \text{ and } \tau_{n,i-1} \leq t^{(2)}_{k-1} < t^{(2)}_{k} \leq \tau_{n,i} \qquad (2.62)$$

the set of returns $X^{(1)}_{t^{(1)}_{n,j}} - X^{(1)}_{t^{(1)}_{n,j-1}}$ and $X^{(2)}_{t^{(2)}_{n,k}} - X^{(2)}_{t^{(2)}_{n,k-1}}$ are conditionally jointly normal with mean zero and covariances

$$\mathrm{Cov}_{Q_n}((X^{(r)}_{t^{(r)}_{n,j}} - X^{(r)}_{t^{(r)}_{n,j-1}}, X^{(s)}_{t^{(s)}_{n,k}} - X^{(s)}_{t^{(s)}_{n,k-1}}) \mid \mathcal{F}_{\tau_{n,i-1}})$$
$$= (\zeta_{\tau_{n,i-1}})_{r,s} d\{(t^{(r)}_{n,j-1}, t^{(r)}_{n,j}) \cap (t^{(s)}_{n,k-1}, t^{(s)}_{n,k})\}$$

where d is length (Lebesgue measure). Set

$$\kappa_{r,s;j,k} = \zeta d\{(t^{(r)}_{n,j-1}, t^{(r)}_{n,j}) \cap (t^{(s)}_{n,k-1}, t^{(s)}_{n,k})\}.$$

The Q_n log likelihood ratio based on observations fully in block $(\tau_{n,i-1}, \tau_{n,i}]$ is therefore given as

$$\begin{aligned} \ell(\zeta) = \quad & -\frac{1}{2} \ln \det(\kappa) \\ & -\frac{1}{2} \sum_{r,s,j,k} \kappa^{r,s;j,k} (X^{(r)}_{t^{(r)}_{n,j}} - X^{(r)}_{t^{(r)}_{n,j-1}})(X^{(s)}_{t^{(s)}_{n,k}} - X^{(s)}_{t^{(s)}_{n,k-1}}) \\ & -\frac{N_i}{2} \ln(2\pi), \end{aligned}$$

where $\kappa^{r,s;j,k}$ are the elements of the matrix inverse of $(\kappa_{r,s;j,k})$, and N_i is a measure of block sample size. The sum in (j,k) is over all intersections $(t^{(r)}_{n,j-1}, t^{(r)}_{n,j}) \cap (t^{(s)}_{n,k-1}, t^{(s)}_{n,k})$ with positive length satisfying (2.62). Call the number of such terms

$$\begin{aligned} m^{(r,s)}_{n,i} = \quad & \# \text{ nonempty intersections } (t^{(r)}_{n,j-1}, t^{(r)}_{n,j}) \cap (t^{(s)}_{n,k-1}, t^{(s)}_{n,k}) \\ & \text{satisfying (2.62)} . \end{aligned}$$

The "parameter" ζ corresponds to $\zeta_{\tau_{n,i-1}}$. The block MLE is thus given as

$$\hat{\zeta}^{(r,s)}_{\tau_{n,i-1}} = \frac{1}{m^{(r,s)}_{n,i}} \sum_{j,k} \frac{(X^{(r)}_{t^{(r)}_{n,j}} - X^{(r)}_{t^{(r)}_{n,j-1}})(X^{(s)}_{t^{(s)}_{n,k}} - X^{(s)}_{t^{(s)}_{n,k-1}})}{d\{(t^{(r)}_{n,j-1}, t^{(r)}_{n,j}) \cap (t^{(s)}_{n,k-1}, t^{(s)}_{n,k})\}} \qquad (2.63)$$

where the sum is over j, k satisfying (2.62) for which the denominator in the summand is nonzero. The overall estimate of covariation is thus

$$\widehat{\langle X^{(r)}, X^{(s)} \rangle}_T = \sum_i \hat{\zeta}^{(r,s)}_{\tau_{n,i-1}} (\tau_{n,i} - \tau_{n,i-1}).$$

We suppose, of course, that each block is large enough for $m^{(r,s)}_{n,i}$ to be always greater than zero.

Under Q_n, $E_{Q_n}(\hat{\zeta}_{\tau_{n,i-1}} | \mathcal{F}_{\tau_{n,i-1}}) = \zeta_{\tau_{n,i-1}}$, and

$$\mathrm{Var}_{Q_n}(\hat{\zeta}^{(r,s)}_{\tau_{n,i-1}} | \mathcal{F}_{\tau_{n,i-1}})$$

$$= \left(\frac{1}{m^{(r,s)}_{n,i}}\right)^2 \left(\zeta^{(r,r)}_{\tau_{n,i-1}} \zeta^{(s,s)}_{\tau_{n,i-1}} \sum_{j,k} \frac{(t^{(r)}_{n,j} - t^{(r)}_{n,j-1})(t^{(s)}_{n,k} - t^{(s)}_{n,k-1})}{d\{(t^{(r)}_{n,j-1}, t^{(r)}_{n,j}) \cap (t^{(s)}_{n,k-1}, t^{(s)}_{n,k})\}^2} \right.$$

$$+ (\zeta^{(r,s)}_{\tau_{n,i-1}})^2 \sum_{j_1, j_2, k_1, k_2} \left(\frac{d\{(t^{(r)}_{n,j_1-1}, t^{(r)}_{n,j_1}) \cap (t^{(s)}_{n,k_2-1}, t^{(s)}_{n,k_2})\}}{d\{(t^{(r)}_{n,j_1-1}, t^{(r)}_{n,j_1}) \cap (t^{(s)}_{n,k_1-1}, t^{(s)}_{n,k_1})\}} \right.$$

$$\left. \left. \times \frac{d\{(t^{(r)}_{n,j_2-1}, t^{(r)}_{n,j_2}) \cap (t^{(s)}_{n,k_1-1}, t^{(s)}_{n,k_1})\}}{d\{(t^{(r)}_{n,j_2-1}, t^{(r)}_{n,j_2}) \cap (t^{(s)}_{n,k_2-1}, t^{(s)}_{n,k_2})\}} \right) \right). \qquad (2.64)$$

The first sum is over the same (j, k) as in (2.63), and the second sum is over all j_1, j_2, k_1, k_2 satisfying (2.62), again for which the denominator in the summand is nonzero.

It is therefore easy to see that subject to conditions on the observation times $t^{(r)}_{n,i}$ and $t^{(s)}_{n,i}$, $n^{1/2}(\widehat{\langle X^{(r)}, X^{(s)} \rangle}_T - \langle X^{(r)}, X^{(s)} \rangle_T)$ converges stably (under Q_n), to a mixed normal distribution with variance as the limit of

$$n \sum_i \mathrm{Var}_{Q_n}(\hat{\zeta}^{(r,s)}_{\tau_{n,i-1}} | \mathcal{F}_{\tau_{n,i-1}})(\tau_{n,i} - \tau_{n,i-1})^2. \qquad (2.65)$$

It is straightforward to see that there is no adjustment from Q_n to P. A formal asymptotic analysis would be tedious, and has therefore been omitted. In any case, to estimate the asymptotic variance, one would use (2.64) – (2.65), with $\hat{\zeta}_{\tau_{n,i-1}}$ replacing $\zeta_{\tau_{n,i-1}}$ in (2.64).

Remark 2.46 *An important difference from the Hayashi-Yoshida estimator is that (2.63) depends on the observation times. This is in many instances undesirable, and the choice of estimator will depend on the degree to which these times are trusted. The Hayashi-Yoshida estimator is also aesthetically more*

pleasing. We note, however, that from likelihood considerations, the estima-
tor (2.63) will have an asymptotic variance which, as the block size tends to
infinity, converges to a limit which corresponds to the efficient minimum for
constant volatility matrix.

This phenomenon can be best illustrated for a scalar process (so there is no
asynchronicity). In this case, our estimator (2.63) of $\langle X, X \rangle_T$ becomes (for
block size M fixed)

$$\widehat{\langle X, X \rangle}_T = \sum_i (\tau_{n,i} - \tau_{n,i-1}) \frac{1}{M} \sum_{j: \, \tau_{n,i-1} < t_{n,j} \leq \tau_{n,i}} \frac{\Delta X_{t_{n,j}}^2}{\Delta t_{n,j}}. \qquad (2.66)$$

It is easy to see, by the methods in this paper, or directly, that for this esti-
mator, the asymptotic variance is $2T \int_0^T \sigma_t^4 dt$, while for the standard realized
volatility, the corresponding expression is $2T \int_0^T \sigma_t^4 H'(t) dt$, where $H(t)$ is
the asymptotic quadratic variation of time (2.39). It is always the case that
$H'(t) \geq 1$, and when observations are sufficiently irregular (under, say, Pois-
son sampling), the inequality is strict, cf. Section 2.7.2 below. Thus, (2.66) is
more efficient than regular realized volatility, but since the times can in many
cases not be trusted, the realized volatility remains a main tool for estimating
volatility.

2.6.4 More complicated data generating mechanisms

Jumps

We only consider the case of finitely many jumps (compound Poisson processes, and similar). The conceptually simplest approach is to remove these
jumps using the kind of procedure described in Mancini (2001) and Lee and
Mykland (2006). The procedure will detect all intervals $(t_{n,j-1}, t_{n,j}]$, with
probability tending to one (exponentially fast) as $n \to \infty$. If one simply removes the detected intervals from the analysis, it is easy to see that our asymptotic results go through unchanged.

The case of infinitely many jumps is more complicated, and beyond the scope
of this paper.

Note that there is a range of approaches for estimating the continuous part
of volatility in such data. Methods include bi- and multi-power (Barndorff–
Nielsen and Shephard (2004b)). Other devices are considered by Aït-Sahalia
(2004), and Aït-Sahalia and Jacod (2007). One can use our method of analysis
for all of these approaches.

Microstructure noise

The presence of noise does not alter the analysis in any major way. Suppose one observes

$$Y_{t_{n,j}} = X_{t_{n,j}} + \epsilon_{n,j}$$

where the $\epsilon_{n,j}$'s are independent of the (X_t) process. The latter still follows (2.25). We take the σ-field $\mathcal{X}_{n,n}$ to be generated by $\{X_{t_{n,j}}, \epsilon_{n,j}, 0 \le j \le n\}$. Suppose that P_1 and P_2 are two measures on $\mathcal{X}_{n,n}$ for which: (1) the variables $\{\epsilon_{n,j}, 0 \le j \le n\}$ are independent of $\{X_{t_{n,j}}, 0 \le j \le n\}$, and (2) the variables $\{\epsilon_{n,j}, 0 \le j \le n\}$ have the same distribution under P_1 and P_2. Then, from standard results in measure theory,

$$\frac{dP_2}{dP_1}\left((X_{t_{n,j}}, \epsilon_{n,j}), 0 \le j \le n\right) = \frac{dP_2}{dP_1}\left(X_{t_{n,j}}, 0 \le j \le n\right).$$

The results in our theorems are therefore unchanged in the case of microstructure noise (unless one also wants to change the probability distribution of the noise). We note that this remains the case irrespective of the internal dependence structure of the noise.

The key observation which leads to this easy extension is that it is not required for our results to work that the observables $Y_{t_{n,j}}$ generate the σ-field $\mathcal{X}_{n,n}$. It is only required that the observables be measurable with respect to this σ-field. The same principle was invoked in Section 2.6.3.

The extension does not, obviously, solve all problems relating to microstructure noise, since this type of data generating mechanism is best treated with an asymptotics where $M \to \infty$ as $n \to \infty$. This is currently under investigation. For one approach, see Mykland and Zhang (2011).

2.7 Irregularly spaced data

2.7.1 A second block approximation.

The approximation in Section 2.6.1 reduces the problem (in each block) to a case of independent (but not identically distributed) increments. Can we do better than this, and go to iid observations? We here give a criterion for this to be the case.

Consider yet another approximate probability measure R_n, under which $X_0 = x_0$, and

$$\text{for each } i = 1, ..., K_n : \quad \Delta X_{t_{j+1}} = \sigma_{\tau_{n,i-1}}\left(\frac{\Delta\tau_i}{\Delta t_{j+1}M_i}\right)^{1/2}\Delta W^*_{t_{j+1}}$$

$$\text{for } t_{n,j+1} \in (\tau_{n,i-1}, \tau_{n,i}]. \tag{2.67}$$

Formally, we define the approximation as follows.

Definition 2.47 R_n is defined as Q_n in Definition 2.42, but with (2.67) replacing (2.53).

The crucial fact will be that under R_n, the observables $\Delta X_{t_{j+1}}$ are conditionally iid $N(0, \zeta_{\tau_{n,i-1}} \Delta \tau_i / M_i)$ for $t_{n,j+1} \in (\tau_{n,i-1}, \tau_{n,i}]$.

So that the following can be used together with the results in Mykland and Zhang (2009), we will in the following let the process X be multivariate, and we make the following assumption

Condition 2.48 *(Structure of the instantaneous volatility). We assume that the matrix process σ_t is itself an Itô process, and that if $\lambda_t^{(p)}$ is the smallest eigenvalue of σ_t, then $\inf_t \lambda_t^{(p)} > 0$ a.s.*

The contiguity question is then addressed as follows. Let P_n^* be the measure from Remark 2.45 (corresponding to block length $M = 1$). Recall that

$$\log \frac{dR_n}{dP} = \log \frac{dR_n}{dQ_n} + \log \frac{dQ_n}{dP_n^*} + \log \frac{dP_n^*}{dP}.$$

Define

$$B_{n,j} = \left(\Delta t_{n,j+1} \left(\frac{\Delta \tau_{n,i}}{M_i} \right)^{-1} - 1 \right).$$

Theorem 2.49 *(Asymptotic relationship between P_n^*, Q_n and R_n). Assume the conditions of Theorem 4 in Mykland and Zhang (2009), and let $Z_n^{(1)}$ and $M_n^{(1)}$ be as in that theorem (see (2.71) and (2.74) in Section 2.7.3). Assume that the following limits exist:*

$$\Gamma_2 = \frac{p}{2} \lim_{n \to \infty} \sum_j B_{n,j}^2 \text{ and } \Gamma_3 = \frac{p}{2} \lim_{n \to \infty} \sum_j \log(1 + B_{n,j}).$$

Set

$$Z_n^{(2)} = \frac{1}{2} \sum_i \sum_{t_{n,j} \in (\tau_{n,i-1}, \tau_{n,i}]} \Delta X_{t_{n,j}}^T ((\sigma \sigma^T)_{\tau_{n,i-1}}^{-1}) \Delta X_{t_{n,j}}$$

$$\times \left(\Delta t_{n,j+1}^{-1} - \left(\frac{\Delta \tau_{n,i}}{M_i} \right)^{-1} \right),$$

and let $M_n^{(2)}$ be the end point of the martingale part of $Z_n^{(2)}$ (see (2.72) and (2.74) in Section 2.7.3 for the explicit formula). Then, as $n \to \infty$, $(M_n^{(1)}, M_n^{(2)})$ converges stably in law under P^* to a normal distribution with mean zero and diagonal variance matrix with diagonal elements Γ_1 and Γ_2. Also, under P^*,

$$\log \frac{dR_n}{dQ_n} = M_n^{(2)} + \Gamma_3 + o_p(1).$$

The theorem can be viewed from the angle of contiguity:

Corollary 2.50 *Under regularity conditions, the following statements are equivalent, as $n \to \infty$:*
(i) R_n is contiguous to P.
(ii) R_n is contiguous to Q_n.
(iii) The following relationship holds:

$$\Gamma_3 = -\frac{1}{2}\Gamma_2. \tag{2.68}$$

As we shall see, the requirement (2.68) is a substantial restriction. Corollary 2.50 says that unlike the case of Q_n, inference under R_n may not give rise to desired results. Part of the probability mass under Q_n (and hence P) is not preserved under R_n.

To understand the requirement (2.68), note that

$$\frac{p}{2}\sum_j \log(1 + B_{n,j}) = -\frac{p}{4}\sum_j B_{n,j}^2 + \frac{p}{6}\sum_j B_{n,j}^3 - \cdots$$

since $\sum_j B_{n,j} = 0$. Hence, (2.68) will, for example, be satisfied if $\max_j |B_{n,j}| \to 0$ as $n \to \infty$. One such example is

$$t_{n,j} = f(j/n) \text{ and } f \text{ is continuously differentiable.} \tag{2.69}$$

However, (2.69) will not hold in more general settings, as we shall see from the following examples.

Example 2.51 (POISSON SAMPLING.) *Suppose that the sampling time points follow a Poisson process with parameter λ. If one conditions on the number of sampling points n, these points behave like the order statistics of n uniformly distributed random variables (see, for example, Chapter 2.3 in S. Ross (1996)). Consider the case where $M_i = M$ for all but (possibly) the last interval in $\mathcal{H}_n$. In this case, K_n is the smallest integer larger than or equal to n/M. Let Y_i be the M-tuple $(B_j, \tau_{i-1} \le t_j < \tau_i)$.*

We now obtain, by passing between the conditional and unconditional, that $Y_1, ..., Y_{K_n-1}$ are iid, and the distribution can be described by

$$Y_1 = M(U_{(1)}, U_{(2)} - U_{(1)}, ..., U_{(M-1)} - U_{(M-2)}, 1 - U_{(M-1)}) - 1,$$

where $U_{(1)}, ..., U_{(M-1)}$ is the order statistic of $M - 1$ independent uniform random variables on $(0, 1)$. It follows that

$$\sum_j B_{n,j}^2 = \frac{n}{M}(M^2 E U_{(1)}^2 - 1) + o_p(n)$$

$$\sum_j \log(1 + B_{n,j}) = \frac{n}{M} E \log(M U_{(1)}) + o_p(n)$$

since $EU_{(1)}^2 = 2/(M+1)(M+2)$. *Hence, both Γ_2 and Γ_3 are infinite. The contiguity between R_n and the other probabilities fails. On the other hand, all our assumptions up to Section 2.6 are satisfied, and so P, P_n^* and Q_n are all contiguous. The AQVT (equation (2.39)) is given by $H(t) = 2t$. Also, if the block size is constant (size M), the ADD is $K(t) = (M-1)t$.*

Example 2.52 (SYSTEMATIC IRREGULARITY.) *Let ϵ be a small positive number, and let $\Delta t_{n,j} = (1+\epsilon)T/n$ for odd j and $\Delta t_{n,j} = (1-\epsilon)T/n$ for even j (with $\Delta t_{n,n} = T/n$ for odd n). Again, all our assumptions up to Section 2.6 are satisfied. The AQVT is given by $H(t) = t(1+\epsilon^2)$. If we suppose that all $M_i = 2$, the ADD becomes $K(t) = t$. On the other hand, $B_{n,j} = \pm\epsilon$, so that, again, both Γ_2 and Γ_3 are infinite. The contiguity between R_n and the other probabilities thus fails in the same radical fashion as in the case of Poisson sampling.*

2.7.2 Irregular spacing and subsampling

We here return to a more direct study of the effect of irregular spacings. We put ourselves in the situation from Section 2.4.3, where observation times are independent of the process. As stated in equation (2.44), the limit law for the realized volatility (for $\sqrt{n}\left([X,X]_t^{\mathcal{G}_n} - [X,X]_t\right)$) is mixed normal with (random) variance

$$2T \int_0^t \sigma_s^4 dH_s,$$

where H is the asymptotic quadratic variation of time (AQVT). When observations are equidistant, $H'(t) \equiv 1$. From the preceeding section, we also know that if times are on the form (2.69), the asymptotic variance is unaffected. It is worth elaborating on this in direct computation. Set

$$F(t) = \lim_{n\to\infty} \frac{1}{n}\#\{t_{n,i+1} \le t\}.$$

This quantity exists, if necessary by going through subsequences (Helly's Theorem, see, for example, p. 336 in Billingsley (1995)). Set

$$u_{n,i} = F(t_{n,i}). \tag{2.70}$$

Asymptotically, the $u_{n,i}$ are equispaced:

$$\frac{1}{n}\#\{u_{n,i+1} \le t\} = \frac{1}{n}\#\{t_{n,i+1} \le F^{(-1)}(t)\} \to F(F^{(-1)}(t)) = t$$

Inference is invariant to this transformation: Observing the process X_t at times $t_{n,i}$ is the same as observing the process $Y_t = X_{F^{(-1)}(t)}$ at times $u_{n,i}$. If we set $\mathcal{U} = \{u_{n,j}, j=0,...,n\}$, then $[X,X]_T^{\mathcal{G}} = [Y,Y]_T^{\mathcal{U}}$. Also, in the limit,

$[X, X]_T = [Y, Y]_T$. Finally, the asymptotic distribution is the same in these two cases.

If the $u_{n,i}$ have AQVT $U(t)$, the mixed normal variance transforms

$$2T \int_0^T H'(u)(\langle X, X \rangle_t')^2 dt = 2 \int_0^1 U'(u)(\langle Y, Y \rangle_t')^2 dt.$$

The transformation (2.70) regularizes spacing. It means that without loss of generality, one can take $T = 1$, $F' = 1$ and $U = H$. Also, the transformation (2.70) regularizes spacing defined by (2.69), and in this case, $U'(t) \equiv 1$.

Example 2.53 *On the other hand, it is clear from Example 2.51 that it is possible for $U'(t)$ to take other values than 1. The example shows that for Poisson distributed observation times, $H' = U' \equiv 2$, while, indeed $F'(t) \equiv 1/T$.*

The general situation can be expressed as follows:

Proposition 2.54 *Assume that F exists and is monotonely increasing. Also assume that H exists. Then U exists. For all $s \leq t$, $U(t) - U(s) \geq t - s$. In particular, if $U'(t)$ exists, then $U'(t) \geq 1$. The following statements are equivalent:*
(i) $U(1) = 1$
(ii) $U' \equiv 1$
(iii) $\sum_{j=0}^n \left(u_{n,j+1} - u_{n,j} - \frac{1}{n} \right)^2 = o_p(n^{-1})$.

Proof of Proposition 2.54. The first statement uses a standard property of the variance: if $\Delta t_{n,j+1} = t_{n,j+1} - t_{n,j}$, and $\overline{\Delta}_n = T/n$, then

$$\frac{n}{T} \sum_{t_{n,j+1} \leq t} (\Delta t_{n,j+1})^2$$

$$= \frac{n}{T} \sum_{t_{n,j+1} \leq t} (\Delta t_{n,j+1} - \overline{\Delta}_n)^2 + \frac{n}{T} \#\{t_{n,i+1} \leq t\}(\overline{\Delta}_n)^2$$

$$\geq \frac{n}{T} \#\{t_{n,i+1} \leq t\}(\overline{\Delta}_n)^2.$$

By taking limits as $n \to \infty$ under $F'(t) \equiv 1/T$, we get that $H(t) - H(s) \geq t - s$. In particular, the same will be true for U.

The equivalence between (i) and (iii) follows from the proof of Lemma 2 (p. 1029) in Zhang (2006). (The original lemma uses slightly different assumptions). □

The implication of the proposition is that under the scenario $U(1) = 1$, observation times are "almost" equidistant. In particular, subsampling does not change the structure of the spacings. On the other hand, when $U(1) > 1$, there is scope for subsampling to regularize the times.

Example 2.55 *Suppose that the times are Poisson distributed. Instead of picking every observation, we now pick every M'th observation. By the same methods as in Example 2.51, we obtain that*

$$U(t) = \frac{M+1}{M}t.$$

Hence the sparser the subsampling, the more regular the times will be. This is an additional feature of subsampling that remains to be exploited.

2.7.3 Proof of Theorem 2.49

We begin by describing the relationship between R_n and P_n^*. In analogy with Proposition 2 of Mykland and Zhang (2009), we obtain that

Lemma 2.56

$$\log \frac{dR_n}{dP_n^*}(U_{t_0}, ..., U_{t_{n,j}}, ..., U_{t_{n,n}})$$

$$= \sum_i \sum_{\tau_{i-1} \le t_j < \tau_i} \left\{ \ell(\Delta X_{t_{j+1}}; \zeta_{\tau_{n,i-1}} \Delta \tau_i / M_i) - \ell(\Delta X_{t_{j+1}}; \zeta_{t_{n,j}} \Delta t_{j+1}) \right\}.$$

Now set $\zeta_t = \sigma_t \sigma_t^T$ (superscript "T" meaning transpose).

PROOF OF THEOREM 2.49. Let $Z_n^{(1)}$ and $Z_n^{(2)}$ be as in the statement of the theorem. Set

$$\Delta Z_{n,t_{n,j+1}}^{(1)} = \frac{1}{2} \Delta X_{t_{n,j+1}}^T (\zeta_{t_{n,j}}^{-1} - \zeta_{\tau_{n,i-1}}^{-1}) \Delta X_{t_{n,j+1}} \Delta t_{n,j+1}^{-1}$$

$$\Delta Z_{n,t_{n,j+1}}^{(2)} = \frac{1}{2} \Delta X_{t_{n,j+1}}^T (\zeta_{\tau_{n,i-1}}^{-1}) \Delta X_{t_{n,j+1}} \left(\Delta t_{n,j+1}^{-1} - \left(\frac{\Delta \tau_{n,i}}{M_i} \right)^{-1} \right)$$

$$\hspace{10cm} (2.71)$$

and note that $Z_n^{(v)} = \sum_j \Delta Z_{n,t_{n,j+1}}^{(v)}$ for $v = 1, 2$. Set $A_j = \zeta_{t_{n,j}}^{1/2} \zeta_{\tau_{n,i-1}}^{-1} \zeta_{t_{n,j}}^{1/2} - I$ and $B_j = \left(\Delta t_{n,j+1} \left(\frac{\Delta \tau_{n,i}}{M_i} \right)^{-1} - 1 \right)$ (the latter is a scalar). Set $C_j = \zeta_{t_{n,j}}^{1/2} \zeta_{\tau_{n,i-1}}^{-1} \zeta_{t_{n,j}}^{1/2} \left(\Delta t_{n,j+1} \left(\frac{\Delta \tau_{n,i}}{M_i} \right)^{-1} - 1 \right) = (I + A_j) B_j$.

Since $\Delta X_{t_{n,j}}$ is conditionally Gaussian, we obtain (under P_n^*)

$$E_{P_n^*}(\Delta Z_{n,t_{n,j+1}}^{(1)} | \mathcal{X}_{n,t_{n,j}}) = -\frac{1}{2} \text{tr}(A_j)$$

$$E_{P_n^*}(\Delta Z_{n,t_{n,j+1}}^{(2)} | \mathcal{X}_{n,t_{n,j}}) = -\frac{1}{2} \text{tr}(C_j) = -\frac{1}{2}(p + \text{tr}(A_j)) B_j \hspace{1cm} (2.72)$$

and

$$\text{conditional covariance of } \Delta Z^{(1)}_{n,t_n,j+1} \text{ and } \Delta Z^{(2)}_{n,t_n,j+1} = \frac{1}{2}\begin{pmatrix} \text{tr}(A_j^2) & \text{tr}(A_jC_j) \\ \text{tr}(A_jC_j) & \text{tr}(C_j^2) \end{pmatrix}. \quad (2.73)$$

Finally, let $M_n^{(v)}$ be the (end point of the) martingale part (under P^*) of $Z_n^{(v)}$ ($v = 1, 2$), so that

$$M_n^{(1)} = Z^{(1)} + (1/2)\sum_j \text{tr}(A_j) \text{ and } M_n^{(2)} = Z^{(2)} + (1/2)\sum_j \text{tr}(C_j). \quad (2.74)$$

If $\langle \cdot, \cdot \rangle^{\mathcal{G}}$ represents discrete time predictable quadratic variation on the grid $\mathcal{G}$, then equation (2.73) yields

$$\begin{pmatrix} \langle M_n^{(1)}, M_n^{(1)}\rangle^{\mathcal{G}} & \langle M_n^{(1)}, M_n^{(2)}\rangle^{\mathcal{G}} \\ \langle M_n^{(1)}, M_n^{(2)}\rangle^{\mathcal{G}} & \langle M_n^{(2)}, M_n^{(2)}\rangle^{\mathcal{G}} \end{pmatrix} = \frac{1}{2}\sum_j \begin{pmatrix} \text{tr}(A_j^2) & \text{tr}(A_jC_j) \\ \text{tr}(A_jC_j) & \text{tr}(C_j^2) \end{pmatrix}. \quad (2.75)$$

The following is shown in ibid., Appendix B:

$$\langle M_n^{(1)}, M_n^{(1)}\rangle^{\mathcal{G}} = \Gamma_1 + o_p(1), \quad (2.76)$$

where K is the ADD given by equation (2.54),

$$\sup_j \text{tr}(A_j^2) \to 0 \text{ as } n \to \infty, \quad (2.77)$$

$$\text{for } r > 2, |\text{tr}(A_j^r)| \le \text{tr}(A_j^2)^{r/2}, \quad (2.78)$$

and

$$\log \frac{dQ_n}{dP_n^*} = M_n^{(1)} - \frac{1}{2}\langle M_n^{(1)}, M_n^{(1)}\rangle^{\mathcal{G}} + o_p(1).$$

Now observe that by (2.76) – (2.78),

$$\sum_j \text{tr}(C_j^2) = \sum_j \text{tr}(I_p)B_j^2 + \sum_j \text{tr}(A_j)B_j^2 + \sum_j \text{tr}(A_j^2)B_j^2$$
$$= p\sum_j B_j^2 + o_p(1)$$
$$= 2\Gamma_2 + o_p(1),$$

and

$$\sum_j \text{tr}(A_jC_j) = \sum_j \text{tr}(A_j)B_j + \sum_j \text{tr}(A_j^2)B_j$$
$$= \sum_j \text{tr}(A_j)B_j + o_p(1)$$
$$= o_p(1)$$

where the last transition follows by Condition 2.48. Meanwhile, since

$$\log \frac{dR_n}{dP_n^*} = \log \frac{dR_n}{dQ_n} + \log \frac{dQ_n}{dP_n^*},$$

we obtain similarly that

$$\log \frac{dR_n}{dQ_n} = Z_n^{(2)} + \frac{p}{2} \sum_j \log(1 + B_j)$$

$$= M_n^{(2)} + \Gamma_3 + o_p(1).$$

At this point, let $\langle M_n, M_n \rangle$ be the quadratic variation of the continuous martingale that coincides at points $t_{n,j}$ with the discrete time martingale leading up to the end point $M_n^{(1)}$. By a standard quarticity argument (as in the proof of Remark 2 in Mykland and Zhang (2006)), (2.75) – (2.76) – (2.78) and the conditional normality of $(\Delta Z_{n,t_{n,j+1}}^{(1)}, \Delta Z_{n,t_{n,j+1}}^{(2)})$ yield that $\langle M_n, M_n \rangle = \langle M_n, M_n \rangle^{\mathcal{G}} + o_p(1)$. The stable convergence to a normal distribution with covariance matrix

$$\begin{pmatrix} \Gamma_1 & 0 \\ 0 & \Gamma_2 \end{pmatrix}$$

then follows by the same methods as in Zhang et al. (2005). The result is thus proved. □

Acknowledgments

We would like to thank Mathieu Kessler, Alexander Lindner, and Michael Sørensen, and the anonymous referee, for encouragement and constructive comments. We would also like to thank Eric Renault and D. Christina Wang for a careful reading of the manuscript. We are also grateful to seminar participants in Hong Kong and Bergen for terrific and useful feedback, in particular Knut Aase, Jonas Andersson, Lancelot James, Bingyi Jing, Jostein Lillestøl, Albert Lo, Trygve Nilsen, and Bård Støve. Financial support from the National Science Foundation under grants DMS 06-04758, SES 06-31605, and SES 11-24526 is gratefully acknowledged. We would like to thank Hong Kong University of Science and Technology, where part of the manuscript was written.

References

Aalen, O. (1976). Nonparametric inference in connection with multiple decrement models. *Scand. J. Statist.*, *3*, 15–27.

Aalen, O. (1978). Nonparametric inference for a family of counting processes. *Ann. Statist.*, *6*, 701–726.

Aït-Sahalia, Y. (2004). Disentangling diffusion from jumps. *J. Finan. Econ.*, *74*, 487–528.

Aït-Sahalia, Y., & Jacod, J. (2007). Volatility estimators for discretely sampled Lévy processes. *Ann. Statist.*, *35*, 335–392.

Aït-Sahalia, Y., Mykland, P. A., & Zhang, L. (2011). Ultra high frequency volatility estimation with dependent microstructure noise. *J. Econometrics*, *160*, 160–165.

Aldous, D. J., & Eagleson, G. K. (1978). On mixing and stability of limit theorems. *Ann. Probab.*, *6*, 325–331.

Andersen, P. K., Borgan, Ø., Gill, R. D., & Keiding, N. (1992). *Statistical Models Based on Counting Processes*. New York: Springer.

Andersen, T., Bollerslev, T., & Meddahi, N. (2005). Correcting the errors: Volatility forecast evaluation using high frequency data and realized volatilities. *Econometrica*, *73*, 279–296.

Andersen, T. G., & Bollerslev, T. (1998). Answering the skeptics: Yes, standard volatility models do provide accurate forecasts. *Int. Econ. Rev.*, *39*, 885–905.

Andersen, T. G., Bollerslev, T., Diebold, F. X., & Labys, P. (2000). Great realizations. *Risk*, *13*, 105–108.

Andersen, T. G., Bollerslev, T., Diebold, F. X., & Labys, P. (2001). The distribution of realized exchange rate volatility. *J. Amer. Statistical Assoc.*, *96*, 42–55.

Andersen, T. G., Bollerslev, T., Diebold, F. X., & Labys, P. (2003). Modeling and forecasting realized volatility. *Econometrica*, *71*, 579–625.

Barndorff–Nielsen, O. E., Hansen, P. R., Lunde, A., & Shephard, N. (2008). Designing realized kernels to measure ex-post variation of equity prices in the presence of noise. *Econometrica*, *76*, 1481–1536.

Barndorff–Nielsen, O. E., Hansen, P. R., Lunde, A., & Shephard, N. (2011). Multivariate realised kernels: Consistent positive semi-definite estimators of the covariation of equity prices with noise and non-synchronous trading. *J. Econometrics, 162,* 149–169.

Barndorff–Nielsen, O. E., & Shephard, N. (2001). Non-Gaussian Ornstein-Uhlenbeck-based models and some of their uses in financial economics. *J. R. Stat. Soc. Ser. B Stat. Methodol., 63,* 167–241.

Barndorff–Nielsen, O. E., & Shephard, N. (2002). Econometric analysis of realized volatility and its use in estimating stochastic volatility models. *J. R. Stat. Soc. Ser. B Stat. Methodol., 64,* 253–280.

Barndorff–Nielsen, O. E., & Shephard, N. (2004a). Econometric analysis of realised covariation: high frequency based covariance, regression and correlation in financial economics. *Econometrica, 72,* 885–925.

Barndorff–Nielsen, O. E., & Shephard, N. (2004b). Power and bipower variation with stochastic volatility and jumps (with discussion). *J. Finan. Econ., 2,* 1–48.

Bibby, B. M., Jacobsen, M., & Sørensen, M. (2009). Estimating functions for discretely sampled diffusion-type models. In Y. Aït-Sahalia & L. P. Hansen (Eds.), *Handbook of Financial Econometrics* (pp. 203–268). North Holland, Oxford.

Bibby, B. M., Skovgaard, I. M., & Sørensen, M. S. (2005). Diffusion-type models with given marginal distribution and autocorrelation function. *Bernoulli, 11,* 191–220.

Billingsley, P. (1995). *Probability and Measure* (Third ed.). New York: Wiley.

Black, F., & Scholes, M. (1973). The pricing of options and corporate liabilities. *J. Polit. Economy, 81,* 637–654.

Cochran, W. (1934). The distribution of quadratic forms in a normal system, with applications to the analysis of variance. *Proc. Camb. Phil. Soc., 30,* 178–191.

Comte, F., & Renault, E. (1998). Long memory in continuous-time stochastic volatility models. *Math. Finance, 8,* 291–323.

Dacorogna, M. M., Gençay, R., Müller, U., Olsen, R. B., & Pictet, O. V. (2001). *An Introduction to High-Frequency Finance.* San Diego: Academic Press.

Dambis, K. (1965). On the decomposition of continuous sub-martingales. *Theory Probab. Appl., 10,* 401–410.

Delattre, S., & Jacod, J. (1997). A central limit theorem for normalized functions of the increments of a diffusion process, in the presence of round-off errors. *Bernoulli, 3,* 1–28.

Delbaen, F., & Schachermayer, W. (1995). The existence of absolutely continuous local martingale measures. *Ann. Appl. Probab., 5,* 926–945.

Dellacherie, C., & Meyer, P. (1982). *Probabilities and Potential B.* Amsterdam: North-Holland.

Ditlevsen, P., Ditlevsen, S., & Andersen, K. (2002). The fast climate fluctuations during the stadial and interstadial climate states. *Annals of Glaciology*, *35*, 457–462.

Ditlevsen, S., & Sørensen, M. (2004). Inference for observations of integrated diffusion processes. *Scand. J. Statist.*, *31*, 417–429.

Dubins, L., & Schwartz, G. (1965). On continuous martingales. *Proc. Nat. Acad. Sci. U.S.A.*, *53*, 913–916.

Duffie, D. (1996). *Dynamic Asset Pricing Theory*. Princeton, N.J.: Princeton University Press.

Elerian, O., Siddhartha, C., & Shephard, N. (2001). Likelihood inference for discretely observed nonlinear diffusions. *Econometrica*, *69*, 959–993.

Feller, W. (1951). Two singular diffusion problems. *Ann. Math.*, *54*, 173–182.

Fleming, J., Kirby, C., & Ostdiek, B. (2001). The economic value of volatility timing. *J. Finance*, *56*, 329–352.

Fleming, J., Kirby, C., & Ostdiek, B. (2003). The economic value of volatility timing using realized volatility. *J. Finan. Econ.*, *67*, 473–509.

Foster, D., & Nelson, D. (1996). Continuous record asymptotics for rolling sample variance estimators. *Econometrica*, *64*, 139–174.

Genon-Catalot, V., & Jacod, J. (1994). Estimation of the diffusion coefficient of diffusion processes: Random sampling. *Scand. J. Statist.*, *21*, 193–221.

Genon-Catalot, V., Jeantheau, T., & Larédo, C. (1999). Parameter estimation for discretely observed stochastic volatility models. *Bernoulli*, *5*, 855–872.

Genon-Catalot, V., Jeantheau, T., & Larédo, C. (2000). Stochastic volatility models as hidden markov models and statistical applications. *Bernoulli*, *6*, 1051–1079.

Gikhman, I. I., & Skorohod, A. V. (1969). *Introduction to the Theory of Random Processes*. Philadelphia, P.A.: W. B. Sauders Company.

Gloter, A. (2000). Discrete sampling of an integrated diffusion process and parameter estimation of the diffusion coefficient. *ESAIM Probab. Statist.*, *4*, 205–227.

Gloter, A., & Jacod, J. (2001a). Diffusions with measurement errors. II - optimal estimators. *ESAIM Probab. Statist.*, *5*, 243–260.

Gloter, A., & Jacod, J. (2001b). Diffusions with measurement errors. I - local asymptotic normality. *ESAIM Probab. Statist.*, *5*, 225–242.

Hájek, J., & Sidak, Z. (1967). *Theory of Rank Tests*. New York: Academic Press.

Hall, P., & Heyde, C. C. (1980). *Martingale Limit Theory and Its Application*. Boston: Academic Press.

Harrison, M., & Kreps, D. (1979). Martingales and arbitrage in multiperiod securities markets. *J. Econ. Theory*, *20*, 381–408.

Harrison, M., & Pliska, S. (1981). Martingales and stochastic integrals in the theory of continuous trading. *Stochastic Process. Appl.*, *11*, 215–260.

Hayashi, T., & Mykland, P. (2005). Evaluating hedging errors: An asymptotic approach. *Math. Finance*, *15*, 1931–1963.

Hayashi, T., & Yoshida, N. (2005). On covariance estimation of non-synchronously observed diffusion processes. *Bernoulli*, *11*, 359–379.

Heston, S. (1993). A closed-form solution for options with stochastic volatility with applications to bonds and currency options. *Rev. Finan. Stud.*, *6*, 327–343.

Hoffmann, M. (2002). Rate of convergence for parametric estimation in a stochastic volatility model. *Stochastic Process. Appl.*, *97*, 147–170.

Jacobsen, M. (2001). Discretely observed diffusions: Classes of estimating functions and small δ-optimality. *Scand. J. Statist.*, *28*, 123–150.

Jacod, J. (1994). *Limit of random measures associated with the increments of a Brownian semimartingale* (Tech. Rep.). Université de Paris VI.

Jacod, J., Li, Y., Mykland, P. A., Podolskij, M., & Vetter, M. (2009). Microstructure noise in the continuous case: The pre-averaging approach. *Stochastic Process. Appl.*, *119*, 2249–2276.

Jacod, J., & Protter, P. (1998). Asymptotic error distributions for the Euler method for stochastic differential equations. *Ann. Probab.*, *26*, 267–307.

Jacod, J., & Shiryaev, A. N. (2003). *Limit Theorems for Stochastic Processes* (Second ed.). New York: Springer.

Kaplan, E., & Meier, P. (1958). Nonparametric estimation from incomplete observations. *J. Amer. Statistical Assoc.*, *53*, 457–481.

Karatzas, I., & Shreve, S. E. (1991). *Brownian Motion and Stochastic Calculus*. New York: Springer.

LeCam, L. (1986). *Asymptotic Methods in Statistical Decision Theory*. New York: Springer.

LeCam, L., & Yang, G. (2000). *Asymptotics in Statistics: Some Basic Concepts* (Second ed.). New York: Springer.

Lee, S. Y., & Mykland, P. A. (2006). Jumps in financial markets: A new nonparametric test and jump dynamics. *Rev. Finan. Stud.*, *21*, 2535–2563.

Lehmann, E. (1983). *Theory of Point Estimation*. New York: Wiley.

Li, Y., Mykland, P., Renault, E., Zhang, L., & Zheng, X. (2009). *Realized volatility when endogeniety of time matters* (Tech. Rep.). University of Chicago.

Li, Y., & Mykland, P. A. (2007). Are volatility estimators robust with respect to modeling assumptions? *Bernoulli*, *13*, 601–622.

Linton, O. (2007). *Notes about Gaussian calculus.* (Presented in Imperical College Financial Econometrics Conference)

Mancini, C. (2001). Disentangling the jumps of the diffusion in a geometric jumping Brownian motion. *Giornale dell'Istituto Italiano degli Attuari, LXIV*, 19–47.

Mardia, K. V., Kent, J., & Bibby, J. (1979). *Multivariate Analysis*. London: Academic Press.

Meddahi, N. (2001). *An eigenfunction approach for volatility modeling* (Tech. Rep.). Université de Montréal.

Merton, R. C. (1973). The theory of rational option pricing. *Bell J. Econ. Manag. Sci., 4*, 141–183.

Mykland, P. A. (1995). Embedding and asymptotic expansions for martingales. *Probab. Theory Rel. Fields, 103*, 475–492.

Mykland, P. A. (2000). Conservative delta hedging. *Ann. Appl. Probab., 10*, 664–683.

Mykland, P. A. (2003a). Financial options and statistical prediction intervals. *Ann. Statist., 31*, 1413–1438.

Mykland, P. A. (2003b). The interpolation of options. *Finance Stochastics, 7*, 417–432.

Mykland, P. A. (2005). *Combining statistical intervals and market proces: The worst case state price distribution* (Tech. Rep.). University of Chicago.

Mykland, P. A. (2010a). A Gaussian calculus for inference from high frequency data. *Ann. Finance (to appear)*.

Mykland, P. A. (2010b). Option pricing bounds and statistical uncertainty: Using econometrics to find an exit strategy in derivatives trading. In Y. Aït-Sahalia & L. P. Hansen (Eds.), *Handbook of Financial Econometrics* (Vol. II, pp. 135–196). North Holland, Oxford.

Mykland, P. A., & Zhang, L. (2006). ANOVA for diffusions and Itô processes. *Ann. Statist., 34*, 1931–1963.

Mykland, P. A., & Zhang, L. (2008). Inference for volatility type objects and implications for hedging. *Statistics and its Interface, 1*, 255–278.

Mykland, P. A., & Zhang, L. (2009). Inference for continuous semimartingales observed at high frequency. *Econometrica, 77*, 1403–1455.

Mykland, P. A., & Zhang, L. (2011). The double Gaussian approximation for high frequency data. *Scand. J. Statist., 38*, 215–236.

Neftci, S. N. (2000). *Introduction to the Mathematics of Financial Derivatives* (Second ed.). Amsterdam: Academic Press.

Nelson, W. (1969). Hazard plotting for incomplete failure data. *J. Qual. Tech., 1*, 27–52.

Øksendal, B. (2003). *Stochastic Differential Equations: An Introduction with Applications* (Sixth ed.). Berlin: Springer.

Podolskij, M., & Vetter, M. (2009). Estimation of volatility functionals in the simultaneous presence of microstructure noise and jumps. *Bernoulli, 15*, 634–658.

Pollard, D. (1984). *Convergence of Stochastic Processes*. New York: Springer.

Protter, P. (2004). *Stochastic Integration and Differential Equations* (Second ed.). New York: Springer.

Rényi, A. (1963). On stable sequences of events. *Sankyā Series A, 25*, 293–302.

Rice, J. (2006). *Mathematical Statistics and Data Analysis* (Third ed.). Duxbury Press.

Rootzén, H. (1980). Limit distributions for the error in approximations of stochastic integrals. *Ann. Probab., 8*, 241–251.

Ross, S. (1996). *Stochastic Processes* (Second ed.). New York, N.Y.: Wiley.

Ross, S. M. (1976). The arbitrage theory of capital asset pricing. *J. Econ. Theory, 13*, 341–360.

Shreve, S. E. (2004). *Stochastic Calculus for Finance II*. New York: Springer-Verlag.

Sørensen, H. (2001). Discretely observed diffusions: Approximation of the continuous-time score function. *Scand. J. Statist., 28*, 113–121.

Valdés-Sosa, P. A., Bornot-Sánchez, J. M., Melie-García, L., Lage-Castellanos, A., & Canales-Rodriguez, E. (2007). Granger causality on spatial manifolds: Applications to neuroimaging. In B. Schelter, M. Winterhalder, & J. Timmer (Eds.), *Handbook of Time Series Analysis: Recent Theoretical Developments* (pp. 461–485). New York: Wiley.

Xiu, D. (2010). Quasi-maximum likelihood estimation of volatility with high frequency data. *J. Econometrics, 159*, 235–250.

Zhang, L. (2001). *From martingales to ANOVA: Implied and realized volatility*. Unpublished doctoral dissertation, The University of Chicago, Department of Statistics.

Zhang, L. (2006). Efficient estimation of stochastic volatility using noisy observations: A multi-scale approach. *Bernoulli, 12*, 1019–1043.

Zhang, L. (2009). Implied and realized volatility: empirical model selection. *Ann. Finance (to appear)*.

Zhang, L. (2011). Estimating covariation: Epps effect and microstructure noise. *J. Econometrics, 160*, 33–47.

Zhang, L., Mykland, P. A., & Aït-Sahalia, Y. (2005). A tale of two time scales: Determining integrated volatility with noisy high-frequency data. *J. Amer. Statistical Assoc., 100*, 1394–1411.

CHAPTER 3

Statistics and high-frequency data

Jean Jacod

Institut de mathématiques de Jussieu
Université Pierre et Marie Curie (Paris-6) and CNRS, UMR 7586
4 place Jussieu, 75252 Paris, France

3.1 Introduction

This short course is devoted to a few statistical problems related to the observation of a given process on a fixed time interval, when the observations occur at regularly spaced discrete times. These kinds of observations may occur in many different contexts, but they are particularly relevant in finance: we do have now huge amounts of data on the prices of various assets, exchange rates, and so on, typically "tick data" which are recorded at every transaction time. So we are mainly concerned with the problems which arise in this context, and the concrete applications we will give are all pertaining to finance.

In some sense they are not "standard" statistical problems, for which we want to estimate some unknown parameter. We are rather concerned with the "estimation" of some random quantities. This means that we would like to have procedures that are as model-free as possible, and also that they are in some sense more akin to non-parametric statistics.

Let us describe the general setting in more detail. We have an underlying process $X = (X_t)_{t \geq 0}$, which may be multi-dimensional (its components are then denoted by $X^1, X^2, \cdots$). This process is defined on some probability space $(\Omega, \mathcal{F}, \mathbb{P})$. We observe this process at discrete times, equally spaced, over some fixed finite interval $[0, T]$, and we are concerned with asymptotic properties as the time lag, denoted by Δ_n, goes to 0. In practice, this means that we are in the context of *high-frequency data*.

The objects of interest are various quantities related to the particular outcome ω which is (partially) observed. The main object is the *volatility*, but other quantities or features are also of much interest for modeling purposes, for example whether the observed path has jumps and, when this is the case, whether several components may jump at the same times or not.

All these quantities are related in some way to the probabilistic model which is assumed for X: we do indeed need some model assumption, otherwise nothing can be said. In fact, any given set of observed values $X_0, X_{\Delta_n}, \cdots, X_{i\Delta_n}, \cdots$, with Δ_n fixed, is of course compatible with many different models for the continuous time process X: for example we can suppose that X is piecewise constant between the observation times, or that it is piecewise linear between these times. Of course neither one of these two models is in general compatible with the observations if we modify the frequency of the observations.

So in the sequel we will always assume that X is an Itô semimartingale, that is a semimartingale whose characteristics are absolutely continuous with respect to Lebesgue measure. This is compatible with virtually all semimartingale models used for modeling quantities like asset prices or log-prices, although it rules out some non-semimartingale models sometimes used in this context, like the fractional Brownian motion.

Before stating more precisely the questions which we will consider, and in order to be able to formulate them in precise terms, we recall the structure of Itô semimartingales. We refer to Jacod and Shiryaev (2003), Chapter I, for more details.

Semimartingales: We start with $(\Omega, \mathcal{F}, (\mathcal{F}_t)_{t\geq 0}, \mathbb{P})$, a basic filtered probability space where the family of sub-σ-fields $(\mathcal{F}_t)$ of $\mathcal{F}$ is increasing and right-continuous in t. A semimartingale is simply the sum of a local martingale on this space, plus an adapted process of finite variation (meaning, its paths are right-continuous, with finite variation on any finite interval). In the multidimensional case it means that each component is a real-valued semimartingale.

Any multidimensional semimartingale can be written as

$$X_t = X_0 + B_t + X_t^c + \int_0^t \int_{\mathbb{R}^d} \kappa(x)(\mu - \nu)(ds, dx) + \int_0^t \int_{\mathbb{R}^d} \kappa'(x)\mu(ds, dx).$$
$$(3.1)$$

In this formula we use the following notation:

- μ is the "jump measure" of X: if we denote by $\Delta X_t = X_t - X_{t-}$ the size of the jump of X at time t (recall that X is right-continuous with left limits), then the set $\{t : \Delta X_t(\omega) \neq 0\}$ is at most countable for each ω, and μ is the random measure on $(0, \infty) \times \mathbb{R}^d$ defined by

$$\mu(\omega; dt, dx) = \sum_{s>0:\ \Delta X_s(\omega) \neq 0} \varepsilon_{(s, \Delta X_s(\omega))}(dt, dx),$$

where ε_a is the Dirac measure sitting at a.

- ν is the "compensator" (or, predictable compensator) of μ. This is the unique random measure on $(0, \infty) \times \mathbb{R}^d$ such that, for any Borel subset A of $\mathbb{R}^d$ at a positive distance of 0, the process $\nu((0, t] \times A)$ is predictable and the difference $\mu((0, t] \times A) - \nu((0, t] \times A)$ is a local martingale.

- κ is a "truncation function," that is a function: $\mathbb{R}^d \to \mathbb{R}^d$, bounded with compact support, such that $\kappa(x) = x$ for all x in a neighborhood of 0. This function is fixed throughout, and we choose it to be *continuous* for convenience.

- κ' is the function $\kappa'(x) = x - \kappa(x)$.

- B is a predictable process of finite variation, with $B_0 = 0$.

- X^c is a continuous local martingale with $X_0^c = 0$, called the "continuous martingale part" of X.

With this notation, the decomposition (3.1) is unique (up to null sets), but the process B depends on the choice of the truncation function κ. The continuous martingale part does *not* depend on the choice of κ. Note that the first integral in (3.1) is a stochastic integral (in general), whereas the second one is a pathwise integral (in fact for any t is is simply the finite sum $\sum_{s \le t} \kappa'(\Delta X_s)$). Of course (3.1) should be read "componentwise" in the multidimensional setting.

In the sequel we use the shorthand notation $\star$ to denote the (possibly stochastic) integral w.r.t. a random measure, and also $\bullet$ for the (possibly stochastic) integral of a process w.r.t. a semimartingale. For example, (3.1) may be written more shortly as

$$X = X_0 + B + X^c + \kappa \star (\mu - \nu) + \kappa' \star \mu. \qquad (3.2)$$

The "*" symbol will also be used, as a superscript, to denote the transpose of a vector or matrix (no confusion may arise).

Another process is of great interest, namely the quadratic variation of the continuous martingale part X^c, which is the following $\mathbb{R}^d \otimes \mathbb{R}^d$-valued process:

$$C = \langle X^c, X^{c\star} \rangle, \quad \text{that is, componentwise, } C^{ij} = \langle X^{i,c}, X^{j,c} \rangle. \qquad (3.3)$$

This is a continuous adapted process with $C_0 = 0$, which further is increasing in the set $\mathcal{M}_d^+$ of symmetric nonnegative matrices, that is $C_t - C_s$ belongs to $\mathcal{M}_d^+$ for all $t > s$.

The triple (B, C, ν) is called the *triple of characteristics* of X, this name coming from the fact that in "good cases" it completely determines the law of X.

The fundamental example of semimartingales is the case of Lévy processes. We say that X is a *Lévy process* if it is adapted to the filtration, with right-continuous and left-limited paths and $X_0 = 0$, and such that $X_{t+s} - X_t$ is independent of $\mathcal{F}_t$ and has the same law as X_s for all $s, t \geq 0$. Such a process is always a semimartingale, and its characteristics (B, C, ν) are of the form

$$B_t(\omega) = bt, \quad C_t = ct, \quad \nu(\omega; dt, dx) = dt \otimes F(dx). \tag{3.4}$$

Here $b \in \mathbb{R}^d$ and $c \in \mathcal{M}_d^+$ and F is a measure on $\mathbb{R}^d$ which does not charge 0 and integrates the function $x \mapsto \|x\|^2 \wedge 1$. The triple (b, c, F) is connected with the law of the variables X_t by the formula (for all $u \in \mathbb{R}^d$)

$$\mathbb{E}(e^{i\langle u, X_t\rangle}) = \tag{3.5}$$

$$\exp t\Big(i\langle u, b\rangle - \frac{1}{2}\langle u, cu\rangle + \int F(dx)\Big(e^{i\langle u, x\rangle} - 1 - i\langle u, \kappa(x)\rangle\Big)\Big),$$

called Lévy-Khintchine's formula. So we sometimes call (b, c, F) the characteristics of X as well, and it is the Lévy-Khintchine characteristics of the law of X_1 in the context of infinitely divisible distributions. b is called the drift, c is the covariance matrix of the Gaussian part, and F is called the Lévy measure.

As seen above, for a Lévy process the characteristics (B, C, ν) are deterministic, and they do characterize the law of the process. Conversely, if the characteristics of a semimartingale X are deterministic one can show that X has independent increments, and if they are of the form (3.4) then X is a Lévy process.

Itô semimartingales. By definition, an *Itô semimartingale* is a semimartingale whose characteristics (B, C, ν) are absolutely continuous with respect to Lebesgue measure, in the following sense:

$$B_t(\omega) = \int_0^t b_s(\omega)ds, \quad C_t(\omega) = \int_0^t c_s(\omega)ds, \quad \nu(\omega; dt, dx) = dt\, F_{\omega,t}(dx). \tag{3.6}$$

Here we can always choose a version of the processes b or c which is optional, or even predictable, and likewise choose F in such a way that $F_t(A)$ is optional, or even predictable, for all Borel subsets A of $\mathbb{R}^d$.

It turns out that Itô semimartingales have a nice representation in terms of a Wiener process and a Poisson random measure, and this representation will be very useful for us. Namely, it can be written as follows (where for example $\kappa'(\delta) * \underline{\mu}_t$ denotes the value at time t of the integral process $\kappa'(\delta) * \underline{\mu}$):

$$X_t = X_0 + \int_0^t b_s ds + \int_0^t \sigma_s dW_s + \kappa(\delta) \star (\underline{\mu} - \underline{\nu})_t + \kappa'(\delta) \star \underline{\mu}_t. \tag{3.7}$$

In this formula W is a standard d'-dimensional Wiener process and $\underline{\mu}$ is a Poisson random measure on $(0, \infty) \times E$ with intensity measure $\underline{\nu}(dt, \overline{dx}) = dt \otimes \lambda(dx)$, where λ is a σ-finite and infinite measure without atom on an auxiliary measurable set $(E, \mathcal{E})$.

The process b_t is the same in (3.6) and in (3.7), and $\sigma = (\sigma^{ij})_{1 \le i \le d, 1 \le j \le d'}$ is an $\mathbb{R}^d \otimes \mathbb{R}^{d'}$-valued optional (or predictable, as one wishes to) process such that $c = \sigma\sigma^\star$, and $\delta = \delta(\omega, t, x)$ is a predictable function on $\Omega \times [0, \infty) \times E$ (that is, measurable with respect to $\mathcal{P} \otimes \mathcal{E}$, where $\mathcal{P}$ is the predictable σ-field of $\Omega \times [0, \infty)$). The connection between δ above and F in (3.6) is that $F_{t,\omega}$ is the image of the measure λ by the map $x \mapsto \delta(\omega, t, x)$, and restricted to $\mathbb{R}^d \backslash \{0\}$.

Remark 3.1 One should be a bit more precise in characterizing W and μ: W is an $(\mathcal{F}_t)$-Wiener process, meaning it is $\mathcal{F}_t$ adapted and $W_{t+s} - W_t$ is independent of $\mathcal{F}_t$ (on top of being Wiener, of course). Likewise, μ is an $(\mathcal{F}_t)$-Poisson measure, meaning that $\mu((0, t] \times A)$ is $\mathcal{F}_t$-measurable and $\mu((t, t + s] \times A)$ is independent of $\mathcal{F}_t$, for all $A \in \mathcal{E}$. □

Remark 3.2 The original space $(\Omega, \mathcal{F}, \mathbb{P})$ on which X is defined may be too small to accommodate a Wiener process and a Poisson measure, so we may have to enlarge the space. Such an enlargement is always possible. □

Remark 3.3 When the matrix $c_t(\omega)$ is of full rank for all (ω, t) and $d' = d$, then it has "square-roots" $\sigma_t(\omega)$ which are invertible, and we may take $W = (\sigma)^{-1} \bullet X^c$. However, even in this case, there are many ways of choosing σ such that $\sigma\sigma^\star = c$, hence many ways of choosing W, and to begin with of choosing its dimension d' (which can always be taken such that $d' \le d$).

In a similar way, we have a lot of freedom for the choice of μ. In particular we can choose at will the space $(E, \mathcal{E})$ and the measure λ, subject to the above conditions, and for example we can always take $E = \mathbb{R}$ with λ the Lebesgue measure, although in the d-dimensional case it is somewhat more intuitive to take $E = \mathbb{R}^d$. □

Of course a Lévy process is an Itô semimartingale (compare (3.2) and (3.6)). In this case the two representations (3.2) and (3.7) coincide if we take $E = \mathbb{R}^d$ and $\lambda = F$ (the Lévy measure) and $\mu = \mu$ (the jump measure of a Lévy process is a Poisson measure) and $\delta(\omega, t, x) = x$, and also if we recall that in this case the continuous martingale (or "Gaussian") part of X is always of the form $X^c = \sigma W$, with $\sigma\sigma^\star = c$.

The setting of Itô semimartingales encompasses most processes used for modeling purposes, at least in mathematical finance. For example, solutions of stochastic differential equations driven by a Wiener process, or a by a Lévy process, or by a Wiener process plus a Poisson random measure, are all Itô semimartingales. Such solutions are obtained directly in the form (3.7), which of course implies that X is an Itô semimartingale.

The volatility. In a financial context, the process c_t is called the volatility (sometimes it is σ_t which is thus called). This is by far the most important quantity which needs to be estimated, and there are many ways to do so. A very widely spread way of doing so consists in using the so-called "implied volatility," and it is performed by using the observed current prices of options drawn on the stock under consideration, by somehow inverting the Black–Scholes equation or extensions of it.

However, this way usually assumes a given type of model, for example that the stock price is a diffusion process of a certain type, with unknown coefficients. Among the coefficients there is the volatility, which further may be "stochastic," meaning that it depends on some random inputs other than the Wiener process which drives the price itself. But then it is of primary importance to have a sound model, and this can be checked only by statistical means. That is, we have to make a statistical analysis, based on series of (necessarily discrete) observations of the prices.

In other words, there is a large body of work, essentially in the econometrical literature, about the (statistical) estimation of the volatility. This means finding good methods for estimating the path $t \mapsto c_t(\omega)$ for $t \in [0, T]$, on the basis of the observation of $X_{i\Delta_n}(\omega)$ for all $i = 0, 1, \cdots, [T/\Delta_n]$.

In a sense this is very similar to the non-parametric estimation of a function $c(t)$, say in the 1-dimensional case, when one observes the Gaussian process

$$Y_t = \int_0^t \sqrt{c(s)}\, dW_s$$

(here W is a standard 1-dimensional Wiener process) at the time $i\Delta_n$, and when Δ_n is "small" (that is, we consider the asymptotic $\Delta_n \to 0$). As is well known, this is possible only under some regularity assumptions on the function $c(t)$, whereas the "integrated" value $\int_0^t c(s)ds$ can be estimated as in parametric statistics, since it is just a number. On the other hand, if we know $\int_0^t c(s)ds$ for all t, then we also know the function $c(t)$, up to a Lebesgue-null set, of course: it should be emphasized that if we modify c on such a null set, we do not change the process Y itself; the same comment applies to the volatility process c_t in (3.6).

This is why we mainly consider, as in most of the literature, the problem of estimating the *integrated volatility*, which with our notation is the process C_t. One has to be aware of the fact that in the case of a general Itô semimartingale, this means "estimating" the random number or matrix $C_t(\omega)$, for the observed ω, although of course ω is indeed not "fully" observed.

Let us consider for simplicity the 1-dimensional case, when further X is continuous, that is

$$X_t = X_0 + \int_0^t b_s ds + \int_0^t \sigma_s dW_s, \qquad (3.8)$$

and σ_t (equivalently, $c_t = \sigma_t^2$) is random. It may be of the form $\sigma_t(\omega) = \sigma(X_t(\omega))$, it can also be by itself the solution of another stochastic differential equation, driven by W and perhaps another Wiener process W', and perhaps also some Poisson measures if it has jumps (even though X itself does not jump).

By far, the simplest thing to do is to consider the "realized" integrated volatility, or "approximate quadratic variation," that is the process

$$B(2, \Delta_n)_t = \sum_{i=1}^{[t/\Delta_n]} |\Delta_i^n X|^2, \qquad \text{where } \Delta_i^n X = X_{i\Delta_n} - X_{(i-1)\Delta_n}. \tag{3.9}$$

Then if (3.8) holds, well known results on the quadratic variation (going back to Itô in this case), we know that

$$B(2, \Delta_n)_t \xrightarrow{\mathbb{P}} C_t \tag{3.10}$$

(convergence in probability), and this convergence is even uniform in t over finite intervals. Further, as we will see later, we have a rate of convergence (namely $1/\sqrt{\Delta_n}$) under some appropriate assumptions.

Now what happens when X is discontinuous? We no longer have (3.10), but rather

$$B(2, \Delta_n)_t \xrightarrow{\mathbb{P}} C_t + \sum_{s \leq t} |\Delta X_s|^2 \tag{3.11}$$

(the right side above is always finite, and is the quadratic variation of the semi-martingale X, also denoted $[X, X]_t$). Nevertheless we do want to estimate C_t; a good part of these notes is devoted to this problem. For example, we will show that the two quantities

$$\left. \begin{array}{l} B(1, 1, \Delta_n)_t = \sum_{i=1}^{[t/\Delta_n]} |\Delta_i^n X| |\Delta_{i+1}^n X| \\[2mm] B(2, \varpi, \alpha)_t = \sum_{i=1}^{[t/\Delta_n]} |\Delta_i^n X|^2 1_{\{|\Delta_i^n X| \leq \alpha \Delta_n^\varpi\}} \end{array} \right\} \tag{3.12}$$

converge in probability to $\frac{2}{\pi} C_t$ and C_t respectively, provided that $\varpi \in (0, 1/2)$ and $\alpha > 0$ for the second one.

Inference for jumps. Now, when X is discontinuous, there is also a lot of interest about jumps and, to begin with, are the observations compatible with a model without jumps, or should we use a model with jumps? More complex questions may be posed: for a 2-dimensional process, do the jumps occur at the same times for the two components or not? Are there infinitely many (small) jumps? In this case, what is the "concentration" of the jumps near 0?

Here again, the analysis is based on the asymptotic behavior of quantities involving sums of functions of the increments $\Delta_i^n X$ of the observed process.

So, before going to the main results in a general situation, we consider first two very simple cases: when $X = \sigma W$ for a constant $\sigma > 0$, and when $X = \sigma W + Y$ when Y is a compound Poisson process. It is also of primary importance to determine which quantities can be "consistently estimated" when $\Delta_n \to 0$, and which ones cannot. We begin with the latter question.

3.2 What can be estimated?

Recall that our underlying process X is observed at discrete times $0, \Delta_n, 2\Delta_n,$ $\cdots$, up to some fixed time T. Obviously, we cannot have consistent estimators, as $\Delta_n \to 0$, for quantities which cannot be retrieved when we observe the whole path $t \mapsto X_t(\omega)$ for $t \in [0, T]$, a situation referred to below as the "complete observation scheme."

We begin with two simple observations:

1) The drift b_t can *never* be identified in the complete observation scheme, except in some very special cases, like when $X_t = X_0 + \int_0^t b_s ds$.

2) The quadratic variation of the process is fully known in the complete observation scheme, up to time T, of course. This implies in particular that the integrated volatility C_t is known for all $t \leq T$, hence also the process c_t (this is of course up to a $\mathbb{P}$-null set for C_t, and a $\mathbb{P}(d\omega) \otimes dt$-null set for $c_t(\omega)$).

3) The jumps are fully known in the complete observation scheme, up to time T again.

Now, the jumps are not so interesting by themselves. More important is the "law" of the jumps in some sense. For Lévy processes, the law of jumps is in fact determined by the Lévy measure. In a similar way, for a semimartingale the law of jumps can be considered as known if we know the measures $F_{t,\omega}$, since these measures specify the jump coefficient δ in (3.7). (Warning: this specification is in a "weak" sense, exactly as c specifies σ; we may have several square-root of c, as well as several δ such that F_t is the image of λ, but all choices of σ_t and δ which are compatible with a given c_t and F_t give rise to equations that have exactly the same weak solutions).

Consider Lévy processes first. Basically, the restriction of F to the complement of any neighborhood of 0, after normalization, is the law of the jumps of X lying outside this neighborhood. Hence to consistently estimate F we need potentially infinitely many jumps far from 0, and this possible only if $T \to \infty$. In our situation with T fixed there is no way of consistently estimating F.

We can still say something in the Lévy case: for the complete observation scheme, if there is a jump then F is not the zero measure; if we have infinitely many jumps in $[0, T]$ then F is an infinite measure; in this case, we can also determine for which $r > 0$ the sum $\sum_{s \leq T} |\Delta X_s|^r$ is finite, and this is also the set of r's such that $\int_{\{|x| < \infty\}} |x|^r F(dx) < \infty$.

The same statements also hold for more general semimartingales: we can decide for which r's the sum $\sum_{s \leq T} |\Delta X_s|^r$ is finite, and also if we have zero, or finitely many, or infinitely many jumps. Those are "characteristics" of the model which are of much interest for modelling purposes.

Hence we will be interested, when coming back to the actual discrete observation scheme, in estimating C_t for $t \leq T$, and whether there are zero or finitely many or infinitely many jumps in $[0, T]$.

3.3 Wiener plus compound Poisson processes

This section is about a very particular case: the underlying process is $X = \sigma X + Y$ for some $\sigma > 0$, and Y a compound Poisson process independent of W. And in the first subsection we even consider the most elementary case of $X = \sigma W$. In these two cases we state all limit theorems that are available about sums of a function of the increments. We do not give the full proofs, but heuristic reasons for the results to be true. The reason for devoting a special section to this simple case is to show the variety of results that can be obtained, whereas the full proofs can be easily reconstructed without annoying technical details.

Before getting started, we introduce some notation, to be used also for a general d-dimensional semimartingale X later on. Recall the increments $\Delta_i^n X$ in (3.9). First for any $p > 0$ and $j \leq d$ we set

$$B(p, j, \Delta_n)_t = \sum_{i=1}^{[t/\Delta_n]} |\Delta_i^n X^j|^p. \tag{3.13}$$

In the 1-dimensional case this is written simply $B(p, \Delta_n)_t$. Next if f is a function on $\mathbb{R}^d$, the state space of X in general, we set

$$\left. \begin{array}{l} V(f, \Delta_n)_t = \sum_{i=1}^{[t/\Delta_n]} f(\Delta_i^n X), \\ V'(f, \Delta_n)_t = \sum_{i=1}^{[t/\Delta_n]} f(\Delta_i^n X/\sqrt{\Delta_n}). \end{array} \right\} \tag{3.14}$$

The reason for introducing the normalization $1/\sqrt{\Delta_n}$ will be clear below. These functionals are related one to the other by trivial identity $V'(f, \Delta_n) = V(f_n, \Delta_n)$ with $f_n(x) = f(x/\sqrt{\Delta_n})$. Moreover, with the notation

$$y \in \mathbb{R} \mapsto h_p(y) = |y|^p, \qquad x = (x_j) \in \mathbb{R}^d \mapsto h_p^j(x) = |x_j|^p, \tag{3.15}$$

we also have $B(p, j, \Delta_n) = V(h_p^j, \Delta_n) = \Delta_n^{-p/2} V'(h_p^j, \Delta_n)$. Finally, if we need to emphasize the dependency on the process X, we write these functionals as $B(X; p, j, \Delta_n)$ or $V(X; f, \Delta_n)$ or $V'(X; f, \Delta_n)$.

3.3.1 The Wiener case

Here we suppose that $X = \sigma W$ for some constant $\sigma > 0$, so $d = 1$. Among all the previous functionals, the simplest ones to study are the functionals $V'(f, \Delta_n)$ with f a fixed function on $\mathbb{R}$. We need f to be Borel, of course, and "not too big," for example with polynomial growth, or even with exponential growth. In this case, the results are straightforward consequences of the usual law of large numbers (LNN) and central limit theorem (CLT).

Indeed, for any n the variables $(\Delta_i^n X / \sqrt{\Delta_n} : i \geq 1)$ are i.i.d. with law $\mathcal{N}(0, \sigma^2)$. In the formulas below we write ρ_σ for the law $\mathcal{N}(0, \sigma^2)$ and also $\rho_\sigma(g)$ the integral of a function g with respect to it. Therefore, with f as above, the variables $f(\Delta_i^n X / \sqrt{\Delta_n})$ when i varies are i.i.d. with moments of all orders, and their first and second moments equal $\rho_\sigma(f)$ and $\rho_\sigma(f^2)$ respectively. Then the classical LLN and CLT give us that

$$
\left.
\begin{array}{l}
\Delta_n\, V'(f, \Delta_n)_t \;\xrightarrow{\;\mathbb{P}\;}\; t\rho_\sigma(f) \\[2mm]
\frac{1}{\sqrt{\Delta_n}} \left(\Delta_n\, V'(f, \Delta_n)_t - t\rho_\sigma(g) \right) \;\xrightarrow{\;\mathcal{L}\;}\; \mathcal{N}\!\left(0, t(\rho_\sigma(f^2) - \rho_\sigma(f)^2) \right).
\end{array}
\right\}
$$
(3.16)

We clearly see here why we have put the normalizing factor $1/\sqrt{\Delta_n}$ inside the function f.

The reader will observe that, contrary to the usual LNN, we get convergence in probability but *not* almost surely in the first part of (3.16). The reason is as follows: let ζ_i be a sequence of i.i.d. variables with the same law as $f(X_1)$. The LLN implies that $Z_n = \frac{t}{[t/\Delta_n]} \sum_{i=1}^{[t/\Delta_n]} \zeta_i$ converges a.s. to $t\rho_\sigma(f)$. Since $\Delta_n V'(f, \Delta_n)_t$ has the same law as Z_n we deduce the convergence in probability in (3.16) because, for a deterministic limit, convergence in probability and convergence in law are equivalent. However the variables $V'(f, \Delta_n)_t$ are connected one with the others in a way we do not really control when n varies, so we cannot conclude to $\Delta_n V'(f; \Delta_n)_t \to t\rho_\sigma(f)$ a.s.

(3.9) gives us the convergence for any time t, but we also have functional convergence:

1) First, recall that a sequence g_n of nonnegative increasing functions on $\mathbb{R}_+$ converging pointwise to a *continuous* function g also converges locally uniformly; then, from the first part of (3.9) applied separately for the positive and negative parts f^+ and f^- of f and using a "subsequence principle" for the convergence in probability, we obtain

$$\Delta_n V'(f, \Delta_n)_t \xrightarrow{\text{u.c.p.}} t\rho_\sigma(f) \tag{3.17}$$

where $Z_t^n \xrightarrow{\text{u.c.p.}} Z_t$ means "convergence in probability, locally uniformly in time:" that is, $\sup_{s \le t} |Z_s^n - Z_s| \xrightarrow{\mathbb{P}} 0$ for all t finite.

2) Next, if instead of the 1-dimensional CLT we use the "functional CLT," or Donsker's Theorem, we obtain

$$\left(\frac{1}{\sqrt{\Delta_n}} \left(\Delta_n V'(f, \Delta_n)_t - t\rho_\sigma(f) \right) \right)_{t \ge 0} \xrightarrow{\mathcal{L}} \sqrt{\rho_\sigma(f^2) - \rho_s(f)^2}\, W' \tag{3.18}$$

where W' is another standard Wiener process, and $\xrightarrow{\mathcal{L}}$ stands for the convergence in law of processes (for the Skorokhod topology). Here we see a new Wiener process W' appear. What is its connection with the basic underlying Wiener process W? To study that, one can try to prove the "joint convergence" of the processes on the left side of (3.18) together with W (or equivalently X) itself.

This is an easy task: consider the 2-dimensional process Z^n whose first component is the left side of (3.18) and second component is $X_{\Delta_n[t/\Delta_n]}$ (the discretized version of X, which converges pointwise to X). Then Z^n takes the form $Z_t^n = \sqrt{\Delta_n} \sum_{i=1}^{[t/\Delta_n]} \zeta_i^n$, where the ζ_i^n are 2-dimensional i.i.d. variables as i varies, with the same distribution as $(g_1(X_1), g_2(X_1))$, where $g_1(x) = f(x) - \rho_\sigma(f)$ and $g_2(x) = x$. Then the 2-dimensional version of Donsker's Theorem gives us that

$$\left(\frac{1}{\sqrt{\Delta_n}} \left(\Delta_n V'(f; \Delta_n)_t - t\rho_\sigma(f) \right), X_t \right)_{t \ge 0} \xrightarrow{\mathcal{L}} \left(B, X \right) \tag{3.19}$$

and the pair (B, X) is a 2-dimensional (correlated) Wiener process, characterized by its variance-covariance at time 1, which is the following matrix:

$$\begin{pmatrix} \rho_\sigma(f^2) - \rho_s(f)^2 & \rho_\sigma(fg_2) \\ \rho_\sigma(fg_2) & \sigma^2 \end{pmatrix} \tag{3.20}$$

(note that $\sigma^2 = \rho_\sigma(g_2^2)$ and also $\rho_\sigma(g_2) = 0$, so the above matrix is semi-definite positive). Equivalently, we can write B as $B = \sqrt{\rho_\sigma(f^2) - \rho_s(f)^2}\, W'$ with W' a standard Brownian motion (as in (3.19))) which is correlated with W, the correlation coefficient being $\rho_\sigma(fg_2)/\sigma\sqrt{\rho_\sigma(f^2) - \rho_s(f)^2}$.

Now we turn to the processes $B(p, \Delta_n)$. Since $B(p, \Delta_n) = \Delta_n^{-p/2} V'(h_p, \Delta_n)$ this is just a particular case of (3.17) and (3.19), which we reformulate below (m_p denotes the pth absolute moment of the normal law $\mathcal{N}(0, 1)$):

$$\Delta_n^{1-p/2} B(p, \Delta_n) \xrightarrow{\text{u.c.p.}} t\sigma^p m_p, \tag{3.21}$$

$$\left(\frac{1}{\sqrt{\Delta_n}}\left(\Delta_n^{1-p/2}B(p,\Delta_n)_t - t\sigma^p m_p\right), X_t\right)_{t\geq 0} \xrightarrow{\mathcal{L}} \left(B, X\right), \quad (3.22)$$

with B a Wiener process unit variance $\sigma^{2p}(m_{2p} - m_p^2)$, independent of X (the independence comes from that fact that $\rho_\sigma(g) = 0$, where $g(x) = x|x|^p$).

Finally for the functionals $V(f, \Delta_n)$, the important thing is the behavior of f near 0, since the increments $\Delta_i^n X$ are all going to 0 as $\Delta_n \to 0$. In fact, $\sup_{i\leq[t/\Delta_n]}|\Delta_i^n X| \to 0$ pointwise, so when the function f vanishes on a neighborhood of 0, for all n bigger than some (random) finite number N depending also on t we have

$$V(f, \Delta_n)_s = 0 \qquad \forall s \leq t. \qquad (3.23)$$

For a "general" function f we can combine (3.21) with (3.23): we easily obtain that (3.21) holds with $V(f, \Delta_n)$ instead of $B(p, \Delta_n)$ as soon as $f(x) \sim |x|^p$ as $x \to 0$, and the same holds for (3.22) if we further have $f(x) = |x|^p$ on a neighborhood of 0.

Of course these results do not exhaust all possibilities for the convergence of $V(f; \Delta_n)$. For example one may prove the following:

$$f(x) = |x|^p \log|x| \quad \Rightarrow \quad \frac{\Delta_n^{1-p/2}}{\log(1/\Delta_n)} V(f, \Delta_n) \xrightarrow{\text{u.c.p.}} -\frac{1}{2} t\sigma^p m_p, \quad (3.24)$$

and a CLT is also available in this situation. Or, we could consider functions f which behave like x^p as $x \downarrow\downarrow 0$ and like $(-x)^{p'}$ as $x \uparrow\uparrow 0$, with $p \neq p'$. However, we essentially restrict our attention to functions behaving like h_p: for simplicity first, and since more general functions do not really occur in the applications we have in mind, and also because the extension to processes X more general than the Brownian motion is not easy for other functions.

3.3.2 The Wiener plus compound Poisson case

Our second example is when the underlying process X has the form $X = \sigma W + Y$, where as before $\sigma > 0$ and W is a Brownian motion, and Y is a compound Poisson process independent of W. We will write $X' = \sigma W$. Recall that Y has the form

$$Y_t = \sum_{p\geq 1} \Phi_p 1_{\{T_p \leq t\}}, \qquad (3.25)$$

where the T_p's are the successive arrival times of a Poisson process, say with parameter 1 (they are finite stopping times, positive, strictly increasing with p and going to ∞), and the Φ_p's are i.i.d. variables, independent of the T_p's, and with some law G. Note that in (3.25) the sum, for any given t, is actually a finite sum.

The processes $V'(f, \Delta_n)$, which were particularly easy to study when X was a Wiener process, are not so simple to analyze now. This is easy to understand: let us fix t; at stage n, we have $\Delta_i^n X = \Delta_i^n X'$ for all $i \leq [t/\Delta_n]$, except for those finitely many i's corresponding to an interval $((i-1)\Delta_n, i\Delta_n]$ containing at least one of the T_p's. Furthermore, all those exceptional intervals contain exactly one T_p, as soon as n is large enough (depending on (ω, t)). Therefore for n large we have

$$V'(f, \Delta_n)_t = V'(X'; f, \Delta_n)_t + A_t'^n, \quad \text{where}$$

$$A_t'^n = \sum_{i=1}^{[t/\Delta_n]} \sum_{p \geq 1} 1_{\{(i-1)\Delta_n < T_p \leq i\Delta_n\}}$$

$$\left(f((\Phi_p + \Delta_i^n X')\sqrt{\Delta_n}) - f(\Delta_i^n X'/\sqrt{\Delta_n}) \right). \quad \left.\right\} \quad (3.26)$$

The double sum in $A_t'^n$ is indeed a finite sum, with as many non-zero entries as the number of T_p's less than $\Delta_n[t/\Delta_n]$.

Therefore the behavior of $V'(f, \Delta_n)$ depends in an essential way on the behavior of f near infinity. There are essentially two possibilities:

1) The function f is bounded, or more generally satisfies $|f(x)| \leq K(1+|x|^p)$ for some $p < 2$. Then $|A_t'^n|$ above is "essentially" smaller than $K \sum_{q:T_q \leq t}(1 + |\Phi_p|^r \Delta_n^{-p/2})$ for some constant K, and thus $\Delta_n A_t'^n \to 0$. So obviously the convergence (3.17) holds.

If further $p < 1$ we even have $\sqrt{\Delta_n}\, A_t'^n \to 0$. Therefore (3.19) holds. Observe that in this situation, the presence of the jumps does *not* modify the results that held for the Brownian case; this will be the rule for more general processes X as well.

2) The function f is equivalent to $|x|^p$ at infinity, for some $p > 2$. Then in (3.26) the leading term becomes $A_t'^n$, which is approximately equal to the quantity $\Delta_n^{-p/2} \sum_{s \leq t} |\Delta X_s|^p$. So $\Delta_n^{p/2} V'(f, \Delta_n)_t$ converges in probability to the variable

$$B(p)_t = \sum_{s \leq t} |\Delta X_s|^p \quad (3.27)$$

(we have just "proved" the convergence for any given t, but it is also a functional convergence, for the Skorokhod topology, in probability).

Again, these cases do not exhaust the possible behaviors of f, and further, we have not given a CLT in the second situation above. But, when f is not bounded it looks a bit strange to impose a specific behavior at infinity, and without this there is simply *no convergence result* for $V'(f, \Delta_n)_t$, not to speak about CLTs.

Now we turn to the processes $V(f, \Delta_n)$. To begin with, we observe that, similar to (3.26), we have

$$\left.\begin{aligned}
V(f, \Delta_n)_t &= V(X', f, \Delta_n)_t + A_t^n, \quad \text{where} \\
A_t^n &= \sum_{i=1}^{[t/\Delta_n]} \sum_{p \geq 1} 1_{\{(i-1)\Delta_n < T_p \leq i\Delta_n\}} \left(f(\Phi_p + \Delta_i^n X') - f(\Delta_i^n X') \right).
\end{aligned}\right\}$$
(3.28)

The first — fundamental — difference with the continuous case is that (3.23) fails now when f vanishes in a neighborhood of 0. In this case, though, for each given t and all n bigger than some number depending on (ω, t), we have $V(X'; f, \Delta_n)_s = 0$ for all $s \leq t$ by (3.23), hence

$$V(f, \Delta_n)_s = \sum_{i=1}^{[s/\Delta_n]} \sum_{p \geq 1} 1_{\{(i-1)\Delta_n < T_p \leq i\Delta_n\}} f(\Phi_p + \Delta_i^n X'), \qquad \forall s \leq t.$$
(3.29)

Then, as soon as f is continuous and vanishes on a neighborhood of 0, we get

$$V(f, \Delta_n)_t \xrightarrow{\text{Sk}} V(f)_t := \sum_{s \leq t} f(\Delta X_s).$$
(3.30)

Here $\xrightarrow{\text{Sk}}$ means "convergence for the Skorokhod topology", pointwise in ω (the reason for which we have convergence in the Skorokhod sense will be explained later; what is clear at this point is that we have the — pointwise in ω — convergence for all t such that X is continuous at t; we also have for each t an almost sure convergence above).

Next, we consider the case where f is still continuous and, say, coincides with h_p for some $p > 0$ on a neighborhood of 0. For any given $\varepsilon > 0$ we can write $f = f_\varepsilon + \widehat{f}_\varepsilon$ with f_ε and $\widehat{f}_\varepsilon$ continuous, and $f_\varepsilon(x) = h_p(x)$ if $|x| \leq \varepsilon/2$ and $f_\varepsilon(x) = 0$ if $|x| \geq \varepsilon$ and $|f_\varepsilon| \leq h_p$ everywhere. Since $\widehat{f}_\varepsilon$ vanishes around 0, we have $V(\widehat{f}_\varepsilon, \Delta_n)_t \to V(\widehat{f}_\varepsilon)_t$ by (3.30), and $V(\widehat{f}_\varepsilon)_t$ converges to $V(f)_t$ as $\varepsilon \to 0$. On the other hand the process A^n associated with f_ε by (3.28) is the sum of summands smaller than $2\varepsilon^p$, the number of them being bounded for each (ω, t) by a number independent of ε: hence A_t^n is negligible and $V(f_\varepsilon, \Delta_n)$ and $V(X'; f_\varepsilon, \Delta_n)$ behave essentially in the same way. This means heuristically that, with the symbol $\asymp$ meaning "approximately equal to," we have

$$V(\widehat{f}_\varepsilon, \Delta_n)_t \asymp V(f)_t, \qquad V(f_\varepsilon, \Delta_n)_t \asymp \Delta_n^{p/2-1} t \sigma^p m_p.$$
(3.31)

Adding these two expressions, we get

$$\left.\begin{aligned}
V(f, \Delta_n)_t &\xrightarrow{\mathbb{P}-\text{Sk}} V(f)_t && \text{if } p > 2 \\
V(f, \Delta_n)_t &\xrightarrow{\mathbb{P}-\text{Sk}} V(f)_t + t\sigma^2 && \text{if } p = 2 \\
\Delta_n^{1-r/2} V(f, \Delta_n)_t &\xrightarrow{\text{u.c.p.}} t\sigma^{p/2} m_p && \text{if } p < 2.
\end{aligned}\right\}$$
(3.32)

This type of LLN, which shows a variety of behaviors according to how f behaves near 0, will be found for much more general processes later, in (almost) exactly the same terms.

Now we turn to the CLT. Here again we single out first the case where f vanishes in a neighborhood of 0. We need to find out what happens to the difference $V(f, \Delta_n) - V(f)$. It is easier to evaluate the difference $V(f, \Delta_n)_t - V(f)_{\Delta_n[t/\Delta_n]}$, since by (3.29) we have

$$V(f, \Delta_n)_s - V(f)_{\Delta_n[s/\Delta_n]} = \qquad\qquad\qquad (3.33)$$

$$\sum_{i-1}^{[s/\Delta_n]} \sum_{p\geq 1} 1_{\{(i-1)\Delta_n < T_p \leq i\Delta_n\}} \left(f(\Phi_p + \Delta_i^n X') - f(\Phi_p) \right)$$

for all $s \leq t$, as soon as n is large enough. Provided f is C^1, with derivative f', the pth summand above is approximately $f'(\Phi_p)\Delta_i^n X'$. Now the normalized increment $\Delta_i^n X'/\sqrt{\Delta_n}$, for the value of i such that $(i-1)\Delta_n < T_p \leq i\Delta_n$, has the law $\mathcal{N}(0, \sigma^2)$ (because X' and Y are independent), and it is asymptotically independent of the process X (more details are to be found later). Thus if $(U_p)_{p\geq 1}$ denotes a sequence of i.i.d. $\mathcal{N}(0, 1)$ variables, independent of X, it is not difficult to see that

$$\frac{1}{\sqrt{\Delta_n}} \left(V(f, \Delta_n)_t - V(f)_{\Delta_n[t/\Delta_n]} \right) \overset{\mathcal{L}}{\Longrightarrow} \overline{B}(f)_t := \sum_{p:T_p \leq t} f'(\Phi_p)\sigma U_p,$$

$$(3.34)$$

and in fact, this convergence in law (for the Skorokhod topology) is even stable (denoted $\overset{\mathcal{L}-s}{\Longrightarrow}$), a stronger property than the mere convergence in law, which will be defined later but nevertheless is used in the statements below.

When f coincides with h_p for some $p > 0$ in a neighborhood of 0 and is still C^1 outside 0, exactly as for (3.31) we obtain heuristically that

$$V(\widehat{f_\varepsilon}, \Delta_n)_t \;\asymp\; V(f)_{\Delta_n[t/\Delta_n]} + \sqrt{\Delta_n}\, U_t^n,$$
$$V(f_\varepsilon, \Delta_n)_t \;\asymp\; \Delta_n^{p/2-1} t\sigma^p m_p + \Delta_n^{p/2-1/2} U_t'^n,$$

where U^n and U'^n converge stably in law to the right side of (3.34) and to the process B of (3.22), respectively. We then have two conflicting rates, and we can indeed prove that, with $\overline{B}(f)$ as in (3.34) and B as in (3.22) (thus depending on r):

$$\frac{1}{\sqrt{\Delta_n}}\left(V(f,\Delta_n)_t - V(f)_{\Delta_n[t/\Delta_n]}\right)\overset{\mathcal{L}-s}{\Longrightarrow}\overline{B}(f)_t \qquad\qquad \text{if } p > 3$$

$$\frac{1}{\sqrt{\Delta_n}}\left(V(f,\Delta_n)_t - V(f)_{\Delta_n[t/\Delta_n]}\right)\overset{\mathcal{L}-s}{\Longrightarrow}t\sigma^3 m_3 + \overline{B}(f)_t \quad \text{if } p = 3$$

$$\frac{1}{\Delta_n^{p/2-1}}\left(V(f,\Delta_n)_t - V(f)_{\Delta_n[t/\Delta_n]}\right)\overset{\text{u.c.p.}}{\longrightarrow}t\sigma^p m_p \qquad \text{if } 2 < p < 3$$

$$\frac{1}{\sqrt{\Delta_n}}\left(V(f,\Delta_n)_t - V(f)_{\Delta_n[t/\Delta_n]} - t\sigma^2\right)\overset{\mathcal{L}-s}{\Longrightarrow}B_t + \overline{B}(f)_t \text{ if } p = 2$$

$$\frac{1}{\Delta_n^{1-p/2}}\left(\Delta_n^{1-p/2}V(f;\Delta_n)_t - t\sigma^p m_p\right)\overset{\mathbb{P}-\text{Sk}}{\longrightarrow}V(f)_t \qquad \text{if } 1 < p < 2$$

$$\frac{1}{\sqrt{\Delta_n}}\left(\sqrt{\Delta_n}\,V(f,\Delta_n)_t - t\sigma m_1\right)\overset{\mathcal{L}-s}{\Longrightarrow}V(f)_t + B_t \qquad \text{if } p = 1$$

$$\frac{1}{\sqrt{\Delta_n}}\left(\Delta_n^{1-p/2}V(f,\Delta_n)_t - t\sigma^p m_p\right)\overset{\mathcal{L}-s}{\Longrightarrow}B_t \qquad\qquad \text{if } p < 1.$$

$$\left.\right\} \qquad (3.35)$$

Hence we obtain a genuine CLT, relative to the LLN (3.32), in the cases $p > 3$, $p = 2$ and $p < 1$. When $p = 3$ and $p = 1$ we still have a CLT, with a bias. When $2 < p < 3$ or $1 < p < 2$ we have a "second order LNN," and the associated genuine CLTs run as follows:

$$\frac{1}{\sqrt{\Delta_n}}\left(V(f,\Delta_n)_t - V(f)_{\Delta_n[t/\Delta_n]} - \Delta_n^{p/2-1}t\sigma^p m_p\right)$$

$$\overset{\mathcal{L}-s}{\Longrightarrow} \overline{B}(f)_t \quad \text{if } 2 < p < 3$$

$$\frac{1}{\Delta_n^{p/2-1/2}}\left(V(f,\Delta_n)_t - V(f)_{\Delta_n[t/\Delta_n]} - \Delta_n^{p/2-1}t\sigma^p m_p\right)$$

$$\overset{\mathcal{L}-s}{\Longrightarrow} B_t \quad \text{if } 1 < p < 2$$

$$\left.\right\} \qquad (3.36)$$

We see that these results again exhibit a large variety of behavior. This will be encountered also for more general underlying processes X, with of course more complicated statements and proofs (in the present situation we have not really given the complete proof, of course, but it is relatively easy along the lines outlined above). However, in the general situation we will not give such a complete picture, which is useless for practical applications. Only (3.32) and the cases $r > 2$ in (3.35) will be given.

3.4 Auxiliary limit theorems

The aims of this section are twofold: first we define the stable convergence in law, already mentioned in the previous section. Second, we recall a number of limit theorems for partial sums of triangular arrays of random variables.

1) Stable convergence in law. This notion has been introduced by Rényi in

Rényi (1963), for the very same reasons as we need it here. We refer to Aldous and Eagleson (1978) for a very simple exposition and to Jacod and Shiryaev (2003) for more details.

It often happens that a sequence of statistics Z_n converges in law to a limit Z which has, say, a mixed centered normal distribution: that is, $Z = \Sigma U$ where U is an $\mathcal{N}(0, 1)$ variable and Σ is a positive variable independent of U. This poses no problem other than computational when the law of Σ is known. However, in many instances the law of Σ is unknown, but we can find a sequence of statistics Σ_n such that the pair (Z_n, Σ_n) converges in law to (Z, Σ); so although the law of the pair (Z, Σ) is unknown, the variable Z_n/Σ_n converges in law to $\mathcal{N}(0, 1)$ and we can base estimation or testing procedures on this new statistic Z_n/Σ_n. This is where the stable convergence in law comes into play.

The formal definition is a bit involved. It applies to a sequence of random variables Z_n, all defined on the same probability space $(\Omega, \mathcal{F}, \mathbb{P})$, and taking their values in the same state space $(E, \mathcal{E})$, assumed to be Polish (= metric complete and separable). We say that Z_n *stably converges in law* if there is a probability measure η on the product $(\Omega \times E, \mathcal{F} \otimes \mathcal{E})$, such that $\eta(A \times E) = \mathbb{P}(A)$ for all $A \in \mathcal{F}$ and

$$\mathbb{E}(Yf(Z_n)) \rightarrow \int Y(\omega)f(x)\eta(d\omega, dx) \qquad (3.37)$$

for all bounded continuous functions f on E and bounded random variables Y on $(\Omega, \mathcal{F})$.

This is an "abstract" definition, similar to the definition of the convergence in law which says that $\mathbb{E}(f(Z_n)) \rightarrow \int f(x)\rho(dx)$ for some probability measure ρ. Now for the convergence in law we usually want a limit, that is, we say $Z_n \xrightarrow{\mathcal{L}} Z$, and the variable Z is *any* variable with law ρ, of course. In a similar way it is convenient to "realize" the limit Z for the stable convergence in law.

We can always realize Z in the following way: take $\widetilde{\Omega} = \Omega \times E$ and $\widetilde{\mathcal{F}} = \mathcal{F} \otimes \mathcal{E}$ and endow $(\widetilde{\Omega}, \widetilde{\mathcal{F}})$ with the probability η, and put $Z(\omega, x) = x$. But, as for the simple convergence in law, we can also consider other extensions of $(\Omega, \mathcal{F}, \mathbb{P})$: that is, we have a probability space $(\widetilde{\Omega}, \widetilde{\mathcal{F}}, \widetilde{\mathbb{P}})$, where $\widetilde{\Omega} = \Omega \times \widetilde{\Omega}'$ and $\widetilde{\mathcal{F}} = \mathcal{F} \otimes \mathcal{F}'$ for some auxiliary measurable space $(\Omega', \mathcal{F}')$ and $\widetilde{\mathbb{P}}$ is a probability measure on $(\widetilde{\Omega}, \widetilde{\mathcal{F}})$ whose first marginal is $\mathbb{P}$, and we also have a random variable Z on this extension. Then in this setting, (3.37) is equivalent to saying (with $\widetilde{\mathbb{E}}$ denoting the expectation w.r.t. $\widetilde{\mathbb{P}}$)

$$\mathbb{E}(Yf(Z_n)) \rightarrow \widetilde{\mathbb{E}}(Yf(Z)) \qquad (3.38)$$

for all f and Y as above, as soon as $\widetilde{\mathbb{P}}(A \cap \{Z \in B\}) = \eta(A \times B)$ for all $A \in \mathcal{F}$

and $B \in \mathcal{E}$. We then say that Z_n converges stably to Z, and this convergence is denoted by $\xrightarrow{\mathcal{L}-s}$.

Clearly, when η is given, the property $\widetilde{\mathbb{P}}(A \cap \{Z \in B\}) = \eta(A \times B)$ for all $A \in \mathcal{F}$ and $B \in \mathcal{E}$ simply amounts to specifying the law of Z, conditionally on the σ-field $\mathcal{F}$. Therefore, saying $Z_n \xrightarrow{\mathcal{L}-s} Z$ amounts to saying that we have the stable convergence in law towards a variable Z, defined on any extension $(\widetilde{\Omega}, \widetilde{\mathcal{F}}, \widetilde{\mathbb{P}})$ of $(\Omega, \mathcal{F}, \mathbb{P})$, and with a specified conditional law knowing $\mathcal{F}$.

Obviously, the stable convergence in law implies the convergence in law. But it implies much more, and in particular the following crucial result: if $Z_n \xrightarrow{\mathcal{L}-s} Z$ and if Y_n and Y are variables defined on $(\Omega, \mathcal{F}, \mathbb{P})$ and with values in the same Polish space F, then

$$Y_n \xrightarrow{\mathbb{P}} Y \quad \Rightarrow \quad (Y_n, Z_n) \xrightarrow{\mathcal{L}-s} (Y, Z). \tag{3.39}$$

On the other hand, there are criteria for stable convergence in law of a given sequence Z_n. The σ-field generated by all Z_n is necessarily separable, that is generated by a countable algebra, say $\mathcal{G}$. Then if for any finite family $(A_p : 1 \leq p \leq q)$ in $\mathcal{G}$, the sequence $(Z_n, (1_{A_p})_{1 \leq p \leq q})$ of $E \times \mathbb{R}^q$-valued variables converges in law as $n \to \infty$, then necessarily Z_n converges stably in law.

2) Convergence of triangular arrays. Our aim is to prove the convergence of functionals like those in (3.13) and (3.14), which appear in a natural way as partial sums of triangular arrays. We really need the convergence for the terminal time T, but in most cases the available convergence criteria also give the convergence as processes, for the Skorokhod topology. So now we provide a set of conditions implying the convergence of partial sums of triangular arrays, all results being in Jacod and Shiryaev (2003).

We are not looking for the most general situation here, and we restrict our attention to the case where the filtered probability space $(\Omega, \mathcal{F}, (\mathcal{F}_t)_{t \geq 0}, \mathbb{P})$ is fixed. For each n we have a sequence of $\mathbb{R}^d$-valued variables $(\zeta_i^n : i \geq 1)$, the components being denoted by $\zeta_i^{n,j}$ for $j = 1, \cdots, d$. The *key assumption* is that for all n, i the variable ζ_i^n is $\mathcal{F}_{i\Delta_n}$-measurable, and this assumption is in force in the remainder of this section.

Conditional expectations w.r.t. $\mathcal{F}_{(i-1)\Delta_n}$ will play a crucial role, and to simplify notation we write it $\mathbb{E}_{i-1}^n$ instead of $\mathbb{E}(. \mid \mathcal{F}_{(i-1)\Delta_n})$, and likewise $\mathbb{P}_{i-1}^n$ is the conditional probability.

Lemma 3.4 *If we have*

$$\sum_{i=1}^{\lfloor t/\Delta_n \rfloor} \mathbb{E}_{i-1}^n (\|\zeta_i^n\|) \xrightarrow{\mathbb{P}} 0 \qquad \forall t > 0, \tag{3.40}$$

then $\sum_{i=1}^{[t/\Delta_n]} \zeta_i^n \xrightarrow{u.c.p.} 0$. *The same conclusion holds under the following two conditions:*

$$\sum_{i=1}^{[t/\Delta_n]} \mathbb{E}_{i-1}^n(\zeta_i^n) \xrightarrow{u.c.p.} 0, \tag{3.41}$$

$$\sum_{i=1}^{[t/\Delta_n]} \mathbb{E}_{i-1}^n(\|\zeta_i^n\|^2) \xrightarrow{\mathbb{P}} 0 \qquad \forall t > 0. \tag{3.42}$$

In particular when ζ_i^n *is a martingale difference, that is* $\mathbb{E}_{i-1}^n(\zeta_i^n) = 0$, *then (3.42) is enough to imply* $\sum_{i=1}^{[t/\Delta_n]} \zeta_i^n \xrightarrow{u.c.p.} 0$.

Lemma 3.5 *If we have*

$$\sum_{i=1}^{[t/\Delta_n]} \mathbb{E}_{i-1}^n(\zeta_i^n) \xrightarrow{u.c.p.} A_t \tag{3.43}$$

for some continuous adapted $\mathbb{R}^d$-*valued process of finite variation* A, *and if further (3.42) holds, then we have* $\sum_{i=1}^{[t/\Delta_n]} \zeta_i^n \xrightarrow{u.c.p.} A_t$.

Lemma 3.6 *If we have (3.43) for some (deterministic) continuous* $\mathbb{R}^d$-*valued function of finite variation* A, *and also the following two conditions:*

$$\sum_{i=1}^{[t/\Delta_n]} \left(\mathbb{E}_{i-1}^n(\zeta_i^{n,j} \zeta_i^{n,k}) - \mathbb{E}_{i-1}^n(\zeta_i^{n,j}) \mathbb{E}_{i-1}^n(\zeta_i^{n,k}) \right) \xrightarrow{\mathbb{P}} C_t^{\prime jk} \tag{3.44}$$

$$\forall t > 0, \; j, k = 1, \cdots, d,$$

$$\sum_{i=1}^{[t/\Delta_n]} \mathbb{E}_{i-1}^n(\|\zeta_i^n\|^4) \xrightarrow{\mathbb{P}} 0 \qquad \forall t > 0, \tag{3.45}$$

where $C' = (C'^{jk})$ *is a (deterministic) function, continuous and increasing in* $\mathcal{M}_d^+$, *then the processes* $\sum_{i=1}^{[t/\Delta_n]} \zeta_i^n$ *converge in law to* $A + B$, *where* B *is a continuous centered Gaussian* $\mathbb{R}^d$-*valued process with independent increments with* $\mathbb{E}(B_t^j B_t^k) = C_t^{\prime jk}$.

(3.45) is a conditional *Lindeberg condition*, whose aims is to ensure that the limiting process is continuous; other, weaker, conditions of the same type are available, but not needed here. The conditions given above completely characterize, of course, the law of the process B. Equivalently we could say that B is a Gaussian martingale (relative to the filtration it generates), starting from 0, and with quadratic variation process C'.

3) Stable convergence of triangular arrays. The reader will have observed that the conditions (3.43) and (3.44) in Lemma 3.6 are very restrictive, because the limits are non-random. In the sequel, such a situation rarely occurs, and typically these conditions are satisfied with A and C' random. But then we need an additional condition, under which it turns out that the convergence holds not only in law, but even stably in law.

Note that the stable convergence in law has been defined for variables taking values in a Polish space, so it also applies to right-continuous and left limited d-dimensional processes: such a process can be viewed as a variable taking its values in the Skorokhod space $\mathbb{D}(\mathbb{R}^d)$ of all functions from $\mathbb{R}_+$ into $\mathbb{R}^d$ which are right-continuous with left limits, provided we endow this space with the Skorokhod topology which makes it a Polish space. See Billingsley (1999) or Chapter VI of Jacod and Shiryaev (2003) for details on this topology. In fact, in Lemma 3.6 the convergence in law is also relative to this Skorokhod topology. The stable convergence in law for processes is denoted as $\overset{\mathcal{L}-s}{\Longrightarrow}$ below.

In the previous results the fact that all variables were defined on the same space $(\Omega, \mathcal{F}, (\mathcal{F}_t)_{t\geq 0}, \mathbb{P})$ was essentially irrelevant. This is no longer the case for the next result, for which this setting is fundamental.

Below we single out, among all martingales on $(\Omega, \mathcal{F}, (\mathcal{F}_t)_{t\geq 0}, \mathbb{P})$, a possibly multidimensional Wiener process W. The following lemma holds for any choice of W, and even with no W at all (in which case a martingale "orthogonal to W" below means any martingale) but we will use it mainly with the process W showing in (3.7). The following is a particular case of Theorem IX.7.28 of Jacod and Shiryaev (2003).

Lemma 3.7 *Assume (3.43) for some continuous adapted $\mathbb{R}^d$-valued process of finite variation A, and (3.44) with some continuous adapted process $C' = (C'^{jk})$ with values in $\mathcal{M}_d^+$ and increasing in this set, and also (3.45). Assume also*

$$\sum_{i=1}^{[t/\Delta_n]} \mathbb{E}_{i-1}^n(\zeta_i^n \Delta_i^n N) \overset{\mathbb{P}}{\longrightarrow} 0 \qquad \forall t > 0 \tag{3.46}$$

whenever N is one of the components of W or is a bounded martingale orthogonal to W. Then the processes $\sum_{i=1}^{[t/\Delta_n]} \zeta_i^n$ converge stably in law to $A + B$, where B is a continuous process defined on an extension $(\widetilde{\Omega}, \widetilde{\mathcal{F}}, \widetilde{\mathbb{P}})$ of the space $(\Omega, \mathcal{F}, \mathbb{P})$ and which, conditionally on the σ-field $\mathcal{F}$, is a centered Gaussian $\mathbb{R}^d$-valued process with independent increments satisfying $\widetilde{\mathbb{E}}(B_t^j B_t^k \mid \mathcal{F}) = C_t'^{jk}$.

The conditions stated above completely specify the conditional law of B, knowing $\mathcal{F}$, so we are exactly in the setting explained in §1 above and the stable convergence in law is well defined. However, one can say even more: letting $(\widetilde{\mathcal{F}}_t)$ be the smallest filtration on $\widetilde{\Omega}$ which make B adapted and which contains $(\mathcal{F}_t)$

(that is, $A \times \Omega' \in \widetilde{\mathcal{F}}_t$ whenever $A \in \mathcal{F}_t$), then B is a continuous local martingale on $(\widetilde{\Omega}, \widetilde{\mathcal{F}}, (\widetilde{\mathcal{F}}_t)_{t \geq 0}, \widetilde{\mathbb{P}})$ which is orthogonal in the martingale sense to any martingale on the space $(\Omega, \mathcal{F}, (\mathcal{F}_t)_{t \geq 0}, \mathbb{P})$, and whose quadratic variation process is C'. Of course, on the extended space B is no longer Gaussian.

The condition (3.46) could be substituted with weaker ones. For example if it holds when N is orthogonal to W, whereas $\sum_{i=1}^{[t/\Delta_n]} \mathbb{E}_{i-1}^n(\zeta_i^n \Delta_i^n W^j)$ converges in probability to a continuous process for all indices j, we still have the stable convergence in law of $\sum_{i=1}^{[t/\Delta_n]} \zeta_i^n$, but the limit has the form $A + B + M$, where the process M is a stochastic integral with respect to W. See Jacod and Shiryaev (2003) for more details.

3.5 A first LNN (Law of Large Numbers)

At this stage we start giving the basic limit theorems which are used later for statistical applications. Perhaps giving first all limit theorems in a purely probabilistic setting is not the most pedagogical way of proceeding, but it is the most economical in terms of space...

We are in fact going to provide a version of the results of Section 3.3, and other connected results, when the basic process X is an Itô semimartingale. There are two kinds of results: first some LNNs similar to (3.17), (3.21), (3.30) or (3.32); second, some "central limit theorems" (CLT) similar to (3.22) or (3.35). We will not give a complete picture, and rather restrict ourselves to those results which are used in the statistical applications.

Warning: Below, and in all these notes, the proofs are often sketchy and sometimes absent; for the full proofs, which are sometimes a bit complicated, we refer essentially to Jacod (2008) (which is restricted to the 1-dimensional case for X, but the multidimensional extension is straightforward).

In this section, we provide some general results, valid for any d-dimensional semimartingale $X = (X^j)_{1 \leq j \leq d}$, not necessarily Itô. We also use the notations (3.13) and (3.14). We start by recalling the fundamental result about quadratic variation, which says that for any indices j, k, and as $n \to \infty$ (recall $\Delta_n \to 0$):

$$\sum_{i=1}^{[t/\Delta_n]} \Delta_i^n X^j \Delta_i^n X^k \xrightarrow{\mathbb{P}-\text{sk}} [X^j, X^k]_t = C_t^{jk} + \sum_{s \leq t} \Delta X_s^j \Delta X_s^k. \quad (3.47)$$

This is the convergence in probability, for the Skorokhod topology, and we even have the joint convergence for the Skorokhod topology for the d^2-dimensional processes, when $1 \leq j, k \leq d$. When further X has no fixed times of discontinuity, for example when it is an Itô semimartingale, we also have the convergence in probability for any fixed t.

Theorem 3.8 *Let f be a continuous function from $\mathbb{R}^d$ into $\mathbb{R}^{d'}$.*

a) If $f(x) = \mathrm{o}(\|x\|^2)$ as $x \to 0$, then

$$V(f, \Delta_n)_t \xrightarrow{\mathbb{P}-\text{Sk}} f \star \mu_t = \sum_{s \leq t} f(\Delta X_s). \qquad (3.48)$$

b) If f coincides in a neighborhood of 0 with the function $g(x) = \sum_{j,k=1}^{d} \gamma_{jk} x_j x_k$ (here each γ_{jk} is a vector in $\mathbb{R}^{d'}$), then

$$V(f, \Delta_n)_t \xrightarrow{\mathbb{P}-\text{Sk}} \sum_{j,k=1}^{d} \gamma_{jk} C_t^{jk} + f \star \mu_t. \qquad (3.49)$$

Moreover both convergences above also hold in probability for any fixed t such that $\mathbb{P}(\Delta X_t = 0) = 1$ (hence for all t when X is an Itô semimartingale).

Proof. 1) Suppose first that $f(x) = 0$ when $\|x\| \leq \varepsilon$, for some $\varepsilon > 0$. Denote by $S_1, S_2, \cdots$ the successive jump times of X corresponding to jumps of norm bigger than $\varepsilon/2$, so $S_p \to \infty$. Fix $T > 0$. For each $\omega \in \Omega$ there are two integers $Q = Q(T, \omega)$ and $N = N(T, \omega)$ such that $S_Q(\omega) \leq T < S_{Q+1}(\omega)$ and for all $n \geq N$ and for any interval $(i-1)\Delta_n, i\Delta_n]$ in $[0, T]$ then either there is no S_q in this interval and $\|\Delta_i^n X\| \leq \varepsilon$, or there is exactly one S_q in it and then we set $\alpha_q^n = \Delta_i^n X - \Delta X_{S_q}$. Since $f(x) = 0$ when $\|x\| \leq \varepsilon$ we clearly have for all $t \leq T$ and $n \geq N$:

$$\left\| V(f, \Delta_n)_t - \sum_{q:\, S_q \leq \Delta_n[t/\Delta_n]} f(\Delta X_{S_q}) \right\| \leq \sum_{q=1}^{Q} |f(\Delta X_{S_q} + \alpha_q^n) - f(\Delta X_{S_q})|.$$

Then the continuity of f yields (3.48), because $\alpha_q^n \to 0$ for all q.

2) We now turn to the general case in (a). For any $\eta > 0$ there is $\varepsilon > 0$ such that we can write $f = f_\varepsilon + f_\varepsilon'$, where f_ε is continuous and vanishes for $\|x\| \leq \varepsilon$, and where $\|f_\varepsilon'(x)\| \leq \eta \|x\|^2$. By virtue of (3.47) and the first part of the proof, we have

$$\begin{cases} \|V(f_\varepsilon', \Delta_n)\| \leq \eta \sum_{i=1}^{[t/\Delta_n]} \|\Delta_i^n X\|^2 \xrightarrow{\mathbb{P}-\text{Sk}} \eta \sum_{j=1}^{d} [X^j, X^j], \\ V(f_\varepsilon, \Delta_n) \xrightarrow{\mathbb{P}-\text{Sk}} f_\varepsilon \star \mu \end{cases}$$

Moreover, $f_\varepsilon \star \mu \xrightarrow{\text{u.c.p.}} f \star \mu$ as $\varepsilon \to 0$ follows easily from Lebesgue convergence theorem and the property $f(x) = \mathrm{o}(\|x\|^2)$ as $x \to 0$, because $\|x\|^2 \star \mu_t < \infty$ for all t. Since $\eta > 0$ and $\varepsilon > 0$ are arbitrarily small, we deduce (3.48) from $V(f, \Delta_n) = V(f_\varepsilon, \Delta_n) + V(f_\varepsilon', \Delta_n)$.

3) Now we prove (b). Let $f' = f - g$, which vanishes on a neighborhood of 0. Then if we combine (3.47) and (3.48), plus a classical property of the

Skorokhod convergence, we obtain that the pair $(V(g, \Delta_n), V(f', \Delta_n))$ converges (for the $2d'$-dimensional Skorokhod topology, in probability) to the pair $\left(\sum_{j,k=1}^{d} \gamma_{jk} C^{jk} + g \star \mu, f' \star \mu \right)$, and by adding the two components we obtain (3.49).

Finally, the last claim comes from a classical property of the Skorokhod convergence, plus the fact that an Itô semimartingale has no fixed time of discontinuity.   □

In particular, in the 1-dimensional case we obtain (recall (3.13)):

$$p > 2 \quad \Rightarrow \quad B(p, \Delta_n) \overset{\mathbb{P}-\text{sk}}{\longrightarrow} B(p)_t := \sum_{s \leq t} |\Delta X_s|^p. \qquad (3.50)$$

This result is due to Lépingle (1976), who even proved the almost sure convergence. It completely fails when $r \leq 2$ except under some special circumstances.

3.6 Some other LNNs

3.6.1 Hypotheses

So far we have generalized (3.30) to any semimartingale, under appropriate conditions on f. If we want to generalize (3.17) or (3.26) we need X to be an Itô semimartingale, plus the fact that the processes (b_t) and (σ_t) and the function δ in (3.7) are locally bounded and (σ_t) is either right-continuous or left-continuous.

When it comes to the CLTs we need even more. So for a clearer exposition we gather all hypotheses needed in the sequel, either for LNNs or CLTs, in a single assumption.

Assumption (H): The process X has the form (3.7), and the volatility process σ_t is also an Itô semimartingale of the form

$$\sigma_t = \sigma_0 + \int_0^t \widetilde{b}_s \, ds + \int_0^t \widetilde{\sigma} \, dW_s + \widetilde{\kappa}(\widetilde{\delta}) \star (\underline{\mu} - \underline{\nu})_t + \widetilde{\kappa}'(\widetilde{\delta}) \star \underline{\mu}_t. \qquad (3.51)$$

In this formula, σ_t (a $d \times d'$ matrix) is considered as an $\mathbb{R}^{dd'}$-valued process; $\widetilde{b}_t(\omega)$ and $\widetilde{\sigma}_t(\omega)$ are optional processes, respectively dd' and dd'^2-dimensional, and $\widetilde{\delta}(\omega, t, x)$ id a dd'-dimensional predictable function on $\Omega \times \mathbb{R}_+ \times E$; finally $\widetilde{\kappa}$ is a truncation function on $\mathbb{R}^{dd'}$ and $\widetilde{\kappa}'(x) = x - \widetilde{\kappa}(x)$.

Moreover, we have:

(a) The processes $\widetilde{b}_t(\omega)$ and $\sup_{x \in E} \frac{\|\delta(\omega, t, x)\|}{\gamma(x)}$ and $\sup_{x \in E} \frac{\|\widetilde{\delta}(\omega, t, x)\|}{\widetilde{\gamma}(x)}$ are locally bounded, where γ and $\widetilde{\gamma}$ are (non-random) nonnegative functions satisfying $\int_E (\gamma(x)^2 \wedge 1)\, \lambda(dx) < \infty$ and $\int_E (\widetilde{\gamma}(x)^2 \wedge 1)\, \lambda(dx) < \infty$.

(b) All paths $t \mapsto b_t(\omega)$, $t \mapsto \widetilde{\sigma}_t(\omega)$, $t \mapsto \delta(\omega, t, x)$ and $t \mapsto \widetilde{\delta}(\omega, t, x)$ are left-continuous with right limits.   □

Recall that "$(\widetilde{b}_t)$ is locally bounded," for example, means that there exists an increasing sequence (T_n) of stopping times, with $T_n \to \infty$, and such that each stopped process $\widetilde{b}_t^{T_n} = \widetilde{b}_{t \wedge T_n}$ is bounded by a constant (depending on n, but not on (ω, t)).

Remark 3.9 For the LNNs, and also for the CLTs in which there is a discontinuous limit below, we need a weaker form of this assumption, namely *Assumption (H′)*: this is as (H), except that we do not require σ_t to be an Itô semimartingale but only to be càdlàg (then of course $\widetilde{b}$, $\widetilde{\sigma}$, $\widetilde{\delta}$ are not present), and b_t is only locally bounded.

As a rule, we will state the results with the mention of this assumption (H′), when the full force of (H) is not needed. However, all proofs will be made assuming (H), because it simplifies the exposition, and because the most useful results need it anyway.   □

Apart from the regularity and growth conditions (a) and (b), this assumption amounts to saying that both X and the process σ in (3.7) are Itô semimartingales: since the dimension d' is arbitrary large (and in particular may be bigger than d), this accommodates the case where in (3.7) only the first d components of W occur (by taking $\sigma_t^{ij} = 0$ when $j > d$), whereas in (3.51) other components of W come in, thus allowing σ_t to be driven by the same Wiener process as X, plus an additional multidimensional process. In the same way, it is no restriction to assume that both X and σ are driven by the same Poisson measure $\underline{\mu}$.

So in fact, this hypothesis accommodates virtually all models of stock prices or exchange rates or interest rates, with stochastic volatility, including those with jumps, and allows for correlation between the volatility and the asset price processes. For example, if we consider a q-dimensional equation

$$dY_t = f(Y_{t-})dZ_t \qquad (3.52)$$

where Z is a multi-dimensional Lévy process, and f is a C^2 function with at most linear growth, then if X consists in a subset of the components of Y, it satisfies Assumption (H). The same holds for more general equations driven by a Wiener process and a Poisson random measure.

3.6.2 The results

Now we turn to the results. The first, and most essential, result is the following; recall that we use the notation ρ_σ for the law $\mathcal{N}(0, \sigma\sigma^\star)$, and $\rho_\sigma^{\otimes k}$ denotes the k-fold tensor product. We also write $\rho_\sigma^{\otimes k}(f) = \int f(x)\rho_\sigma^{\otimes k}(dx)$ if f is a (Borel) function on $(\mathbb{R}^d)^k$. With such a function f we also associate the following processes

$$V'(f, k, \Delta_n)_t = \sum_{i=1}^{[t/\Delta_n]} f\left(\Delta_i^n X/\sqrt{\Delta_n}, \cdots, \Delta_{i+k-1}^n X/\sqrt{\Delta_n}\right). \quad (3.53)$$

Of course when f is a function on $\mathbb{R}^d$, then $V'(f, 1, \Delta_n) = V'(f, \Delta_n)$, as defined by (3.14).

Theorem 3.10 *Assume (H) (or (H') only, see Remark 3.9)), and let f be a continuous function on $(\mathbb{R}^d)^k$ for some $k \geq 1$, which satisfies*

$$|f(x_1, \cdots, x_k))| \leq K_0 \prod_{j=1}^{k}(1 + \|x_j\|^p) \quad (3.54)$$

for some $p \geq 0$ and K_0. If either X is continuous, or if $p < 2$, we have $\Delta_n V'^m(f, k, \Delta_n)_t \xrightarrow{u.c.p.} \int_0^t \rho_{\sigma_u}^{\otimes k}(f)du$.

In particular, if X is continuous and the function f on $\mathbb{R}^d$ satisfies $f(\lambda x) = \lambda^p f(x)$ for all $x \in \mathbb{R}^d$ and $\lambda \geq 0$, then

$$\Delta_n^{1-p/2} V(f, \Delta_n)_t \xrightarrow{u.c.p.} \int_0^t \rho_{\sigma_u}(f)du. \quad (3.55)$$

The last claim above may be viewed as an extension of Theorem 3.8 in the case when the limit in (3.48) vanishes. The continuity of f can be somehow relaxed. The proof will be given later, after we state some other LLNs, of two kinds, to be proved later also.

Recalling that one of our main objectives is to estimate the integrated volatility C_t^{jk}, we observe that Theorem 3.8 does not provide "consistent estimators" for C_t when X is discontinuous. There are two ways to solve this problem, and the first one is as follows: when X has jumps, (3.47) does not give information on C_t because of the jumps, essentially the "big" ones. However, a big jump gives rise to a big increment $\Delta_i^n X$. So the idea, following Mancini (2001, 2009), consists in throwing away the big increments. The cutoff level has to be chosen carefully, so as to eliminate the jumps but keeping the increments which are "mainly" due to the continuous martingale part X^c, and those are of order $\sqrt{\Delta_n}$. So we choose two numbers $\varpi \in (0, 1/2)$ and $\alpha > 0$, and for all indices $j, k \leq d$ we set

$$V^{jk}(\varpi, \alpha, \Delta_n)_t = \sum_{i=1}^{[t/\Delta_n]} (\Delta_i^n X^j \Delta_i^n X^k) 1_{\{\|\Delta_i^n X\| \leq \alpha \Delta_n^\varpi\}}. \qquad (3.56)$$

More generally one can consider the truncated analogue of $V'(f, k, \Delta_n)$ of (3.53). With ϖ and α as above, and if f is a function on $(\mathbb{R}^d)^k$, we set

$$V'(\varpi, \alpha; f, k, \Delta_n)_t = \qquad\qquad\qquad\qquad\qquad\qquad (3.57)$$

$$\sum_{i=1}^{[t/\Delta_n]} f\left(\Delta_i^n X / \sqrt{\Delta_n}, \cdots, \Delta_{i+k-1}^n X / \sqrt{\Delta_n}\right) 1_{\cap_{j=1,\cdots,k} \{\|\Delta_{i+j-1}^n X\| \leq \alpha \Delta_n^\varpi\}}.$$

Theorem 3.11 *Assume (H) (or (H') only), and let f be a continuous function on $(\mathbb{R}^d)^k$ for some $k \geq 1$, which satisfies (3.54) for some $p \geq 0$ and some $K_0 > 0$. Let also $\varpi \in (0, \frac{1}{2})$ and $\alpha > 0$. If either X is continuous, or X is discontinuous and $p \leq 2$ we have $\Delta_n V'^n(\varpi, \alpha; f, k, \Delta_n)_t \xrightarrow{u.c.p.} \int_0^t \rho_{\sigma_u}^{\otimes k}(f) du.$*

In particular, $V^{jk}(\varpi, \alpha, \Delta_n) \xrightarrow{u.c.p.} C_t^{jk}$.

This result has no real interest when X is continuous. When X jumps, and at the expense of a more complicated proof, one could show that the result holds when $p \leq 4$, and also when $p > 4$ and $\varpi \geq \frac{p-4}{2p-2r-4}$ when additionally we have $\int (\gamma(x)^r \wedge 1)\lambda(dz) < \infty$ for some $r \in [0, 2)$ (where γ is the function occurring in (H)).

The (slight) improvement on the condition on p, upon the previous theorem, allows to easily estimate not only C_t, but also the integral $\int_0^t g(c_s) ds$ for any polynomial g on the set of $d \times d$ matrices. For example, if we take

$$f(x_1, \cdots, x_k) = \prod_{j=1}^{k} (x_j^{m_j} x_j^{n_j}), \qquad\qquad (3.58)$$

for arbitrary indices m_j and n_j in $\{1, \cdots, d\}$, then we get

$$\Delta_n V'^n(\varpi, \alpha; f, k, \Delta_n)_t \xrightarrow{u.c.p.} \int_0^t \prod_{j=1}^{k} c_s^{m_j n_j} ds. \qquad (3.59)$$

The problem with this method is that we do not really know how to choose ϖ and α *a priori*: empirical evidence from simulation studies leads to choosing ϖ to be very close to $1/2$, like $\varpi = 0.47$ or 0.48, whereas α for estimating C_t^{jj}, say, should be chosen between 2 and 5 times the "average $\sqrt{c^{jj}}$ " (recall $c = \sigma\sigma^\star$). So this requires a preliminary rough estimate of the order of magnitude of c^{jj}; of course, for financial data this order of magnitude is usually pretty much well known.

Another way, initiated by Barndorff–Nielsen and Shephard (2003, 2004) consists in using the so-called bipower, or more generally multipower, variations. This is in fact a particular case of the Theorem 3.10. Indeed, recalling that m_r is the rth absolute moment of $\mathcal{N}(0,1)$, we set for any $r_1, \cdots, r_l \in (0,2)$ with $r_1 + \cdots + r_l = 2$ (hence $l \geq 2$):

$$V^{jk}(r_1, \cdots, r_l, \Delta_n)_t = \tag{3.60}$$

$$\frac{1}{4m_{r_1} \cdots m_{r_l}} \sum_{i=1}^{[t/\Delta_n]} \left(\prod_{v=1}^{l} |\Delta_{i+v-1}^n (X^j + X^k)|^{r_v} - \prod_{v=1}^{l} |\Delta_{i+v-1}^n (X^j - X^k)|^{r_v} \right)$$

Then obviously this is equal to $\frac{1}{\Delta_n} V'(f, l, \Delta_n)$, where

$$f(x_1, \cdots, x_l) = \frac{1}{4m_{r_1} \cdots m_{r_l}} \left(\prod_{v=1}^{l} |x_v^j + x_v^k|^{r_v} - \prod_{v=1}^{l} |x_v^j - x_v^k|^{r_v} \right),$$

and $\rho_\sigma^{\otimes l}(f) = (\sigma\sigma^\star)^{jk}$ by a simple calculation. Then we deduce from Theorem 3.10 the following result:

Theorem 3.12 *Assume (H) (or (H') only)), and let $r_1, \cdots, r_l \in (0,2)$ be such that $r_1 + \cdots + r_l = 2$. Then $V^{jk}(r_1, \cdots, r_l, \Delta_n) \xrightarrow{u.c.p.} C_t^{jk}$.*

Now, the previous LNNs are not enough for the statistical applications we have in mind. Indeed, we need consistent estimators for a few other processes than C_t, and in particular for the following one which appears as a conditional variance in some of the forthcoming CLTs:

$$D^{jk}(f)_t = \sum_{s \leq t} f(\Delta X_s)(c_{s-}^{jk} + c_s^{jk}) \tag{3.61}$$

for indices $j, k \leq d$ and a function f on $\mathbb{R}^d$ with $|f(x)| \leq K\|x\|^2$ for $\|x\| \leq 1$, so the summands above are non-vanishing only when $\Delta X_s \neq 0$ and the process $D^{jk}(f)$ is finite-valued.

To do this we take any sequence k_n of integers satisfying

$$k_n \to \infty, \qquad k_n \Delta_n \to 0, \tag{3.62}$$

and we let $I_{n,t}(i) = \{j \in N : j \neq i : 1 \leq j \leq [t/\Delta_n], |i - j| \leq k_n\}$ define a local window in time of length $k_n \Delta_n$ around time $i\Delta_n$. We also choose $\varpi \in (0, 1/2)$ and $\alpha > 0$ as in (3.56). We will consider two distinct cases for f and associate with it the functions f_n:

- $f(x) = o(\|x\|^2)$ as $x \to 0$, $\qquad f_n(x) = f(x)$

- $f(x) = \sum_{v,w=1}^d \gamma_{vw} x_v x_w \qquad f_n(x) = f(x)1_{\{\|x\| > \alpha \Delta_n^\varpi\}}.$
  on a neighborhood of 0, $\qquad\qquad\qquad\qquad\qquad\qquad$ $\left.\begin{array}{c} \\ \\ \\ \end{array}\right\}$ (3.63)

Finally, we set

$$D^{jk}(f, \varpi, \alpha, \Delta_n)_t = \tag{3.64}$$

$$\frac{1}{k_n \Delta_n} \sum_{i=1+k_n}^{[t/\Delta_n]-k_n} f_n(\Delta_i^n X) \sum_{l \in I_{n,t}(i)} (\Delta_l^n X^j \, \Delta_l^n X^k) 1_{\{\|\Delta_l^n X\| \le \alpha \Delta_n^\varpi\}}.$$

Theorem 3.13 *Assume (H) (or (H') only), and let f be a continuous function on $\mathbb{R}^d$ satisfying (3.63), and $j, k \le d$ and $\varpi \in (0, 1/2)$ and $\alpha > 0$. Then*

$$D^{jk}(f, \varpi, \alpha, \Delta_n) \xrightarrow{\mathbb{P}-\text{sk}} D^{jk}(f). \tag{3.65}$$

If further X is continuous and $f(\lambda x) = \lambda^p f(x)$ for all $\lambda > 0$ and $x \in \mathbb{R}^d$, for some $p > 2$ (hence we are in the first case of (3.63)), then

$$\Delta_n^{1-p/2} D^{jk}(f, \varpi, \alpha, \Delta_n) \xrightarrow{u.c.p.} 2 \int_0^t \rho_{\sigma_u}(f) c_u^{jk} ds. \tag{3.66}$$

Before proceeding to the proof of all those results, we give some preliminaries.

3.6.3 A localization procedure

The localization is a simple but very important tool for proving limit theorems for discretized processes, over a finite time interval. We describe it in detail in the setting of the previous theorems, but it will also be used later for the CLTs.

The idea is that, for those theorems, we can replace the local boundedness assumptions in (H-r), for example, by boundedness (by a constant), which is a much stronger assumption. More precisely, we set

Assumption (SH): We have (H) and also, for some constant Λ and all (ω, t, x):

$$\left. \begin{array}{ll} \|b_t(\omega)\| \le \Lambda, \quad \|\sigma_t(\omega)\| \le \Lambda, \quad \|X_t(\omega)\| \le \Lambda, \\ \|\tilde{b}_t(\omega)\| \le \Lambda, \quad \|\tilde{\sigma}_t(\omega)\| \le \Lambda, \\ \|\delta(\omega, t, x)\| \le \Lambda(\gamma(x) \wedge 1), \quad \|\tilde{\delta}(\omega, t, x)\| \le \Lambda(\tilde{\gamma}(x) \wedge 1) \end{array} \right\} \tag{3.67}$$

If these are satisfied, we can of course choose γ and $\tilde{\gamma}$ smaller than 1.

Lemma 3.14 *If X satisfies (H) we can find a sequence of stopping times R_p increasing to $+\infty$ and a sequence of processes $X(p)$ satisfying (SH) and with volatility process $\sigma(p)$, such that*

$$t < R_p \quad \Rightarrow \quad X(p)_t = X_t, \quad \sigma(p)_t = \sigma_t. \tag{3.68}$$

Proof. Let X satisfy assumption (H). The processes b_t, $\widetilde{b}_t$, $\widetilde{\sigma}_t$, $\sup_{x \in E} \frac{\|\delta(t,x)\|}{\gamma(x)}$ and $\sup_{x \in E} \frac{\|\widetilde{\delta}(t,x)\|}{\widetilde{\gamma}(x)}$ are locally bounded, so we can assume the existence of a "localizing sequence" of stopping times T_p (i.e. this sequence is increasing, with infinite limit) such that for $p \geq 1$:

$$t \leq T_p(\omega) \quad \Rightarrow \quad \left\{ \begin{array}{ll} \|b_t(\omega)\| \leq p, & \|\widetilde{b}_t(\omega)\| \leq p, \quad \|\widetilde{\sigma}_t(\omega)\| \leq p, \\ \|\delta(\omega,t,x)\| \leq p\gamma(x), & \|\widetilde{\delta}(\omega,t,x)\| \leq p\widetilde{\gamma}(x). \end{array} \right. \tag{3.69}$$

We also set $S_p = \inf(t : \|X_t\| \geq p \text{ or } \|\sigma_t\| \geq p)$, so $R_p = T_p \wedge S_p$ is again a localizing sequence, and we have (3.69) for $t \leq R_p$ and also $\|X_t\| \leq p$ and $\|\sigma_t\| \leq p$ for $t < R_p$. Then we set

$$b(p)_t = \left\{ \begin{array}{ll} b_t & \text{if } t \leq R_p \\ 0 & \text{otherwise,} \end{array} \right. \qquad \widetilde{b}(p)_t = \left\{ \begin{array}{ll} \widetilde{b}_t & \text{if } t \leq R_p \\ 0 & \text{otherwise,} \end{array} \right.$$

$$\widetilde{\sigma}(p)_t = \left\{ \begin{array}{ll} \widetilde{\sigma}_t & \text{if } t \leq R_p \\ 0 & \text{otherwise,} \end{array} \right.$$

$$\delta(p)(\omega,t,x) = \left\{ \begin{array}{ll} \delta(\omega,t,x) & \text{if } \|\delta(\omega,t,x)\| \leq 2p \text{ and } t \leq R_p \\ 0 & \text{otherwise,} \end{array} \right.$$

$$\widetilde{\delta}(p)(\omega,t,x) = \left\{ \begin{array}{ll} \widetilde{\delta}(\omega,t,x) & \text{if } \|\widetilde{\delta}(\omega,t,x)\| \leq 2p \text{ and } t \leq R_p \\ 0 & \text{otherwise,} \end{array} \right.$$

At this stage we define the process $\sigma(p)$ by (3.51) with the starting point $\sigma(p)_0 = \sigma_0$ if $\|\sigma_0\| < p$ and $\sigma(p)_0 = 0$ otherwise, and the coefficients $\widetilde{b}(p)$ and $\widetilde{\sigma}(p)$ and $\widetilde{\delta}(p)$, and we define the process $X(p)$ by (3.7) with the starting point $X(p)_0 = X_0$ if $\|X_0\| < p$ and $X(p)_0 = 0$ otherwise, and the coefficients $b(p)$ and $\sigma(p)$ (as defined just above) and $\delta(p)$.

We can write μ as $\mu = \sum_{t>0} 1_D(t)\,\varepsilon_{(t,\beta_t)}$ where D is the countable (random) support of μ and β_t is E-valued. Outside a $\mathbb{P}$-null set N we have $\Delta X_t = 1_D(t)\,\delta(t,\beta_t)$ and $\Delta X(p)_t = 1_D(t)\,\delta(p)(t,\beta_t)$, and since $\|\Delta X_t\| \leq 2p$ when $t < R_p$ we deduce $\Delta X_t = \Delta X(p)_t$ if $t < R_p$, which implies that $\kappa'(\delta) * \underline{\mu}_t = \kappa'(\delta(p)) * \underline{\mu}_t$ for $t < R_p$. As for the two local martingales $\kappa(\delta) * (\underline{\mu} - \underline{\nu})$ and $\kappa(\delta(p)) * (\underline{\mu} - \underline{\nu})$, they have (a.s.) the same jumps on the predictable interval $[0, R_p]$ as soon as $\kappa(x) = 0$ when $\|x\| > 2p$ (this readily follows from the definition of $\delta(p)$), so they coincide a.s. on $[0, R_p]$.

The same argument shows that $\widetilde{\kappa}'(\widetilde{\delta}) * \underline{\mu}_t = \widetilde{\kappa}'(\widetilde{\delta}(p)) * \underline{\mu}_t$ for $t < R_p$, and $\widetilde{\kappa}(\widetilde{\delta}) * (\underline{\mu} - \underline{\nu})_t = \widetilde{\kappa}(\widetilde{\delta}(p)) * (\underline{\mu} - \underline{\nu})_t$ for $t \leq R_p$. It first follows in an obvious way that $\sigma(p)_t = \sigma_t$ for all $t < R_p$, and then $X(p)_t = X_t$ for all $t < R_p$, that is (3.68) holds.

Finally by definition the coefficients $b(p)$, $\widetilde{b}(p)$, $\widetilde{\sigma}(p)$, $\delta(p)$ and $\widetilde{\delta}(p)$ satisfy

(3.67) with $\Lambda = 2p$. Moreover, the processes $\tilde{\sigma}(p)$ and $X(p)$ are constant after time R_p, and they have jumps bounded by $2P$, so they satisfy (3.67) with $\Lambda = 3p$, and thus (SH) holds for $X(p)$.   □

Now, suppose that, for example, Theorem 3.10 has been proved when X satisfies (SH). Let X satisfy (H) only, and $(X(p), R_p)$ be as above. We then know that, for all p, T and all appropriate functions f,

$$\sup_{t \leq T} \left| \Delta_n V'^n(X(p); f, k, \Delta_n)_t - \int_0^t \rho_{\sigma(p)_u}^{\otimes k}(f) du \right| \overset{\mathbb{P}}{\longrightarrow} 0. \qquad (3.70)$$

On the set $\{R_p > T + 1\}$, and if $k\Delta_n \leq 1$, we have $V'^n(X(p); f, k, \Delta_n)_t = V'^n(X; f, k, \Delta_n)_t$ and $\sigma(p)_t = \sigma_t$ for all $t \leq T$, by (3.68). Since $\mathbb{P}(R_p > T + 1) \to 1$ as $p \to \infty$, it readily follows that $\Delta_n V'^n(X; f, k, \Delta_n)_t \overset{\text{u.c.p.}}{\longrightarrow} \int_0^t \rho_{\sigma_u}^{\otimes k}(f) du$. This proves Theorem 3.10 under (H).

This procedure works in exactly the same way for all the theorems below, LNNs or CLTs, and we will call this the "localization procedure" without further comment.

Remark 3.15 If we assume (SH), and if we choose the truncation functions κ and $\tilde{\kappa}$ in such a way that they coincide with the identity on the balls centered at 0 and with radius 2Λ, in $\mathbb{R}^d$ and $\mathbb{R}^{dd'}$ respectively, then clearly (3.7) and (3.51) can be rewritten as follows:

$$\left. \begin{array}{l} X_t = X_0 + \int_0^t b_s\, ds + \int_0^t \sigma_s\, dW_s + \delta \star (\mu - \nu)_t, \\[2mm] \sigma_t = \sigma_0 + \int_0^t \tilde{b}_s\, ds + \int_0^t \tilde{\sigma}\, dW_s + \tilde{\delta} \star (\mu - \nu)_t. \end{array} \right\} \qquad (3.71)$$

3.6.4 Some estimates

Below, we assume (SH), and we use the form (3.71) for X and σ. We will give a number of estimates, to be used for the LLNs and also for the CLTs, and we start with some notation. We set

$$\left. \begin{array}{l} \chi_{i,l}'^n = \frac{1}{\sqrt{\Delta_n}} \int_{(i+l-1)\Delta_n}^{(i+l)\Delta_n} \left(b_s\, ds + (\sigma_s - \sigma_{(i-1)\Delta_n})\, dW_s \right) \\[2mm] \beta_{i,l}^n = \sigma_{(i-1)\Delta_n} \Delta_{i+l}^n W / \sqrt{\Delta_n}, \qquad \chi_{i,l}''^n = \frac{1}{\sqrt{\Delta_n}} \Delta_{i+l}^n (\delta \star (\mu - \nu)), \\[2mm] \chi_{i,l}^n = \chi_{i,l}'^n + \chi_{i,l}''^n, \qquad \beta_i^n = \beta_{i,0}^n, \qquad \chi_i^n = \chi_{i,0}^n, \qquad \chi_i'^n = \chi_{i,0}'^n. \end{array} \right\} \qquad (3.72)$$

In particular, $\Delta_{i+l}^n X = \sqrt{\Delta_n}(\chi_{i,l}^n + \beta_{i,l}^n)$. It is well known that the boundedness of the coefficients in (SH) yields, through a repeated use of Doob and Davis-Burkholder-Gundy inequalities, for all $q > 0$ (below, K denotes a constant which varies from line to line and may depend on the constants occurring in (SH); we write it K_p if we want to emphasize its dependency on another parameter p):

$$\left. \begin{array}{l} \mathbb{E}_{i-1}^n(\|\Delta_i^n X^c\|^q) \le K_q \Delta_n^{q/2}, \quad \mathbb{E}(\|\sigma_{t+s} - \sigma_t\|^q \mid \mathcal{F}_t) \le K_q s^{1 \wedge (q/2)}, \\[2mm] \mathbb{E}_{i+l-1}^n(\|\beta_{i,l}^n\|^q) \le K_q, \qquad \mathbb{E}_{i+l-1}^n(\|\chi_{i,l}'^n\|^q) \le K_{q,l} \Delta_n^{1 \wedge (q/2)}, \\[2mm] \mathbb{E}_{i+l-1}^n(\|\chi_{i,l}^n\|^q + \|\chi_{i,l}''^n\|^q) \le \left\{ \begin{array}{ll} K_{q,l} \Delta_n^{-(1-q/2)^-} & \text{in general} \\[1mm] K_{q,l} \Delta_n^{1 \wedge (q/2)} & \text{if } X \text{ is continuous} \end{array} \right. \end{array} \right\} \quad (3.73)$$

We also use the following notation, for $\eta > 0$:

$$\psi_\eta(x) = \psi(x/\eta), \tag{3.74}$$

where ψ is a C^∞ function on $\mathbb{R}^d$ with $1_{\{\|x\| \le 1\}} \le \psi(x) \le 1_{\{\|x\| \le 2\}}$.

Lemma 3.16 *Assume (SH) and let $r \in [0,2]$ be such that $\int (\gamma(x)^r \wedge 1) \lambda(dx) < \infty$, and α_n a sequence of numbers with $\alpha_n \ge 1$ and $\alpha_n \sqrt{\Delta_n} \to 0$. Then*

$$\lim_{n \to \infty} \sup_{i \ge 1, \, \omega \in \Omega} \Delta_n^{r/2-1} \alpha_n^{r-2} \, \mathbb{E}_{i+l-1}^n(\|\chi_{i,l}''^n\|^2 \wedge \alpha_n^2) = 0, \tag{3.75}$$

$$r \le 1 \quad \Rightarrow \tag{3.76}$$

$$\lim_{n \to \infty} \sup_{i \ge 1, \, \omega \in \Omega} \Delta_n^{r/2-1} \alpha_n^{r-1} \, \mathbb{E}_{i+l-1}^n \left(\left| \frac{1}{\sqrt{\Delta_n}} \Delta_{i+l}^n(\delta \star \underline{\mu}) \right| \wedge \alpha_n \right) = 0,$$

$$\lim_{\eta \to 0} \limsup_{n \to \infty} \sup_{i \ge 1, \, \omega \in \Omega} \frac{1}{\Delta_n} \mathbb{E}_{i+l-1}^n(\|\sqrt{\Delta_n} \, \chi_{i,l}^n\|^2 \wedge \eta^2) = 0. \tag{3.77}$$

(When $r \le 1$ above, the two integral processes $\delta \star \underline{\mu}$ and $\delta \star \underline{\nu}$ are well defined, and of finite variation).

Proof. It is enough to consider the 1-dimensional case. For any $\varepsilon \in (0,1]$ we have $\delta \star (\mu - \nu) = N(\varepsilon) + M(\varepsilon) + B(\varepsilon)$, where ($\kappa$ is the truncation function in (3.1)).

$$N(\varepsilon) = (\delta 1_{\{|\delta| > \varepsilon\}}) \star \mu, \quad M(\varepsilon) = (\delta 1_{\{|\delta| \le \varepsilon\}}) \star (\mu - \nu), \quad B(\varepsilon) = -(\delta 1_{\{|\delta| > \varepsilon\}}) \star \underline{\nu}.$$

Then if $\gamma_\varepsilon = \int_{\{\gamma(x) \le \varepsilon\}} \gamma(x)^r \lambda(dx)$, we have by (SH):

$$\left. \begin{array}{l} \mathbb{P}_{i+l-1}^n(\Delta_{i+l}^n N(\varepsilon) \ne 0) \le \mathbb{E}_{i+l-1}^n(\Delta_{i+l}^n(1_{\{\gamma > \varepsilon\}} \star \mu)) = \\[2mm] \hspace{4cm} \Delta_n \lambda(\{\gamma > \varepsilon\}) \le K \Delta_n \varepsilon^{-r} \\[3mm] \mathbb{E}_{i+l-1}^n((\Delta_{i+l}^n M(\varepsilon))^2) \le \Delta_n \int_{\{\gamma(x) \le \varepsilon\}} \gamma(x)^2 \lambda(dx) \le \Delta_n \varepsilon^{2-r} \gamma_\varepsilon, \\[3mm] |\Delta_{i+l}^n B(\varepsilon)| \le K \Delta_n \left(1 + \int_{\{\gamma(x) > \varepsilon\}} (\gamma(x) \wedge 1) \lambda(dx) \right) \le K \Delta_n \varepsilon^{-(r-1)^+}. \end{array} \right\}$$

We also trivially have

$$|\chi_{i,l}^n|^2 \wedge \alpha_n \leq$$
$$\alpha_n^2 1_{\{\Delta_{i+l}^n N(\varepsilon) \neq 0\}} + 3|\chi_{i,l}'^n|^2 + 3\Delta_n^{-1}|\Delta_{i+l}^n M(\varepsilon)|^2 + 3\Delta_n^{-1}|\Delta_{i+l}^n B(\varepsilon)|^2.$$

Therefore, using (3.73), we get

$$\mathbb{E}_{i+l-1}^n(\|\chi_{i,l}^n\|^2 \wedge \alpha_n^2) \leq K\left(\frac{\alpha_n^2 \Delta_n}{\varepsilon^r} + \Delta_n + \varepsilon^{2-r}\gamma_\varepsilon + \Delta_n \varepsilon^{-(r-1)^+}\right).$$

Then since $\gamma_\varepsilon \to 0$ as $\varepsilon \to 0$, (3.75) follows by taking $\varepsilon = \varepsilon_n = u_n^2(u_n^{-1} \wedge (\gamma_{u_n})^{-1/4})$, where $u_n = \alpha_n^{1/2}\Delta_n^{1/4} \to 0$ (note that $\varepsilon_n \leq u_n$, hence $\gamma_{\varepsilon_n} \leq \gamma_{u_n}$).

Next, suppose $r \leq 1$. Then $\delta \star \mu = N(\varepsilon) + A(\varepsilon)$, where $A(\varepsilon) = (\delta 1_{\{|\delta|\leq\varepsilon\}}) \star \mu$, and obviously $\mathbb{E}_{i+l-1}^n(|\Delta_{i+l}^n \bar{A}(\varepsilon)|) \leq K\Delta_n \varepsilon^{1-r}\gamma_\varepsilon$. Moreover

$$\left|\frac{1}{\sqrt{\Delta_n}}\Delta_{i+l}^n(\delta \star \mu)\right| \wedge \alpha_n \leq \alpha_n 1_{\{\Delta_{i+l}^n N(\varepsilon)\neq 0\}} + \frac{1}{\sqrt{\Delta_n}}|\Delta_{i+l}^n A(\varepsilon)|.$$

Therefore

$$\mathbb{E}_{i+l-1}^n\left(\left|\frac{1}{\sqrt{\Delta_n}}\Delta_{i+l}^n(\delta \star \mu)\right| \wedge 1\right) \leq K\left(\frac{\alpha_n \Delta_n}{\varepsilon^r} + \sqrt{\Delta_n}\,\varepsilon^{1-r}\gamma_\varepsilon\right),$$

and the same choice as above for $\varepsilon = \varepsilon_n$ gives (3.76).

Finally, we have for any $\eta > 0$:

$$|\sqrt{\Delta_n}\,\chi_{i,l}^n|^2 \wedge \eta^2 \leq$$
$$\eta^2 1_{\{\Delta_{i+l}^n N(\varepsilon)\neq 0\}} + 3\Delta_n|\chi_{i,l}'^n|^2 + 3|\Delta_{i+l}^n M(\varepsilon)|^2 + 3|\Delta_{i+l}^n B(\varepsilon)|^2,$$

hence if we take $\varepsilon = \sqrt{\eta}$ above we get

$$\mathbb{E}_{i+l-1}^n\left(|\sqrt{\Delta_n}\,\chi_{i,l}^n|^2 \wedge \eta^2\right) \leq K\Delta_n g_n'(\eta),$$

where

$$g_n'(\eta) = \eta^{2-r/2} + \Delta_n + \eta^{1-r/2}\gamma_{\sqrt{\eta}} + \Delta_n \eta^{-(r-1)^+}.$$

Since $g_n'(\eta) \to g'(\eta) := \eta^{2-r/2} + \eta^{1-r/2}\gamma_{\sqrt{\eta}}$ and $\gamma_\varepsilon \to 0$ as $\varepsilon \to 0$, we readily get (3.77). $\qquad\square$

Lemma 3.17 *Assume (SH). Let $k \geq 1$ and $l \geq 0$ be integers and let $q > 0$. Let f be a continuous function on $(\mathbb{R}^d)^k$, satisfying (3.54) for some $p \geq 0$ and $K_0 > 0$.*

a) If either X is continuous or if $qp < 2$, we have as $n \to \infty$:

$$\sup_{i\geq l,\,\omega\in\Omega} \mathbb{E}_{i-l-1}^n\left(\left|f\left(\frac{\Delta_i^n X}{\sqrt{\Delta_n}}, \cdots, \frac{\Delta_{i+k-1}^n X}{\sqrt{\Delta_n}}\right) - \right.\right. \tag{3.78}$$

$$\left.\left. f\left(\beta_{i-l,l}^n, \cdots, \beta_{i-l,l+k-1}^n\right)\right|^q\right) \to 0.$$

b) If $qp \leq 2$, and if α_n is like in the previous lemma, we have as $n \to \infty$:

$$\sup_{i \geq l, \, \omega \in \Omega} \mathbb{E}^n_{i-l-1} \left(\left| f \left(\frac{\Delta^n_i X}{\sqrt{\Delta_n}}, \cdots, \frac{\Delta^n_{i+k-1} X}{\sqrt{\Delta_n}} \right) 1_{\cap_{1 \leq j \leq k} \{ \| \Delta^n_{i+j-1} X \| \leq \alpha_n \}} \right. \right. \tag{3.79}$$

$$\left. \left. - f \left(\beta^n_{i-l,l}, \cdots, \beta^n_{i-l,l+k-1} \right) \right|^q \right) \to 0.$$

Proof. For any $A > 0$, the supremum $G_A(\varepsilon)$ of $|f(x_1 + y_1, \cdots, x_k + y_k) - f(x_1, \cdots, x_k)|$ over all $\|x_j\| \leq A$ and $\|y_j\| \leq \varepsilon$ goes to 0 as $\varepsilon \to 0$. We set $g(x, y) = 1 + \|x\|^{qp} + \|y\|^{qp}$. If we want to prove (3.79), the sequence α_n is of course as above, whereas if we want to prove (3.78), we put $\alpha_n = \infty$ for all n. Then for all $A > 1$ and $s \geq 0$ and $\varepsilon > 0$ we have, by a (tedious) calculation using (3.54), the constant K depending on K_0, q, k:

$$|f(x_1 + y_1, \cdots, x_k + y_k) 1_{\cap_{j=1,\cdots,k} \{ \|x_j\| \leq \alpha_n \}} - f(x_1, \cdots, x_k)|^q \tag{3.80}$$

$$\leq G_A(\varepsilon)^q + K \sum_{m=1}^k \left(h_{\varepsilon,s,A,n}(x_m, y_m) \prod_{j=1,\cdots,k, \, j \neq m} g(x_j, y_j) \right),$$

where

$$h_{\varepsilon,s,A,n}(x, y) = \frac{\|x\|^{pq+1}}{A} + \|x\|^{pq}(\|y\| \wedge 1) + A^{pq} \frac{\|y\|^2 \wedge 1}{\varepsilon^2} + \frac{\|y\|^{pq+s} \wedge \alpha_n^{pq+s}}{A^s}.$$

We apply these estimates with $x_j = \beta^n_{i-l,l+j-1}$ and $y_j = \chi^n_{i-l,l+j-1}$. In view of (3.73) we have if X is continuous or if $pq \leq 2$:

$$\mathbb{E}^n_{i+j-2}(g(\beta^n_{i-l,l+j-1}, \chi^n_{i-l,l+j-1})) \leq K. \tag{3.81}$$

Next consider $\zeta^n_{i,j,\varepsilon,A} = \mathbb{E}^n_{i+j-2}(h_{\varepsilon,s,A,n}(\beta^n_{i-l,l+j-1}, \chi^n_{i-l,l+j-1}))$ for an adequate choice of s, to be done below. When X is continuous we take $s = 1$, and (3.73) and Cauchy-Schwarz inequality yield $\zeta^n_{i,j,\varepsilon,A} \leq K(1/A + \sqrt{\Delta_n} + \Delta_n A^{pq}/\varepsilon^2)$. In the discontinuous case when $pq < 2$ and $\alpha_n = \infty$ we take $s = 2 - pq > 0$ and by (3.73) and Cauchy-Schwarz again, plus (3.75) with $r = 2$, we get the existence on a sequence $\delta_n \to 0$ such that $\zeta^n_{i,j,\varepsilon,A} \leq K(1/A + 1/A^s + A^{pq}\delta_n/\varepsilon^2)$. Finally in the discontinuous case when $\alpha_n < \infty$ we have $pq \leq 2$ and we take $s = 0$ and we still obtain $\zeta^n_{i,j,\varepsilon,A} \leq K(1/A + A^{pq}\delta_n/\varepsilon^2)$ by the same argument. To summarize, in all cases we have for all $\varepsilon > 0$:

$$\sup_{\omega,i,j} \zeta^n_{i,j,\varepsilon,A}(\omega) \leq \psi_n(A, \varepsilon), \text{ where } \lim_{A \to \infty} \limsup_{n \to \infty} \psi_n(A, \varepsilon) = 0. \tag{3.82}$$

At this stage, we make use of (3.80) and use the two estimates (3.81) and (3.82) and take successive downward conditional expectations to get the left sides of (3.78) and (3.79) are smaller than $G_A(\varepsilon)^q + K\psi_n(A, \varepsilon)$. This holds for all $A > 1$ and $\varepsilon > 0$. Then by using $G_A(\varepsilon) \to 0$ as $\varepsilon \to 0$ and the last part of (3.82), we readily get the results. $\qquad \square$

Lemma 3.18 *Under (SH), for any function $(\omega, x) \mapsto g(\omega, x)$ on $\Omega \times \mathbb{R}^d$ which is $\mathcal{F}_{(i-1)\Delta_n} \otimes \mathcal{R}^d$-measurable and even, and with polynomial growth in x, we have*

$$\mathbb{E}_{i-1}^n \left(\Delta_i^n N \, g(., \beta_i^n) \right) = 0 \tag{3.83}$$

for N being any component of W, or being any bounded martingale orthogonal to W.

Proof. When $N = W^j$ we have $\Delta_i^n N g(\beta_i^n)(\omega) = h(\sigma_{(i-1)\Delta_n}, \Delta_i^n W)(\omega)$ for a function $h(\omega, x, y)$ which is odd and with polynomial growth in y, so obviously (3.83) holds.

Next assume that N is bounded and orthogonal to W. We consider the martingale $M_t = \mathbb{E}(g(., \beta_i^n) | \mathcal{F}_t)$, for $t \geq (i-1)\Delta_n$. Since W is an $(\mathcal{F}_t)$-Brownian motion, and since β_i^n is a function of $\sigma_{(i-1)\Delta_n}$ and of $\Delta_i^n W$, we see that $(M_t)_{t \geq (i-1)\Delta_n}$ is also, conditionally on $\mathcal{F}_{i-1)\Delta_n}$, a martingale w.r.t. the filtration which is generated by the process $W_t - W_{(i-1)\Delta_n}$. By the martingale representation theorem the process M is thus of the form $M_t = M_{(i-1)\Delta_n} + \int_{(i-1)\Delta_n}^t \eta_s dW_s$ for an appropriate predictable process η. It follows that M is orthogonal to the process $N_t' = N_t - N_{(i-1)\Delta_n}$ (for $t \geq (i-1)\Delta_n$), or in other words the product MN' is an $(\mathcal{F}_t)_{t \geq (i-1)\Delta_n}$-martingale. Hence

$$\mathbb{E}_{i-1}^n (\Delta_i^n N \, g(., \sqrt{\Delta_n} \, \sigma_{(i-1)\Delta_n} \Delta_i^n W)) = \mathbb{E}_{i-1}^n (\Delta_i^n N' M_{i\Delta_n}) =$$
$$\mathbb{E}_{i-1}^n \Delta_i^n N' \Delta_i^n M) = 0,$$

and thus we get (3.83). $\qquad \qquad \square$

3.6.5 Proof of Theorem 3.10

When $f(\lambda x) = \lambda^p f(x)$ we have $V(f, \Delta_n) = \Delta_n^{p/2} V'(f, \Delta_n)$, hence (3.55) readily follows from the first claim. For this first claim, and as seen above, it is enough to prove it under the stronger assumption (SH).

If we set

$$V''(f, k, \Delta_n)_t = \sum_{i=1}^{[t/\Delta_n]} f(\beta_{i,0}^n, \cdots, \beta_{i,k-1}^n),$$

we have $\Delta_n(V'(f, k, \Delta_n) - V''(f, k, \Delta_n)) \xrightarrow{\text{u.c.p.}} 0$ by Lemma 3.17-(a) applied with $l = 0$ and $q = 1$. Therefore it is enough to prove that

$$\Delta_n V''(f, k, \Delta_n)_t \xrightarrow{\text{u.c.p.}} \int_0^t \rho_{\sigma_v}^{\otimes k}(f_v) dv.$$

For this, with $I(n, t, l)$ denoting the set of all $i \in \{1, \cdots, [t/\Delta_n]\}$ which are equal to l modulo k, it is obviously enough to show that for $l = 0, 1, \cdots, k-1$:

$$\sum_{i \subset I(n,t,l)} \eta_i^n \xrightarrow{\text{u.c.p.}} \frac{1}{k} \int_0^t \rho_{\sigma_v}^{\otimes k}(f_v)dv, \qquad \text{where } \eta_i^n = \Delta_n f(\beta_{i,0}^n, \cdots, \beta_{i,k-1}^n).$$

(3.84)

Observe that η_i^n is $\mathcal{F}_{(i+k-1)\Delta_n}$-measurable, and obviously

$$\mathbb{E}_{i-1}^n(\eta_i^n) = \Delta_n \rho_{\sigma_{(i-1)\Delta_n}}^{\otimes k}(f), \qquad \mathbb{E}_{i-1}^n(|\eta_i^n|^2) \leq K\Delta_n^2.$$

By Riemann integration, we have $\sum_{i \in I(n,t,l)} \mathbb{E}_{i-1}^n(\eta_i^n) \xrightarrow{\text{u.c.p.}} \frac{1}{k} \int_0^t \rho_{\sigma_v}^{\otimes k}(f_v)dv$, because $t \mapsto \rho_{\sigma_t}^{\otimes k}(f)$ is right-continuous with left limits. Hence (3.84) follows from Lemma 3.4.

3.6.6 Proof of Theorem 3.11

The proof is exactly the same as for Theorem 3.10, once noticed that in view of Lemma 3.17-(b) applied with $\alpha_n = \alpha\Delta_n^{\varpi-1/2}$ we have

$$\Delta_n(V'(\varpi, \alpha; f, k, \Delta_n)_t - V''(f, k, \Delta_n)_t)) \xrightarrow{\text{u.c.p.}} 0.$$

3.6.7 Proof of Theorem 3.13

Once more we may assume (SH). Below, j, k are fixed, as well as ϖ and α and the function f, satisfying (3.63), and for simplicity we write $D = D^{jk}(f)$ and $D^n = D^{jk}(f, \varpi, \alpha, \Delta_n)$. Set also

$$\left.\begin{array}{l} \widehat{D}_t^n = \frac{1}{k_n} \sum_{i=1+k_n}^{[t/\Delta_n]-k_n} f_n(\Delta_i^n X) \sum_{l \in I_{n,t}(i)} \beta_l^{n,j} \beta_l^{n,k}, \\[2mm] \widehat{D}_t'^n = \frac{1}{k_n} \sum_{i=1+k_n}^{[t/\Delta_n]-k_n} f(\sqrt{\Delta_n}\,\beta_i^n) \sum_{l \in I_{n,t}(i)} \beta_l^{n,j} \beta_l^{n,k}. \end{array}\right\}$$

(3.85)

Lemma 3.19 *We have* $\widehat{D}^n \xrightarrow{\mathbb{P}-Sk} D$.

Proof. a) Let ψ_ε be as in (3.74) and

$$Y(\varepsilon)_t^n = \frac{1}{k_n} \sum_{i=1+k_n}^{[t/\Delta_n]-k_n} (f_n\psi_\varepsilon)(\Delta_i^n X) \sum_{l \in I_{n,t}(i)} \beta_l^{n,j} \beta_l^{n,k},$$

$$Z(\varepsilon)_t^n = \widehat{D}_t^n - Y(\varepsilon)_t^n.$$

It is obviously enough to show the following three properties, for some suitable processes $Z(\varepsilon)$:

$$\lim_{\varepsilon \to 0} \limsup_n \mathbb{E}(\sup_{s \leq t} |Y(\varepsilon)_s^n|) = 0,$$

(3.86)

$$\varepsilon \in (0,1), \ n \to \infty \quad \Rightarrow \quad Z(\varepsilon)^n \xrightarrow{\mathbb{P}-\text{sk}} Z(\varepsilon), \qquad (3.87)$$

$$\varepsilon \to 0 \quad \Rightarrow \quad Z(\varepsilon) \xrightarrow{\text{u.c.p.}} D. \qquad (3.88)$$

b) Let us prove (3.86) in the first case of (3.63). We have $|(f\psi_\varepsilon)(x)| \leq \phi(\varepsilon)\|x\|^2$ for some function ϕ such that $\phi(\varepsilon) \to 0$ as $\varepsilon \to 0$. Hence (3.73) yields $\mathbb{E}_{i-1}^n(|(f\psi_\varepsilon)(\Delta_i^n X)|) \leq K\phi(\varepsilon)\Delta_n$. Now, $Y(\varepsilon)_t^n$ is the sum of at most $2k_n \cdot [t/\Delta_n]$ terms, all smaller in absolute value than $\frac{1}{k_n} |(f\psi_\varepsilon)(\Delta_i^n X)|\|\beta_j^n\|^2$ for some $i \neq j$. By taking two successive conditional expectations and by using again (3.73) the expectation of such a term is smaller than $K\phi(\varepsilon)\Delta_n/k_n$, hence the expectation in (3.86) is smaller than $Kt\phi(\varepsilon)$ and we obtain (3.86).

Next, consider the second case of (3.63). Then $(f_n\psi_\varepsilon)(x) = g(x)1_{\{\alpha\Delta_n^\varpi < \|x\| \leq \varepsilon\}}$, where g is an homogeneous polynomial of degree 2. Then if $\alpha\Delta_n^\varpi < \varepsilon < 1/2$ we have

$$|(f_n\psi_\varepsilon)(x+y)| \leq K\left(\|x\|^4\Delta_n^{-2\varpi} + \|y\|^2 \wedge \varepsilon^2\right).$$

Using this with $x = \sqrt{\Delta_n} \ \beta_i^n$ and $y = \Delta_i^n X'$, we deduce from (3.73) and (3.77) that

$$\mathbb{E}_{i-1}^n(|(f_n\psi_\varepsilon)(\Delta_i^n X)|) \leq K\Delta_n\left(\Delta_n^{1-2\varpi} + \alpha(n,\varepsilon)\right),$$

where $\lim_{\varepsilon \to 0} \limsup_{n \to \infty} \alpha(n,\varepsilon) = 0$. Then exactly as for the first case, we deduce that the expectation in (3.86) is smaller than $Kt(\Delta_n^{1-2\varpi} + \alpha(n,\varepsilon))$, and we obtain again (3.86).

c) Now we define $Z(\varepsilon)$. Let us call $T_q(\varepsilon)$ for $q = 1, 2, \cdots$ the successive jump times of the Poisson process $\underline{\mu}([0,t] \times \{x : \gamma(x) > \varepsilon/2\})$, and set

$$Z(\varepsilon)_t = \sum_{q:T_q(\varepsilon) \leq t} (f(1-\psi_\varepsilon))(\Delta X_{T_q(\varepsilon)}) \, (c_{T_q(\varepsilon)-}^{jk} + c_{T_q(\varepsilon)}^{jk}).$$

For all $\omega \in \Omega$, $q \geq 1$, $\varepsilon' \in (0,\varepsilon)$ there is q' such that $T_q(\varepsilon)(\omega) = T_{q'}(\varepsilon')(\omega)$, whereas $1 - \psi_\varepsilon$ increases to the indicator of $\mathbb{R}^d\backslash\{0\}$. Thus we obviously have (3.88).

d) It remains to prove (3.87). Fix $\varepsilon \in (0,1)$ and write $T_q = T_q(\varepsilon)$. Recall that for u different from all T_q's, we have $\|\Delta X_u\| \leq \varepsilon/2$. Hence, for each ω and each $t > 0$, we have the following properties for all n large enough: there is no T_q in $(0, k_n\Delta_n]$, nor in $(t - (k_n+1)\Delta_n, t]$; there is at most one T_q in an interval $((i-1)\Delta_n, i\Delta_n]$ with $i\Delta_n \leq t$, and if this is not the case we have $\psi_\varepsilon(\Delta_i^n X) = 1$. Hence for n large enough we have

$$Z(\varepsilon)_t^n = \sum_{q: \ k_n\Delta_n < T_q \leq t-(k_n+1)\Delta_n} \zeta_q^n,$$

where

$$\zeta_q^n = \frac{1}{k_n}(f(1-\psi_\varepsilon))(\Delta_{i(n,q)}^n X) \sum_{l \in I'(n,q)} \beta_l^{n,j}\beta_l^{n,k},$$

and $i(n,q) = \inf(i : i\Delta_n \geq T_q)$ and $I'(n,q) = \{l : l \neq i(n,q), |l - i(n,q)| \leq k_n\}$.

To get (3.87) it is enough that $\zeta_q^n \xrightarrow{\mathbb{P}} (f(1-\psi_\varepsilon))(\Delta X_{T_q})(c_{T_q-}^{jk} + c_{T_q}^{jk})$ for any q. Since $(f(1-\psi_\varepsilon))(\Delta_{i(n,q)}^n X) \to (f(1-\psi_\varepsilon))(\Delta X_{T_q})$ pointwise, it remains to prove that

$$\frac{1}{k_n} \sum_{l \in I'_-(n,q)} \beta_l^{n,j} \beta_l^{n,k} \xrightarrow{\mathbb{P}} c_{T_q-}^{jk}, \quad \frac{1}{k_n} \sum_{l \in I'_+(n,q)} \beta_l^{n,j} \beta_l^{n,k} \xrightarrow{\mathbb{P}} c_{T_q}^{jk}. \quad (3.89)$$

where $I'_-(n,q)$ and $I'_+(n,q)$ are the subsets of $I'(n,q)$ consisting in those l smaller, respectively bigger, than $i(n,q)$. Letting $l(n,q)$ be the smallest l in $I'_-(n,q)$, we see that the left side of the first expression in (3.89) is $U_q^n + U_q'^n$, where

$$U_q^n = \sum_{r,s=1}^{d'} \sigma_{l(n,q)\Delta_n}^{jr} \sigma_{l(n,q)\Delta_n}^{ks} \overline{U}_n^q(r,s),$$

$$\overline{U}_q^n(r,s) = \frac{1}{k_n \Delta_n} \sum_{l \in I'_-(n,q)} \Delta_l^n W^r \Delta_l^n W^s,$$

$$U_q'^n = \sum_{r,s=1}^{d'} \frac{1}{k_n \Delta_n} \sum_{l \in I'(n,q)} \left((\sigma_{(l-1)\Delta_n}^{jr} - \sigma_{i(n,q)\Delta_n}^{jr}) \sigma_{(l-1)\Delta_n}^{ks} \right.$$

$$\left. + (\sigma_{(l-1)\Delta_n}^{ks} - \sigma_{i(n,q)\Delta_n}^{ks}) \sigma_{(l-1)\Delta_n}^{jr} \right) \Delta_l^n W^r \Delta_l^n W^s.$$

On the one hand, the variables $\Delta_l^n W$ are i.i.d. $N(0, \Delta_n I_{d'})$, so $\overline{U}_q^n(r,s)$ is distributed as $1/k_n$ times the sum of k_n i.i.d. variables with the same law as $W_1^r W_1^s$, hence obviously $\overline{U}_q^n(r,s)$ converges in probability to 1 if $r = s$ and to 0 otherwise. Since $\sigma_{l(n,q)\Delta_n} \to \sigma_{T_q-}$, we deduce that $U_q^n \xrightarrow{\mathbb{P}} c_{T_q-}^{jk}$.

On the other hand, due to (3.73) and by successive integrations we obtain

$$\mathbb{E}(|U_q'^n|) \leq \frac{1}{k_n} \sum_{l \in I'_-(n,q)} \mathbb{E}(\|\sigma_{(l-1)\Delta_n} - \sigma_{l(n,q)\Delta_n}\|^2) \leq K k_n \Delta_n$$

which goes to 0 by virtue of (3.62). Therefore we have proved the first part of (3.89), and the second part is proved in a similar way. $\qquad\square$

Lemma 3.20 *If f is continuous and $f(\lambda x) = \lambda^p f(x)$ for all $\lambda > 0$, $x \in \mathbb{R}^d$ and some $p \geq 2$, we have $\Delta_n^{1-p/2} \widehat{D}_t'^n \xrightarrow{u.c.p.} 2 \int_0^t \rho_{\sigma_u}(f) c_u^{jk} du$.*

Proof. First we observe that by polarization, and exactly as in the proof of Theorem 3.11, it is enough to show the result when $j = k$, and of course when $f \geq 0$: then $\widehat{D}_t'^n$ is increasing in t, and $\int_0^t \rho_{\sigma_u}(f) c_u^{jk} du$ is also increasing and

continuous. Then instead of proving the local uniform convergence it is enough
to prove the convergence (in probability) for any given t.

With our assumptions on f, we have

$$\Delta_n^{1-p/2}\widehat{D}_t^{\prime n} = \frac{\Delta_n}{k_n}\sum_{i=1+k_n}^{[t/\Delta_n]-k_n}\sum_{l\in I_{n,t}(i)}f(\beta_i^n)\beta_i^{n,j}\beta_i^{n,k}.$$

Moreover, $\Delta_n\sum_{i=1+k_n}^{[t/\Delta_n]-k_n}\rho_{\sigma_{(i-1)\Delta_n}}(f)c_{(i-1-k_n)\Delta_n}^{jk}\xrightarrow{\mathbb{P}}\int_0^t\rho_{\sigma_u}(f)c_u^{jk}\,du$ by
Riemann integration. Therefore, it is enough to prove the following two prop-
erties:

$$\sum_{i=1+k_n}^{[t/\Delta_n]-k_n}\Delta_n(f(\beta_i^n)-\rho_{\sigma_{(i-1)\Delta_n}}(f))\,c_{(i-1-k_n)\Delta_n}^{jk}\xrightarrow{\mathbb{P}}0,\qquad(3.90)$$

$$Y_t^n:=\frac{\Delta_n}{k_n}\sum_{i=1+k_n}^{[t/\Delta_n]-k_n}\sum_{l\in I_{n,t}(i)}\zeta_{i,l}^n\xrightarrow{\mathbb{P}}0,\qquad(3.91)$$

where

$$\zeta_{i,l}^n=f(\beta_i^n)(\beta_l^{n,j}\beta_l^{n,k}-c_{(i-1-k_n)\Delta_n}^{jk}).$$

Each summand, say ζ_i^n, in the left side of (3.90) is $\mathcal{F}_{i\Delta_n}$-measurable with
$\mathbb{E}_{i-1}^n(\zeta_i^n)=0$ and $\mathbb{E}_{i-1}^n((\zeta_i^n)^2)\le K\Delta_n^2$ (apply (3.73) and recall that $|f(x)|\le$
$K\|x\|^r$ with our assumptions on f), so (3.90) follows from Lemma 3.4.

Proving (3.91) is a bit more involved. We set

$$Y_t^{\prime n}=\frac{\Delta_n}{k_n}\sum_{i=1+k_n}^{[t/\Delta_n]-k_n}\sum_{l\in I_{n,t}(i)}\zeta_{i,l}^{\prime n},$$

where

$$\zeta_{i,l}^{\prime n}=f(\sigma_{(i-1-k_n)\Delta_n}\Delta_i^n W/\sqrt{\Delta_n})\,(\beta_l^{n,j}\beta_l^{n,k}-c_{(i-1-k_n)\Delta_n}^{jk}).$$

On the one hand, for any $l\in I_{n,t}(i)$ (hence either $l<i$ or $l>i$) and by
successive integration we have

$$|\mathbb{E}_{i-1-k_n}^n(\zeta_{i,l}^{\prime n})|=$$
$$|\rho_{\sigma_{(i-1-k_n)\Delta_n}}(f)\mathbb{E}_{i-1-k_n}^n(c_{(l-1)\Delta_n}^{jk}-c_{(i-1-k_n)\Delta_n}^{jk})|\le K\sqrt{k_n\Delta_n}$$

by (3.73), the boundedness of σ and $|f(x)|\le K\|x\|^r$. Moreover, the inequality
$\mathbb{E}_{i-1-k_n}^n((\zeta_{i,l}^{\prime n})^2)\le K$ is obvious. Therefore, since $\mathbb{E}((Y_t^{\prime n})^2)$ is Δ^2/k_n^2 times
the sum of all $\mathbb{E}(\zeta_{i,l}^{\prime n}\zeta_{i',l'}^{\prime n})$ for all $1+k_n\le i,i'\le[t/\Delta_n]-k_n$ and $l\in I_{n,t}(i)$
and $l'\in I_{n,t}(i')$, by singling out the cases where $|i-i'|>2k_n$ and $|i-i'|\le$
$2k_n$ and in the first case by taking two successive conditional expectations, and

in the second case by using Cauchy–Schwarz inequality, we obtain that

$$\mathbb{E}((Y_t'^n)^2) \leq K\frac{\Delta_n^2}{k_n^2}\left(4k_n^2[t/\Delta_n]^2(k_n\Delta_n) + 4k_n^2[t/\Delta_n]\right) \leq$$

$$K(t^2 k_n\Delta_n + t\Delta_n) \rightarrow 0.$$

In order to get (3.91) it remains to prove that $Y_t^n - Y_t'^n \xrightarrow{\mathbb{P}} 0$. By Cauchy-Schwarz inequality and (3.73), we have

$$\mathbb{E}(|\zeta_{i,l}^n - \zeta_{i,l}'^n|) \leq K\left(\mathbb{E}(|\rho_{\sigma_{(i-1)\Delta_n}}(f) - \rho_{\sigma_{(i-1-k_n)\Delta_n}}(f)|^2)\right)^{1/2}.$$

Then another application of Cauchy-Schwarz yields that $\mathbb{E}(|Y_t^n - Y_t'^n|) \leq K_t\sqrt{\alpha_n(t)}$, where

$$\alpha_n(t) = \frac{\Delta_n}{k_n}\sum_{i=1+k_n}^{[t/\Delta_n]-k_n}\sum_{l\in I_{n,t}(i)}\mathbb{E}(|\rho_{\sigma_{(i-1)\Delta_n}}(f) - \rho_{\sigma_{(i-1-k_n)\Delta_n}}(f)|^2)$$

$$= 2\Delta_n\sum_{i=1+k_n}^{[t/\Delta_n]-k_n}\mathbb{E}(|\rho_{\sigma_{(i-1)\Delta_n}}(f) - \rho_{\sigma_{(i-1-k_n)\Delta_n}}(f)|^2)$$

$$\leq 2\int_0^t g_n(s)ds,$$

with the notation $g_n(s) = \mathbb{E}((\rho_{\sigma_{\Delta_n(k_n+[s\Delta_n])}}(f) - \rho_{\sigma_{\Delta_n[s/\Delta_n]}}(f))^2)$. Since c_t is bounded and f is with polynomial growth, we first have $g_n(s) \leq K$. Since further $t \mapsto \sigma_t$ has no fixed time of discontinuity and f is continuous and $\Delta_n k_n \rightarrow 0$, we next have $g_n(s) \rightarrow 0$ pointwise: hence $\alpha_n(t) \rightarrow 0$ and we have the result.   □

Proof of (3.65). In view of Lemma 3.19 it is enough to prove that $\widehat{D}^n - D^n \xrightarrow{u.c.p.} 0$, and this will obviously follow if we prove that

$$\sup_{i\neq l}\frac{1}{\Delta_n^2}\mathbb{E}(|f_n(\Delta_i^n X)\zeta_i^n|) \rightarrow 0 \qquad \text{as } n \rightarrow \infty, \tag{3.92}$$

where $\zeta_i^n = \Delta_l^n X^j \Delta_l^n X^k 1_{\{\|\Delta_l^n X\| \leq \alpha\Delta_n^\varpi\}} - \Delta_n\beta_l^{n,j}\beta_l^{n,k}$.

A simple computation shows that for $x, y \in \mathbb{R}^d$ and $\varepsilon > 0$, we have

$$|(x_j+y_j)(x_k+y_k)1_{\{\|x+y\|\leq\varepsilon\}} - x_j x_k| \leq K\left(\frac{1}{\varepsilon}\|x\|^3 + \|x\|(\|y\|\wedge\varepsilon) + \|y\|^2\wedge\varepsilon^2\right).$$

We apply this to $x = \sqrt{\Delta_n}\,\beta_l^n$ and $y = \sqrt{\Delta_n}\,\chi_l^n$ and $\varepsilon = \alpha\Delta_n^\varpi$, and (3.73) and (3.77) with $\eta = \varepsilon$ and Cauchy-Schwarz inequality, to get

$$\mathbb{E}_{l-1}^n(|\zeta_l^n|) \leq K\Delta_n(\Delta_n^{1/2-\varpi} + \alpha_n)$$

for some α_n going to 0. On the other hand, (SH) implies that $\Delta_i^n X$ is bounded by a constant, hence (3.63) yields $|f_n(\Delta_i^n X)| \leq K \|\Delta_i^n X\|^2$ and (3.73) again gives $\mathbb{E}_{i-1}^n(|f_n(\Delta_i^n X)|) \leq K\Delta_n$. Then, by taking two successive conditional expectations, we get $\mathbb{E}(|f_n(\Delta_i^n X)\zeta_l^n|) \leq K\Delta_n^2(\Delta_n^{1/2-\varpi} + \alpha_n)$ as soon as $l \neq i$, and (3.92) follows. $\qquad\qquad\square$

Proof of (3.66). In view of Lemma 3.20 it is enough to prove that $\Delta_n^{1-r/2}(\widehat{D}'^n - D^n) \overset{\text{u.c.p.}}{\longrightarrow} 0$, when X is continuous and $f(\lambda x) = \lambda^r f(x)$ for some $r > 2$. With the notation ζ_i^n of the previous proof, this amounts to prove the following two properties:

$$\sup_{i \neq l} \frac{1}{\Delta_n^{1+r/2}} \, \mathbb{E}(|f(\Delta_i^n X)\zeta_l^n|) \to 0 \qquad \text{as } n \to \infty, \qquad (3.93)$$

$$\sup_{i \neq l} \frac{1}{\Delta_n^{r/2}} \, \mathbb{E}\Big(|f(\Delta_i^n X)|1_{\{\|\Delta_i^n X\| > \alpha\Delta_n^\varpi\}}\|\beta_l^n\|\Big) \to 0 \qquad \text{as } n \to \infty. \qquad (3.94)$$

Since X is continuous and $|f(x)| \leq K\|x\|^r$, we have $\mathbb{E}_{i-1}^n(|f_n(\Delta_i^n X)|) \leq K\Delta_n^{r/2}$, hence the proof of (3.93) is like in the previous proof. By Bienaymé–Tchebycheff inequality and (3.73) we also have

$$\mathbb{E}_{i-1}^n(|f(\Delta_i^n X)|1_{\{\|\Delta_i^n X\| > \alpha\Delta_n^\varpi\}}) \leq K_q\Delta_n^q$$

for any $q > 0$, hence (3.94) follows. $\qquad\qquad\square$

3.7 A first CLT

As we have seen after (3.26), we have the CLT (3.19) when X is the sum of a Wiener process and a compound Poisson process, as soon as the function f in $V'(f, \Delta_n)$ satisfies $f(x)/|x|^p \to 0$ as $|x| \to \infty$, for some $p < 1$. In this section we prove the same result, and even a bit more (the stable convergence in law) when X satisfies (H).

In other words, we are concerned with the CLT associated with Theorem 3.10. For statistical purposes we need a CLT when the function $f = (f_1, \cdots, f_q)$ is multidimensional: in this case, $V'(f, k, \Delta_n)$ is also multidimensional, with components $V'(f_j, k, \Delta_n)$. On the other hand, we will strongly restrict the class of functions f for which we give a CLT: although much more general situations are available, they also are much more complicated and will not be used in the sequel. Let us mention, however, that the present setting does not allow to consider the CLT for multipower variations in the interesting cases as in (3.60); for this, we refer to Barndorff–Nielsen, Graversen, Jacod, Podolskij, and Shephard (2006) when X is continuous, and to Barndorff–Nielsen, Shephard, and Winkel (2006) when X is a discontinuous Lévy process. For discon-

tinuous semimartingales which are not Lévy processes, essentially nothing is
known as far as CLTs are concerned, for processes like (3.60).

One of the difficulties of this question is to characterize the limit, and more
specifically the quadratic variation of the limiting process. To do this, we consider a sequence $(U_i)_{i \geq 1}$ of independent $\mathcal{N}(0, I_d)$ variables (they take values
in $\mathbb{R}^d$, and I_d is the unit $d \times d$ matrix). Recall that ρ_σ, defined before (3.53), is
also the law of σU_1, and so $\rho_\sigma(g) = \mathbb{E}(g(\sigma U_1))$. In a similar way, for any q-
dimensional function $f = (f_1, \cdots, f_q)$ on $(\mathbb{R}^d)^k$, say with polynomial growth,
we set for $i, j = 1, \cdots, q$:

$$
\begin{aligned}
R_\sigma^{ij}(f, k) \quad &= \quad \sum_{l=-k+1}^{k-1} \mathbb{E}\Big(f_i(\sigma U_k, \cdots, \sigma U_{2k-1}) f_j(\sigma U_{l+k}, \cdots, \sigma U_{l+2k-1}) \Big) \\
&\quad - (2k-1)\mathbb{E}(f_i(\sigma U_1, \cdots, \sigma U_k))\mathbb{E}(f_j(\sigma U_1, \cdots, \sigma U_k)). \quad (3.95)
\end{aligned}
$$

One can of course express this in terms of integrals of f with respect to the
measures ρ_σ and their tensor powers, but this is very complicated. Let us just
mention the special case where $k = 1$:

$$
R_\sigma^{ij}(f, 1) \quad = \quad \rho_\sigma(f_i f_j) - \rho_\sigma(f_i)\rho_\sigma(f_j). \quad (3.96)
$$

The main result goes as follows:

Theorem 3.21 *Assume (H). Let f be a q-dimensional function on $(\mathbb{R}^d)^k$ for
some $k \geq 1$, which is even in each argument, that is*

$$
f(x_1, \cdots, x_{l-1}, -x_l, x_{l+1}, \cdots, x_k) = f(x_1, \cdots, x_{l-1}, x_l, x_{l+1}, \cdots, x_k)
$$

identically for all l. In the following two cases:

a) X is continuous, and f is C^1 with derivatives having polynomial growth,

*b) f is C_b^1 (bounded with first derivatives bounded), and $\int (\gamma(x) \wedge 1)\lambda(dx)
< \infty$ (hence the jumps of X are summable over each finite interval),*

the q-dimensional processes

$$
\frac{1}{\sqrt{\Delta_n}} \left(\Delta_n V'(f, k, \Delta_n)_t - \int_0^t \rho_{\sigma_u}^{\otimes k}(f) du \right)
$$

*converge stably in law to a continuous process $V'(f, k)$ defined on an extension
$(\widetilde{\Omega}, \widetilde{\mathcal{F}}, \widetilde{\mathbb{P}})$ of the space $(\Omega, \mathcal{F}, \mathbb{P})$, which conditionally on the σ-field $\mathcal{F}$ is a
centered Gaussian $\mathbb{R}^q$-valued process with independent increments, satisfying*

$$
\widetilde{\mathbb{E}}(V'(f_i, k)_t V'(f_j, k)_t) \quad = \quad \int_0^t R_{\sigma_u}^{ij}(f, k) du. \quad (3.97)
$$

Another, equivalent, way to characterize the limiting process $V'(f, k)$ is as follows, see Jacod and Shiryaev (2003): for each σ, the matrix $R_\sigma(f, k)$ is symmetric nonnegative definite, so we can find a square-root $S_\sigma(f, k)$, that is $S_\sigma(f, k)S_\sigma(f, k)^* = R_\sigma(f, k)$, which as a function of σ is measurable. Then there exists a q-dimensional Brownian motion $B = (B^i)_{i \leq q}$ on an extension of the space $(\Omega, \mathcal{F}, \mathbb{P})$, independent of $\mathcal{F}$, and $V'(f, k)$ is given componentwise by

$$V'(f_i, k)_t = \sum_{j=1}^{q} \int_0^t S_{\sigma_u}^{ij}(f, k)dB_u^j. \tag{3.98}$$

As a consequence we obtain a CLT for estimating C_t^{jk} when X is continuous. It suffices to apply the theorem with $k = 1$ and the $d \times d$-dimensional function f with components $f_{jl}(x) = x^j x^k$. Upon a simple calculation using (3.96) in this case, we obtain:

Corollary 3.22 *Assume (H) (or (H′) only, although it is not then a consequence of the previous theorem) and that X is continuous. Then the $d \times d$-dimensional process with components*

$$\frac{1}{\sqrt{\Delta_n}} \Big(\sum_{i=1}^{[t/\Delta_n]} \Delta_i^n X^j \Delta_i^n X^k - C_t^{jk} \Big)$$

converge stably in law to a continuous process $(V^{jk})_{1 \leq j, k \leq d}$ defined on an extension $(\widetilde{\Omega}, \widetilde{\mathcal{F}}, \widetilde{\mathbb{P}})$ of the space $(\Omega, \mathcal{F}, \mathbb{P})$, which conditionally on the σ-field $\mathcal{F}$ is a centered Gaussian $\mathbb{R}^q$-valued process with independent increments, satisfying

$$\widetilde{\mathbb{E}}(V_t^{jk} V_t^{j'k'}) = \int_0^t (c_u^{jk'} c_u^{j'k} + c_u^{jj'} c_u^{kk'})du. \tag{3.99}$$

It turns out that this result is very special: Assumption (H) is required for Theorem 3.21, essentially because one needs that $\Delta_n \sum_{i=1}^{[t/\Delta_n]} \rho_{\sigma_{(i-1)\Delta_n}}(g)$ converges to $\int_0^t \rho_{\sigma_s}(g)ds$ at a rate faster than $1/\sqrt{\Delta_n}$, and this necessitates strong assumptions on σ (instead of assuming that it is an Itô semimartingale, as in (H), one could require some Hölder continuity of its paths, with index bigger than $1/2$). However, for the corollary, and due to the quadratic form of the test function, some cancelations occur which allow to obtain the result under the weaker assumption (H′) only. Although this is a theoretically important point, it is not proved here.

There is a variant of Theorem 3.21 which concerns the case where in (3.53) one takes the sum over the i's that are multiples of k. More precisely, we set

$$V''(f, k, \Delta_n)_t = \sum_{i=1}^{[t/k\Delta_n]} f\Big(\Delta_{(i-1)k+1}^n X/\sqrt{\Delta_n}, \cdots, \Delta_{ik}^n X/\sqrt{\Delta_n}\Big). \tag{3.100}$$

The LLN is of course exactly the same as Theorem 3.10, except that the limit should be divided by k in (3.55). As for the CLT, it runs as follows (and although similar to Theorem 3.21 it is not a direct consequence):

Theorem 3.23 *Under the same assumptions as in Theorem 3.21, the q-dimensional processes*

$$\frac{1}{\sqrt{\Delta_n}} \left(\Delta_n V''(f, k, \Delta_n)_t - \frac{1}{k} \int_0^t \rho_{\sigma_u}^{\otimes k}(f) du \right)$$

converge stably in law to a continuous process $V''(f, k)$ defined defined on an extension $(\widetilde{\Omega}, \widetilde{\mathcal{F}}, \widetilde{\mathbb{P}})$ of the space $(\Omega, \mathcal{F}, \mathbb{P})$, which conditionally on the σ-field $\mathcal{F}$ is a centered Gaussian $\mathbb{R}^q$-valued process with independent increments, satisfying

$$\widetilde{\mathbb{E}}(V''(f_i, k)_t V''(f_j, k)_t \mid \mathcal{F}) = \frac{1}{k} \int_0^t \left(\rho_{\sigma_u}^{\otimes k}(f_i f_j) - \rho_{\sigma_u}^{\otimes k}(f_i) \rho_{\sigma_u}^{\otimes k}(f_j) \right) du.$$

$$(3.101)$$

Theorem 3.21 does not allow one to deduce a CLT associated with Theorem 3.12, since the function f which is used in (3.60) cannot meet the assumptions above. Nevertheless such a CLT is available when X is continuous: see Barndorff–Nielsen, Graversen, et al. (2006), under the (weak) additional assumption that $\sigma_t \sigma_t^\star$ is everywhere invertible. When X is discontinuous and with the additional assumption that $\int (\gamma(x) \wedge 1) \lambda(dx) < \infty$, it is also available, see Barndorff–Nielsen, Shephard, and Winkel (2006) for the case when X is a Lévy process.

We do however give the CLT associated with Theorem 3.11, although it is not a direct consequence of the previous one.

Theorem 3.24 *Assume (H), and also that X is continuous or that $\int (\gamma(x)^r \wedge 1) \lambda(dx) < \infty$ for some $r \in [0, 1)$. Then for all $\varpi \in [\frac{1}{2(2-r)}, \frac{1}{2})$ and $\alpha > 0$ the $d \times d$-dimensional process with components*

$$\frac{1}{\sqrt{\Delta_n}} \left(V^{jk}(\varpi, \alpha, \Delta_n)_t - C_t^{jk} \right)$$

converge stably in law to the continuous process $(V^{jk})_{1 \le j, k \le d}$ defined in Corollary 3.22.

In the discontinuous case, this is not fully satisfactory since we need the assumption about $r < 1$, which we *a priori* do not know to hold, and further ϖ has to be bigger than $\frac{1}{2(2-r)}$. In the continuous case for X the assumption is simply (H), but of course in this case there is no reason to prefer the estimators $V^{jk}(\varpi, \alpha, \Delta_n)_t$ to $\sum_{i=1}^{[t/\Delta_n]} \Delta_i^n X^j \Delta_i^n X^k$.

3.7.1 The scheme of the proof of Theorem 3.21

This theorem is rather long to prove, and quite technical. We first describe here the main steps. Note that the localization argument expounded earlier works here as well, so we can and will assume (SH) instead of (H), without special mention. Also, the multidimensional case for f reduces to the 1-dimensional one by polarization, as in the proof of Theorem 3.11, so below we suppose that f is 1-dimensional (that is, $q = 1$). These assumptions are in force through the remainder of this section. We also denote by $\mathcal{M}'$ the set of all $d' \times d'$ matrices bounded by K where K is a bound for the process $\|\sigma_t\|$.

We use the notation

$$
\begin{aligned}
\zeta_i^n &= f(\Delta_i^n X/\sqrt{\Delta_n}, \cdots, \Delta_{i-k}^n X/\sqrt{\Delta_n}), \\
\zeta_i'^n &= f(\beta_{i,0}^n, \beta_{i,1}^n, \cdots, \beta_{i,k-1}^n), \\
\zeta_i''^n &= \zeta_i^n - \zeta_i'^n.
\end{aligned}
$$

First, we replace each normalized increment $\Delta_{i+l}^n X/\sqrt{\Delta_n}$ in (3.53) by $\beta_{i,l}^n$ (notation (3.72)): this is of course much simpler, and we have the following:

Proposition 3.25 *The processes*

$$
\overline{V}_t^n = \sqrt{\Delta_n} \sum_{i=1}^{[t/\Delta_n]} \left(\zeta_i'^n - \rho_{\sigma(i-1)\Delta_n}^{\otimes k}(f) \right) \tag{3.102}
$$

converge stably in law to the process $V'(f, k)$, as defined in Theorem 3.21.

Next, we successively prove the following three properties:

$$
\sqrt{\Delta_n} \sum_{i=1}^{[t/\Delta_n]} \mathbb{E}_{i-1}^n(\zeta_i''^n) \xrightarrow{\text{u.c.p.}} 0, \tag{3.103}
$$

$$
\sqrt{\Delta_n} \sum_{i=1}^{[t/\Delta_n]} \left(\zeta_i''^n - \mathbb{E}_{i-1}^n(\zeta_i''^n) \right) \xrightarrow{\text{u.c.p.}} 0, \tag{3.104}
$$

$$
\frac{1}{\sqrt{\Delta_n}} \left(\Delta_n \sum_{i=1}^{[t/\Delta_n]} \rho_{\sigma(i-1)\Delta_n}^{\otimes k}(f) - \int_0^t \rho_{\sigma_u}^{\otimes k}(f) du \right) \xrightarrow{\text{u.c.p.}} 0. \tag{3.105}
$$

Obviously our theorem is a consequence of these three properties and of Proposition 3.25. Apart from (3.104), which is a simple consequence of Lemma 3.17, all these steps are non trivial, and the most difficult is (3.103).

3.7.2 Proof of the convergence result 3.104

We use the notation $I(n, t, l)$ of the proof of Theorem 3.10, and it is of course enough to prove

$$\sqrt{\Delta_n} \sum_{i \in I(n,t,l)} \left(\zeta_i''^n - \mathbb{E}_{i-1}^n(\zeta_i''^n) \right) \xrightarrow{\text{u.c.p.}} 0$$

for each $l = 0, \cdots, k - 1$. Since each $\zeta_i''^n$ is $\mathcal{F}_{(i+k-1)\Delta_n}$-measurable, by Lemma 3.4 it is even enough to prove that

$$\Delta_n \sum_{i \in I(n,t,l)} \mathbb{E}_{i-1}^n \left((\zeta_i''^n)^2 \right) \xrightarrow{\text{u.c.p.}} 0.$$

But this is a trivial consequence of Lemma 3.17 applied with $q = 2$ and $l = 0$: in case (a) the function f obviously satisfies (3.54) for some $r \geq 0$ and X is continuous, whereas in case (b) it satisfies (3.54) with $r = 0$.

3.7.3 Proof of the convergence result 3.105

Let us consider the function $g(\sigma) = \rho_\sigma^{\otimes k}(f)$, defined on the set $\mathcal{M}'$. (3.105) amounts to

$$\sum_{i=1}^{[t/\Delta_n]} \eta_i^n \xrightarrow{\text{u.c.p.}} 0, \quad \text{where} \quad \eta_i^n = \frac{1}{\sqrt{\Delta_n}} \int_{(i-1)\Delta_n}^{i\Delta_n} (g(\sigma_u) - g(\sigma_{(i-1)\Delta_n}))du.$$

(3.106)

Since f is at least C^1 with derivatives having polynomial growth, the function g is C_b^1 on $\mathcal{M}$. However, the problem here is that σ may have jumps, and even when it is continuous its paths are typically Hölder with index $\alpha > 1/2$, but not $\alpha = 1/2$: so (3.106) is not trivial.

With ∇g denoting the gradient of g (a $d \times d'$-dimensional function), we may write $\eta_i^n = \eta_i'^n + \eta_i''^n$ where (with matrix notation)

$$\eta_i'^n = \frac{1}{\sqrt{\Delta_n}} \nabla g(\sigma_{(i-1)\Delta_n}) \int_{(i-1)\Delta_n}^{i\Delta_n} (\sigma_u - \sigma_{(i-1)\Delta_n}) \, du,$$

$$\eta_i''^n = \frac{1}{\sqrt{\Delta_n}} \int_{(i-1)\Delta_n}^{i\Delta_n} \big(g(\sigma_u) - g(\sigma_{(i-1)\Delta_n}) - $$
$$\nabla g(\sigma_{(i-1)\Delta_n})(\sigma_u - \sigma_{(i-1)\Delta_n}) \big) \, du.$$

In view of (3.71) we can decompose further $\eta_i'^n$ as $\eta_i'^n = \mu_i^n + \mu_i'^n$, where

$$\mu_i^n = \frac{1}{\sqrt{\Delta_n}} \nabla g(\sigma_{(i-1)\Delta_n}) \int_{(i-1)\Delta_n}^{i\Delta_n} du \int_{(i-1)\Delta_n}^{u} \widetilde{b}_s ds,$$

$$\mu_i'^n = \frac{1}{\sqrt{\Delta_n}} \nabla g(\sigma_{(i-1)\Delta_n}) \int_{(i-1)\Delta_n}^{i\Delta_n} du \left(\int_{(i-1)\Delta_n}^{u} \tilde{\sigma}_s dW_s \right.$$
$$\left. + \int_{(i-1)\Delta_n}^{u} \int \tilde{\delta}(s,x)(\underline{\mu} - \underline{\nu})(ds,dx) \right).$$

On the one hand, we have $|\mu_i^n| \le K\Delta_n^{3/2}$ (recall that g is C_b^1 and $\tilde{b}$ is bounded), so $\sum_{i=1}^{[t/\Delta_n]} \mu_i^n \xrightarrow{\text{u.c.p.}} 0$. On the other hand, we have $\mathbb{E}_{i-1}^n(\mu_i'^n) = 0$ and $\mathbb{E}_{i-1}^n((\mu_i'^n)^2) \le K\Delta_n^2$ by Doob and Cauchy–Schwarz inequalities, hence by Lemma 3.4 $\sum_{i=1}^{[t/\Delta_n]} \mu_i'^n \xrightarrow{\text{u.c.p.}} 0$.

Finally, since g is C_b^1 on the compact set $\mathcal{M}$, we have $|g(\sigma')-g(\sigma)-\nabla g(\sigma)(\sigma'-\sigma)| \le K\|\sigma' - \sigma\|h(\|\sigma' - \sigma\|)$ for all $\sigma, \sigma' \in \mathcal{M}$, where $h(\varepsilon) \to 0$ as $\varepsilon \to 0$. Therefore

$$|\eta_i''^n| \le \frac{1}{\sqrt{\Delta_n}} \int_{(i-1)\Delta_n}^{i\Delta_n} h(\|\sigma_u - \sigma_{(i-1)\Delta_n}\|) \|\sigma_u - \sigma_{(i-1)\Delta_n}\| \, du$$
$$\le \frac{1}{\sqrt{\Delta_n}} h(\varepsilon) \int_{(i-1)\Delta_n}^{i\Delta_n} \|\sigma_u - \sigma_{(i-1)\Delta_n}\| \, du$$
$$+ \frac{K}{\varepsilon\sqrt{\Delta_n}} \int_{(i-1)\Delta_n}^{i\Delta_n} \|\sigma_u - \sigma_{(i-1)\Delta_n}\|^2 \, du.$$

Since $h(\varepsilon)$ is arbitrarily small we deduce from the above and from (3.73) that $\sum_{i=1}^{[t/\Delta_n]} \mathbb{E}(|\eta_i''^n|) \to 0$. This clearly finishes to prove (3.106).

3.7.4 Proof of Proposition 3.25

We prove the result when $k = 2$ only. The case $k \ge 3$ is more tedious but similar.

Letting $g_t(x) = \int \rho_{\sigma_t}(dy)f(x,y)$, we have $\overline{V}^n(f)_t = \sum_{i=2}^{[t/\Delta_n]+1} \eta_i^n + \gamma_1'^n - \gamma_{[t/\Delta_n]+1}'^n$, where $\eta_i^n = \gamma_i^n + \gamma_i'^n$ and

$$\gamma_i^n = \sqrt{\Delta_n} \left(f(\beta_{i-1,0}^n, \beta_{i-1,1}^n) - \int \rho_{\sigma(i-2)\Delta_n}(dx)f(\beta_{i-1,0}^n, x) \right),$$
$$\gamma_i'^n = \sqrt{\Delta_n} \left(\int \rho_{\sigma(i-1)\Delta_n}(dx)f(\beta_{i,0}^n, x) - \rho_{\sigma(i-1)\Delta_n}^{\otimes 2}(f) \right).$$

Since obviously $\mathbb{E}(|\gamma_i'^n|) \le K\sqrt{\Delta_n}$, it is enough to prove that $\overline{V}'^n(f)_t = \sum_{i=2}^{[t/\Delta_n]+1} \eta_i^n$ converges stably in law to the process $V'(f,2)$.

Note that η_i^n is $\mathcal{F}_{i\Delta_n}$-measurable, and a (tedious) calculation yields

$$\mathbb{E}_{i-1}^n(\eta_i^n) = 0, \quad \mathbb{E}_{i-1}^n((\eta_i^n)^2) = \Delta_n \phi_i^n, \quad \mathbb{E}_{i-1}^n(|\eta_i^n|^4) \le K\Delta_n^2, \quad (3.107)$$

where $\phi_i^n = g((i-2)\Delta_n, (i-1)\Delta_n, \beta_{i-1}^n)$, and

$$g(s,t,x) = \int \rho_{\sigma_t}(dy)\left(f(x,y)^2 + \left(\int \rho_{\sigma_t}(dz)f(y,z)\right)^2\right)$$

$$- \left(\int \rho_{\sigma_t}(dy)f(x,y)\right)^2 - \left(\rho_{\sigma_t}^{\otimes 2}(f)\right)^2 - 2\rho_{\sigma_t}^{\otimes 2}(f)\int \rho_{\sigma_s}(dy)f(x,y)$$

$$+ 2\int \rho(dy)\rho(dz)f(x,\sigma_s y)f(\sigma_t y, \sigma_t z)$$

(here, ρ is the law $\mathcal{N}(0, I_{d'})$). Then if we can prove the following two properties:

$$\sum_{i=2}^{[t/\Delta_n]+1} \mathbb{E}_{i-1}^n(\Delta_i^n N\, \eta_i^n) \xrightarrow{\mathbb{P}} 0 \qquad\qquad (3.108)$$

for any N which is a component of W or is a bounded martingale orthogonal to W, and

$$\Delta_n \sum_{i=2}^{[t/\Delta_n]+1} \phi_i^n \xrightarrow{\mathbb{P}} \int_0^t R_{\sigma_u}(f,2)du \qquad\qquad (3.109)$$

(with the notation (3.95); here f is 1-dimensional, so $R_\sigma(f,2)$ is also 1-dimensional), then Lemma 3.7 will yield the stable convergence in law of $\overline{V}'^n$ to $V'(f,2)$.

Let us prove first (3.108). Recall $\eta_i^n = \gamma_i^n + \gamma_i'^n$, and observe that

$$\gamma_i^n = \sqrt{\Delta_n}\, h(\sigma_{(i-2)\Delta_n}, \Delta_{i-1}^n W/\sqrt{\Delta_n}, \Delta_i^n W/\sqrt{\Delta_n})$$
$$\gamma_i'^n = \sqrt{\Delta_n}\, h'(\sigma_{(i-1)\Delta_n}, \Delta_i^n W/\sqrt{\Delta_n}),$$

where $h(\sigma, x, y)$ and $h'(\sigma, x)$ are continuous functions with polynomial growth in x an y, uniform in $\sigma \in \mathcal{M}'$. Then (3.108) when N is a bounded martingale orthogonal to W readily follows from Lemma 3.18.

Next, suppose that N is a component of W, say W^1. Since f is globally even and ρ_{σ_s} is a measure symmetric about the origin, the function $h'(\sigma, x)$ is even in x, so $h'(\sigma, x)x^1$ is odd in x and obviously $\mathbb{E}_{i-1}^n(\gamma_i'^n \Delta_i^n W^1) = 0$. So it remains to prove that

$$\sum_{i=2}^{[t/\Delta_n]+1} \zeta_i^n \xrightarrow{\mathbb{P}} 0, \qquad \text{where } \zeta_i^n = \mathbb{E}_{i-1}^n(\gamma_i^n \Delta_i^n W^1). \qquad (3.110)$$

An argument similar to the previous one shows that $h(\sigma, x, y)$ is globally even in (x,y), so ζ_i^n has the form $\Delta_n\, k(\sigma_{(i-2)\Delta_n}, \Delta_{i-1}^n W/\sqrt{\Delta_n})$ where $k(\sigma, x) = \int \rho_\sigma(dy)h(\sigma, x, y)y^1$ is odd in x, and also C^1 in x with derivatives with polynomial growth, uniformly in $\sigma \in \mathcal{M}'$. Then $\mathbb{E}_{i-2}^n(\zeta_i^n) = 0$ and $\mathbb{E}_{i-2}^n(|\zeta_i^n|^2) \le K\Delta_n^2$. Since ζ_i^n is also $\mathcal{F}_{(i-1)\Delta_n}$-measurable, we deduce (3.110) from Lemma 3.4, and we have finished the proof of (3.108).

Now we prove (3.109). Observe that ϕ_i^n is $\mathcal{F}_{(i-1)\Delta_n}$-measurable and

$$\mathbb{E}_{i-2}^n(\phi_i^n) = h((i-2)\Delta_n, (i-1)\Delta_n), \qquad \mathbb{E}_{i-2}^n(|\phi_i^n|^2) \leq K,$$

where $h(s,t) = \int \rho_{\sigma_s}(dx)g(s,t,x)$. Then, by Lemma 3.4, the property (3.109) follows from

$$\Delta_n \sum_{i=1}^{[t/\Delta_n]} h((i-1)\Delta_n, i\Delta_n) \xrightarrow{\mathbb{P}} \int_0^t R_{\sigma_u}(f,2)du, \qquad (3.111)$$

so it remains to show (3.111). On the one hand, we have $|h(s,t)| \leq K$. On the other hand, since f is continuous with polynomial growth and σ_t is bounded we clearly have $h(s_n, t_n) \to h(t,t)$ for any sequences $s_n, t_n \to t$ which are such that σ_{s_n} and σ_{t_n} converge to σ_t: since the later property holds, for $\mathbb{P}$-almost all ω and Lebesgue-almost all t, for all sequences $s_n, t_n \to t$, we deduce that

$$\Delta_n \sum_{i=1}^{[t/\Delta_n]} h((i-1)\Delta_n, i\Delta_n) \xrightarrow{\mathbb{P}} \int_0^t h(u,u)du.$$

Since

$$h(t,t) = \rho_{\sigma_t}^{\otimes 2}(f^2) - 3\left(\rho_{\sigma_t}^{\otimes 2}(f)\right)^2 + 2\int \rho_{\sigma_t}(dx)\rho_{\sigma_t}(dy)\rho_{\sigma_t}(dz)f(x,y)f(y,z),$$

is trivially equal to $R_{\sigma_t}(f,2)$, as given by (3.95). Hence we have (3.111).

3.7.5 Proof of the convergence result 3.103

As said before, this is the hard part, and it is divided into a number of steps.

Step 1. For $l = 0, \cdots, k-1$ we define the following (random) functions on $\mathbb{R}^d$:

$$g_{i,l}^n(x) =$$

$$\int f\left(\frac{\Delta_i^n X}{\sqrt{\Delta_n}}, \cdots, \frac{\Delta_{i+l-1}^n X}{\sqrt{\Delta_n}}, x, x_{l+1}, \cdots, x_{k-1}\right) \rho_{\sigma_{(i-1)\Delta_n}}^{\otimes(k-l-1)}(dx_{l+1}, \cdots, x_{k-1})$$

(for $l = 0$ we simply integrate $f(x, x_{l+1}, \cdots, x_{k-1})$, whereas for $l = k-1$ we have no integration). As a function of ω this is $\mathcal{F}_{(i+l-1)\Delta_n}$-measurable. As a function of x it is C^1, and further it has the following properties, according to the case (a) or (b) of Theorem 3.21 (we heavily use the fact that σ_t is bounded, and also (3.73)):

$$\left.\begin{array}{c} |g_{i,l}^n(x)| + \|\nabla g_{i,l}^n(x)\| \leq KZ_{i,l}^n(1 + \|x\|^r) \qquad \text{where} \\ \text{in case (a): } r \geq 0, \quad \mathbb{E}_{i-1}^n(|Z_{i,l}^n|^p) \leq K_p \;\; \forall p > 0, \\ Z_{i,l}^n \text{ is } \mathcal{F}_{(i+l-2)\Delta_n}\text{-measurable} \\ \text{in case (b): } r = 0, \quad Z_{i,l}^n = 1. \end{array}\right\} \qquad (3.112)$$

For all $A \geq 1$ there is also a positive function $G_A(\varepsilon)$ tending to 0 as $\varepsilon \to 0$, such that with $Z_{i,l}^n$ as above:

$$\|x\| \leq A, \ Z_{i,l}^n \leq A, \ \|y\| \leq \varepsilon \quad \Rightarrow \tag{3.113}$$
$$\|\nabla g_{i,l}^n(x+y) - \nabla g_{i,l}^n(x)\| \ \leq \ G_A(\varepsilon).$$

Observing that $\zeta_i^{\prime\prime n}$ is the sum over l from 0 to $k-1$ of

$$f\left(\frac{\Delta_i^n X}{\sqrt{\Delta_n}}, \cdots, \frac{\Delta_{i+l}^n X}{\sqrt{\Delta_n}}, \beta_{i,l+1}^n, \cdots, \beta_{i,k-1}^n\right) -$$
$$f\left(\frac{\Delta_i^n X}{\sqrt{\Delta_n}}, \cdots, \frac{\Delta_{i+l-1}^n X}{\sqrt{\Delta_n}}, \beta_{i,l}^n, \cdots, \beta_{i,k-1}^n\right),$$

we have

$$\mathbb{E}_{i-1}^n(\zeta_i^{\prime\prime n}) \ = \ \sum_{l=0}^{k-1} \mathbb{E}_{i-1}^n\left(g_{i,l}^n(\Delta_{i+l}^n X/\sqrt{\Delta_n}) - g_{i,l}^n(\beta_{i,l}^n)\right).$$

Therefore it is enough to prove that for any $l \geq 0$ we have

$$\sqrt{\Delta_n} \sum_{i=1}^{[t/\Delta_n]} \mathbb{E}_{i-1}^n\left(g_{i,l}^n(\Delta_{i+l}^n X/\sqrt{\Delta_n}) - g_{i,l}^n(\beta_{i,l}^n)\right) \xrightarrow{\text{u.c.p.}} 0. \tag{3.114}$$

Step 2. In case (b) the process X has jumps, but we assume that $\int(\gamma(x) \wedge 1)\lambda(dx) < \infty$, hence the two processes $\delta \star \mu$ and $\delta \star \underline{\nu}$ are well defined. Moreover (3.112) readily gives $|g_{i,l}^n(x+y) - g_{i,l}^n(x)| \leq K(\|y\| \wedge 1)$. Hence it follows from (3.76) with $\alpha_n = 1$ that

$$\sqrt{\Delta_n} \sum_{i=1}^{[t/\Delta_n]} \mathbb{E}_{i-1}^n\left(g_{i,l}^n(\Delta_{i+l}^n X/\sqrt{\Delta_n}) - \right.$$
$$\left. g_{i,l}^n(\Delta_{i+l}^n X/\sqrt{\Delta_n} - \Delta_{i+l}^n(\delta \star \mu)/\sqrt{\Delta_n})\right) \xrightarrow{\text{u.c.p.}} 0.$$

Therefore if we put

$$\xi_{i,l}^n \ = \ \begin{cases} \Delta_{i+l}^n X/\sqrt{\Delta_n} - \beta_{i,l}^n & \text{in case (a)} \\ \Delta_{i+l}^n X/\sqrt{\Delta_n} - \Delta_{i+l}^n(\delta \star \mu)/\sqrt{\Delta_n} - \beta_{i,l}^n & \text{in case (b)}, \end{cases} \tag{3.115}$$

(3.114) amounts to

$$\sqrt{\Delta_n} \sum_{i=1}^{[t/\Delta_n]} \mathbb{E}_{i-1}^n\left(g_{i,l}^n(\beta_{i,l}^n + \xi_{i,l}^n) - g_{i,l}^n(\beta_{i,l}^n)\right) \xrightarrow{\text{u.c.p.}} 0. \tag{3.116}$$

Step 3. At this stage, we set (for simplicity, in the forthcoming formulas we write $S = S(i,l,n) = (i+l-1)\Delta_n$ and $T = T(i,l,n) = (i+l)\Delta_n$; recall that $x \mapsto \delta(s, x)$ is λ-integrable (in case (a) because then $\delta \equiv 0$, in case (b) because $|\delta(s, .)| \leq K(\gamma \wedge 1)$):

$$\widehat{\xi}_{i,l}^n = \int_S^T \left(b_s - b_S + \int_E (\delta(s, x) - \delta(S, x))\lambda(dx) \right) ds$$

$$+ \int_S^T \left(\int_S^s (\widetilde{b}_u du + (\widetilde{\sigma}_u - \widetilde{\sigma}_S) dW_u) \right.$$

$$\left. + \int_S^s \int_E (\widetilde{\delta}(u, x) - \widetilde{\delta}(S, x))(\underline{\mu} - \underline{\nu})(du, dx) \right) dW_s$$

$$\widetilde{\xi}_{i,l}^n = \left(b_S + \int_E \delta(S, x)\lambda(dx) \right) \Delta_n$$

$$+ \int_S^T \left(\widetilde{\sigma}_S \int_S^s dW_u + \int_S^s \int \widetilde{\delta}(S, x)(\underline{\mu} - \underline{\nu})(du, dx) \right) dW_s$$

In view of (3.115), we obviously have $\xi_{i,l}^n = \left(\widehat{\xi}_{i,l}^n + \widetilde{\xi}_{i,l}^n \right)/\sqrt{\Delta_n}$.

Consider the process $Y = (\widetilde{\gamma}^2 \wedge 1) \star \mu$. This is an increasing pure jump Lévy process, whose Laplace transform is

$$u \mapsto \mathbb{E}(e^{-u(Y_{s+t} - Y_s)}) = \exp t \int \left(e^{-u(\widetilde{\gamma}(x)^2 \wedge 1)} - 1 \right) \lambda(dx).$$

If q is a non zero integer, we compute the qth moment of $Y_{s+t} - Y_s$ by differentiating q times its Laplace transform at 0: this is the sum, over all choices $p_1, \ldots, p_k$ of positive integers with $\sum_{i=1}^k p_i = q$, of suitable constants times the product for all $i = 1, \ldots, k$ of the terms $t \int (\widetilde{\gamma}(x)^{2p_i} \wedge 1)\lambda(dx)$, each one being smaller than Kt. Then we deduce that $\mathbb{E}((Y_{s+t} - Y_s)^q \mid \mathcal{F}_s) \leq K_q t$, and by interpolation this also holds for any real $q \geq 1$.

Then, coming back to the definition of $\widehat{\xi}_{i,l}^n$ and $\widetilde{\xi}_{i,l}^n$, and using the properties $\|\delta(t, x)\| \leq K(\gamma(x) \wedge 1)$ and $\|\widetilde{\delta}(t, x)\| \leq K(\widetilde{\gamma}(x) \wedge 1)$, plus the fact that $\int (\gamma(x) \wedge 1)\lambda(dx) < \infty$ when δ is not identically 0, and the boundedness of $b, \widetilde{b}, \sigma, \widetilde{\sigma}$, we deduce from Burkholder-Davis-Gundy and Hölder inequalities that

$$q \geq 2 \quad \Rightarrow \quad \mathbb{E}_{i+l-1}^n(|\widehat{\xi}_{i,l}^n|^q) + \mathbb{E}_{i+l-1}^n(|\widetilde{\xi}_{i,l}^n|^q) \leq K\Delta_n^{1+q/2}. \quad (3.117)$$

The same arguments, plus Cauchy-Schwarz inequality, yield that with notation

$$\alpha_{i,l}^n = \mathbb{E}_{i+l-1}^n \left(\int_S^T \left(\|b_s - b_S\|^2 + \|\widetilde{\sigma}_s - \widetilde{\sigma}_S\|^2 \right. \right.$$

$$\left. \left. + \int \|\widetilde{\delta}(s, x) - \widetilde{\delta}(S, x)\|^2 \lambda(dx) + \int \|\delta(s, x) - \delta(S, x)\|\lambda(dx) \right) ds \right),$$

then

$$\mathbb{E}_{i+l-1}^n(|\widehat{\xi}_{i,l}^n|^2) \leq K\Delta_n \left(\Delta_n^2 + \alpha_{i,l}^n \right). \quad (3.118)$$

Next, since the restriction of μ to $(S, \infty) \times E$ and the increments of W after time S are independent, conditionally on $\mathcal{F}'_S = \mathcal{F}_S \vee \sigma(W_t : t \geq 0)$, we get

$$\mathbb{E}(\widetilde{\xi}^n_{i,l} \mid \mathcal{F}'_S) = \left(b_S + \int_E \delta(S, x)\lambda(dx)\right)\Delta_n + \widetilde{\sigma}_S \int_S^T \left(\int_S^s dW_u\right)dW_s.$$

Hence the product of the right side above with $h(\beta^n_{i,l})$, where h is an odd function on $\mathbb{R}^d$ with polynomial growth, is a function of the form $Y(\omega, (W_{S+t} - W_S)_{t\geq 0})$ on $\Omega \times C(\mathbb{R}_+, \mathbb{R}^{d'})$ which is $\mathcal{F}_S \otimes \mathcal{C}$-measurable ($\mathcal{C}$ is the Borel σ-field on $C(\mathbb{R}_+, \mathbb{R}^{d'})$), and such that $Y(\omega, w) = Y(\omega, -w)$. Therefore we deduce

$$\mathbb{E}^n_{i+l-1}(\widetilde{\xi}^n_{i,l} h(\beta^n_{i,l})) = 0. \tag{3.119}$$

Step 4. Here we prove the following auxiliary result:

$$\sqrt{\Delta_n} \sum_{i=1}^{[t/\Delta_n]} \sqrt{\mathbb{E}(\alpha^n_{i,l})} \to 0. \tag{3.120}$$

Indeed, by Cauchy-Schwarz inequality the square of the left side of (3.120) is smaller than

$$t \sum_{i=1}^{[t/\Delta_n]} \mathbb{E}(\alpha^n_{i,l}) = t\mathbb{E}\left(\int_{l\Delta_n}^{\Delta_n(l+[t/\Delta_n])} \left(\|b_s - b_{\Delta_n[s/\Delta_n]}\|^2\right.\right.$$

$$+ \|\widetilde{\sigma}_s - \widetilde{\sigma}_{\Delta_n[s/\Delta_n]}\|^2 + \int \|\widetilde{\delta}(s, x) - \widetilde{\delta}(\Delta_n[s/\Delta_n], x)\|^2\lambda(dx)$$

$$\left.\left.+ \int \|\delta(s, x) - \delta(\Delta_n[s/\Delta_n], x)\|\lambda(dx)\right)ds\right),$$

which goes to 0 by the dominated convergence theorem and the bounds given in (SH) and $\int(\gamma(x) \wedge 1)\lambda(dx) < \infty$.

Step 5. By a Taylor expansion we can write

$$g^n_{i,l}(\beta^n_{i,l} + \xi^n_{i,l}) - g^n_{i,l}(\beta^n_{i,l}) = \nabla g^n_{i,l}(\beta^n_{i,l})\xi^n_{i,l} + (\nabla g^n_{i,l}(\beta'^n_{i,l}) - \nabla g^n_{i,l}(\beta^n_{i,l}))\xi^n_{i,l},$$

where $\beta'^n_{i,l}$ is some (random) vector lying on the segment between $\beta^n_{i,l}$ and $\beta^n_{i,l} + \xi^n_{i,l}$. Therefore we can write $g^n_{i,l}(\beta^n_{i,l} + \xi^n_{i,l}) - g^n_{i,l}(\beta^n_{i,l}) = \sum_{j=1}^3 \zeta^n_{i,l}(j)$, where

$$\zeta^n_{i,l}(1) = \frac{1}{\sqrt{\Delta_n}} \nabla g^n_{i,l}(\beta^n_{i,l})\widetilde{\xi}^n_{i,l}, \qquad \zeta^n_{i,l}(2) = \frac{1}{\sqrt{\Delta_n}} \nabla g^n_{i,l}(\beta^n_{i,l})\widehat{\xi}^n_{i,l},$$

$$\zeta^n_{i,l}(3) = (\nabla g^n_{i,l}(\beta'^n_{i,l}) - \nabla g^n_{i,l}(\beta^n_{i,l}))\xi^n_{i,l}.$$

Then at this point it remains to prove that we have, for $j = 1, 2, 3$:

$$\sqrt{\Delta_n} \sum_{i=1}^{[t/\Delta_n]} \mathbb{E}^n_{i-1}(\zeta^n_{i,l}(j)) \xrightarrow{\text{u.c.p.}} 0. \tag{3.121}$$

For $j = 1$ this is obvious: indeed f is even in each of its (d-dimensional) arguments, so the functions $g^n_{i,l}$ are even as well, hence $\nabla g^n_{i,l}$ is odd and by (3.119) the left side of (3.121) is equal to 0.

Step 6) Now we prove (3.121) for $j = 3$. By (3.112) and (3.113) we have for all $A \geq 1$ and $\varepsilon > 0$:

$$
\begin{aligned}
|\zeta^n_{i,l}(3)| &\leq G_A(\varepsilon)\|\xi^n_{i,l}\| + KZ^n_{i,l}(1 + \|\beta^n_{i,l}\|^r + \|\xi^n_{i,l}\|^r)\|\xi^n_{i,l}\| \cdot \\
&\qquad (1_{\{Z^n_{i,l}>A\}} + 1_{\{\|\beta^n_{i,l}\|>A\}} + 1_{\{\|\xi^n_{i,l}\|>\varepsilon\}}) \\
&\leq G_A(\varepsilon)\|\xi^n_{i,l}\| + KZ^n_{i,l}(1 + Z^n_{i,l}) \cdot \\
&\qquad \left(\frac{1 + \|\beta^n_{i,l}\|)^{r+1}}{A} + \frac{(1 + \|\beta^n_{i,l}\|)^r \|\xi^n_{i,l}\|}{\varepsilon} + \right. \\
&\qquad\qquad \left. \frac{(1 + \|\beta^n_{i,l}\|)\|\xi^n_{i,l}\|^{r+1}}{A} + \frac{\|\xi^n_{i,l}\|^{r+2}}{\varepsilon} \right).
\end{aligned}
$$

By (3.117) we have $\mathbb{E}^n_{i+l-1}(\|\xi^n_{i,l}\|^q) \leq K_q \Delta_n$ if $q \geq 2$. Then in view of (3.73) we get by Hölder inequality:

$$
\mathbb{E}^n_{i+l-1}(|\zeta^n_{i,l}(3)|) \leq K\sqrt{\Delta_n} \left(G_A(\varepsilon) + Z^n_{i,l}(1 + Z^n_{i,l})\Big(\frac{1}{A} + \frac{\Delta_n^{1/6}}{\varepsilon}\Big) \right).
$$

Then since $\mathbb{E}((Z^n_{i,l})^q) \leq K_q$ for all $q > 0$ we have

$$
\mathbb{E}\Big(\sqrt{\Delta_n} \sum_{i=1}^{[t/\Delta_n]} \big|\mathbb{E}^n_{i+l-1}(\zeta^n_{i,l}(3))\big|\Big) \leq Kt\Big(G_A(\varepsilon) + \frac{1}{A} + \frac{\Delta_n^{1/6}}{\varepsilon}\Big),
$$

and (3.121) for $j = 3$ follows (choose A big and then ε small).

Step 7) It remains to prove (3.121) for $j = 2$. By (3.112) we have

$$
|\zeta^n_{i,l}(2)| \leq \frac{K}{\sqrt{\Delta_n}} Z^n_{i,l}(1 + \|\beta^n_{i,l}\|^r)\|\widehat{\xi}^n_{i,l}\|.
$$

Hence by Cauchy-Schwarz inequality and (3.73),

$$
\mathbb{E}\Big(\big|\mathbb{E}^n_{i+l-1}(\zeta^n_{i,l}(2))\big|\Big) \leq K\mathbb{E}\Big(Z^n_{i,l}(\Delta_n + \sqrt{\alpha^n_{i,l}})\Big) \leq K\Big(\Delta_n + \sqrt{\mathbb{E}(\alpha^n_{i,l})}\Big).
$$

Then, in view of (3.120), the result is obvious.

3.7.6 Proof of Theorem 3.23

The proof is exactly the same as above, with the following changes:

1) In Proposition 3.25 we substitute $\overline{V}^n$ and $V'(f, k)$ with

$$\overline{V}_t''^n = \sqrt{\Delta_n} \sum_{i=1}^{[t/k\Delta_n]} \left(\zeta_{(i-1)k+1}'^n - \rho_{\sigma_{(i-1)k\Delta_n}}^{\otimes k}(f) \right)$$

and $V''(f, k)$ respectively. The proof is then much shorter, because $\eta_i^n = \sqrt{\Delta_n}(\zeta_{(i-1)k+1}'^n - \rho_{\sigma_{(i-1)k\Delta_n}}^{\otimes k}(f))$ is $\mathcal{F}_{ik\Delta_n}$-measurable. We have

$$\mathbb{E}_{(i-1)k}^n(\eta_i^n) = 0, \quad \mathbb{E}_{(i-1)k}^n((\eta_i^n)^2) = \Delta_n \phi_i^n, \quad \mathbb{E}_{(i-1)k}^n((\eta_i^n)^4) \leq K\Delta_n^2,$$

with $\phi_i^n = \rho_{\sigma_{(i-1)k\Delta_n}}^{\otimes k}(f^2) - \rho_{\sigma_{(i-1)k\Delta_n}}^{\otimes k}(f)^2$, and (3.111) is replaced by the obvious convergence of $\sum_{i=1}^{[t/k\Delta_n]} \mathbb{E}_{(i-1)k}^n((\eta_i^n)^2)$ to the right side of (3.101) (recall that we assumed $q = 1$ here). We also have

$$\mathbb{E}_{(i-1)k}^n((N_{ik\Delta_n} - N_{(i-1)k\Delta_n})\zeta_i'^n) = 0$$

when N is a bounded martingale orthogonal to W by Lemma 3.18, and if N is one of the components of W because then this conditional expectation is the integral of a globally odd function, with respect to a measure on $(\mathbb{R}^{d'})^k$ which is symmetric about 0. So Lemma 3.7 readily applies directly, and the proposition is proved.

2) Next, we have to prove the analogues of (3.103), (3.104) and (3.105), where we only take the sum for those i of the form $i = (j-1)k + 1$, and where in (3.105) we divide the integral by k. Proving the new version of (3.104) is of course simpler than the old one; the new version of (3.105) is the old one for Δ_n, whereas for (3.103) absolutely nothing is changed. So we are done.

3.7.7 Proof of Theorem 3.24

For this theorem again we can essentially reproduce the previous proof, with $k = 1$, and with the function f with components $f_{jm}(x) = x^j x^m$ (here m replaces the index k in the theorem). Again it suffices by polarization to prove the result for a single pair (j, m).

Below we set $\alpha_n = \alpha \Delta_n^{\varpi - 1/2}$, which goes to ∞. Introduce the function on $\mathbb{R}^d$ defined by $g_n(x) = x^j x^m \psi_{\alpha_n}(x)$ (recall (3.74)), and set

$$\eta_i^n = \frac{\Delta_i^n X^j \Delta_i^n X^k}{\Delta_n} 1_{\{\|\Delta_i^n X\| \leq \alpha \Delta_n^\varpi\}} - g_n(\Delta_i^n X/\sqrt{\Delta_n}),$$

$$\eta_i'^n = g_n(\Delta_i^n X/\sqrt{\Delta_n}) - \beta_i^{n,m}\beta_i^{n,k}.$$

Proposition 3.25 implies that the processes

$$\sqrt{\Delta_n} \sum_{i=1}^{[t/\Delta_n]} \left(\beta_i^{n,j}\beta_i^{n,m} - c_{(i-1)\Delta_n}^{jm} \right)$$

converges stably in law to V^{jm}, and we also have (3.105), which here reads as

$$\frac{1}{\sqrt{\Delta_n}} \left(\Delta_n \sum_{i=1}^{[t/\Delta_n]} c_{(i-1)\Delta_n}^{jj} - \int_0^t ((c_u^{jm})^2 + c_u^{jj}c_u^{mm}) \, du \right) \xrightarrow{\text{u.c.p.}} 0.$$

Therefore it remains to prove the following three properties:

$$\sqrt{\Delta_n} \sum_{i=1}^{[t/\Delta_n]} \eta_i^n \xrightarrow{\text{u.c.p.}} 0, \tag{3.122}$$

$$\sqrt{\Delta_n} \sum_{i=1}^{[t/\Delta_n]} \mathbb{E}_{i-1}^n(\eta_i'^n) \xrightarrow{\text{u.c.p.}} 0, \tag{3.123}$$

$$\sqrt{\Delta_n} \sum_{i=1}^{[t/\Delta_n]} \left(\eta_i'^n - \mathbb{E}_{i-1}^n(\eta_i'^n) \right) \xrightarrow{\text{u.c.p.}} 0. \tag{3.124}$$

Proof of (3.122). Observe that $|\eta_i^n| \le (\|\Delta_i^n X\|^2/\Delta_n) 1_{\{\alpha\Delta_n^\varpi < \|\Delta_i^n X\| \le 2\alpha\Delta_n^\varpi\}}$, hence

$$\begin{aligned} |\eta_i^n| &\le 2\alpha_n 1_{\{\|\beta_i^n\| > \alpha_n/2\}} + 4\|\chi_i^n\|^2 1_{\{\alpha_n/2 < \|\chi_i^n\| \le 3\alpha_n\}} \\ &\le 2\alpha_n^{1-q}\|\beta_i^n\|^q + +36(\|\chi_i^n\|^2 \wedge \alpha_n^2) \end{aligned}$$

for any $q > 0$ (recall that $\Delta_i^n X/\sqrt{\Delta_n} = \beta_i^n + \chi_i^n$). Then we take q such that $(q-1)(1/2 - \varpi) > 2\varpi - \varpi r$, and we apply (3.73) and (3.75) with α_n as above (so $\alpha_n \ge 1$ for n large enough, and $\alpha_n\sqrt{\Delta_n} \to 0$), to get $\mathbb{E}(|\eta_i^n|) \le K\Delta_n^{2\varpi - \varpi r}u_n$, where $u_n \to 0$. Hence

$$\sqrt{\Delta_n} \sum_{i=1}^{[t/\Delta_n]} \mathbb{E}(|\eta_i^n|) \le Kt\Delta_n^{2\varpi - \varpi r - 1/2}u_n,$$

which goes to 0 if $\varpi \ge \frac{1}{2(2-r)}$. Hence we have (3.122).

Proof of (3.123). From the properties of ψ, the function ψ_{α_n} is differentiable and $\|\nabla\psi_{\alpha_n}(x)\| \le (K/\alpha_n) 1_{\{\|x\| \le 2\alpha_n\}}$. Hence we clearly have $\|\nabla g_n(x)\| \le K(\|x\| \wedge \alpha_n)$, and thus

$$\left. \begin{aligned} |g_n(x+y) - g_n(x))| &\le K\alpha_n(\|y\| \wedge \alpha_n), \\ |g_n(x+y) - g_n(x) - \nabla g_n(x)y| &\le K\|y\|^2. \end{aligned} \right\} \tag{3.125}$$

If we use the first estimate above and (3.76) we obtain, as in Step 2 of the previous proof (we use again $\varpi \geq \frac{1}{2(2-r)}$ here), that

$$\sqrt{\Delta_n} \sum_{i=1}^{[t/\Delta_n]} \mathbb{E}_{i-1}^n \Big(g_n(\Delta_{i+l}^n X/\sqrt{\Delta_n})$$

$$- g_n(\Delta_{i+l}^n X/\sqrt{\Delta_n} - \Delta_{i+l}^n(\delta \star \underline{\mu})/\sqrt{\Delta_n}) \Big) \xrightarrow{\text{u.c.p.}} 0.$$

Then with the notation of (3.115), in order to prove (3.123) it is enough to prove (3.106) with g_n instead of $g_{i,l}^n$, and $l = 0$. Then the second estimate in (3.125) allows to write $g_n(\beta_{i,l}^n + \xi_{i,l}^n) - g_n(\beta_{i,l}^n) = \sum_{j=1}^3 \zeta_i^n(j)$, where

$$\zeta_i^n(1) = \frac{1}{\sqrt{\Delta_n}} \nabla g_n(\beta_i^n) \widetilde{\xi}_{i,0}^n, \qquad \zeta_i^n(2) = \frac{1}{\sqrt{\Delta_n}} \nabla g_{i,l}^n(\beta_i^n) \widehat{\xi}_{i,0}^n,$$

$$|\zeta_i^n(3)| \leq K \|\xi_{i,0}^n\|^2.$$

Then it remains to prove (3.121) for $j = 1, 2, 3$, with $\zeta_i^n(j)$ instead of $\zeta_{i,l}^n(j)$.

Since g_n is even, this property for $j = 1$ follows from (3.119) exactly as in the previous proof. The proof for $j = 2$ is the same as in Step 7 of the previous proof (here $Z_{l,i}^n = 1$ and $r = 2$). Finally by (3.117) we have $\mathbb{E}(\|\xi_{i,0}^n\|^2) \leq K\Delta_n$, so the result for $j = 3$ is immediate.

Proof of (3.124). Exactly as for (3.104) it is enough to prove that

$$\Delta_n \sum_{i=1}^{[t/\Delta_n]} \mathbb{E}_{i-1}^n ((\eta_i'^n)^2) \xrightarrow{\text{u.c.p.}} 0. \tag{3.126}$$

First, we have

$$\sum_{i=1}^{[t/\Delta_n]} \mathbb{E}_{i-1}^n \left(\left(\beta_i^{n,j} \beta_i^{n,m} - g_n(\beta_i^n) \right)^2 \right) \leq \sum_{i=1}^{[t/\Delta_n]} \mathbb{E}_{i-1}^n \left(\|\beta_i^n\|^4 1_{\{\|\beta_i^n\| > \alpha_n\}} \right)$$

$$\leq K\Delta_n^{q((1/2-\varpi))} \sum_{i=1}^{[t/\Delta_n]} \mathbb{E}_{i-1}^n (\|\beta_i^n\|^{4+q}) \leq Kt\Delta_n,$$

by choosing appropriately q for the last inequality.

Second, since $\Delta_i^n X = \sqrt{\Delta_n}(\beta_i^n + \xi_{i,0}^n) + \Delta_i^n(\delta \star \underline{\mu})$, we deduce from (3.125) that

$$\left(g_n(\Delta_i^n X/\sqrt{\Delta_n}) - g_n(\beta_i^n) \right)^2 \leq$$

$$K\alpha_n^2 \|\xi_{i,0}^n\|^2 + K\alpha_n^3 \left((|\Delta_i^n(\delta \star \underline{\mu})|/\sqrt{\Delta_n}) \bigwedge \alpha_n \right).$$

Then by (3.76) and (3.117) again, we get

$$\mathbb{E}\left(\left(g_n(\Delta_i^n X/\sqrt{\Delta_n}) - g_n(\beta_i^n)\right)^2\right) \leq$$
$$K(\alpha_n^2\Delta_n + \alpha_n^{4-r}\Delta_n^{1-r/2}) \leq K\Delta_n^{\varpi(4-r)-1}.$$

If we put together these estimates, we find that

$$\Delta_n \sum_{i=1}^{[t/\Delta_n]} \mathbb{E}((\eta_i^{\prime n})^2) \leq Kt(\Delta_n + \Delta_n^{\varpi(4-r)-1}),$$

which goes to 0 because $\varpi \geq \frac{1}{2(2-r)}$. Hence we have (3.126).

3.8 CLT with discontinuous limits

So far we have been concerned with CLTs associated with Theorems 3.10, 3.11 and 3.12, in which the limiting processes are always continuous. Now, as seen in the case $r = 3$ of (3.35) there are cases where the limit is a sum of jumps, and we are looking at this kind of question here. In case $r = 2$ of (3.35) we even have a "mixed" limit with a continuous and a purely discontinuous parts: this has less statistical interest, and we will state the result without proof.

Here, more than in the continuous case, it is important and not completely trivial to define the limiting processes. This is the aim of the first subsection below. Throughout, we assume (H), and we also fix an integer $k \geq 2$.

3.8.1 The limiting processes

As for the case of continuous limits, we will have stable convergence in law, and the limiting processes will be defined on an extension of the space $(\Omega, \mathcal{F}, \mathbb{P})$. To do this, it is convenient to introduce another probability space $(\Omega', \mathcal{F}', \mathbb{P}')$. We assume that this space supports the following variables:

- four sequences (U_p), (U_p'), $(\overline{U}_p)$, $(\overline{U}_p')$ of d'-dimensional $N(0, I_{d'})$ variables;
- a sequence (κ_p) of uniform variables on $[0, 1]$;
- a sequence (L_p) of uniform variables on the finite set $\{0, 1, \cdots, k-1\}$, where $k \geq 2$ is some fixed integer;

and all these variables are mutually independent. Then we put

$$\widetilde{\Omega} = \Omega \times \Omega', \qquad \widetilde{\mathcal{F}} = \mathcal{F} \otimes \mathcal{F}', \qquad \widetilde{\mathbb{P}} = \mathbb{P} \otimes \mathbb{P}'. \qquad (3.127)$$

We extend the variables X_t, b_t, ... defined on Ω and U_p, κ_p,... defined on Ω' to the product $\widetilde{\Omega}$ in the obvious way, without changing the notation. We write $\widetilde{\mathbb{E}}$ for the expectation with respect to $\widetilde{\mathbb{P}}$.

Next, we need a filtration $(\widetilde{\mathcal{F}}_t)_{t\geq 0}$ on our extension. To this effect, we first denote by $(S_p)_{p\geq 1}$ a sequence of stopping times which exhausts the "jumps" of the Poisson measure μ: this means that for each ω we have $S_p(\omega) \neq S_q(\omega)$ if $p \neq q$, and that $\mu(\omega, \{t\} \times E) = 1$ if and only if $t = S_p(\omega)$ for some p. There are many ways of constructing those stopping times, but it turns out that what follows does not depend on the specific description of them.

With a given choice of the above stopping times S_p, we let $(\widetilde{\mathcal{F}}_t)$ be the smallest (right-continuous) filtration of $\widetilde{\mathcal{F}}$ containing the filtration $(\mathcal{F}_t)$ and such that $U_p, U'_p, \overline{U}_p, \overline{U}'_p, \kappa_p$ and L_p are $\widetilde{\mathcal{F}}_{S_p}$-measurable for all p. Obviously, μ is still a Poisson measure with compensator ν, and W is still a Wiener process on $(\widetilde{\Omega}, \widetilde{\mathcal{F}}, (\widetilde{\mathcal{F}}_t)_{t\geq 0}, \widetilde{\mathbb{P}})$. Finally we define the q-dimensional variables

$$\left.\begin{aligned}
R_p &= \sqrt{\kappa_p}\, \sigma_{S_p-}U_p + \sqrt{1 - \kappa_p}\, \sigma_{S_p}\overline{U}_p \\
R''_p &= \sqrt{L_p}\, \sigma_{S_p-}U'_p + \sqrt{k - 1 - L_p}\, \sigma_{S_p}\overline{U}'_p \\
R'_p &= R_p + R''_p.
\end{aligned}\right\} \tag{3.128}$$

If f is a C^q function on $\mathbb{R}^d$ we denote by $\nabla^r f$ for $r \leq q$ the tensor of its rth derivatives, and if we want to be more specific, we write $\partial^r_{i_1,\cdots,i_r} f$ the rh partial derivative with respect to the components $x^{i_1},\cdots,x^{j_r}$, and simply ∇f and $\partial_i f$ when $r = 1$. If f and g are two C^1 functions we set

$$C(f,g)_t := \sum_{s\leq t} \sum_{i,j=1}^d (\partial_i f \partial_j g)(\Delta X_s)\, (c^{ij}_{s-} + c^{ij}_s). \tag{3.129}$$

This makes sense (that is, the series above converges for all t) as soon as $f(0) = 0$, because then $\|\nabla f(x)\| \leq K\|x\|$ for $\|x\| \leq 1$ and the process c is locally bounded and $\sum_{s\leq t}\|\Delta X_s\|^2 < \infty$; the process $C(f,g)$ is then of finite variation, and even increasing when $g = f$. In the same way, if f is C^2 and $\|\nabla f(x)\| \leq K\|x\|^2$ when $\|x\| \leq 1$, the following defines a process of finite variation:

$$\overline{C}(f)_t := \sum_{s\leq t} \sum_{i,j=1}^d \partial^2_{ij} f(\Delta X_s)\, (c^{ij}_{s-} + c^{ij}_s). \tag{3.130}$$

In the following lemma we simultaneously define and prove the existence of our limiting processes. We do it for a q-dimensional function $f = (f_1,\cdots,f_q)$, since it costs us nothing.

Lemma 3.26 *a) Let f be a q-dimensional C^1 function on $\mathbb{R}^d$, vanishing at 0.
The formulas*

$$\left.\begin{array}{l} Z(f_l)_t = \sum_{p:\, S_p \le t} \sum_{i=1}^d \partial_i f_l(\Delta X_{S_p})\, R_p^i, \\[2mm] Z'(f_l)_t = \sum_{p:\, S_p \le t} \sum_{i=1}^d \partial_i f_l(\Delta X_{S_p})\, R_p'^i \end{array}\right\} \qquad (3.131)$$

*define q-dimensional processes $Z(f) = (Z(f_l))_{l \le q}$ and $Z'(f) = (Z'(f_l))_{l \le q}$,
and conditionally on $\mathcal{F}$ the pair $(Z(f), Z'(f))$ is a square-integrable martingale with independent increments, zero mean and variance-covariance given
by*

$$\left.\begin{array}{l} \widetilde{\mathbb{E}}(Z(f_l)_t Z(f_{l'})_t \mid \mathcal{F}) = \widetilde{\mathbb{E}}(Z(f_l)_t Z'(f_{l'})_t \mid \mathcal{F}) = \tfrac{1}{2}\, C(f_l, f_{l'})_t, \\[2mm] \widetilde{\mathbb{E}}(Z'(f_l)_t Z'(f_{l'})_t \mid \mathcal{F}) = \tfrac{k}{2}\, C(f_l, f_{l'})_t. \end{array}\right\} \qquad (3.132)$$

*Moreover, if X and c have no common jumps, conditionally on $\mathcal{F}$ the process
$(Z(f), Z'(f))$ is a Gaussian martingale.*

*b) Let f be a q-dimensional C^2 function on $\mathbb{R}^d$, with $\|\nabla^2 f(x)\| \le \|x\|^2$ for
$\|x\| \le 1$. The formulas*

$$\left.\begin{array}{l} \overline{Z}(f_l)_t = \sum_{p:\, S_p \le t} \sum_{i,j=1}^d \partial_{ij}^2 f_l(\Delta X_{S_p})\, R_p^i, R_p^j \\[2mm] \overline{Z}'(f_l)_t = \sum_{p:\, S_p \le t} \sum_{i,j=1}^d \partial_{ij}^2 f_l(\Delta X_{S_p})\, R_p'^i R_p'^j \end{array}\right\} \qquad (3.133)$$

*define q-dimensional processes $\overline{Z}(f) = (\overline{Z}(f_l))_{l \le q}$ and $\overline{Z}'(f) = (\overline{Z}'(f_l))_{l \le q}$
of finite variation, and with $\mathcal{F}$-conditional expectations given by*

$$\left.\begin{array}{l} \widetilde{\mathbb{E}}(\overline{Z}(f_l)_t \mid \mathcal{F}) = \tfrac{1}{2}\, \overline{C}(f_l)_t, \\[2mm] \widetilde{\mathbb{E}}(\overline{Z}'(f_l)_t \mid \mathcal{F}) = \tfrac{k}{2}\, \overline{C}(f_l)_t. \end{array}\right\} \qquad (3.134)$$

*c) The processes $(Z(f), Z'(f))$ and $(\overline{Z}(f), \overline{Z}'(f))$ above depend on the choice
of the sequence (S_p) of stopping times exhausting the jumps of μ, but their $\mathcal{F}$-
conditional laws do not.*

Proof. a) Among several natural proofs, here is an "elementary" one. We
set $\alpha_p(l, l') = \sum_{i,j=1}^d (\partial_i f_l \partial_j f_{l'})(\Delta X_{S_p})(c_{S_p-}^{ij} + c_{S_p}^{ij})$, so $C(f_l, f_{l'})_t = \sum_{p:\, S_p \le t} \alpha_p(l, l')$. We fix $\omega \in \Omega$, and we consider the q-dimensional variables
$\Phi_p(\omega, .)$ and $\Phi_p'(\omega, .)$ on $(\Omega', \mathcal{F}')$ with components

$$\Phi_p^l(\omega, \omega') = \sum_{i=1}^d \partial_i f_l(\Delta X_{S_p}(\omega)) R_p^i(\omega, \omega'),$$

$$\Phi_p'^l(\omega, \omega') = \sum_{i=1}^d \partial_i f_l(\Delta X_{S_p}(\omega)) R_p'^i(\omega, \omega').$$

The variables $(\Phi_p(\omega,.), \Phi'_p(\omega,.))$ on $(\Omega', \mathcal{F}', \mathbb{P}')$ are independent as p varies, and a simple calculation shows that they have zero mean and variance-covariance given by

$$\left.\begin{array}{l} \mathbb{E}'(\Phi^l_p(\omega,.)\Phi^{l'}_p(\omega,.)) = \mathbb{E}'(\Phi'^l_p(\omega,.)\Phi'^{l'}_p(\omega,.)) = \frac{1}{2}\alpha_p(l,l';\omega) \\ \mathbb{E}'(\Phi^l_p(\omega,.)\Phi'^{l'}_p(\omega,.)) = \frac{k}{2}\alpha_p(l,l';\omega) \end{array}\right\} \quad (3.135)$$

Since $\sum_{p:\, S_p(\omega)\leq t}\alpha_p(l,l';\omega) < \infty$, a standard criterion for convergence of series of independent variables yields that the formulas

$$Z(f_l)_t(\omega,.) = \sum_{p:\, S_p(\omega)\leq t} \Phi^l_p(\omega,.),$$

$$Z'(f_l)_t(\omega,.) = \sum_{p:\, S_p(\omega)\leq t} \Phi'^l_p(\omega))R'_p(\omega,.)$$

define a $2q$-dimensional process $(\omega',t) \mapsto (Z(f)(\omega,\omega')_t, Z'(f)(\omega,\omega')_t)$. Obviously this is a martingale with independent increments, and with $((2q) \times 2$-dimensional) predictable bracket being deterministic (that is, it does not depend on ω') and equal at time t to the sum over all p with $S_p(\omega) \leq t$ of the right sides of (3.135). That is, we can consider $(Z(f), Z'(f))$ as a process on the extended space, and it satisfies (3.132). Since the law of a centered martingale with independent increments depends only on its predictable bracket we see that the law of $(Z(f), Z'(f))$, conditional on $\mathcal{F}$, only depends on the processes $C(f_l, f_{l'})$ and thus does not depend on the particular choice of the sequence (S_p).

Moreover, this martingale is purely discontinuous and jumps at times $S_p(\omega)$, and if X and c have no common jumps, the jump of $(Z(f)(\omega,.), Z'(f)(\omega,.))$ at $S_p(\omega)$ equals

$$\left(\nabla f(\Delta X_{S_p})\sigma_{S_p}(\omega)\left(\sqrt{\kappa_p}\, U_p + \sqrt{1-\kappa_p}\, \overline{U}_p\right), \nabla f(\Delta X_{S_p})\sigma_{S_p}(\omega)\cdot\right.$$
$$\left.\left(\sqrt{\kappa_p}\, U_p + \sqrt{1-\kappa_p}\, \overline{U}_p + \sqrt{L_p}\, U'_p + \sqrt{k-1-L_p}\, \overline{U}'_p\right)\right)$$

(we use here product matrix notation); this 2-dimensional variable is $\mathcal{F}$-conditionally Gaussian and centered, so in this case the pair $(Z(f), Z'(f))$ is $\mathcal{F}$-conditionally a Gaussian process.

b) Since $\widetilde{\mathbb{E}}(|R^i_p R^j_p| \mid \mathcal{F}) \leq K(\|c_{S_p-}\| + \|c_{S_p}\|)$, and the same with R'_p, it is obvious in view of our assumption on f that the $\mathcal{F}$-conditional expectation of the two variables

$$\sum_{p:\, S_p\leq t} \left|\sum_{i,j=1}^{d} \partial^2_{ij}f_l(\Delta X_{S_p})\, R^i_p, R^j_p\right|, \quad \sum_{p:\, S_p\leq t} \left|\sum_{i,j=1}^{d} \partial^2_{ij}f_l(\Delta X_{S_p})\, R'^i_p R'^j_p\right|$$

is finite for all t. Then all claims are obvious.

It remains to prove (c) for the process $(\overline{Z}(f), \overline{Z}'(f))$. For this, we observe that conditionally on $\mathcal{F}$ this process is the sum of its jump and it has independent increments. Moreover, it jumps only when X jumps, and if T is a finite $(\mathcal{F}_t)$-stopping time such that $\Delta X_T \neq 0$, then its jump at time T is

$$\left(\sum_{i,j=1}^{d} \partial_{ij}^2 f(\Delta X_T) \widetilde{R}^{ij}, \sum_{i,j=1}^{d} \partial_{ij}^2 f(\Delta X_T) \widetilde{R}'^{ij} \right),$$

where $WR^{ij} = \sum_{p \geq 1} R_p^i R_p^j 1_{\{S_p = T\}}$ and a similar expression for $\widetilde{R}'^{ij}$. But the $\mathcal{F}$-conditional law of $(\widetilde{R}^{ij}, WR'^{ij})$ clearly depends only on σ_{T-} and σ_T, but not on the particular choice of the sequence (S_p). This proves the result. $\square$

3.8.2 The results

Now we proceed to giving a CLT associated with the convergence in (3.48), and as seen already in (3.35) we need some smoothness for the test function f, and also that $f(x)$ goes to 0 faster than $\|x\|^3$ instead of $\|x\|^2$ as $x \to 0$. As in Theorem 3.21 we also consider a q-dimensional function $f = (f_1, \cdots, f_q)$.

Theorem 3.27 *Assume (H) (or (H') only), and let f be a q-dimensional C^2 function on $\mathbb{R}^d$ satisfying $f(0) = 0$ and $\nabla f(0) = 0$ and $\nabla^2 f(x) = o(\|x\|)$ as $x \to 0$. The pair of q-dimensional processes*

$$\left(\frac{1}{\sqrt{\Delta_n}} (V(f, \Delta_n)_t - f \star \mu_{\Delta_n[t/\Delta_n]}), \frac{1}{\sqrt{\Delta_n}} (V(f, k\Delta_n)_t - f \star \mu_{k\Delta_n[t/k\Delta_n]}) \right)$$

(3.136)

converges stably in law, on the product $\mathbb{D}(\mathbb{R}_+, \mathbb{R}^q) \times \mathbb{D}(\mathbb{R}_+, \mathbb{R}^q)$ of the Skorokhod spaces, to the process $(Z(f), Z'(f))$.

We have the (stable) convergence in law of the above processes, as elements of the product functional space $\mathbb{D}(\mathbb{R}_+, \mathbb{R}^q)^2$, but usually not as elements of the space $\mathbb{D}(\mathbb{R}_+, \mathbb{R}^{2q})$ with the ($2q$-dimensional) Skorokhod topology, because a jump of X at time S, say, entails a jump for both components above at two times S_n and S_n' which both converge to S but are in general different (with a probability close to $(k-1)/k$, in fact): this prevents the $2q$-dimensional Skorokhod convergence. In the same way, although $S_n \to S$, we have $S_n \neq S$ and $V(f, \Delta_n)$ jumps at S_n whereas $f \star \mu$ jumps at S: this is why, if we want Skorokhod convergence, we have to center $V(f, \Delta_n)$ around the discretized version of $f \star \mu$.

However, in most applications we are interested in the convergence at a given fixed time t. Since $\mathbb{P}(\Delta X_t \neq 0) = 0$ for all t, in view of the properties of the Skorokhod convergence we immediately get the following corollary:

Corollary 3.28 *Under the assumptions of the previous theorem, for any fixed $t > 0$ the $2q$-dimensional variables*

$$\Big(\frac{1}{\sqrt{\Delta_n}} \big(V(f, \Delta_n)_t - f \star \mu_t\big), \frac{1}{\sqrt{\Delta_n}} \big(V(f, k\Delta_n)_t - f \star \mu_t\big)\Big)$$

converges stably in law to the variable $(Z(f)_t, Z'(f)_t)$.

Now, it may happen that f is such that $f \star \mu = 0$, and also $(\nabla f) \star \mu = 0$: this is the case when X is continuous, of course, but it may also happen when X is discontinuous, as we will see in some statistical applications later. Then the above result degenerates, and does not give much insight. So we need a further CLT, which goes as follows. There is a general result in the same spirit as Theorem 3.27, but here we consider a very special situation, which is enough for the applications we have in mind:

Theorem 3.29 *Assume (H) (or (H') only), and suppose that the two components X^1 and X^2, say, never jump at the same times. Let f be the function $f(x) = (x^1 x^2)^2$. Then the 2-dimensional processes*

$$\Big(\frac{1}{\Delta_n} V(f, \Delta_n), \frac{1}{\Delta_n} V(f, k\Delta_n)\Big) \tag{3.137}$$

converge stably in law, on the product $\mathbb{D}(\mathbb{R}_+, \mathbb{R}) \times \mathbb{D}(\mathbb{R}_+, \mathbb{R})$ of the Skorokhod spaces, to the process

$$\Big(\frac{1}{2} \overline{Z}(f)_t + \int_0^t (c_u^{ii} c_u^{jj} + 2(c_u^{ij})^2) du, \frac{1}{2} \overline{Z}'(f)_t + k \int_0^t (c_u^{ii} c_u^{jj} + 2(c_u^{ij})^2) du\Big) \tag{3.138}$$

Of course the same result holds for any two other components. More generally a similar result holds when f is a homogeneous polynomial of degree 4, which satisfies outside a $\mathbb{P}$-null set:

$$f \star \mu = 0, \qquad (\nabla f) \star \mu = 0. \tag{3.139}$$

Finally, as said before, we also state, without proof, the result about the quadratic variation itself. Although not so important for statistical applications, it is of great theoretical significance. Exactly as in Corollary 3.22, only the weak assumption (H') is required here (see Jacod (2008) for a proof, and Jacod and Protter (1998) for an early version stated somewhat differently).

Theorem 3.30 *Assume (H'), Then the $d \times d$-dimensional process with components*

$$\frac{1}{\sqrt{\Delta_n}} \Big(\sum_{i=1}^{[t/\Delta_n]} \Delta_i^n X^j \Delta_i^n X^k - [X^j, X^k]_{[t/\Delta_n]\Delta_n} \Big)$$

converge stably in law to $V + Z(f)$, where V is a s described in Corollary 3.22 and $Z(f)$ is as above with $f_{jk}(x) = x_j x_k$, and conditionally on $\mathcal{F}$ the processes V and $Z(f)$ are independent.

3.8.3 Some preliminary on stable convergence

Once more, for the above results it is enough to prove them under (SH), which we assume henceforth. The basis of the proof is a rather general result of stable convergence about discontinuous processes, which cannot be found in a book form so far.

Although what follows does not depend on the choice of the sequence (S_p), for convenience we make a specific choice. For any $m \geq 1$ we denote by $(T(m,r) : r \geq 1)$ the successive jump times of the process

$$N^m = 1_{\{1/m < \gamma \leq 1/(m-1)\}} \star \underline{\mu}$$

(note that N^m is an homogeneous Poisson process with intensity $\lambda(\{z : \frac{1}{m} < \gamma(z) < \frac{1}{m-1}\}))$. Then (S_p) is a reordering of the double sequence $(T(m,r) : r, m \geq 1)$ into a single sequence.

Next we introduce some notation. For any $p \geq 1$ the time S_p is in one and only one interval $((ik+j)\Delta_n, (ik+j+1)\Delta_n]$, for some $i \geq 0$ and $j = 0, \cdots, k-1$. So, we can define a number of quantities by setting their values on each set $\{(ik + j)\Delta_n < S_p \leq (ik + j + 1)\Delta_n\}$:

$$
\left.
\begin{aligned}
&L(n,p) = j, \qquad\qquad K(n,p) = \frac{S_p}{\Delta_n} - (ik + j) \\
&\alpha_-(n,p) = \tfrac{1}{\sqrt{\Delta_n}} \left(W_{S_p} - W_{(ik+j)\Delta_n} \right) \\
&\alpha_+(n,p) = \tfrac{1}{\sqrt{\Delta_n}} \left(W_{(ik+j+1)\Delta_n} - W_{S_p} \right) \\
&\beta_-(n,p) = \tfrac{1}{\sqrt{\Delta_n}} \left(W_{(ik+j)\Delta_n} - W_{ik\Delta_n} \right) \\
&\beta_+(n,p) = \tfrac{1}{\sqrt{\Delta_n}} \left(W_{(i+1)k\Delta_n} - W_{(ik+j+1)\Delta_n} \right) \\
&A(n,p) = (\alpha_-(n,p), \alpha_+(n,p), \beta_-(n,p), \beta_+(n,p)) \\
&\widehat{R}_p^n = \sigma_{(ik+j)\Delta_n} \alpha_-(n,p) + \sigma_{T_p} \alpha_+(n,p) \\
&\widehat{R}_p'''^n = \sigma_{ik\Delta_n} \beta_-(n,p) + \sigma_{T_p} \beta_+(n,p) \\
&R_p^n = \tfrac{1}{\sqrt{\Delta_n}} \left(X_{(ik+j+1)\Delta_n} - X_{(ik+j)\Delta_n} - \Delta X_{S_p} \right) \\
&R_p'''^n = \tfrac{1}{\sqrt{\Delta_n}} \left(X_{(i+1)k\Delta_n} - X_{(ik+j+1)\Delta_n} + X_{(ik+j)\Delta_n} - X_{ik\Delta_n} \right).
\end{aligned}
\right\} \quad (3.140)
$$

In the next lemma, we consider the variables $\Theta_n = A(n,p)_{p \geq 1}$ taking values in the Polish space $F = (\mathbb{R}^4)^{N^*}$, and also the variable $\Theta = (A_p)_{p \geq 1}$ taking values in F as well, where

$$A_p = (\sqrt{\kappa_p}\, U_p, \sqrt{1 - \kappa_p}\, \overline{U}_p, \sqrt{L_p}\, U'_p, \sqrt{k-1-L_p}\, \overline{U}'_p)$$

uses the variables introduced at the beginning of this section.

Lemma 3.31 *The sequence* (Θ_n) *of variables stably converges in law to* Θ.

Proof. We need to prove that

$$\mathbb{E}(Zh(\Theta_n)) \to \widetilde{\mathbb{E}}(Zh(\Theta)) \tag{3.141}$$

for any bounded $\mathcal{F}$-measurable variable Z and any bounded continuous function h on F.

Let $\mathcal{G}$ be the σ-field of Ω generated by the process W and the random measure μ. Each Θ_n is $\mathcal{G}$-measurable and Θ is $\mathcal{G} \otimes \mathcal{F}'$-measurable, so $\mathbb{E}(Zh(\Theta_n)) = \mathbb{E}(Z'h(\Theta_n))$ and $\widetilde{\mathbb{E}}(Zh(\Theta)) = \widetilde{\mathbb{E}}(Z'h(\Theta))$, where $Z' = \mathbb{E}(Z \mid \mathcal{G})$. Hence it suffices to prove (3.141) when Z is $\mathcal{G}$-measurable.

We can go further: recalling that μ has the form $\mu = \sum_{p\geq1} \varepsilon_{(S_p, V_p)}$ for suitable E-valued variables V_p (ε_a = Dirac mass at a), then $\mathcal{G}$ is generated by W and the variables (S_p, V_p). Then by a density argument it is enough to prove (3.141) when

$$Z = f(W) \prod_{p=1}^{P} g_p(S_p)g'_p(V_p), \qquad h((z_p)_{p\geq1}) = \prod_{p=1}^{P} h_p(z_p)$$

where f is continuous and bounded on the space of all continuous $\mathbb{R}^{d'}$-valued functions, and the g_p's are continuous and bounded on R_+ and the g'_p's are continuous and bounded on E, and the h_p's are continuous and bounded on $\mathbb{R}^4$, and P is an integer.

Let $W^n_t = W_t - \sum_{p=1}^{P}(W_{S_p+2k\Delta_n} - W_{(S_p-2k\Delta_n)^+})$. Clearly $W^n \to W$ uniformly (for each ω), hence $f(W^n) \to f(W)$. If

$$\Omega(n, P) = \cap_{p,p'\in\{1,\cdots,P\}, \, p\neq p'} \{|S_p - S_{p'}| > k\Delta_n\},$$

we also have $\Omega(n, P) \to \Omega$ as $n \to \infty$. Therefore by Lebesgue's theorem,

$$\mathbb{E}\left(f(W) \prod_{p=1}^{P} g_p(S_q)g'_p(V_p)h_p(A(n,p)) \, 1_{\Omega(n,P)} \right)$$

$$-\mathbb{E}\left(f(W^n) \prod_{p=1}^{P} g_p(S_p)g'_p(V_p)h_p(A(n,p)) \right)$$

goes to 0, and we are left to prove that

$$\mathbb{E}\Big(f(W^n) \prod_{p=1}^{P} g_p(S_p)g'_p(V_p)h_p(A(n,p)) \, 1_{\Omega(n,P)}\Big) \ \to$$

$$\widetilde{\mathbb{E}}\Big(f(W) \prod_{p=1}^{P} g_p(S_p)g'_q(V_p)h_p(A_p)\Big).$$

Now, W and $\underline{\mu}$ are independent, and with our choice of the sequence (S_p) the two sequences (S_p) and (V_p) are also independent. This implies that W^n, the family (V_p) and the family $(A(n,p))_{p\leq P}$ are independent as well. Therefore the left side above equals the product of $\mathbb{E}\Big(f(W^n) \prod_{p=1}^{P} g'_p(V_p)\Big)$ with $\mathbb{E}\Big(\prod_{p=1}^{P} g_p(T_p)h_p(A(n,p)) \, 1_{\Omega(n,P)}\Big)$, and likewise for the right side. So finally it remains to prove that

$$\mathbb{E}\Big(\prod_{p=1}^{P} g_p(S_p)h_p(A(n,p)) \, 1_{\Omega(n,P)}\Big) \ \to \ \widetilde{\mathbb{E}}\Big(\prod_{p=1}^{P} g_p(S_p)h_p(A_p)\Big). \quad (3.142)$$

At this stage, and by another application of the independence between W and $\underline{\mu}$, we observe that in restriction on the set $\Omega(n,P)$, the sequence $(A(n,p) : p = 1, \cdots, P)$ has the same law as the sequence $(A'(n,p) : p = 1, \cdots, P)$, where

$$A'(n,p) =$$
$$(\sqrt{K(n,p)} \, U_p, \sqrt{1 - K(n,p)} \, \overline{U}_q, \sqrt{L(n,p)} \, U'_p, \sqrt{k - 1 - L(n,p)} \, \overline{U}'_p).$$

Therefore (3.142) amounts to proving that $((S_p, K(n,p), L(n,p)) : p = 1, \cdots, P)$ converges in law to $((S_p, \kappa_p, L_p) : p = 1, \cdots, P)$.

To see this, one may introduce the fractional part $G(n,p)$ of $[S_p/k\Delta_n]$, which equals $\frac{1}{\Delta_n}(S_p - ik\Delta_n)$ on the set $\{ik\Delta_n \leq S_p < (i+1)k\Delta_n\}$. Since the family $(S_p : p = 1, \cdots, P)$ admits a smooth density on its support in $\mathbb{R}_+^P$ (again because of our choice of (S_p)), an old result of Tukey (1939) shows that the sequences $((S_p, G(n,p)) : p = 1, \cdots, P)$ converge in law, as $n \to \infty$, to $((S_p, G_p) : p = 1, \cdots, P)$ where the G_p's are independent one from the other and from the S_p's and uniformly distributed on $[0,1]$ (Tukey's result deals with 1-dimensional variables, but the multidimensional extension is straightforward). Since $K(n,p)$ and $L(n,p)$ are respectively the fractional part and the integer part of $G(n,p)/k$, and since the fractional part and the integer part of G_p/k are independent and respectively uniform on $[0,1]$ and uniform on $\{0, \cdots, k-1\}$, the desired result is now obvious. $\qquad\square$

Lemma 3.32 *The sequence of $(\mathbb{R}^{2d})^{\mathcal{N}^*}$-valued variables $((R_p^n, R_p''^n) : p \geq 1)$ stably converges in law to $((R_p, R_p'') : p \geq 1)$ (see (3.128)).*

Proof. This result is a consequence of one of the basic properties of the stable convergence in law. Namely, if a sequence Y_n of E-variables defined on the space $(\Omega, \mathcal{F}, \mathbb{P})$ stably converges in law to Y (defined on an extension), and if a sequence Z_n of F-variables defined on $(\Omega, \mathcal{F}, \mathbb{P})$ again converges in probability to Z, then for any continuous function f on $E \times F$ the variables $f(Y_n, Z_n)$ stably converge in law to $f(Y, Z)$.

A first application of this property allows to deduce from the previous lemma and from the fact that σ_t is right continuous with left limits is that $((\widehat{R}_p^n, \widehat{R}_p^{\prime\prime n}) : p \geq 1)$ stably converges in law to $((R_p, R_p^{\prime\prime}) : p \geq 1)$. A second application of the same shows that, in order to get our result, it is enough to prove that for each $p \geq 1$ we have

$$R_p^n - \widehat{R}_p^n \xrightarrow{\mathbb{P}} 0, \qquad R_p^{\prime\prime n} - \widehat{R}_p^{\prime\prime n} \xrightarrow{\mathbb{P}} 0. \tag{3.143}$$

We will prove the first part of (3.143), the proof of the second part being similar. Recall that $S_p = T(m, r)$ for some $r, m \geq 1$, and set $X' = X - X^c - (\delta 1_{\{\gamma > 1/m\}}) \star \mu$ and

$$\begin{cases} \zeta_i^n(t) = \frac{1}{\sqrt{\Delta_n}} \Big(\int_{(i-1)\Delta_n}^t (\sigma_u - \sigma_{(i-1)\Delta_n}) dW_u \\ \qquad\qquad + \int_t^{i\Delta_n} (\sigma_u - \sigma_t) dW_u \Big) 1_{((i-1)\Delta_n, i\Delta_n]}(t) \\ \zeta_i^{\prime n} = \frac{1}{\sqrt{\Delta_n}} \Delta_i^n X', \end{cases}$$

and observe that

$$R_p^n - \widehat{R}_p^n = \sum_{i \geq 1} \Big(\zeta_i^n(S_p) + \zeta_i^{\prime n} \Big) 1_{D_i^n}, \tag{3.144}$$

where $D_i^n = \{(i-1)\Delta_n < S_p \leq i\Delta_n\}$.

There is a problem here: it is easy to evaluate the conditional expectations of $|\zeta_i^n(t)|$ and $|\zeta_i^{\prime n}|$ w.r.t. $\mathcal{F}_{(i-1)\Delta_n}$ and to check that they go to 0, uniformly in i, but the set D_i^n is not $\mathcal{F}_{(i-1)\Delta_n}$-measurable. To overcome this difficulty we denote by $(\mathcal{G}_t)_{t \geq 0}$ the smallest filtration such that $\mathcal{G}_t$ contains $\mathcal{F}_t$ and $\sigma(S_p)$. Then W and the restriction μ' of μ to the set $\mathbb{R}_+ \times \{z : \gamma(z) \leq 1/m\}$ are still a Wiener process and a Poisson random measure relative to this bigger filtration $(\mathcal{G}_t)$, and X' is driven by μ'.

Therefore applying (3.75) with $\alpha_n = 1$ and $r = 2$ and to the process X' instead of X, we get $\mathbb{E}(|\Delta_i^n X'| \wedge 1 \mid \mathcal{G}'_{(i-1)\Delta_n}) \leq \varepsilon_n$, where $\varepsilon_n \to 0$. Since $D_i^n \in \mathcal{G}'_{(i-1)\Delta_n}$ we then have

$$\mathbb{E}\Big(\sum_{i \geq 1} (|\zeta_i^{\prime n}| \wedge 1) 1_{D_i^n} \Big) = \tag{3.145}$$

$$\mathbb{E}\Big(\sum_{i \geq 1} 1_{D_i^n} \mathbb{E}(|\zeta_i^{\prime n}| \wedge 1 \mid \mathcal{G}'_{(i-1)\Delta_n}) \Big) \leq \varepsilon_n \mathbb{E}\Big(\sum_{i \geq 1} 1_{D_i^n} \Big) = \varepsilon_n.$$

By Doob inequality and the fact that W is an $(\mathcal{F}'_t)$-Wiener process, for any $t \in ((i-1)\Delta_n, i\Delta_n]$, the conditional expectation $\mathbb{E}(|\zeta_i'^n(t)|^2 \mid \mathcal{G}'_{(i-1)\Delta_n})$ is smaller than

$$K\mathbb{E}\left(\int_{(i-1)\Delta_n}^{t} \|\sigma_u - \sigma_{(i-1)\Delta_n}\|^2 du + \int_t^{i\Delta_n} \|\sigma_u - \sigma_t\|^2 du \mid \mathcal{G}'_{(i-1)\Delta_n}\right).$$

Then the same argument as above yields

$$\mathbb{E}\left(\sum_{i\geq 1} |\zeta_i^n(T_p)|^2 1_{D_i^n}\right) \leq K\mathbb{E}\left(\int_{\Delta_n[T_p/\Delta_n]-\Delta_n}^{T_p} \|\sigma_u - \sigma_{(i-1)\Delta_n}\|^2 du \right.$$
$$\left. + \int_{T_p}^{\Delta_n[T_p/\Delta_n]} \|\sigma_u - \sigma_t\|^2 du \mid \mathcal{F}'_{(i-1)\Delta_n}\right).$$

This quantity goes to 0 by Lebesgue's theorem, because σ is right continuous with left limit, so this together with (3.144) and (3.145) gives us the first part of (3.143). □

3.8.4 Proof of Theorem 3.27

Step 1) We begin with some preliminaries, to be used also for the next theorem. We fix $m \geq 1$ and let P_m be the set of all p such that $S_p = T(m', r)$ for some $r \geq 1$ and some $m' \leq m$ (see the previous subsection). We also set

$$X(m)_t = X_t - \sum_{p\in P_m:\, T_p\leq t} \Delta X_{S_p} = X_t - (\delta 1_{\{\gamma>1/m\}}) \star \underline{\mu}_t. \quad (3.146)$$

Observe that, due to (3.71), and with the notation

$$b(m)_t = b_t - \int_{\{z:\gamma(z)>1/m\}} \delta(t,z)\lambda(dz),$$

we have $X(m) = X'(m) + X''(m)$, where

$$\left.\begin{array}{rcl} X'(m)_t &=& X_0 + \int_0^t b(m)_s ds + \int_0^t \sigma_s dW_s \\ X''(m) &=& (\delta 1_{\{\gamma\leq 1/m\}}) \star (\mu - \nu). \end{array}\right\} \quad (3.147)$$

Then we denote by $\Omega_n(t,m)$ the set of all ω satisfying the following for all $p \geq 1$:

$$\left.\begin{array}{l} p, p' \in P_m,\ S_p(\omega) \leq t \ \Rightarrow\ |S_p(\omega) - S_{p'}(\omega)| > k\Delta_n, \\ 0 \leq s \leq t,\ 0 \leq u \leq k\Delta_n \ \Rightarrow\ \|X(m)_{s+u}(\omega) - X(m)_s(\omega)\| \leq 2/m. \end{array}\right\} \quad (3.148)$$

Since $\|\delta\| \leq \gamma$, implying $\|\Delta X(m)_s\| \leq 1/m$, we deduce that for all $t > 0$ and $m \geq 1$:

$$\Omega_n(t,m) \rightarrow \Omega \quad \text{a.s. as } n \rightarrow \infty. \quad (3.149)$$

If g is C^2 with $g(0) = 0$ and $\nabla g(0) = 0$, for any integer $l \geq 1$ and any d-dimensional semimartingale Z, we write $G^n(Z, g, l)_t = V(Z, g, l\Delta_n)_t - \sum_{s \leq l\Delta_n[t/l\Delta_n]} g(\Delta Z_s)$. Observe that on the set $\Omega_n(t, m)$ we have for all $s \leq t$ and $l = 1$ or $l = k$:

$$G^n(X, g, l)_t = G^n(X(m), g, l) + Y^n(m, g, l), \qquad (3.150)$$

where

$$\left.\begin{aligned}
Y^n(m, g, l)_t &= \textstyle\sum_{p \in P_m : \, S_p \leq l\Delta_n[t/l\Delta_n]} \zeta(g, l)_p^n, \\
\zeta(g, 1)_p^n &= g(\Delta X_{S_p} + \sqrt{\Delta_n}\, R_p^n) - g(\Delta X_{S_p}) - g(\sqrt{\Delta_n}\, R_p^n) \\
\zeta(g, k)_p^n &= g(\Delta X_{S_p} + \sqrt{\Delta_n}\,(R_p^n + R_p''^n)) - g(\Delta X_{S_p}) \\
&\qquad - g(\sqrt{\Delta_n}\,(R_p^n + R_p''^n)).
\end{aligned}\right\} \qquad (3.151)$$

Step 2) Now we turn to the proof itself, with a function f satisfying the relevant assumptions. Recall in particular that $f(x)/\|x\| \to 0$ as $x \to 0$. A Taylor expansion in the expressions giving $\zeta(m, f, l)_q^n$ and Lemma 3.32 readily gives

$$\left(\frac{1}{\sqrt{\Delta_n}}\, \zeta(f, 1)_p^n, \frac{1}{\sqrt{\Delta_n}}\, \zeta(f, k)_p^n\right)_{p \geq 1} \xrightarrow{\mathcal{L} - s}$$

$$\left(\nabla f(\Delta X_{S_p}) R_p, \nabla f(\Delta X_{S_p}) R_p'\right)_{p \geq 1}$$

(here, $\nabla f(\Delta X_{S_p}) R_p$ for example stands for the q-dimensional vector with components $\sum_{i=1}^d \partial_i f_l(\Delta X_{S_p}) R_p^i$). Since the sum giving $Y^n(m, f, l)_t$ has in fact finitely many entries, we deduce from well known properties of the Skorokhod topology that, as $n \to \infty$:

$$\left.\begin{aligned}
&\text{the processes } \left(\tfrac{1}{\sqrt{\Delta_n}}\, Y^n(m, f, 1), \tfrac{1}{\sqrt{\Delta_n}}\, Y^n(m, f, k)\right) \\
&\text{converges stably in law, in } \mathbb{D}(\mathbb{R}_+, \mathbb{R}^q) \times \mathbb{D}(\mathbb{R}_+, \mathbb{R}^q) \\
&\text{to the process } (Z^m(f), Z'^m(f)),
\end{aligned}\right\} \qquad (3.152)$$

where $(Z^m(f), Z'^m(f))$ is defined componentwise by (3.131), except that the sum is taken over all $p \in P_m$ only.

If we consider, say, the first component, we have by (3.132) and Doob's inequality:

$$\widetilde{\mathbb{E}}\left(\sup_{s \leq t} |Z^m(f_1)_s - Z(f_1)_s|^2\right) = \widetilde{\mathbb{E}}\left(\widetilde{\mathbb{E}}\left(\sup_{s \leq t} |Z^m(f_1)_s - Z(f_1)_s|^2 \mid \mathcal{F}\right)\right)$$

$$\leq 4\mathbb{E}\left(\sum_{p \notin P_m, \, S_p \leq t} \sum_{i,j=1}^d (\partial_i f_1 \partial_j f_1)(\Delta X_{S_p})(c_{S_p-}^{ij} + c_{S_p}^{ij})\right).$$

The variable of which the expectation is taken in the right side above is smaller than $K \sum_{s \leq t} \|\Delta X_s\|^2 1_{\{\|\Delta X_s\| \leq 1/m\}}$ (because c_t is bounded and if $p \notin P_m$ then $\|\Delta X_s\| \leq 1/m$), so by Lebesgue theorem this expectation goes to 0 as

$m \to \infty$. The same argument works for the other components, and thus we have proved that

$$(Z^m(f), Z'^m(f)) \xrightarrow{\text{u.c.p.}} (Z(f), Z'(f)). \qquad (3.153)$$

Hence, in view of (3.152) and (3.153), and also of (3.149) and (3.150), it remains to prove

$$\lim_{m\to\infty} \limsup_{n\to\infty} \mathbb{P}\Big(\Omega_n(t,m) \cap \Big\{ \sup_{s\le t} \frac{1}{\sqrt{\Delta_n}} |G^n(X(m), f_r, l)_s| > \eta \Big\}\Big) = 0. \qquad (3.154)$$

Step 3) Now we proceed to proving (3.154), and we drop the index r, pretending that f is 1-dimensional. It is also enough to consider the case $l = 1$ (the case $l = k$ is the same, upon replacing everywhere Δ_n by $k\Delta_n$). We set

$$k(x,y) = f(x+y) - f(x) - f(y), \qquad g(x,y) = k(x,y) - \nabla f(x)y. \quad (3.155)$$

Recall that f is C^2 and that (3.71) and (3.146) hold. Then we apply Itô's formula to the process $X(m)_s - X(m)_{i\Delta_n}$ and the function f, for $t > i\Delta_n$ to get

$$\frac{1}{\sqrt{\Delta_n}} \Big(G^n(X(m), f, 1)_t - G^n(X(m), f, 1)_{i\Delta_n}\Big) = \qquad (3.156)$$

$$A(n, m, i)_t + M(n, m, i)_t,$$

where $M(n, m, i)$ is a locally square-integrable martingale with predictable bracket $A'(n, m, i)$, and with

$$\left.\begin{array}{l} A(n, m, i)_t = \int_{i\Delta_n}^t a(n, m, i)_u \, du \\[2mm] A'(n, m, i)_t = \int_{i\Delta_n}^t a'(n, m, i)_u \, du, \end{array}\right\} \qquad (3.157)$$

and

$$\begin{cases} a(n, m, i)_t = \frac{1}{\sqrt{\Delta_n}} \Big(\sum_{j=1}^d \partial_j f(X(m)_t - X(m)_{i\Delta_n}) b(m)_t^j \\ \qquad\qquad +\frac{1}{2} \sum_{j,l=1}^d \partial_{jl}^2 f(X(m)_t - X(m)_{i\Delta_n}) c_t^{jl} \\ \qquad\qquad + \int_{\{z:\gamma(z)\le 1/m\}} g(X(m)_t - X(m)_{i\Delta_n}, \delta(t,z)) \lambda(dz)\Big) \\ a'(n, m, i)_t = \frac{1}{\Delta_n}\Big(\sum_{j,l=1}^d (\partial_j f \, \partial_l f)(X(m)_t - X(m)_{i\Delta_n}) c_t^{jl} \\ \qquad\qquad + \int_{\{z:\gamma(z)\le 1/m\}} k(X(m)_t - X(m)_{i\Delta_n}, \delta(t,z))^2 \lambda(dz)\Big). \end{cases}$$

Now we set $T(n, m, i) = \inf(s > i\Delta_n : \|X(m)_s - X(m)_{i\Delta_n}\| > 2/m)$. On the set $\Omega_n(t, m)$ we have by construction $T(n, m, i) > (i+1)\Delta_n$ for all

$i < [t/\Delta_n]$. Therefore in view of (3.156) we have on this set:

$$\frac{1}{\sqrt{\Delta_n}} \sup_{s \leq t} |G^n(X(m), f, 1)_s| \leq \sum_{i=1}^{[t/\Delta_n]} |A(n, m, i-1)_{(i\Delta_n) \wedge T(n,m,i-1)}|$$

$$+ \left| \sum_{i=1}^{[t/\Delta_n]} M(n, m, i-1)_{(i\Delta_n) \wedge T(n,m,i-1)} \right|.$$

Henceforth in order to get (3.154), it is enough to prove the following:

$$\left. \begin{array}{l} \lim_{m \to \infty} \limsup_n \mathbb{E}\left(\sum_{i=1}^{[t/\Delta_n]} |A(n, m, i-1)_{(i\Delta_n) \wedge T(n,m,i-1)}| \right) = 0, \\ \lim_{m \to \infty} \limsup_n \mathbb{E}\left(\sum_{i=1}^{[t/\Delta_n]} A'(n, m, i-1)_{(i\Delta_n) \wedge T(n,m,i-1)} \right) = 0. \end{array} \right\}$$
(3.158)

Recall that $f(0) = 0$ and $\nabla f(0) = 0$ and $\|\nabla^2 f(x)\| = o(\|x\|)$ as $x \to 0$, so we have

$$j = 0, 1, 2, \;\; \|x\| \leq \frac{3}{m} \;\; \Rightarrow \;\; \|\nabla^j f(x)\| \leq \alpha_m \|x\|^{3-j} \qquad (3.159)$$

for some α_m going to 0 as $m \to \infty$, which implies

$$\|x\| \leq \frac{3}{m}, \;\; \|y\| \leq \frac{1}{m} \;\; \Rightarrow \qquad\qquad\qquad\qquad (3.160)$$

$$|k(x,y)| \leq K\alpha_m \|x\| \|y\|, \quad |g(x,y)| \leq K\alpha_m \|x\| \|y\|^2.$$

Observe that $\|X(m)_{s \wedge T(n,m,i)} - X(m)_{i\Delta_n}\| \leq 3/m$ for $s \geq i\Delta_n$ (because the jumps of $X(m)$ are smaller than $1/m$). Then in view of (SH) and (3.160) and of the fact that $\|b(m)_t\| \leq Km$ we obtain for $i\Delta_n \leq t \leq T(n, m, i)$:

$$\left\{ \begin{array}{l} |a(n, m, i)_t| \leq \frac{K\alpha_m}{\sqrt{\Delta_n}} (\|X(m)_t - X(m)_{i\Delta_n}\| \\ \qquad\qquad\qquad\qquad + m\|X(m)_t - X(m)_{i\Delta_n}\|^2), \\ a'(n, m, i)_t \leq \frac{K\alpha_m^2}{\Delta_n} \|X(m)_t - X(m)_{i\Delta_n}\|^2. \end{array} \right.$$

Now, exactly as for (3.73), one has $\mathbb{E}(\|X(m)_{t+s} - X(m)_t\|^p) \leq K_p(s^{p/2} + m^p s^p)$ for all $p \in (0, 2]$ and $s, t \geq 0$, under (SH). Applying this with $p = 1$ and $p = 2$, respectively, gives that the two "lim sup" in (3.158) are smaller than $Kt\alpha_m$ and $Kt\alpha_m^2$ respectively. Then (3.158) holds, and we are finished.

3.8.5 Proof of Theorem 3.29

We essentially reproduce the previous proof, with the same notation. Recall that $f(x) = (x^1 x^2)^2$.

Step 1) The assumption that X^1 and X^2 have no common jumps implies that

$f(\Delta X_{S_p}) = 0$ and $\nabla f(\Delta X_{S_p}) = 0$ for all $p \geq 1$, whereas $f(x)/\|x\|^2 \to 0$ as $x \to 0$. Then a second order Taylor expansion in the expressions giving $\zeta(m, f, l)_q^n$ and Lemma 3.32 gives

$$\left(\frac{1}{\Delta_n}\,\zeta(f,1)_p^n, \frac{1}{\Delta_n}\,\zeta(f,k)_p^n\right)_{p\geq 1} \xrightarrow{\mathcal{L}-s}$$

$$\left(\frac{1}{2}\sum_{i,j=1}^d \partial^2 f(\Delta X_{S_p})R_p^i R_p^j, \frac{1}{2}\sum_{i,j=1}^d \partial^2 f(\Delta X_{S_p})R_p'^i R_p'^j\right)_{p\geq 1}.$$

From this we deduce that, instead of (3.152), and as $n \to \infty$:

$$\left.\begin{array}{l} \text{the processes } \left(\frac{1}{\Delta_n}\,Y^n(m,f,1), \frac{1}{\Delta_n}\,Y^n(m,f,k)\right) \\[4pt] \text{converges stably in law, in } \mathbb{D}(\mathbb{R}_+, \mathbb{R}) \times \mathbb{D}(\mathbb{R}_+, \mathbb{R}), \\[4pt] \text{to the process } \frac{1}{2}(\overline{Z}^m(f), \overline{Z}'^m(f)) \end{array}\right\}$$

where $(\overline{Z}^m(f), \overline{Z}'^m(f))$ is defined componentwise by (3.133), except that the sum is taken over all $p \in P_m$ only. By Lebesgue theorem, we readily obtain

$$(\overline{Z}^m(f), \overline{Z}'^m(f)) \xrightarrow{\text{u.c.p.}} (\overline{Z}(f), \overline{Z}'(f)).$$

Hence, in view of (3.149) and (3.150), and since here

$$G(X(m), f, l) = V(X(m), f, l\Delta_n),$$

it remains to prove that with the notation $\overline{C}_t = \int_0^t (c_u^{ii} c_u^j + 2(c_u^{ij})^2)\,du$ we have for all $t, \eta > 0$ and for $l = 1$ and $l = k$:

$$\lim_{m\to\infty} \limsup_{n\to\infty} \mathbb{P}\left(\sup_{s\leq t} \frac{1}{\Delta_n}\,|V(X(m),f,l\Delta_n)_s - l\overline{C}_{l\Delta_n[s/l\Delta_n]}| > \eta\right) = 0.$$

$$(3.161)$$

Step 2) Recall (3.147), and set $g(x) = \|x\|^4$. By Theorem 3.10 applied to the process $X'(m)$ we have for each $m \geq 1$:

$$\left.\begin{array}{l} \frac{1}{l\Delta_n}\,V(X'(m),f,l\Delta_n) \xrightarrow{\text{u.c.p.}} \overline{C}, \\[6pt] \frac{1}{l\Delta_n}\,V(X'(m),g,l\Delta_n)_t \xrightarrow{\text{u.c.p.}} \int_0^t \rho_{\sigma_u}(g)\,du. \end{array}\right\}$$

$$(3.162)$$

Therefore for getting (3.161) it is enough to prove that

$$\lim_{m\to\infty} \limsup_{n\to\infty} \mathbb{P}\left(\sup_{s\leq t}\frac{1}{\Delta_n}|V(X(m),f,l\Delta_n)_s - V(X'(m),f,l\Delta_n)_s| > \eta\right) = 0.$$

$$(3.163)$$

Here again, it is obviously enough to prove the result for $l = 1$.

Now, the special form of f implies that for each $\varepsilon > 0$ there is a constant K_ε with

$$|f(x+y) - f(x)| \leq \varepsilon\|x\|^4 + K_\varepsilon\|x\|^2\,\|y\|^2 + K_\varepsilon f(y),$$

hence

$$|V(X(m), f, \Delta_n) - V(X'(m), f, \Delta_n)| \leq \varepsilon V(X'(m), g, \Delta_n) + K_\varepsilon (U^n + U'^n),$$

where

$$U_t^n = \sum_{i=1}^{[t/\Delta_n]} (\Delta_i^n X''^1(m))^2 (\Delta_i^n X''^2(m))^2,$$

$$U_t'^n = \sum_{i=1}^{[t/\Delta_n]} \|\Delta_i^n X(m)\|^2 \|\Delta_i^n X'(m)\|^2.$$

$\varepsilon > 0$ being arbitrarily small, by the second part of (3.162) it is then enough to prove

$$\lim_{m\to\infty} \limsup_{n\to\infty} \frac{1}{\Delta_n} \mathbb{E}(U_t^n) = 0, \quad \lim_{m\to\infty} \limsup_{n\to\infty} \frac{1}{\Delta_n} \mathbb{E}(U_t'^n) = 0. \quad (3.164)$$

Step 3) Exactly as in the proof of Lemma 3.16, we have

$$\left.\begin{array}{l} \mathbb{E}(\|X''(m)_{t+s} - X''(m)_t\|^2) \leq \alpha_m s, \\ \text{where} \quad \alpha_m = \int_{\{z:\gamma(z)\leq 1/m\}} \gamma(z)^2 \lambda(dz). \end{array}\right\} \quad (3.165)$$

Now, as for (3.156), we deduce from Itô's formula that

$$(\Delta_i^n X''^1(m))^2 (\Delta_i^n X''^2(m))^2 = M(n, m, i)_{i\Delta_n} + \int_{(i-1)\Delta_n}^{i\Delta_n} a(n, m, i-1)_s ds, \quad (3.166)$$

where $M(n, m, i)$ is a martingale,

$$a(n, m, i)_t = H_m(X''(m)_t - X''(m)_{(i-1)\Delta_n})$$

and

$$H_m(x) = \int_{\{z:\gamma(z)\leq 1/m\}} \left(f(x + \delta(t, z)) - f(x) - \nabla f(x)\delta(t, z) \right) \lambda(dz).$$

Now, since X^1 and X^2 have no common jumps, we have $\delta(\omega, t, z)^1 \delta(\omega, t, z)^2 = 0$ for λ-almost all z. Therefore a simple calculation shows that

$$H(x) = \int_{\{z:\gamma(z)\leq 1/m\}} \left((x^1)^2(\delta(t, z)^2)^2 + (x^2)^2(\delta(t, z)^1)^2 \right) \lambda(dz),$$

and thus

$$0 \leq H_m(x) \leq \alpha_m \|x\|^2.$$

Recall also that $\mathbb{E}(\|X''(m)_{t+s} - X''(m)_t\|^2) \leq Kt$. Then taking the expectation in (3.166) gives us, with

$$\mathbb{E}((\Delta_i^n X''^1(m))^2 (\Delta_i^n X''^2(m))^2) =$$

$$\mathbb{E}\left(\int_{(i-1)\Delta_n}^{i\Delta_n} H_m(X''(m)_s - X''(m)_{(i-1)\Delta_n}) \, ds\right) \leq \alpha_m \Delta_n^2.$$

Then, since $\alpha_m \to 0$ as $m \to \infty$, we readily deduce the first part of (3.164). It remains to prove the second part of (3.164). Itô's formula again yields

$$\|\Delta_i^n X''(m)\|^2 \|\Delta_i^n X'(m)\|^2 = M'(n, m, i)_{i\Delta_n} + \int_{(i-1)\Delta_n}^{i\Delta_n} a'(n, m, i-1)_s ds,$$

$$(3.167)$$

where $M'(n, m, i)$ is a martingale,

$$a'(n, m, i)_t = H'_m(X'(m)_t - X'(m)_{(i-1)\Delta_n}, X''(m)_t - X''(m)_{(i-1)\Delta_n})$$

and

$$H'_m(x, y) = 2\|y\|^2 \sum_{i=1}^d b(m)_t^i x^i + \|y\|^2 \sum_{i=1}^d c_t^{ii}$$

$$+ \|x\|^2 \int_{\{z:\gamma(z)\leq 1/m\}} \|\delta(t, z))\|^2 \, \lambda(dz),$$

and thus

$$|H'_m(x, y)| \leq K\left(\alpha_m\|x\|^2 + \|y\|^2(1 + m\|x\|)\right)$$

because $\|b(m)_t\| \leq Km$ and $\|c\| \leq K$. Then using (3.165) and $\mathbb{E}(\|X'(m)_{t+s} - X'(m)_t\|^p) \leq K_p(s^{p/2} + m^p s^p)$ for all $p > 0$, we deduce from Cauchy-Schwarz inequality, and by taking the expectation in (3.167), that

$$\mathbb{E}(\|\Delta_i^n X''(m)\|^2 \|\Delta_i^n X'(m)\|^2)$$

$$= \mathbb{E}\bigg(\int_{(i-1)\Delta_n}^{i\Delta_n} H'_m(X'(m)_s - X'(m)_{(i-1)\Delta_n},$$

$$X''(m)_s - X''(m)_{(i-1)\Delta_n}) \, ds\bigg)$$

$$\leq K\Delta_n^2\left(\alpha_m(1 + m^2\Delta_n) + m\sqrt{\alpha_m\Delta_n}\,(1 + m\Delta_n)\right).$$

Then again since $\alpha_m \to 0$ as $m \to \infty$, we deduce the second part of (3.164), and the proof is finished.

3.9 Estimation of the integrated volatility

At this point we have established the theoretical results which are needed for the statistical problems we have in mind, and we can turn to these problems. We start with a warning, which applies to all problems studied below:

The underlying process X is observed at times $0, \Delta_n, 2\Delta_n, \cdots$ without measurement errors.

This assumption is clearly *not satisfied* in general in the context of high-frequency data, at least in finance where there is an important microstructure noise. However, dealing with measurement errors involves a lot of complications which would go beyond the scope of this course.

As said before, the first and probably the most important question is the estimation of the integrated volatility, at least when the underlying process is continuous. This is the object of this section.

3.9.1 The continuous case

Here we assume that the underlying process X is a *continuous* Itô semimartingale, i.e. is of the form

$$X_t = X_0 + \int_0^t b_s ds + \int_0^t \sigma_s dW_s. \tag{3.168}$$

Most of the literature is concerned with the 1-dimensional case, but mathematically speaking there is no complication whatsoever in considering the d-dimensional case: so above W is a d'-dimensional Wiener process, and b_t and σ_t are d and $d \times d'$-dimensional (so implicitly in (3.168) the second integral is in fact a sum of stochastic integrals w.r.t. the various components W^j of W).

Our aim is to "estimate" the *integrated volatility*, that is the quadratic variation-covariation process of X:

$$C_t^{jk} = \int_0^t c_s^{jk} ds, \qquad \text{where} \quad c_t = \sigma_t \sigma_t^\star. \tag{3.169}$$

Recall that the process X is observed at the discrete times $0, \Delta_n, 2\Delta_n, \cdots$ over a finite interval $[0, T]$, and one wants to infer C_T, or sometimes the increments $C_t - C_s$ for some pairs (s, t) with $0 \le s \le t \le T$. Each of these increments is a random variable taking values in the set of $d \times d$ symmetric nonnegative matrices.

One point should be mentioned right away, and is in force not only for the integrated volatility but for all quantities estimated in this course: although we speak about estimating the matrix C_T, it is *not* a statistical problem in the usual sense since the quantity to estimate is a *random variable*; so the "estimator," say $\widetilde{C}_T^n$ (the "n" is here to emphasize that it is a function of the observation $(X_{i\Delta_n} : 0 \le i \le [T/\Delta_n])$) does not estimate a parameter, but a variable which depends on the outcome ω, and the quality of this estimator is something which fundamentally depends on ω as well.

Nevertheless we are looking for estimators which behave as in the classical case, asymptotically as $n \to \infty$ (that is, as $\Delta_n \to 0$). We say that $\widetilde{C}_T^n$ is *consistent* if $\widetilde{C}_T^n$ converges in probability to C_T (one should say "weakly" consistent; of course in the present setting, even more than in classical statistics, one would like to have estimators which converge *for all* ω, or at least almost surely, but this is in general impossible to achieve). Then we also aim to a *rate of convergence*, and if possible to a limit theorem, so as to allow for quantitatively asserting the quality of the estimator and for constructing confidence intervals, for example.

Two consistent estimators can be compared on the basis of their rates of convergence and, if those are the same, on their asymptotic variances, for example. However, unlike in classical statistics, we do not have a theory for asymptotic optimality, like the LAN or LAMN theory. The best we can do is to check whether our estimators are asymptotically optimal (in the usual sense) when the problem reduces to a classical parametric problem, that is when C_T is deterministic (this happens when for example the volatility σ_t is not random, like in the Black-Scholes model for the log-returns).

After these lengthy preliminaries we now introduce the estimator. Of course all authors use the approximated quadratic variation given in (3.9), and often called "realized volatility." Since we are in the d-dimensional case, we have a matrix $B(2, \Delta_n)_t$ with components

$$B(2, \Delta_n)_t^{jk} = \sum_{i=1}^{[t/\Delta_n]} \Delta_i^n X^j \Delta_i^n X^k. \tag{3.170}$$

These estimators have the following properties:

Property 3.33 (Consistency) $B(2, \Delta_n)_t \xrightarrow{\mathbb{P}} C_t$.

Property 3.34 (Asymptotic normality-1) $\frac{1}{\sqrt{\Delta_n}} \left(B(2, \Delta_n)_t - C_t \right)$ converges in law to a $d \times d$-dimensional variable which, conditionally on the path of X over $[0, t]$, is centered normal with variance-covariance (Γ_t^{jklm}) (the covariance of the (jk) and the (lm) components) given by

$$\Gamma_t^{jklm} = \int_0^t (c_s^{jl} c_s^{km} + c_s^{jm} c_s^{kl}) ds. \tag{3.171}$$

These are obvious consequences of Theorem 3.8-(b) and Corollary 3.22: for the consistency there is no assumption other than (3.168); for the asymptotic normality we need (H) in these notes, but in fact it is enough that $\int_0^t \|c_s\|^2 ds < \infty$ a.s. (see Jacod and Protter (1998)).

Property 3.34 gives a rate of convergence equal to $1/\sqrt{\Delta_n}$, but the name "asymptotic normality" is not really adequate since the limiting variable after centering and normalization is not unconditionally normal, and indeed it has a law which is essentially unknown. So it is useless in practice. But fortunately we not only have the convergence in law, but also the *stable* convergence in law. That is, as soon as one can find a sequence Γ^n_t of variables, depending on the observations at stage n only, and which converge in probability to the variance given by (3.171), then by normalizing once more by the square-root of the inverse of Γ^n_t (supposed to be invertible), we get a limit which is standard normal.

The "complete" result involving all components of C_t at once is a bit messy to state. In practice one is interested in the estimation of a particular component C^{jk}_t (often with $k = j$ even). So for simplicity we consider below the estimation of a given component C^{jk}_t. The asymptotic variance is $\Gamma^{jkjk}_t = \int_0^t (c^{jj}_s c^{kk}_s + (c^{jk}_s)^2) ds$ and we need an estimator for Γ^{jkjk}_t, which is provided by Theorem 3.10. More specifically, this theorem implies that

$$\Gamma(\Delta_n)^{jkjk}_t = \qquad\qquad (3.172)$$

$$\frac{1}{\Delta_n} \sum_{i=1}^{[t/\Delta_n]} \left((\Delta^n_i X^j)^2 (\Delta^n_{i+1} X^k)^2 + \Delta^n_i X^j \Delta^n_i X^k \Delta^n_{i+1} X^j \Delta^n_{i+1} X^k \right)$$

converges in probability to Γ^{klkl}_t. Therefore we have the following *standardized CLT*:

Theorem 3.35 Asymptotic normality-2 *Assume (H). With the previous notation, and in restriction to the set $\{\Gamma^{jkjk}_t > 0\}$, the variables*

$$\frac{1}{\sqrt{\Delta_n \Gamma(\Delta_n)^{jkjk}_t}} \left(B(2, \Delta_n)^{jk}_t - C^{jk}_t \right) \qquad\qquad (3.173)$$

converge stably in law to an $\mathcal{N}(0, 1)$ random variable independent of $\mathcal{F}$.

The reader will notice the proviso "in restriction to the set $A := \{\Gamma^{jkjk}_t > 0\}$". This set is in fact equal to the set where $s \mapsto c^{jj}_s$ and $s \mapsto c^{kk}_s$ are not Lebesgue-almost surely vanishing on $[0, t]$, and also $\mathbb{P}$-a.s. to the set where neither one of the two paths $s \mapsto X^j_s$ and $s \mapsto X^k_s$ is of finite variation over $[0, t]$. So in practice $A = \Omega$ and the above is the mere (stable) convergence in law.

When $A \neq \Omega$, the stable convergence in law in restriction to A means that $\mathbb{E}(f(T_n)Y) \to \mathbb{E}(Y)\tilde{\mathbb{E}}(f(U))$ for all bounded continuous functions f and all $\mathcal{F}$-measurable bounded variables Y vanishing outside A, and where T_n is the statistics in (3.173) and U is $\mathcal{N}(0, 1)$.

This result is immediately applicable in practice, in contrast to Property 3.34: it may be used to derive confidence intervals for example, in the customary way.

Proof. As above, T_n is the variable (3.173), and we also set

$$S_n = \frac{1}{\sqrt{\Delta_n}} \left(B(2,\Delta_n)_t^{jk} - C_t^{jk} \right).$$

We know that S_n converges stably in law to a variable which can be expressed as the product $\sqrt{\Gamma_t^{jkjk}}\, U$, where U is $\mathcal{N}(0,1)$ and independent of $\mathcal{F}$. By the properties of the stable convergence in law, and since $\Gamma(\Delta_n)_t^{jkjk} \overset{\mathbb{P}}{\longrightarrow} \Gamma_t^{jkjk}$, we also have stable convergence of the pair $(S_n, \Gamma(\Delta_n)_t^{jkjk})$ toward $\left(\sqrt{\Gamma_t^{jkjk}}\, U, \Gamma(\Delta_n)_t^{jkjk}\right)$, Obviously this also holds in restriction to the set A described above. Since $T_n = S_n/\sqrt{\Gamma(\Delta_n)_t^{jkjk}}$ and $\Gamma(\Delta_n)_t^{jkjk} \overset{\mathbb{P}}{\longrightarrow} \Gamma_t^{jkjk} > 0$ on A, the result follows from the continuous mapping theorem. $\qquad\square$

Remark 3.36 When $\sigma_t(\omega) = \sigma$ is a constant matrix, so up to the drift the process X is a Wiener process, then we are in the classical setting of estimation of a matrix-valued parameter $c = \sigma\sigma^*$. In this case we have the LAN property, and it is well known that the estimators $B(2,\Delta_n)_t$ are asymptotically efficient for estimating c in this setting (and when the drift vanishes, it is even the MLE). Note that c is identifiable, but usually not σ itself since there might be many square-roots σ for the matrix c. $\qquad\square$

Remark 3.37 There are many ways, indeed, to find consistent estimators for Γ_t^{jkjk}, and (3.172) is just a possibility. A full set of consistent estimators is provided by the formulas below, where q is a non-zero integer (recall that m_r is the rth absolute moment of $\mathcal{N}(0,1)$):

$$\Gamma(q,\Delta_n)_t^{jkjk} = \frac{1}{8m_{2/q}^{2q}\Delta_n} \sum_{i=1}^{[t/\Delta_n]} g_q^{jk}(\Delta_i^n X, \Delta_{i+1}^n X, \cdots, \Delta_{i+2q-1}^n X),$$

(3.174)

where

$$g_q^{jk}(x_1,\cdots,x_{2q}) = \prod_{i=1}^{2q}|x_i^j + x_i^k|^{2/q} + \prod_{i=1}^{2q}|x_i^j - x_i^k|^{2/q} - 2\prod_{i=1}^{2q}|x_i^j|^{2/q}$$
$$- 2\prod_{i=1}^{2q}|x_i^k|^{2/q} + 4\prod_{i=1}^{q}|x_i^j|^{2/q}\prod_{i=q+1}^{2q}|x_i^k|^{2/q}. \quad (3.175)$$

Indeed a simple computation yields that $\rho_t(g_q^{jk}) = 8m_{2/q}^{2q}\left(c_t^{jj}c_t^{kk} + (c_t^{jk})^2\right)$, so the property $\Gamma(q,\Delta_n)_t^{jkjk} \overset{\mathbb{P}}{\longrightarrow} \Gamma_t^{jkjk}$ again follows from Theorem 3.10. And of course one could make variations on this formula, like taking various powers summing up to 4 instead of the uniform power $2/q$, or varying the order in which the components x_i^j and x_i^k are taken in the last term of (3.175):

for example $2 \prod_{i=1}^{q} |x_i^j|^{2/q} \prod_{i=q+1}^{2q} |x_i^k|^{2/q} + 2 \prod_{i=1}^{q} |x_i^k|^{2/q} \prod_{i=q+1}^{2q} |x_i^j|^{2/q}$ could be taken instead of the last term in (3.175): then, with this substitution, we have in fact $\Gamma(q, \Delta_n)^{jkjk} = \Gamma(\Delta_n)^{jkjk}$ when $q = 2$.

The important fact is that Theorem 3.35 is unchanged, if $\Gamma(\Delta_n)_t^{jkjk}$ is substituted with $\Gamma(q, \Delta_n)_t^{jkjk}$. □

3.9.2 The discontinuous case

Now we come back to the general situation, where X is an Itô semimartingale satisfying (H). In this situation the integrated volatility is probably of less importance than in the continuous case because it captures only a part of the behavior of X and says nothing about jumps, but still many people wish to estimate it.

In this case things are more complicated. For example $B(2, \Delta_n)_t$ is no longer a consistent estimator for C_t, as seen in (3.49). However, we have constructed in Section 3.6 some consistent estimators:

Property 3.38 (Consistency) Assuming (H), the truncated variation $V^{jk}(\varpi, \alpha, \Delta_n)_t$ of (3.56), and the multipower variation $V^{jk}(r_1, \cdots, r_l, \Delta_n)_t$ of (3.60) converge in probability to C_t^{jk}, for all $\alpha > 0$ and $\varpi \in (0, \frac{1}{2})$ for the first one, and for all integer $l \geq 2$ and all $r_1, \cdots, r_l > 0$ with $r_1 + \cdots + r_l = 2$ for the second one.

This is nice enough, but the associated CLTs need some more assumption, as seen in Theorems 3.21 and 3.24. In Theorem 3.21 we need the test function f to be bounded when X jumps, and this precludes the use of multipower variations; hence in these notes we actually have a CLT for truncated powers only, as a consequence of Theorem 3.24 (we do have a CLT for multipower variations as well, under the same assumption $r < 1$ as below, but it is slightly too complicated to prove here; see however, Barndorff–Nielsen, Shephard, and Winkel (2006) for the Lévy case).

Property 3.39 (Asymptotic normality-1) Assume (H) and that $\int (\gamma(z)^r \wedge 1) \lambda(dz) < \infty$ for some $r \in [0, 1)$. If $\alpha > 0$ and $\varpi \in [\frac{1}{2(2-r)}, \frac{1}{2})$ then the $d \times d$-dimensional processes with components $\frac{1}{\sqrt{\Delta_n}} \left(V^{jk}(\varpi, \alpha, \Delta_n)_t - C_t^{jk} \right)$ converge in law to a $d \times d$-dimensional variable which, conditionally on the path of X over $[0, t]$, is centered normal with variance-covariance (Γ_t^{jkjk}) given by (3.171).

The comments made after property 3.34, about the need for a standardized version of the CLT, are in order here. We need a consistent estimator for Γ_t^{jkjk}.

Of course (3.172) does not any longer provide us with such an estimator, but
we can use the "truncated" version

$$\Gamma'(\varpi,\alpha;\Delta_n)_t^{jkjk} = \frac{1}{\Delta_n}\sum_{i=1}^{[t/\Delta_n]}\Big((\Delta_i^n X^j)^2(\Delta_{i+1}^n X^k)^2$$

$$+\Delta_i^n X^j \Delta_i^n X^k \Delta_{i+1}^n X^j \Delta_{i+1}^n X^k\Big)1_{\{\|\Delta_i^n X\|\le\alpha\Delta_n^\varpi,\|\Delta_{i+1}^n X\|\le\alpha\Delta_n^\varpi\}}.$$

By virtue of (3.59), we have $\Gamma'(\varpi,\alpha;\Delta_n)_t^{jkjk}\xrightarrow{\mathbb{P}}\Gamma_t^{jkjk}$. Then the same proof
as for Theorem 3.35 gives:

Theorem 3.40 Asymptotic normality-2 *Assume (H) and that $\int(\gamma(z)^r\wedge 1)\lambda(dz)<\infty$ for some $r\in[0,1)$. If $\alpha>0$ and $\varpi\in[\frac{1}{2(2-r)},\frac12)$. Then
and in restriction to the set $\{\Gamma_t^{jkjk}>0\}$, the variables*

$$\frac{1}{\sqrt{\Delta_n\,\Gamma'(\varpi,\alpha;\Delta_n)_t^{jkjk}}}\Big(V^{jk}(\varpi,\alpha;\Delta_n)_t^{jk}-C_t^{jk}\Big)$$

converge stably in law to an $\mathcal{N}(0,1)$ random variable independent of $\mathcal{F}$.

One could also use multipower variations to estimate Γ_t^{jkjk}.

Remark 3.41 The assumption $\int(\gamma(z)^r\wedge 1)\lambda(dz)<\infty$ for some $r\in[0,1)$
is quite restrictive, but so far there is no known estimator for C_t with a rate
$1/\sqrt{\Delta_n}$, if this fails. However, we do have a (worse) rate in almost every situation. Namely, if $\int(\gamma(z)^r\wedge 1)\lambda(dz)<\infty$ for some $r\in[0,2)$ then the
sequence $\frac{1}{\Delta_n^{(2-r)\varpi}}\Big(V^{jk}(\varpi,\alpha,\Delta_n)_t-C_t^{jk}\Big)$ is tight (or, bounded in probability), see Jacod (2008). This does not give a limit theorem, which we do not
know to exist, but it is a bound for the rate.

Note that the rate gets worse when r approaches 2, and does not exist when
$r=2$ (that is, with no special assumption on the jumps). This is because,
when $r\to 2$, the discontinuous part $\kappa(\delta)\star(\mu-\nu)$ of the process X gets
closer to a Brownian motion in some sense. To take a more specific example,
the symmetric stable processes of index $\alpha\in(0,2)$ (which satisfy the above
assumption for $r>\alpha$ and not for $r\le\alpha$) converge to the Brownian motion as
$\alpha\to 0$. The fact that the rate worsens when r increases is not surprising: it is
more and more difficult to distinguish between the continuous part X^c and the
discontinuous part when r approaches 2.

3.9.3 Estimation of the spot volatility

If one is so much interested in the integrated volatility it is probably because
one does not really know how to estimate the volatility c_t itself. In principle

the knowledge of the process C_t entails the knowledge of its derivative c_t as well. But practically speaking, with discrete observations, the estimation of c_t is quite another matter, and we are not going to give here a serious account on the subject, which still features many open problems.

Let us just say a few words. This is very much like a non-parametric problem for which one wants to estimate an unknown function f, for example the density of a sequence of n i.i.d. variables. In this case, and depending of course on the kind of criterion one chooses (one can consider the estimation error pointwise, or in some $\mathbb{L}^p$), the rate of convergence of the best estimators strongly depends on the smoothness of the estimated function f, although this smoothness is usually not known beforehand. More precisely, if f is "r-Hölder" (that is, Hölder with index r when $r \in (0,1]$, and if $r > 1$ it means that f is $[r]$ times differentiable and its $[r]$th derivative is $(r - [r])$-Hölder), typically the rate of convergence of the best non-parametric estimators is $n^{r/(1+2r)}$, always smaller than $n^{1/2}$.

Here, the unknown function is $t \mapsto c_t(\omega)$, for a given ω. If it were not dependent of ω and if X were simply (say, in the 1-dimensional case) $X_t = \int_0^t \sqrt{c_s}\, dW_s$, the observed increments $\Delta_i^n X$ would be independent, centered, with variances $\int_{(i-1)\Delta_n}^{i\Delta_n} c_s ds$. That is, we would have a genuine non-parametric problem and the rate of convergence of "good" estimators would indeed be $\Delta_n^{-r/(1+2r)}$ with r being the smoothness of the function c_t in the above sense. Now of course c_t is random, and possibly discontinuous, and X has also a drift and possibly jumps.

When σ_t is an Itô semimartingale (hypothesis (H)) and is further continuous, then the path of $t \mapsto c_t$ are a.s. Hölder with any index $r < 1/2$, and not Hölder with index $1/2$. And worse, σ_t can be discontinuous. Nevertheless, one expects estimators which converge at the rate $\Delta_n^{-1/4}$ (the rate when $r = 1/2$). This is what happens for the most elementary kernel estimators which are

$$U_t^{n,jk} = \frac{1}{k_n \Delta_n} \sum_{i \in I_n(t)} \Delta_i^n X^j \Delta_i^n X^k \, 1_{\{\|\Delta_i^n X\| \le \alpha \Delta_n^\varpi\}}, \qquad (3.178)$$

where α and ϖ are as before, and the sequence k_n of integers goes to ∞ with $\Delta_n k_n \to 0$ (as in (3.62)), and $I_n(t)$ is a set of k_n consecutive integers containing $[t/\Delta_n]$. This formula should of course be compared with (3.64). The "optimal" choice, as far as rates are concerned, consists in taking $k_n \sim 1/\sqrt{\Delta_n}$, and it is even possible to prove that the variables $\frac{1}{\Delta_n^{1/4}} \left(U_t^{n,jk} - c_t^{jk} \right)$ converge in law under appropriate conditions (this is *not* a functional CLT, and the limit behaves, as t varies, as a white noise).

3.10 Testing for jumps

This section is about testing for jumps. As before, we observe the process X at discrete times $0, \Delta_n, \cdots$ over a finite interval, and on the basis of these observations we want to decide whether the process has jumps or not. This is a crucial point for modeling purposes, and assuming that there are jumps brings out important mathematical and financial consequences (option pricing and hedging, portfolio optimization).

It would seem that a simple glance at the dataset should be sufficient to decide this issue, and this is correct when a "big" jump occurs. Such big jumps usually do not belong to the model itself, and either they are considered as breakdowns in the homogeneity of the model, or they are dealt with using different methods like risk management. On the other hand, a visual inspection of most time series in finance does not provide a clear evidence for either the presence or the absence of small or medium sized jumps.

Determining whether a process has jumps has been considered by a number of authors. Let us quote for example Aït–Sahalia (2004), Carr and Wu (2003), Barndorff–Nielsen and Shephard (2004), Jiang and Oomen (2008), Huang and Tauchen (2006) and Lee and Mykland (2008). Here we closely follow the approach initiated in Aït–Sahalia and Jacod (2009b).

3.10.1 Preliminary remarks

The present problem is 1-dimensional: if X jumps then at least one of its components jumps, so we can and will assume below that X is 1-dimensional (in the multidimensional case one can apply the forthcoming procedure to each of the components successively). We will also strengthen Hypothesis (H) in a rather innocuous way:

Assumption (K): We have (H); furthermore, with the notation $S = \inf(t : \Delta X_t \neq 0)$, we have:

(a) $C_t > 0$ when $t > 0$,

(b) $t \mapsto \int \kappa(\delta(\omega, t, z))\lambda(dz)$ is left-continuous with right limits on the set $(0, S(\omega)]$. $\qquad \square$

(a) above is a non-degeneracy condition for the continuous martingale part X^c. As for (b), it may appear as a strong assumption because it supposes that $z \mapsto \kappa(\delta(\omega, t, z))$ is λ-integrable if $t < S(\omega)$. However, one may remark that is "empty" on the set where $S(\omega) = 0$, that is where X has infinitely many jumps near the origin. It is also automatically implied by (H) when $\int (\gamma(z) \wedge$

$1)\lambda(dz) < \infty$. Moreover, if $F = \{(\omega, t, z) : \delta(\omega, t, z) \neq 0\}$, the variable $1_F \star \underline{\mu}_t$ is the number of jumps of X on the interval $(0, t]$, so by the very definition of S we have $1_F \star \underline{\mu}_S \leq 1$. Since F is predictable and $\underline{\nu}$ is the predictable compensator of $\underline{\mu}$, we have

$$\mathbb{E}\left(\int_0^S ds \int 1_F(s, z)\lambda(dz)\right) = \mathbb{E}(1_F \star \underline{\nu}_S) = \mathbb{E}(1_F \star \underline{\mu}_S) \leq 1.$$

Therefore outside a $\mathbb{P}$-null set we have $\int_0^S ds \int 1_F(s, z)\lambda(dz) < \infty$ and thus, upon modifying δ on a $\mathbb{P}$-null set, $z \mapsto \kappa(\delta(\omega, t, z))$ is λ-integrable if $t < S(\omega)$. So the condition (b) is really a very mild additional smoothness assumption, of the same nature as (b) of (H).

Before getting started we begin with a very important remark: Suppose that we are in the ideal situation where the path of $t \mapsto X_t(\omega)$ is fully observed over the time interval $[0, T]$. Then we know whether the path jumps or not, but we know nothing about other paths; so, exactly as for the integrated volatility in the previous section we can at the best make an inference about the outcome ω which is (partially) observed. But here there is even more: if we find that there are jumps we should conclude to a model with jumps, of course. But if we find no jump it does not really mean that the model should not have jumps, only that our particular observed path is continuous (and, if jumps occur like for a compound Poisson process, for instance, although the model should include jumps we always have a positive probability that a path does not jump over $[0, T]$).

Therefore, the problem which we really try to solve here is to decide, on the basis of the observations $X_{i\Delta_n}$, in which of the following two complementary sets the path which we have discretely observed falls:

$$\left.\begin{array}{l}\Omega_T^j = \{\omega : s \mapsto X_s(\omega) \text{ is discontinuous on } [0, T]\} \\ \Omega_T^c = \{\omega : s \mapsto X_s(\omega) \text{ is continuous on } [0, T]\}.\end{array}\right\} \quad (3.179)$$

3.10.2 The level and the power function of a test

In view of (3.179) we have two possibilities for the "null hypothesis," namely "there are no jumps" (that is, we are in Ω_T^c), and "there are jumps" (that is, we are in Ω_T^c).

Consider for example the first case where the null hypothesis is "no jump." We are thus going to construct a critical (rejection) region $C_{T,n}^c$ at stage n, which should depend only on the observations $X_0, X_{\Delta_n}, \cdots, X_{\Delta_n[t/\Delta_n]}$. We are not here in a completely standard situation: the problem is asymptotic, and the hypothesis involves the outcome ω.

In a classical asymptotic test problem, the unknown probability measure $\mathbb{P}_\theta$ depends on a parameter $\theta \in \Theta$ (Θ can be a functional space), and the null hypothesis corresponds to θ belonging to some subset Θ_0 of Θ. At stage n one constructs a critical region C_n. The asymptotic level is

$$\alpha = \sup_{\theta \in \Theta_0} \limsup_n \mathbb{P}_\theta(C_n), \qquad (3.180)$$

whereas the asymptotic power function is defined on $\Theta_1 = \Theta \backslash \Theta_0$ as

$$\beta(\theta) = \liminf_n \mathbb{P}_\theta(C_n).$$

Sometimes one exchanges the supremum and the $\limsup$ in (3.180), which is probably more sensible but in general impossible to achieve, in the sense that often $\sup_{\theta \in \Theta_0} \mathbb{P}_\theta(C_n) = 1$. Moreover, usually a prescribed level α_0 is given, and the aim is to construct C_n so that (3.180) holds with $\alpha \le \alpha_0$ (and if possible even, $\alpha = \alpha_0$, which generally increases the power function). Finally a "good" asymptotic critical region satisfies $\beta(\theta) = 1$ for all $\theta \in \Theta_1$ (we cannot hope for $\mathbb{P}_\theta(C_n) = 1$ if $\theta \in \Theta_1$ at any stage n).

In the present situation we have no genuine parameter (although the law of X itself can in a sense be considered as a parameter, or perhaps its characteristics (B, C, ν) can). Rather, the outcome ω, or at least the fact that it lies in Ω_t^c or not, can be considered as a kind of parameter. So, keeping the analogy with (3.180), we are led to consider the following definition for the asymptotic level of our critical region $C_{t,n}^c$:

$$\alpha_t^c = \sup\left(\limsup_{n \to \infty} \mathbb{P}(C_{t,n}^c \mid A) : A \in \mathcal{F}, A \subset \Omega_t^c \right). \qquad (3.181)$$

Here $\mathbb{P}(C_{t,n}^c \mid A)$ is the usual conditional probability with respect to the set A, *with the convention that it vanishes if* $\mathbb{P}(A) = 0$. If $\mathbb{P}(\Omega_t^c) = 0$ then $\alpha_t^c = 0$, which is a rather natural convention. It would seem better to define the level as the essential supremum $\alpha_t^{\prime c}$ (in ω) over Ω_t^c of $\limsup_n \mathbb{P}(C_{t,n}^c \mid \mathcal{F})$; the two notions are closely related and $\alpha_t^{\prime c} \ge \alpha_t^c$, but we cannot exclude a strict inequality here, whereas we have no way (so far) to handle $\alpha_t^{\prime c}$. Note that α_t^c features some kind of "uniformity" over all subsets $A \subset \Omega_t^c$, in the spirit of the uniformity in $\theta \in \Theta_0$ in (3.180).

As for the asymptotic power function, we define it as

$$\beta_t^c = \liminf_n \mathbb{P}(C_{t,n}^c \mid \mathcal{F}) \qquad (3.182)$$

and of course only the restriction of this "power function" (a random variable, indeed) to the alternative set Ω_t^j imports.

When on the opposite we take "there are jumps" as our null hypothesis, that is Ω_t^j, in a similar way we associate to the critical region $C_{t,n}^j$ the asymptotic level α_t^j and the power function β_t^j (simply exchange everywhere Ω_t^c and Ω_t^j).

3.10.3 The test statistics

First we recall the processes (3.13), except that here we do not specify the component since X is 1-dimensional:

$$B(p, \Delta_n)_p = \sum_{i=1}^{[t/\Delta_n]} |\Delta_i^n X|^p. \tag{3.183}$$

The test statistics we will use to construct the critical regions, for both null hypotheses, are the following ones:

$$\widehat{S}(p, k, \Delta_n)_t = \frac{B(p, k\Delta_n)_t}{B(p, \Delta_n)_t}, \tag{3.184}$$

where $k \geq 2$ is an integer, and $p > 3$. Note that the numerator is obtained by considering only the increments of X between successive intervals of length $k\Delta_n$. Then we have (and the assumption (K) is unnecessarily strong for this):

Theorem 3.42 Assume (K). For all $t > 0$ we have the following convergence:

$$\widehat{S}(p, k, \Delta_n)_t \xrightarrow{\mathbb{P}} \begin{cases} 1 & \text{on the set } \Omega_t^j \\ k^{p/2-1} & \text{on the set } \Omega_t^c. \end{cases} \tag{3.185}$$

Proof. By Theorem 3.8 the two variables $B(p, \Delta_n)_t$ and $B(p, k\Delta_n)_t$ both converge in probability to $\sum_{s \leq t} |\Delta X_s|^p$ (this is true as soon as $p > 2$, indeed), and the latter variable is strictly positive on the set Ω_t^j: hence the convergence on the set Ω_t^j is obvious.

When X has no jump, we can apply (3.55) to obtain that $\Delta_n^{1-p/2} B(p, \Delta_n)_t$, and of course $(k\Delta_n)^{1-p/2} B(p, k\Delta_n)_t$ as well, converge to $m_p \int_0^t c_s^{p/2} ds$, which by (H')-(a) is not 0. Then obviously we have the second limit in (3.185) when X is continuous.

This does not end the proof, however, except in the case $\Omega_t^c = \Omega$. It may happen that $0 < \mathbb{P}(\Omega_t^c) < 1$, so X is not (a.s.) continuous on $[0, t]$, but some of its paths are. However, suppose that we have proved the following:

$$X_s = X_s' \text{ for all } s \leq t, \text{ on the set } \Omega_t^c, \text{ where } X' \text{ satisfies (K) and is continuous.} \tag{3.186}$$

Then obviously $B(X, p, \Delta_n)_t = B(X', p, \Delta_n)_t$ and $B(X, p, k\Delta_n)_t = B(X', p, k\Delta_n)_t$ on the set Ω_t^c, and we get the result by applying (3.55) to X' instead of X.

The construction of X' involves the assumption (K)-(b). In fact we set

$$X_t' = X_0 + \int_0^t b_s' ds + \int_0^t \sigma_s dW_s \tag{3.187}$$

where $b'_t = b_t - b''_t$ and $b''_t = \left(\int \kappa(\delta(t,z))\lambda(dz)\right)1_{\{t<S\}}$. Then b'_t is adapted, with left-continuous and right limited paths, so X' satisfies (K), and it is continuous. Now suppose that we are in Ω^c_t. Then $t < S$, hence $\kappa'(\delta) \star \underline{\mu}_s = 0$ for all $s \le t$. As for the stochastic integral $\kappa(\delta) \star (\mu - \nu)_s$ for $s \le t$, we observe that in fact $\kappa(\delta)\star\underline{\nu}_s$ is well-defined as an ordinary integral and equals $\int_0^s b''_u du$; hence $\kappa(\delta) \star \underline{\mu}_s$ is also an ordinary integral, and since $s < S$ it actually vanishes: therefore we deduce that $X_s = X'_s$ if $s \le t$ and we are done. $\qquad\square$

We now turn to the central limit theorem. We introduce two processes, with $q > 0$ and $q \ge 2$ respectively:

$$A(q)_t = \int_0^t c_u^{q/2}du, \qquad D(q)_t = \sum_{s\le t}|\Delta X_s|^q(c_{s-} + c_s). \qquad (3.188)$$

Recalling that $d = 1$ here, these two processes are respectively the right side of (3.55) and the process $D^{11}(f)$ of (3.61), when we take the function $f(x) = |x|^q$. For this function we also write $|x|^q \star \mu$ instead of $f \star \mu$. In addition to the absolute moments m_p used before, we also set

$$m_{2p}(k) = \mathbb{E}\left(|\sqrt{k-1}\,U + V|^p\,|V|^p\right), \qquad (3.189)$$

where U and V are two independent $\mathcal{N}(0,1)$ variables. Finally we set

$$M(p,k) = \frac{1}{m_p^2}\left(k^{p-2}(1+k)(m_{2p} - m_p^2) - 2k^{p/2-1}(m_{2p}(k) - k^{p/2}m_p^2)\right).$$
$$(3.190)$$

When $p = 4$ we get $M(p,k) = 16k(2k^2 - k - 1)/35$, and in particular $M(4,2) = \frac{32}{7}$.

Theorem 3.43 *Assume (K), and let $t > 0$, $p > 3$ and $k \le 2$.*

(a) In restriction to the set Ω^j_t, the variables $\frac{1}{\sqrt{\Delta_n}}\left(\widehat{S}(p,k,\Delta_n)_t - 1\right)$ converge stably in law to a variable $S(p,k)^j_t$ which, conditionally on $\mathcal{F}$, is centered with variance

$$\widetilde{\mathbb{E}}\left((S(p,k)^j_t)^2 \mid \mathcal{F}\right) = \frac{(k-1)p^2}{2}\frac{D(2p-2)_t}{(|x|^p \star \mu)^2_t}. \qquad (3.191)$$

Moreover if the processes σ and X have no common jumps, the variable $S(p,k)^j_t$ is $\mathcal{F}$-conditionally Gaussian.

(b) In restriction to the set Ω^c_t, the variables $\frac{1}{\sqrt{\Delta_n}}\left(\widehat{S}(p,k,\Delta_n)_t - 2\right)$ converge stably in law to a variable $S(p,k)^c_t$ which, conditionally on $\mathcal{F}$, is centered Gaussian with variance

$$\widetilde{\mathbb{E}}\left((S(p,k)^c_t)^2 \mid \mathcal{F}\right) = M(p,k)\frac{A(2p)_t}{(A(p)_t)^2}. \qquad (3.192)$$

We have already encountered and explained after Theorem 3.35 the notion of stable convergence in law in restriction to a subset of Ω. It is also worth noticing that the conditional variances (3.191) and (3.193), although of course random, are more or less behaving in time like $1/t$.

Proof. a) Write $U_n = \frac{1}{\sqrt{\Delta_n}} (B(p, \Delta_n)_t - |x|^p \star \mu_t)$ and $V_n = \frac{1}{\sqrt{\Delta_n}} (B(p, k\Delta_n)_t - |x|^p \star \mu_t)$. Then

$$\widehat{S}(p, k, \Delta_n)_t - 1 = \frac{B(p, k\Delta_n)_t}{B(p, \Delta_n)_t} - 1 = \sqrt{\Delta_n} \frac{V_n - U_n}{B(p, \Delta_n)_t}.$$

Since $p > 3$, Corollary 3.28 yields that $V_n - U_n$ converges stably in law to $Z'(f)_t - Z(f)_t$, and the result readily follows from (3.132), from the fact that $B(p, \Delta_n)_t \xrightarrow{\mathbb{P}} |x|^p \star \mu_t$, and from the last claim in Lemma 3.26.

b) Exactly as for Theorem 3.42 it is enough to prove the result for the process X' of (3.187). This amounts to assuming that the process X itself is continuous, so $\Omega_t^c = \Omega$. Write $U_n' = \frac{1}{\sqrt{\Delta_n}} (\Delta_n^{1-p/2} B(p, \Delta_n) - A(p)_t)$ and $V_n' = \frac{1}{\sqrt{\Delta_n}} (\Delta_n^{1-p/2} B(p, k\Delta_n) - k^{p/2-1} A(p)_t)$. Then

$$\widehat{S}(p, k, \Delta_n)_t - k^{p/2-1} = \frac{B(p, k\Delta_n)_t}{B(p, \Delta_n)_t} - k^{p/2-1} = \sqrt{\Delta_n} \frac{V_n' - k^{p/2-1} U_n'}{\Delta_n^{1-p/2} B(p, \Delta_n)_t}.$$

Now we consider the 2-dimensional function f whose components are $|x_1|^p + \cdots + |x_k|^p$ and $|x_1 + \cdots + x_k|^p$. Recalling (3.100), the two components of $\frac{1}{\sqrt{\Delta_n}} \left(V''(f, k, \Delta_n)_t - \frac{1}{k} \int_0^t \rho_{\sigma_u}^{\otimes k}(f) \right)$ are respectively $U_n' + U_n''$ and V_n', where

$$U_n'' = \sqrt{\Delta_n} \sum_{k[t/k\Delta_n] < i \leq [t/\Delta_n]} |\Delta_i^n X|^p.$$

Obviously $U_n'' \to 0$, hence Theorem 3.23 implies that the pair (U_n', V_n') converges stably in law to a vector which is $\mathcal{F}$-conditionally centered Gaussian, with $\mathcal{F}$-conditional covariance $MA(2p)_t$, where the entries of M are $M_{11} = 1 - m_p^2/m_{2p}$ and $M_{12} = M_{21} = (m_{2p}(k) - m_p^2)/m_{2p}$ and $M_{22} = k^{p-1}(1 - m_p^2/m_{2p})$. Therefore, using also the fact that $\Delta_n^{1-p/2} B(p, \Delta_n)_t \xrightarrow{\mathbb{P}} A(p)_t$, we readily deduce the result. $\qquad\square$

Exactly as for estimating the volatility, (see Theorem 3.35), this CLT is useless in practice and one has to standardize the test statistics so as to obtain a usable result. As usual, the standardization is done by dividing by the square-root of any consistent estimators for the conditional variances in (3.191) and (3.192). For the first one we can again use the fact that $B(p, \Delta_n)_t \xrightarrow{\mathbb{P}} |x|^p \star \mu_t$, plus the following version of (3.64), which by Theorem 3.13 converges to $D(q)_t$ if

$q > 2$:

$$D(q, \varpi, \alpha, \Delta_n)_t = \qquad\qquad (3.193)$$

$$\frac{1}{k_n \Delta_n} \sum_{i=1+k_n}^{[t/\Delta_n]-k_n} |\Delta_i^n X|^q \sum_{j: j \neq i, |j-i| \leq k_n} |\Delta_j^n X|^2 1_{\{|\Delta_j^n X| \leq \alpha \Delta_n^\varpi\}}.$$

where $\alpha > 0$ and $\varpi \in (0, \frac{1}{2})$, and k_n satisfies (3.62).

For the right side of (3.192) we can use estimators of $A(p)_t$, as provided in Theorem 3.11; for example, with ϖ and α as above, we can take

$$A(p, \varpi, \alpha, \Delta_n)_t = \Delta_n^{1-p/2} \sum_{i=1}^{[t/\Delta_n]} |\Delta_i^n X|^p 1_{\{|\Delta_i^n X| \leq \alpha \Delta_n^\varpi\}}, \qquad (3.194)$$

which converges to $A(p)_t$ when X is continuous (and also when X has jumps, in restriction to Ω_t^c, as in the proof of Theorem 3.42). Also the variables $\Delta_n^{1-p/2} B(p, \Delta_n)_t$ converge to $A(p)_t$ on Ω_t^c. Hence the next result follows from Theorem 3.43, with exactly the same proof as for Theorem 3.35:

Theorem 3.44 *Assume (K) and let $t > 0$, $p > 3$ and $k \geq 2$.*

(a) In restriction to the set Ω_t^j, the variables $\frac{1}{\sqrt{\Gamma^j(t,n)}} (\widehat{S}(p, k, \Delta_n)_t - 1)$, where

$$\Gamma^j(t, n) = \frac{\Delta_n(k - 1)p^2 \, D(2p - 2, \varpi, \alpha, \Delta_n)_t}{(B(p, \Delta_n)_t)^2} \qquad (3.195)$$

converge stably in law to a variable which, conditionally on $\mathcal{F}$, is centered with variance 1, and which additionally is $\mathcal{F}$-conditionally normal if the processes σ and X have no common jumps.

(b) In restriction to the set Ω_t^c, the variables $\frac{1}{\sqrt{\Gamma^c(t,n)}} (\widehat{S}(p, k, \Delta_n)_t - k^{p/2-1})$, where either

$$\Gamma^c(t, n) = \frac{\Delta_n M(p, k) \, A(2p, \varpi, \alpha, \Delta_n)_t}{(A(p, \varpi, \alpha, \Delta_n)_t)^2}, \qquad (3.196)$$

or

$$\Gamma^c(t, n) = \frac{M(p, k) \, B(2p, \Delta_n)_t}{(B(p, \Delta_n)_t)^2}, \qquad (3.197)$$

converge stably in law to a variable which, conditionally on $\mathcal{F}$, is $\mathcal{N}(0, 1)$.

We will see later that, although both choice of $\Gamma^c(t, n)$ are asymptotically equivalent for determining the level of our tests, it is no longer the case for the power function: the second choice (3.197) should *never* prevail.

3.10.4 Null hypothesis = no jump

We now use the preceding results to construct actual tests, either for the null hypothesis that there are no jumps, or for the null hypothesis that jumps are present. We start with the first one here. The null hypothesis is then "Ω_t^c," and we are going to construct a critical (rejection) region $C_{t,n}^c$ for it. In view of Theorem 3.42 it is natural to take a region of the form

$$C_{t,n}^c = \{\widehat{S}(p,k,\Delta_n)_t < \gamma_{t,n}^c\} \tag{3.198}$$

for some sequence $\gamma_{t,n}^c > 0$, possibly even a random sequence. What we want, though, is to achieve an asymptotic level α prescribed in advance. For this we need to introduce the α-quantile of $N(0,1)$, that is $\mathbb{P}(U > z_\alpha) = \alpha$ where U is $N(0,1)$.

Theorem 3.45 *Assume (K), and let $t > 0$, $p > 3$ and $k \geq 2$. For any pre-scribed level $\alpha \in (0,1)$ we define the critical region $C_{t,n}^j$ by (3.198), with*

$$\gamma_{t,n}^c = k^{p/2-1} - z_\alpha \sqrt{\Gamma^c(t,n)}, \tag{3.199}$$

where $\Gamma^c(t,n)$ is given either by (3.196) or by (3.197).

(a) The asymptotic level α_t^c for testing the null hypothesis of "no jump" is not bigger than α and equal to α when $\mathbb{P}(\Omega_t^c) > 0$; we even have $\mathbb{P}(C_{t,n}^c \mid A) \to \alpha$ for all $A \subset \Omega_t^c$ with $\mathbb{P}(A) > 0$.

(b) The asymptotic power function β_t^c is a.s. equal to 1 on the complement Ω_t^j if we use (3.196) for $\Gamma^c(t,n)$, with $\varpi \in (\frac{1}{2} - \frac{1}{p}, \frac{1}{2})$, but this fails in general if we use (3.197).

Proof. For (a) it is enough to prove that if $A \in \Omega_t^c$ has $\mathbb{P}(A) > 0$, then $\mathbb{P}(C_{t,n}^c \mid A) \to \alpha$. Let $U_n = \frac{1}{\sqrt{\Gamma^c(t,n)}} (\widehat{S}(p,k,\Delta_n)_t - k^{p/2-1})$. We know that this variable converges stably in law, as $n \to \infty$, and in restriction to Ω_t^c, to an $\mathcal{N}(0,1)$ variable U independent of $\mathcal{F}$. Therefore for A as above we have

$$\mathbb{P}(C_{t,n}^c \cap A) = \mathbb{P}(\{U_n \leq -z_\alpha\} \cap A) \to \mathbb{P}(A)\mathbb{P}(U \leq -z_\alpha) = \alpha\,\mathbb{P}(A),$$

and the result follows.

For (b) we can assume $\mathbb{P}(\Omega_t^j) > 0$, otherwise there is nothing to prove. Theorem 3.42 implies that $\widehat{S}(p,k,\Delta_n)_t \xrightarrow{\mathbb{P}} 1$ on Ω_t^j. If we use the version (3.197) for $\Gamma^c(t,n)$, then Theorem 3.8 implies that $\Gamma^c(t,n)$ converges in probability to a positive finite variable, on Ω_t^j again. Hence on this set the variable U_n converges in probability to a limiting variable U (equal in fact to $(1 - k^{p/2})$-$|x|^p \star \mu_t / \sqrt{M(p,k)|x|^{2p} \star \mu_t}$). In general this variable is not a.s. smaller than $-z_\alpha$ on Ω_t^j, and thus the power function is not equal to 1 on this set.

On the opposite, suppose that we have chosen the version (3.196), with $\varpi \in (\frac{1}{2} - \frac{1}{p}, \frac{1}{2})$. Suppose also that

$$\frac{\Delta_n A(2p, \varpi, \alpha, \Delta_n)t}{(A(p, \varpi, \alpha, \Delta_n)_t)^2} \xrightarrow{\mathbb{P}} 0. \tag{3.200}$$

This means that $\Gamma^c(t, n) \xrightarrow{\mathbb{P}} 0$. Since $1 - k^{p/2-1} < 0$ we deduce that $U_n \xrightarrow{\mathbb{P}} -\infty$ on the set Ω_t^k. Then

$$\mathbb{P}(C_{t,n}^c \cap \Omega_t^j) = \mathbb{P}(\{U_n \leq -z_\alpha\} \cap \Omega_t^j) \to \mathbb{P}(\Omega_t^j).$$

This trivially implies $\mathbb{P}(C_{t,n}^c \mid \mathcal{F}) \xrightarrow{\mathbb{P}} 1$ on the set Ω_t^j.

It remains to prove (3.200), and for this it is no restriction to assume (SH). The reader will observe that when X is continuous this trivially follows from Theorem 3.11, but unfortunately we need this property on Ω_t^j. With the notation of (3.72), one easily check that for all $B > 0$:

$$\left| \Delta_n^{1-p/2} \sum_{i=1}^{[t/\Delta_n]} |\Delta_i^n X|^p 1_{\{|\Delta_i^n X| \leq \sqrt{B\Delta_n}\}} - \Delta_n \sum_{i=1}^{[t/\Delta_n]} |\beta_i^n|^p \right| \leq K Z_n(B), \tag{3.201}$$

where

$$Z_n(B) =$$
$$\Delta_n \sum_{i=1}^{[t/\Delta_n]} \left(|\beta_i^n|^p 1_{\{|\beta_i^n| > \sqrt{B}/2\}} B^{p/2-1}(|\chi_i^n|^2 \wedge B) + |\beta_i^n|^{p-1}(|\chi_i^n| \wedge \sqrt{B}) \right).$$

Equation (3.73) and Bienaymé–Tchebycheff, plus (3.75) and Cauchy-Schwarz give us

$$\limsup_n \mathbb{E}(Z_n(B)) \leq \frac{Kt}{B} \tag{3.202}$$

On the other hand, we know that $\Delta_n^{1-p/2} \sum_{i=1}^{[t/\Delta_n]} |\beta_i^n|^p \xrightarrow{\mathbb{P}} A(p)_t$. Combining this with the above estimates and (3.201), we obtain for all $\eta, B > 0$:

$$\mathbb{P}\left(\Delta_n^{1-p/2} \sum_{i=1}^{[t/\Delta_n]} |\Delta_i^n X|^p 1_{\{|\Delta_i^n X| \leq \sqrt{B\Delta_n}\}} < A(p)_t - Z_n(B) - \eta \right) \to 0.$$

Now, for any $B \geq 1$ we have $\alpha \Delta_n^\varpi > \sqrt{B\Delta_n}$ for all n large enough because $\varpi < 1/2$. Therefore we a fortiori have

$$\mathbb{P}\left(A(p, \varpi, \alpha, \Delta_n)_t < A(p)_t - Z_n(B) - \eta \right) \to 0.$$

Now (3.202) imply that $\lim_{B \to \infty} \limsup_n \mathbb{P}(Z_n(B) > \eta) = 0$, hence

$$\mathbb{P}\left(A(p, \varpi, \alpha, \Delta_n)_t < A(p)_t - 2\eta \right) \to 0.$$

Since $A(p)_t > 0$ a.s., we finally deduce

$$\mathbb{P}\left(A(p, \varpi, \alpha, \Delta_n)_t < \frac{A(p)_t}{2} \right) \; \rightarrow \; 0. \tag{3.203}$$

At this stage, the proof of (3.200) is straightforward: since

$$|\Delta_i^n X|^{2p} \leq \alpha^p \Delta_n^{p\varpi} |\Delta_i^n X|^p$$

when $|\Delta_i^n X| \leq \alpha \Delta_n^\varpi$, one deduces from (3.194) that

$$\frac{\Delta_n \, A(2p, \varpi, \alpha, \Delta_n)_t}{A(p, \varpi, \alpha, \Delta_n)_t^2} \; \leq \; \frac{K \Delta_n^{p\varpi + 1 - p/2}}{A(p, \varpi, \alpha, \Delta_n)_t}.$$

Since $p\varpi + 1 - p/2 > 0$, the result readily follows from (3.203).   □

3.10.5 Null hypothesis = there are jumps

In a second case, we set the null hypothesis to be that there are jumps, that is "Ω_t^j." Then we take a critical region of the form

$$C_{t,n}^j = \{ \widehat{S}(p, k, \Delta_n)_t > \gamma_{t,n}^j \}. \tag{3.204}$$

for some sequence $\gamma_{t,n}^j > 0$. As in (3.181) and (3.182), the asymptotic level and power functions are

$$\alpha_t^j \;\; = \;\; \sup \left(\limsup_n \, \mathbb{P}(C_{t,n}^j \mid A) : \; A \in \mathcal{F}, A \subset \Omega_t^j \right),$$

$$\beta_t^d \;\; = \;\; \liminf_n \mathbb{P}(C_{t,n}^j \mid \mathcal{F}).$$

Theorem 3.46   *Assume (K), and let $t > 0$, $p > 3$ and $k \geq 2$. Define $\Gamma^j(t, n)$ by (3.195), and let $\alpha \in (0, 1)$ be a prescribed level.*

(i) With the critical region $C_{t,n}^j$ given by (3.204), with

$$\gamma_{t,n}^j \;=\; 1 + \frac{1}{\sqrt{\alpha}} \, \sqrt{\Gamma^j(t, n)}, \tag{3.205}$$

the asymptotic level α_t^j for testing the null hypothesis of "jumps" is not bigger than α.

(ii) With the critical region $C_{t,n}^j$ given by (3.204), with

$$\gamma_{t,n}^j \;=\; 1 + z_\alpha \, \sqrt{\Gamma^j(t, n)}, \tag{3.206}$$

and if further the two processes X and σ do not jump at the same times, the asymptotic level α_t^j for testing the null hypothesis of "jumps" is not bigger than α, and equals to α when $\mathbb{P}(\Omega_t^j) > 0$; we even have $\mathbb{P}(C_{t,n}^j \mid A) \rightarrow \alpha$ for all $A \subset \Omega_t^j$ with $\mathbb{P}(A) > 0$.

(iii) In both cases the asymptotic power function β_t^j is a.s. equal to 1 on the complement Ω_t^c of Ω_t^j.

Since $z_\alpha < 1/\sqrt{\alpha}$ the critical region is larger with the version (3.206) than with the version (3.205). Hence, even though asymptotically the two power functions are equal, at any stage n the power is bigger with (3.206) than with (3.205), so one should use (3.206) whenever possible (however, when there are jumps, it is usually the case that the volatility jumps together with X).

Proof. We know that the variables $U_n = \frac{1}{\sqrt{\Gamma^c(t,n)}}\left(\widehat{S}(p,k,\Delta_n)_t - 1\right)$ converges stably in law, as $n \to \infty$, and in restriction to Ω_t^j, to a variable which conditionally on $\mathcal{F}$ is centered with variance 1, and is further $\mathcal{N}(0,1)$ if X and σ do not jump at the same times. Then $\mathbb{P}(U > 1/\sqrt{\alpha}) \leq \alpha$, and also $\mathbb{P}(U > z_\alpha) = \alpha$ in the latter case, the two statements (i) and (ii) follow exactly as in Theorem 3.45.

For (iii) we can assume $\mathbb{P}(\Omega_t^c) > 0$, otherwise there is nothing to prove. Then in restriction to Ω_t^c the statistics $\widehat{S}(p,k,\Delta_n)_t$ converge in probability to $k^{p/2-1} > 1$. Moreover, on this set again, both $D(2p-2,\varpi,\alpha,\Delta_n)_t$ and $B(p,\Delta_n)_t$ are the same as if they were computed on the basis of the continuous process X' of (3.187). Therefore, by virtue of Theorems 3.10 and 3.13 we have that $\Delta_n^{1-p/2}B(p,\Delta_n)_t \xrightarrow{\mathbb{P}} A(p)_t$ and $\Delta_n^{2-p}D(2p-2,\varpi,\alpha,\Delta_n)_t \xrightarrow{\mathbb{P}} \frac{2m_{2p-2}}{m_{2p}}A(2p)_t$ on Ω_t^c. Since by (H') we have $A(p)_t > 0$ it follows that $\Gamma^j(t,n) \xrightarrow{\mathbb{P}} 0$ on Ω_t^j. Therefore we have $U_n \xrightarrow{\mathbb{P}} +\infty$ on Ω_t^c, and as in Theorem 3.45 we conclude that $\mathbb{P}(C_{n,t}^j \cap \Omega_t^c) \to \mathbb{P}(\Omega_t^c)$, hence $\mathbb{P}(C_{t,n}^c \mid \mathcal{F}) \xrightarrow{\mathbb{P}} 1$ on the set Ω_t^c. $\square$

3.11 Testing for common jumps

This section is again about jumps. We suppose here that our underlying process is multidimensional, and that it has jumps, and we want to check whether any two components have jumps occurring at the same time. Below, we follow Jacod and Todorov (2009).

3.11.1 Preliminary remarks

Clearly the problem at hand is 2-dimensional, since in the multidimensional situation one can perform the tests below for any pair of components. So below we assume that $X = (X^1, X^2)$ is 2-dimensional. Exactly as in the previous section, we need a slightly stronger assumption than (H):

Assumption (K'): We have (H); furthermore with the notation $\tau = \inf(t :$ $\Delta X_t^1 \Delta X_t^2 \neq 0)$ (the infimum of all common jump times) and $\Gamma = \{(\omega, t, x) : \delta^1(\omega, t, x) \delta^2(\omega, t, x) \neq 0\}$, we have

(a) $C_t \neq 0$ when $t > 0$

(b) $t \mapsto \int \kappa(\delta(\omega, t, z)) 1_\Gamma(\omega, t, z) \lambda(dz)$ is left-continuous with right limits on the interval $(0, \tau(\omega)]$.   □

(a) above is again a non-degeneracy assumption for X^c, similar in the 2-dimensional case to (a) of (K). As for (b) here, we can state the same remarks as for (b) of (K): it is "empty" on the set $\{\tau = 0\}$, that is where X^1 and X^2 have infinitely many common jumps near the origin. It is implied by (H) when $\int (\gamma(z) \wedge 1) \lambda(dz) < \infty$. Moreover, in all generality, and outside a $\mathbb{P}$-null set, $z \mapsto \kappa(\delta(\omega, t, z))$ is λ-integrable if $t < \tau(\omega)$. So again (b) is a very mild additional smoothness assumption, of the same nature as (b) of (H).

Next, and again as in the previous section, what we can really test on the basis of discrete observations of X over a finite time interval $[0, T]$ is whether the two paths $t \mapsto X_t^1(\omega)$ and $t \mapsto X_t^2(\omega)$ have common jump times or not. That is, we can (hopefully) decide in which one of the following two disjoint subsets of Ω we are:

$$\left. \begin{aligned} \Omega_T^{cj} &= \{\omega : s \mapsto X_s^1(\omega) \text{ and } s \mapsto X_s^2(\omega) \text{ have common} \\ &\qquad\qquad\qquad\qquad\qquad\qquad \text{jumps on } [0, T]\} \\ \Omega_T^{dj} &= \{\omega : \text{ both } s \mapsto X_s^1(\omega) \text{ and } s \mapsto X_s^2(\omega) \text{ have jumps,} \\ &\qquad\qquad\qquad \text{but they have no common jump, on } [0, T]\}. \end{aligned} \right\} \quad (3.207)$$

The union of these two sets is not Ω, but their global complement is

$$\Omega_T^{cc} = \{\omega : \text{ at least one of } X^1(\omega) \text{ and } X^2(\omega) \text{ is continuous on } [0, T]\}. \tag{3.208}$$

All three sets above may have a positive probability. However, we can first perform the tests developed in the previous section, separately on both components, to decide whether both of them jump. Then in this case only, it makes sense to test for joint jumps. That is, we suppose that this preliminary testing has been done and that we have decided that we are *not* in Ω_T^{cc}.

At this point we again have two possible null hypotheses, namely "common jumps" (we are in Ω_T^{cj}) and "disjoint jumps" (we are in Ω_T^{dj}). Exactly as in the previous section we construct at stage n a critical region $C_{T,n}^{cj}$ for the null Ω_T^{cj}, and a critical region $C_{T,n}^{dj}$ for the null Ω_T^{dj}. In the first case, the asymptotic level and power function are respectively

$$\left. \begin{aligned} \alpha_T^{cj} &= \sup \left(\limsup_n \ \mathbb{P}(C_{T,n}^{cj} \mid A) : \ A \in \mathcal{F}, A \subset \Omega_T^{cj} \right), \\ \beta_T^{cj} &= \liminf_n \ \mathbb{P}(C_{T,n}^{cj} \mid \mathcal{F}). \end{aligned} \right\} \quad (3.209)$$

In the second case, they are

$$\left. \begin{array}{l} \alpha_T^{dj} = \sup \left(\limsup_n \; \mathbb{P}(C_{T,n}^{dj} \mid A) : \; A \in \mathcal{F}, A \subset \Omega_T^{dj} \right), \\ \beta_T^{dj} = \liminf_n \; \mathbb{P}(C_{T,n}^{dj} \mid \mathcal{F}). \end{array} \right\} \qquad (3.210)$$

3.11.2 The test statistics

Three functions will be used in the construction of our test statistics (here $x = (x^1, x^2) \in \mathbb{R}^2$):

$$f(x) = (x^1 x^2)^2, \qquad g_1(x) = (x^1)^4, \qquad g_2(x) = (x^2)^4. \qquad (3.211)$$

Then, with $k \geq 2$ being an integer fixed throughout, we put

$$\left. \begin{array}{ll} \widehat{T}^{cj}(k, \Delta_n)_t & = \; \dfrac{V(f, k\Delta_n)_t}{V(f, \Delta_n)_t}, \\[2mm] \widehat{T}^{dj}(\Delta_n)_t & = \; \dfrac{V(f, \Delta_n)_t}{\sqrt{V(g_1, \Delta_n)_t \, V(g_2, \Delta_n)_t}}. \end{array} \right\} \qquad (3.212)$$

These statistics will be used to construct respectively, the two critical regions $C_{t,n}^{cj}$ and $C_{t,n}^{dj}$. Unlike for simply testing jumps, we have to resort to two different statistics to deal with our two cases.

We have now to determine the asymptotic behavior of these statistics, deriving an LLN and a CLT for each one. To prepare for this we need to introduce a number of processes to come in the limiting variables. First we set

$$F_t = \int_0^t (c_s^{11} c_s^{22} + 2(c_s^{12})^2) \, ds. \qquad (3.213)$$

Second, on the extended space described in Subsection 3.8.1 and with the notation S_p, R_p and R'_p of this subsection (recall (3.128), here R_p and R'_p are 2-dimensional), we set

$$\left. \begin{array}{ll} D_t & = \; \sum_{p:S_p \leq t} \left((\Delta X_{S_p}^1 R_p^2))^2 + (\Delta X_{S_p}^2 R_p^1)^2 \right) \\[2mm] D'_t & = \; \sum_{p:S_p \leq t} \left((\Delta X_{S_p}^1 R_p'^2))^2 + (\Delta X_{S_p}^2 R_p'^1)^2 \right). \end{array} \right\} \qquad (3.214)$$

If we are on the set Ω_T^{dj} it turns out (via an elementary calculation) that in fact $D_t = \overline{Z}(f)_t/2$ and $D'_t = \overline{Z}'(f)_t/2$ for all $t \leq T$.

Theorem 3.47 *Assume (K').*

(a) We have

$$\widehat{T}^{cj}(k, \Delta_n)_t \; \overset{\mathbb{P}}{\longrightarrow} \; 1 \qquad \text{on the set } \Omega_t^{cj}, \qquad (3.215)$$

and $\widehat{T}^{cj}(k, \Delta_n)_t$ converges stably in law, in restriction to the set Ω_t^{dj}, to

$$T^{cj}(k) \; = \; \frac{D'_t + kF_t}{D_t + F_t} \qquad (3.216)$$

which is a.s. different from 1.

(b) We have

$$\dot{T}^{dj}(\Delta_n)_t \xrightarrow{\mathbb{P}} \begin{cases} f \star \mu_t / \sqrt{(g_1 \star \mu_t)(g_2 \star \mu_t)} > 0 \text{ on the set } \Omega_t^{cj} \\ 0 \qquad\qquad\qquad\qquad\qquad \text{on the set } \Omega_t^{dj}. \end{cases} \qquad (3.217)$$

The second part of (a) is a kind of LLN because it concerns the behavior of $\widehat{T}^{cj}(k, \Delta_n)_t$ without centering or normalization, but it is also a kind of CLT.

Proof. On both sets Ω_t^{cj} and Ω_t^{dj} both components of X jumps before t, so $g_1 \star \mu_t > 0$ and $g_2 \star \mu_t > 0$, and also $f \star \mu_t > 0$ on Ω_t^{cj}. Then all claims except the second one in (a) are trivial consequences of Theorem 3.8.

Let us now turn to the behavior of $\widehat{T}^{cj}(k, \Delta_n)_t$ on Ω_t^{dj}. If we make the additional assumption that X^1 and X^2 *never* jump at the same time, then the stable convergence in law towards $T^{cj}(k)$, as defined by (3.216), is a trivial consequence of Theorem 3.29 and of the remark which follows (3.214). Moreover, the $\mathcal{F}$-conditional law of the pair of variable (D_t, D_t'), in restriction to Ω_t^{dj}, clearly admits a density, hence $\mathbb{P}(\Omega_t^{dj} \cap \{T^{cj} = 1\}) = 0$ and we have the last claim of (a).

Now, exactly as in Theorem 3.42, this is not quite enough for proving our claim, since it may happen that both Ω_t^{dj} and Ω_t^{cj} have positive probability. However, suppose that

$$\left. \begin{array}{l} X_s = X_s' \text{ for all } s \le t, \text{ on the set } \Omega_t^{dj}, \text{ where } X' \text{ satisfies } (K'), \\ \text{and the two components } X'^1 \text{ and } X'^2 \text{ never jump at the same} \\ \text{times.} \end{array} \right\} \qquad (3.218)$$

Then the above argument applied for X' instead of X yields the result.

The construction of X' involves (K')-(b). We set $b_t' = b_t - b_t''$, where the process $b_t'' = \left(\int \kappa(\delta(t, z)) 1_\Gamma(t, z) \lambda(dz) \right) 1_{\{t < \tau\}}$ is well-defined and left-continuous with right limits everywhere. Set also $\delta' = \delta 1_{\Gamma^c}$. Then the process

$$X_t' = X_0 + \int_0^t b_s'' de + \int_0^t \sigma_s dW_s + \kappa(\delta') \star (\mu - \nu)_t - \kappa'(\delta') \star \underline{\mu}_t$$

satisfies all requirements in (3.218) (we should be more careful here; it satisfies (H), except for one fact, namely we do not know whether $t \mapsto \delta'(\omega, t, z)$ is left-continuous with right limits; however, this particular property plays no role in the proof of Theorem 3.29, so the proof is nevertheless complete.) $\square$

Now we turn to the associated CLTs. Here again we need to complement the notation. Set

$$\overline{D}_t = \frac{1}{2}\sum_{s\leq t}\Big((\Delta X_s^1)^2(c_{s-}^{22}+c_s^{22})+(\Delta X_s^2)^2(c_{s-}^{11}+c_s^{11})\Big), \qquad (3.219)$$

$$\overline{D}_t' = 2\sum_{s\leq t}(\Delta X_s^1\Delta X_s^2)^2\Big((\Delta X_s^2)^2(c_{s-}^{11}+c_s^{11})+ \qquad (3.220)$$

$$(\Delta X_s^1)^2(c_{s-}^{22}+c_s^{22})+2\Delta X_s^1\Delta X_s^2(c_{s-}^{12}+c_s^{12})\Big),$$

In other words, with the notation (3.129) and (3.130), we have $\overline{D}_s = \frac{1}{2}\,\overline{C}(f)_s$ for all $s\leq t$ on the set Ω_t^{dj}, and $\overline{D}' = \frac{1}{2}\,C(f,f)$ everywhere.

Theorem 3.48 *Assume (K').*

(a) In restriction to the set Ω_T^{cj} the sequence $\frac{1}{\sqrt{\Delta_n}}(\widehat{T}^{cj}(k,\Delta_n)_t - 1)$ converges stably in law to a variable $T'^{cj}(k)$ which, conditionally on $\mathcal{F}$, is centered with variance

$$\widetilde{\mathbb{E}}\Big((T'^{cj}(k))^2 \mid \mathcal{F}\Big) = (k-1)\frac{\overline{D}_t'}{(f\star\mu_t)^2}, \qquad (3.221)$$

and is even Gaussian conditionally on $\mathcal{F}$ if the processes X and σ have no common jumps.

(b) In restriction to the set Ω_T^{dj} the sequences $\frac{1}{\Delta_n}\widehat{T}^{dj}(D_n)$ converges stably in law to the positive variable $T^{dj} = (D_t+F_t)/\sqrt{(g_1\star\mu_t)(g_2\star\mu_t)}$ which, conditionally on $\mathcal{F}$, satisfies

$$\widetilde{\mathbb{E}}(T^{dj}\mid\mathcal{F}) = \frac{\overline{D}_t+F_t}{\sqrt{(g_1\star\mu_t)(g_2\star\mu_t)}}. \qquad (3.222)$$

Proof. a) This is the very same proof as for (a) of Theorem 3.43: we write $U_n = \frac{1}{\sqrt{\Delta_n}}\,(V(f,\Delta_n)_t - f\star\mu_t)$ and $V_n = \frac{1}{\sqrt{\Delta_n}}\,(V(f,k\Delta_n)_t - f\star\mu_t)$ and observe that

$$\widehat{T}^{cj}(k,\Delta_n)_t - 1 = \sqrt{\Delta_n}\,\frac{V_n - U_n}{V(f,\Delta_n)_t}.$$

Then we conclude using Corollary 3.28, plus (3.220) and the remark that follows, in exactly the same way.

b) Exactly as in the previous theorem, we can replace X by a process X' satisfying (3.218), or equivalently we can assume that the two X^1 and X^2 never jump at the same times. Then the result immediately derives from Theorem 3.29.   □

Finally, we need to standardize our statistics, and thus to find consistent estimators for the conditional variance in (3.221), and conditional first moment in

(3.222). For the variables $f \star \mu_t$, $g_1 \star \mu_t$ and $g_2 \star \mu_t$ we can use $V(f, \Delta_n)_t$, $V(g_1, \Delta_n)_t$ and $V(g_2, \Delta_n)_t$ respectively. For F_t we can use the truncated powers (see Theorem 3.11; we have to be careful here, because X is discontinuous, whereas f is a polynomial of degree 4; so we choose the version given by (3.58)–(3.59)): we choose $\varpi \in (0, \frac{1}{2})$ and $\alpha > 0$, and we set

$$A(\varpi, \alpha, \Delta_n)_t = \frac{1}{\Delta_n} \sum_{i=1}^{[t/\Delta_n]} \left(|\Delta_i^n X^1|^2 |\Delta_{i+1}^n X^2|^2 \right. \tag{3.223}$$

$$\left. + 2\Delta_i^n X^1 \Delta_i^n X^2 \Delta_{i+1}^n X^1 \Delta_{i+1}^n X^2 \right) 1_{\{\|\Delta_i^n X\| \le \alpha \Delta_n^\varpi\}, \|\Delta_{i+1}^n X\| \le \alpha \Delta_n^\varpi\}}.$$

Finally, by virtue of Theorem 3.13, we can estimate $\overline{D}_t$ and $\overline{D}'_t$ by the following variables, where in addition to ϖ and α we have chosen a sequence k_n of integers satisfying (3.62):

$$\overline{D}(\varpi, \alpha, \Delta_n)_t = \frac{1}{2k_n \Delta_n} \sum_{i=1+k_n}^{[t/\Delta_n]-k_n} 1_{\{\|\Delta_i^n X\| > \alpha \Delta_n^\varpi\}} \tag{3.224}$$

$$\sum_{j \in I_n(i)} \left((\Delta_i^n X^1)^2 (\Delta_j^n X^2)^2 + (\Delta_i^n X^2)^2 (\Delta_j^n X^1)^2 \right) 1_{\{\|\Delta_j^n X\| \le \alpha \Delta_n^\varpi\}},$$

$$\overline{D}'(\varpi, \alpha, \Delta_n)_t = \frac{2}{k_n \Delta_n} \sum_{i=1+k_n}^{[t/\Delta_n]-k_n} \sum_{j \in I_n(i)} (\Delta_i^n X^1)^2 (\Delta_i^n X^2)^2 \tag{3.225}$$

$$\left(\Delta_i^n X^1 \Delta_j^n X^2 + \Delta_i^n X^2 \Delta_j^n X^1 \right)^2 1_{\{\|\Delta_j^n X\| \le \alpha \Delta_n^\varpi\}}.$$

Then we have the following trivial consequence of Theorem 3.48:

Theorem 3.49 *Assume (K').*

(a) In restriction to the set Ω_t^{cj}, the variables $\dfrac{1}{\sqrt{\Gamma^{cj}(t,n)}} \left(\widehat{T}^{cj}(k, \Delta_n)_t - 1 \right)$, *where*

$$\Gamma^{cj}(n, t) = \frac{\Delta_n (k-1) \overline{D}'(\varpi, \alpha, \Delta_n)_t}{(V(f, \Delta_n)_t)^2}, \tag{3.226}$$

converge stably in law to a variable which, conditionally on $\mathcal{F}$, is centered with variance 1, and which additionally is $\mathcal{F}$-conditionally Gaussian if the processes X and σ have no common jumps.

(b) In restriction to the set Ω_t^{dj}, the variables $\dfrac{1}{\Gamma^{dj}(t,n)} \widehat{T}^{dj}(\Delta_n)_t$, where

$$\Gamma^{dj}(t, n) = \frac{\Delta_n (\overline{D}(\varpi, \alpha, \Delta_n)_t + A(\varpi, \alpha, \Delta_n)_t)}{\sqrt{V(g_1, \Delta_n)_t \, V(g_2, \Delta_n)_t}}, \tag{3.227}$$

converge stably in law, in restriction to the set Ω_T^{dj}, to a positive variable which, conditionally on $\mathcal{F}$, has expectation 1.

3.11.3 Null hypothesis = common jumps

Now we are in a position to construct the critical regions we are looking for. We start with the null hypothesis being "there are common jumps," that is, we are in Ω_t^{cj}. In view of Theorem 3.47 it is natural to take a critical region of the form

$$C_{t,n}^{cj} = \{|\widehat{T}^{cj}(k, \Delta_n) - 1| \geq \gamma_{t,n}^{cj}\}. \tag{3.228}$$

For $\alpha \in (0, 1)$ we denote by z_α' the symmetric α-quantile of an $\mathcal{N}(0, 1)$ variable U, that is $\mathbb{P}(|U| \geq z_\alpha) = \alpha$.

Theorem 3.50 *Assume (K'), and let $t > 0$ and $k \geq 2$. Define $\Gamma^{cj}(t, n)$ by (3.226), and let $\alpha \in (0, 1)$ be a prescribed level.*

(i) With the critical region $C_{t,n}^{cj}$ given by (3.228), with

$$\gamma_{t,n}^{cj} = 1 + \frac{1}{\sqrt{\alpha}} \sqrt{\Gamma^{cj}(t, n)}, \tag{3.229}$$

the asymptotic level α_t^{cj} for testing the null hypothesis of "common jumps" is not bigger than α.

(ii) With the critical region $C_{t,n}^{cj}$ given by (3.228), with

$$\gamma_{t,n}^{cj} = 1 + z_\alpha' \sqrt{\Gamma^{cj}(t, n)}, \tag{3.230}$$

and if further the two processes X and σ do not jump at the same times, the asymptotic level α_t^j for testing the null hypothesis of "common jumps" is not bigger than α, and equals to α when $\mathbb{P}(\Omega_t^{cj}) > 0$; we even have $\mathbb{P}(C_{t,n}^{cj} \mid A) \to \alpha$ for all $A \subset \Omega_t^{cj}$ with $\mathbb{P}(A) > 0$.

(iii) In both cases the asymptotic power function β_t^{cj} is a.s. equal to 1 on the set Ω_t^{dj}.

Again $z_\alpha' < 1/\sqrt{\alpha}$, so whenever possible one should choose the critical region defined by (3.230).

Proof. In view of the previous theorem, (i) and (ii) are proved exactly as in Theorem 3.46 for example. For (iii), we observe first that, in view of Theorem 3.47(-a), the variable $\widehat{T}^{cj}(k, \Delta_n)_t$ converges stably in law to $T^j(k) - 1$, which a.s. noon vanishing. On the other hand we have $\overline{D}'(\varpi, \alpha, \Delta_n)_t \xrightarrow{\mathbb{P}} \overline{D}_t$ everywhere and $V(f, \Delta_n)_t \xrightarrow{\mathbb{P}} f \star \mu_t > 0$ on Ω_t^{dj}, hence $\Gamma^{cj}(t, n) \xrightarrow{\mathbb{P}} 0$ on Ω_t^{dj}. That is, $\gamma_{t,n}^{dj} \xrightarrow{\mathbb{P}} 1$ on this set, and this implies the result. $\qquad\square$

3.11.4 Null hypothesis = no common jumps

In a second case, we set the null hypothesis to be "no common jumps," that is, we are in Ω_t^{dj}. We take a critical region of the form

$$C_{t,n}^{dj} = \{\widehat{T}^{dj}(\delta_n)_t \geq \gamma_{t,n}^{dj}\}. \tag{3.231}$$

Theorem 3.51 *Assume (K'), and let $t > 0$. Define $\Gamma^{dj}(t,n)$ by (3.227), and let $\alpha \in (0,1)$ be a prescribed level.*

(a) With the critical region $C_{t,n}^{dj}$ given by (3.231), with

$$\gamma_{t,n}^{dj} = \frac{\Gamma^{dj}(t,n)}{\alpha}, \tag{3.232}$$

the asymptotic level α_t^{dj} for testing the null hypothesis of "common jumps" is not bigger than α.

(b) The asymptotic power function β_t^{dj} is a.s. equal to 1 on the set Ω_t^{cj}.

Proof. The variables $U_n = \widehat{T}^{dj}(\Delta_n)_t / \Gamma^{dj}(t,n)$ converge stably in law to a limit $U > 0$ having $\widetilde{\mathbb{E}}(U \mid \mathcal{F}) = 1$, in restriction to Ω_t^{dj}. Hence if $A \in \mathcal{F}$ is included into Ω_t^{dj} we have

$$\alpha\mathbb{P}(A) \geq \mathbb{P}(A \cap \{U \geq \frac{1}{\alpha}\}) \leq \limsup_n \mathbb{P}(A \cap \{U_n \geq \frac{1}{\alpha}\}) = \mathbb{P}(C_{t,n}^{dj} \cap A).$$

and (a) readily follows.

For (b) one observes that $\Gamma^{dj}(t,n) \xrightarrow{\mathbb{P}} 0$ o Ω_t^{cj}, whereas on this set $\widehat{T}^{dj}(\Delta_n)_t$ converge to a positive variable by Theorem 3.47, hence $U_n \xrightarrow{\mathbb{P}} +\infty$ on Ω_t^{cj} and the result becomes obvious.   □

3.12 The Blumenthal–Getoor index

In the last section of these notes we wish to use the observation already made that if the path $s \mapsto X_s(\omega)$ is fully observed on $[0,t]$, then one also knows the processes

$$H(r)_t = \sum_{s \leq T} \|\Delta X_s\|^r \tag{3.233}$$

for any $r \geq 0$ (with the convention $0^0 = 0$). This is not especially interesting, and it has no predictive value about the laws of the jumps, *but for one point*: we know for which r's we have $H(r)_t < \infty$. We will call the following random number the *Blumenthal–Getoor index up to time T*:

$$R_T = \inf(r : H(r)_T < \infty). \tag{3.234}$$

This is increasing with T, and $0 \leq R_T \leq 2$ always, and we have $H(r)_T = \infty$ for all $r < R_T$, and $H(r)_T < \infty$ for all $r > R_T$, whereas $H(R_T)_T$ may be finite or infinite (except that $H(2)_T < \infty$ always again). We will consider in this section the "estimation" of $R_T(\omega)$, in the same sense as we estimated the integrated volatility above. Clearly, R_T is the maximum of the Blumenthal–Getoor indices $R_t^i(\omega)$ for all components X^i, so this problem is essentially 1-dimensional, and in the sequel we assume X to be 1-dimensional.

To understand why this index is important let us consider the special situation where $X = X' + Y$, where X' is a *continuous* Itô semimartingale and Y is a Lévy process. Of course $H(r)_t = \sum_{s \leq t} |\Delta Y_s|^r$, and the Lévy property yields the following equivalence, which holds for all $t > 0$:

$$H(r)_t < \infty \text{ a.s.} \quad \Longleftrightarrow \quad \int (|x|^r \wedge 1) F(dx) < \infty, \qquad (3.235)$$

where F is the Lévy measure of Y. It is also characterized in the following way: writing

$$x > 0 \quad \mapsto \quad \overline{F}(x) = F([-x, x]^c), \qquad (3.236)$$

for its (symmetrical) tail function (more generally, $\overline{H}(x) = H([-x, x]^c)$ for any measure H on $\mathbb{R}$), then the Blumenthal–Getoor index β is the unique number in $[0, 2]$ such that for all $\varepsilon > 0$ we have

$$\lim_{x \to 0} x^{\beta + \varepsilon} \overline{F}(x) = 0, \qquad \limsup_{x \to 0} x^{\beta - \varepsilon} \overline{F}(x) = \infty. \qquad (3.237)$$

Unfortunately, the "lim sup" above is usually not a limit.

If Y is a stable process, its Blumenthal–Getoor index is the stability index, which is probably the most important parameter in the law of Y (the other three, a scaling constant and a drift and a skewness parameter, are also of course important but not as much; note that here the scaling and skewness parameters can also be in principle estimated exactly, but the drift cannot). More generally, for a Lévy process the observation over $[0, t]$ does not allow to infer the Lévy measure, but one can infer in principle the Blumenthal–Getoor index, which indeed is about the only information which is known about F: this is an essential characteristic of the process, for modeling purposes, for example.

So we are going to estimate R_T. Unfortunately, to do this we need some very restrictive assumptions. We start with the simple case when X is a symmetric stable process plus possibly a Brownian motion. Then we state the results when X is a "general" Itô semimartingale, and we come back to the Lévy process with a slightly different problem. The proofs are mainly given at the end of the subsection, following Aït–Sahalia and Jacod (2009a).

To end these introductory remarks, let us introduce the processes which we will use here. The Blumenthal–Getoor index is related to the behavior of "small jumps," which correspond in our discrete observation scheme to the increments

$\Delta_i^n X$ that are "small"; however, we also have the continuous part X', which plays a preponderant role in those small increments. So we need to "truncate" from below the increments to get rid of the process X'. This leads us to take, as in the previous sections, two numbers $\varpi \in (0, \frac{1}{2})$ and $\alpha > 0$ and, this time, to consider increments bigger than $\alpha \Delta_n^\varpi$ only. We could *a priori* take a "general" test function, but it turns out that simply counting those not too small increments is enough. Hence we set for $u > 0$

$$U(u, \Delta_n)_t = \sum_{i=1}^{[t/\Delta_n]} 1_{\{|\Delta_i^n X| > u\}}, \qquad (3.238)$$

and use in fact the processes $U(\alpha \Delta_n^\varpi, \Delta_n)$ or $U(\alpha \Delta_n^\varpi, 2\Delta_n)$. On the basis of these we introduce two different statistics, which will in fact be our estimators. Below, we choose $\varpi \in (0, \frac{1}{2})$ and two numbers $\alpha' > \alpha > 0$, and we set

$$\widehat{\beta}_n(t, \varpi, \alpha, \alpha') = \frac{\log(U(\alpha \Delta_n^\varpi, \Delta_n)_t / U(\alpha' \Delta_n^\varpi, \Delta_n)_t)}{\log(\alpha'/\alpha)}. \qquad (3.239)$$

Other estimators of the same kind, but involving increments of sizes Δ_n and $k\Delta_n$ and the same cut-off level $\alpha \Delta_n^\varpi$ are possible, in the spirit of the previous two sections, but the results are essentially the same, and in particular the rates.

3.12.1 The stable process case

In this subsection, Y denotes a symmetric stable process with index $\beta \in (0, 2)$. This is a Lévy process whose characteristic function is of the form $\mathbb{E}(e^{iuY_t}) = \exp(-ct|u|^\beta)$ for some constant c, and the Lévy measure is of the form

$$F(dx) = \frac{A\beta}{2|x|^{1+\beta}} dx, \qquad \text{hence} \quad \overline{F}(x) := \frac{A}{x^\beta} \text{ for } x > 0 \qquad (3.240)$$

for some scale parameter $A > 0$, related of course with the c above. The law of Y_1 has an even density g and a tail function $\overline{G}(x) = \mathbb{P}(|Y_1| > x)$ satisfying, as $x \to \infty$ (see Zolotarev (1986), Theorems 2.4.2 and Corollary 2 of Theorem 2.5.1):

$$g(x) = \frac{A\beta}{2|x|^{1+\beta}} + O\left(\frac{1}{x^{1+2\beta}}\right), \quad \overline{G}(x) = \frac{A}{x^\beta} + O\left(\frac{1}{x^{2\beta}}\right). \qquad (3.241)$$

Let us begin with the case $X = Y$. In this case, $U(\alpha \Delta_n^\varpi, \Delta_n)_t$ is the sum of $[t/\Delta_n]$ i.i.d. $\{0, 1\}$-valued variables which, by the scaling property of Y (namely, Y_t has the same law as $t^{1/\beta}Y_1$) have the probability $\overline{G}(\alpha \Delta_n^{\varpi-1/\beta})$ of taking the value 1. Then the following result is completely elementary to prove (it will follows from the more general results proved later):

Theorem 3.52 *Assume that $X = Y$. Let $0 < \alpha < \alpha'$ and $\varpi > 0$ and $t > 0$.*

a) If $\varpi < \frac{1}{\beta}$, the estimators $\widehat{\beta}_n(t, \varpi, \alpha, \alpha')$ converge in probability to β.

b) If $\varpi < \frac{2}{3\beta}$, we have

$$\frac{1}{\Delta_n^{\varpi\beta/2}} \left(\widehat{\beta}_n(t, \varpi, \alpha, \alpha') - \beta\right) \xrightarrow{\mathcal{L}} \mathcal{N}\left(0, \frac{\alpha'^\beta - \alpha^\beta}{At(\log(\alpha'/\alpha))^2}\right), \qquad (3.242)$$

The reader will observe that we do not necessarily assume $\varpi < \frac{1}{2}$, because there is no Brownian part, and the restriction over ϖ will be explained later.

These estimators are not rate-efficient. To see that, one can recall from Aït–Sahalia and Jacod (2008) that the model in which one observes the values $X_{i\Delta_n}$ for $i\Delta_n \le t$ is regular, and its Fisher information (for estimating β) is asymptotically of the form

$$I_n \sim \frac{\log(1/\Delta_n)}{\Delta_n} C_\beta \, t \qquad (3.243)$$

for some constant C_β. So rate-efficient estimators would be such the rate of convergence is $\Delta_n^{-1/2}\sqrt{\log(1/\Delta_n)}$, instead of $\Delta_n^{-\varpi\beta/2}$ found here. With the "optimal" choice of ϖ, namely smaller than but as close as possible to $2/3\beta$, we get a rate which is "almost" $\Delta_n^{-1/3}$ only. In addition β is unknown, so a conservative choice of ϖ is $\varpi = 1/3$ and the rate in (3.242) become $\Delta_n^{-\beta/6}$, quite far from the optimal rate.

The reason for this (huge) lack of optimality is that our method results in discarding a large part of the data. In the absence of a Brownian component this is of course unnecessary, but as seen immediately below the situation is different if a Brownian motion is present.

Now we turn to the situation where $X_t = bt + \sigma W_t + Y_t$, with Y as above.

Theorem 3.53 *Assume that $X_t = bt + \sigma W_t + Y_t$. Let $0 < \alpha < \alpha'$ and $\varpi > 0$ and $t > 0$.*

a) If $\varpi < \frac{1}{2}$, the estimators $\widehat{\beta}_n(t, \varpi, \alpha, \alpha')$ converge in probability to β.

b) If $\varpi < \frac{1}{2+\beta}$, we have (3.242).

These estimators are again not rate-efficient. In fact, one can extend Aït–Sahalia and Jacod (2008) to obtain that in the present situation the Fisher information for estimating β, at stage n, satisfies

$$I_n \sim \frac{A \, (\log(1/\Delta_n))^{2-\beta/2}}{\sigma^\beta \, \Delta_n^{\beta/2}} C_\beta' \, t \qquad (3.244)$$

for another constant C'_β. The discrepancy here comes from the fact that we have absolutely not used the fact that we exactly know the law of X. If one considers the (partial) statistical model where we observe only the increments bigger than $\alpha \Delta_n^\varpi$, the Fisher information becomes

$$I_n \sim \frac{A(1-\varpi)^2 (\log(1/\Delta))^2}{\alpha^\beta \Delta_n^{\varpi\beta}} C'''_\beta \, t. \tag{3.245}$$

This still gives a faster rate than in the theorem, but by a (negligible) factor of $\log(1/\Delta_n)$. There is, however, the restriction $\varpi < \frac{1}{2+\beta}$, which does not appear in (3.245).

3.12.2 The general result

The title of this subsection is rather misleading, since the solution of the problem requires quite strong assumptions. Unfortunately, this seems consubstantial to this problem, as one can see in the next subsection in a much simpler situation. We will assume that X is an Itô semimartingale, with conditions on σ_t even weaker than in (H) or (H'), but the assumptions on the Lévy measures $F_t = F_{\omega,t}(dx)$ of (3.6) are rather strong:

Assumption (L): The process X is a 1-dimensional Itô semimartingale, with b_t and σ_t locally bounded. There are three (non-random) numbers $\beta \in (0,2)$ and $\beta' \in [0, \beta/2)$ and $\gamma > 0$, and a locally bounded process $L_t \geq 1$, such that we have for all (ω, t):

$$F_t = F'_t + F''_t, \tag{3.246}$$

where

a) F'_t has the form

$$F'_t(dx) = \frac{1 + |x|^\gamma f(t,x)}{|x|^{1+\beta}} \left(a_t^{(+)} 1_{\{0 < x \leq z_t\}} + a_t^{(-)} 1_{\{-z_t \leq x < 0\}} \right) dx, \tag{3.247}$$

for some predictable non-negative processes $a_t^{(+)}, a_t^{(-)}, z_t$ and some predictable function $f(\omega, t, x)$, satisfying:

$$\frac{1}{L_t} \leq z_t \leq 1, \; a_t^{(+)} + a_t^{(-)} \leq L_t, \; 1 + |x| f(t,x) \geq 0, \; |f(t,x)| \leq L_t. \tag{3.248}$$

b) F''_t is a measure which is singular with respect to F'_t and satisfies

$$\int_{\mathbb{R}} (|x|^{\beta'} \wedge 1) F''_t(dx) \leq L_t. \tag{3.249}$$
□

This assumption implies in particular that $(|x|^r \wedge 1) * \nu_T$ is finite for all $r > \beta$, and infinite for all $r < \beta$ on the set $\{\overline{A}_T > 0\}$, where we have put

$$A_t = \frac{a_t^{(+)} + a_t^{(-)}}{\beta}, \qquad \overline{A}_t = \int_0^t A_s ds. \tag{3.250}$$

Therefore the Blumenthal–Getoor index R_T satisfies

$$R_T \leq \beta, \qquad A_T > 0 \;\Rightarrow\; R_T = \beta. \tag{3.251}$$

A stable process with index β satisfies (L), and this assumption really means that the small jumps of X behave like the small jumps of such a stable process, on the time set $\{t : A_t > 0\}$, whereas on the complement of this set they are "negligible" in comparison with the small jumps of the stable process. The solution of an equation like (3.52) satisfies (L) when Z is a stable process, and (much) more generally when Z is a Lévy process which itself satisfies (L) (like for example the sum of two stable processes plus a Wiener process, or of a stable process plus a compound Poisson process plus a Wiener process).

Theorem 3.54   *Let $0 < \alpha < \alpha'$ and $0 < \varpi < \frac{1}{2}$ and $t > 0$. Assume (L).*

a) We have $\widehat{\beta}'_n(t, \varpi, \alpha, \varpi') \overset{\mathbb{P}}{\longrightarrow} \beta$ on the set $\{\overline{A}_t > 0\}$.

b) If further $\beta' \in [0, \frac{\beta}{2+\beta})$ and $\gamma > \beta/2$, and if $\varpi < \frac{1}{2+\beta} \bigwedge \frac{1}{3\beta}$, in restriction to the set $\{\overline{A}_t > 0\}$ we have

$$\frac{1}{\Delta_n^{\varpi\beta/2}} \left(\widehat{\beta}_n(t, \varpi, \alpha, \alpha') - \beta\right) \overset{\mathcal{L}-s}{\longrightarrow} U, \tag{3.252}$$

where U is defined on an extension of the original space and is $\mathcal{F}$-conditionally centered Gaussian, with variance:

$$\widetilde{\mathbb{E}}(U^2 \mid \mathcal{F}) = \frac{\alpha'^\beta - \alpha^\beta}{\overline{A}_t (\log(\alpha'/\alpha))^2}. \tag{3.253}$$

At this point, we can replace the variances in (3.253) by estimators for them, to get a standardized CLT:

Theorem 3.55   *Under (L) and the assumptions of (b) of the previous theorem, the variables*

$$\frac{\log(\alpha'/\alpha)}{\sqrt{\frac{1}{U(\alpha'\Delta_n^{\varpi}, \Delta_n)_t} - \frac{1}{U(\alpha\Delta_n^{\varpi}, \Delta_n)_t}}} \left(\widehat{\beta}_n(t, \varpi, \alpha, \alpha') - \beta\right) \tag{3.254}$$

converge stably in law, in restriction to the set $\{\overline{A}_t > 0\}$, to a standard normal variable independent of $\mathcal{F}$.

Despite the strong assumptions, these estimators are thus reasonably good for estimating β on the (random) set $\{\overline{A}_t > 0\}$ on which the Blumenthal–Getoor index is actually β; unfortunately, we do not know how they behave on the complement of this set.

3.12.3 Coming back to Lévy processes

Let us restrict the setting of the previous subsection by assuming that X is a Lévy process, that is an Itô semimartingale with characteristics of the form (3.4). (L) may hold or not, but when it does we have $A_t = at$ for some constant $a > 0$, and so the two theorems 3.54 and 3.55 hold on the whole of Ω.

What is important here, though, is that those results probably fail, even in this simple setting, when (L) fails. We cannot really show this in a serious mathematical way, but we can see on a closely related and even simpler problem why strong assumptions are needed on the Lévy measure. This is what we are going to explain now.

The model is as follows: instead of observing the increments of X, we observe all its jumps (between 0 and t) whose sizes are bigger than $\alpha \Delta_n^\varpi$. A priori, this should give us more information on the Lévy measure than the original observation scheme.

In this setting the estimators (3.239) have no meaning, but may be replaced by

$$\overline{\beta}_n(t, \varpi, \alpha, \alpha') = \frac{\log(\overline{U}(\alpha \Delta_n^\varpi)_t / \overline{U}(\alpha' \Delta_n^\varpi)_t)}{\log(\alpha'/\alpha)}, \qquad (3.255)$$

where

$$\overline{U}(u)_t = \sum_{s \le t} 1_{\{|\Delta X_s| > u\}}.$$

Lemma 3.56 Let $\gamma_n(\alpha) = \overline{F}(\alpha \Delta_n^\varpi)$ and

$$M^n(\alpha)_t = \frac{1}{\sqrt{\gamma_n(\alpha)}} \left(\overline{U}(\alpha \Delta_n^\varpi)_t - \gamma_n(\alpha)\, t \right). \qquad (3.256)$$

a) The processes $M^n(\alpha)$ converge stably in law to a standard Wiener process, independent of $\mathcal{F}$.

b) If $\alpha < \alpha'$ all limit points of the sequence $\frac{\gamma_n(\alpha')}{\gamma_n(\alpha)}$ are in $[0, 1]$. If further this sequence converges to γ then the pairs $(M^n(\alpha), M^n(\alpha'))$ of processes converge stably in law to a process $(\overline{W}, \overline{W}')$, independent of X, where $\overline{W}$ and $\overline{W}'$ are correlated standard Wiener processes with correlation $\sqrt{\gamma}$.

Proof. The processes $M^n = M^n(\alpha)$ and $M'^n = M^n(\alpha')$ are Lévy processes and martingales, with jumps going uniformly to 0, and with predictable brackets

$$\langle M^n, M^n \rangle_t = \langle M'^n, M'^n \rangle_t = t, \qquad \langle M^n, M'^n \rangle_t = \frac{\sqrt{\gamma_n(\alpha')}}{\sqrt{\gamma_n(\alpha)}} t.$$

Observe also that $\alpha' \Delta_n^{\varpi} \geq \alpha \Delta_n^{\varpi}$, hence $\gamma_n(\alpha') \leq \gamma_n(\alpha)$. All results are then obvious (see Jacod and Shiryaev (2003), Chapter VII). □

Theorem 3.57 *If $\alpha' > \alpha$ and if $\frac{\gamma_n(\alpha')}{\gamma_n(\alpha)} \to \gamma \in [0, 1]$, then the sequence*

$$\sqrt{\gamma_n(\alpha')} \left(\overline{B}_n(t, \varpi, \alpha, \alpha') - \frac{\log(\gamma_n(\alpha)/\gamma_n(\alpha'))}{\log(\alpha'/\alpha)} \right) \tag{3.257}$$

converges stably in law to a variable, independent of $\mathcal{F}$ and with the law $\mathcal{N}\left(0, \frac{1-\gamma}{t(\log(\alpha'/\alpha))^2}\right)$.

This result is a simple consequence of the previous lemma, and its proof is the same as for Theorem 3.54 and is thus omitted.

This result shows that in general, that is without specific assumptions on F, the situation is hopeless. These estimators are not even consistent for estimating the Blumenthal–Getoor index β of F, because of a bias, and to remove the bias we have to know the ratio $\gamma_n(\alpha')/\gamma_n(\alpha)$ (or at least its asymptotic behavior in a precise way), and further there is no CLT if this ratio does not converge (a fact which we *a priori* do not know, of course).

The major difficulty comes from the possible erratic behavior of $\overline{F}$ near 0. Indeed, we have (3.237), but there are Lévy measures F satisfying this, and such that for any $r \in (0, \beta)$ we have $x_n^r \overline{F}(x_n) \to 0$ for a sequence $x_n \to 0$ (depending on r, of course). If F is such, the sequence $\gamma_n(\varpi, \alpha')/\gamma_n(\varpi, \alpha)$ may have the whole of $[0, 1]$ as limit points, depending on the parameter values ϖ, α, α', and in a completely uncontrolled way for the statistician.

So we need some additional assumption on F. For the consistency a relatively weak assumption is enough, for the asymptotic normality, we need in fact (L). Recall that under (L) we have necessarily $\overline{A}_t = at$ for some $a \geq 0$, in the Lévy case.

Theorem 3.58 *a) If the tail function $\overline{F}$ is regularly varying at 0, with index $\beta \in (0, 2)$ we have $\overline{\beta}_n(t, \varpi, \alpha, \alpha') \xrightarrow{\mathbb{P}} \beta$.*

b) Under (L) with $a >$, the sequence $\frac{1}{\Delta_n^{\varpi \beta}} \left(\overline{\beta}_n(t, \varpi, \alpha, \alpha') - \beta \right)$ converges stably in law to a variable, independent of $\mathcal{F}$ and with law $\mathcal{N}\left(0, \frac{\alpha'^\beta - \alpha^\beta}{t\,\alpha'^\beta\,(\log(\alpha'/\alpha))^2}\right)$.

Proof. The regular variation implies that $\gamma_n(\alpha) \to \infty$ and $\gamma_n(\alpha')/\gamma_n(\alpha) \to (\alpha/\alpha')^\beta$, so the previous theorem yields (a). (L) clearly implies

$$\sqrt{\gamma_n(\alpha)}\,\frac{\log(\gamma_n(\alpha)/\gamma_n(\alpha'))}{\log(\alpha'/\alpha)} \to \beta,$$

and also $\gamma_n(\varpi, \alpha) \sim a/\alpha^\beta \Delta_n^{\varpi\beta}$, so (b) follows again from the previous theorem.   □

It may of course happen that the regular variation or (L) fails and nevertheless the conclusions of the previous theorem hold for a particular choice of the parameters ϖ, α, α'. But in view of Theorem 3.57 and of the previous proof these assumptions are *necessary* if we want those conclusions to hold *for all choices* of ϖ, α, α'.

Now if we come back to the original problem, for which only increments of X are observed. We have Theorem 3.54 whose part (b) looks like (b) above; however there are restrictions on ϖ, unlike in Theorem 3.58. This is because an increment $\Delta_i^n X$ with size bigger than $\alpha\Delta_n^\varpi$ is, with a high probability, almost equal to a "large" jump only when the cutoff level is higher than a typical Brownian increments, implying at least $\varpi < 1/2$.

3.12.4 Estimates

As all the results in these notes, Theorem 3.54 is "local" in time. So by our usual localization procedure we may assume that (L) is replaced by the stronger assumption below:

Assumption (SL): We have (L), and the process L_t is in fact a constant L, and further $|b_t| \leq L$ and $|\sigma_t| \leq L$ and $|X_t| \leq L$.   □

Before proceeding, we mention a number of elementary consequences of (SL), to be used many times. First, F_t is supported by the interval $[-2L, 2L]$. This and (3.247) and (3.249) imply that for all $u, v, x, y > 0$ we have

$$\left.\begin{array}{l}
\overline{F}_t''(x) \leq \frac{K}{x^{\beta'}}, \quad \left|\overline{F}_t(x) - \frac{A_t}{x^\beta}\right| \leq \frac{K}{x^{(\beta-\gamma)\vee\beta'}}, \quad \overline{F}_t(x) \leq \frac{K}{x^\beta}, \\[2mm]
\int_{\{|x|\leq u\}} x^2 F_t(dx) \leq Ku^{2-\beta}, \qquad \int |x| F_t''(dx) \leq K \\[2mm]
\int_{\{|x|>u\}} (|x|^v \wedge 1) F_t(dx) \leq \begin{cases} K_v & \text{if } v > \beta \\ K_v \log(1/u) & \text{if } v = \beta \\ K_v u^{v-\beta} & \text{if } v < \beta, \end{cases} \\[4mm]
\overline{F}_t(x) - \overline{F}_t(x+y) \leq \frac{K}{x^\beta}\left(1 \wedge \frac{y}{x} + x^{(\beta-\beta')\wedge\gamma}\right).
\end{array}\right\} \quad (3.258)$$

In the next lemma, Y is a symmetric stable process with Lévy measure (3.240), and for $\eta \in (0,1)$ we set

$$Y(\eta)_t = \sum_{s \leq t} \Delta Y_s 1_{\{|\Delta Y_s| > \eta)\}}, \qquad Y'(\eta) = Y - Y(\eta). \qquad (3.259)$$

Lemma 3.59 *There is a constant K depending on (A, β), such that for all $s, \eta \in (0,1)$,*

$$\mathbb{P}(|Y'(\eta)_s| > \eta/2) \leq K s^{4/3}/\eta^{4\beta/3}. \qquad (3.260)$$

Proof. We use the notation (3.240) and (3.241). Set $\eta' = \eta/2$ and $\theta = s\overline{F}(\eta') = sA/\eta'^\beta$, and consider the processes $Y' = Y'(\eta')$ and $Z_t = \sum_{r \leq t} 1_{\{|\Delta Y_r| > \eta'\}}$. Introduce also the sets

$$D = \{|Y_s| > \eta'\}, \quad D' = \{|Y'_s| > \eta'\}, \quad B = \{Z_s = 1\}, \quad B' = \{Z_s = 0\}.$$

It is of course enough to prove the result for s/η^β small, so below we assume $\theta \leq 1/2$.

By scaling, $\mathbb{P}(D) = \overline{G}(\eta' s^{-1/\beta})$, so (3.241) yields

$$|\mathbb{P}(D) - \theta| \leq K\theta^2. \qquad (3.261)$$

On the other hand Z_s is a Poisson variable with parameter $\theta \leq 1/2$, hence

$$|\mathbb{P}(B) - \theta| \leq K\theta^2. \qquad (3.262)$$

Since Y' is a purely discontinuous Lévy process without drift and whose Lévy measure is the restriction of F to $[-\eta', \eta']$, we deduce from (3.240) that

$$\mathbb{E}((Y'_s)^2) = s \int_{\{|x| \leq \eta'\}} x^2 F(dx) \leq K\theta\eta^2. \qquad (3.263)$$

The two processes Y' and Z are independent, and conditionally on B the law of the variable $Y_s - Y'_s$ is the restriction of the measure $\frac{s}{\theta} F$ to $[-\eta', \eta']^c$, and $\mathbb{P}(B) = \theta e^{-\theta}$. Thus

$$\mathbb{P}(B \cap D^c) = e^{-\theta} s \int_{\{|x| > \eta'\}} F(dx)\, \mathbb{P}(|Y'_s + x| \leq \eta')$$

$$\leq s\Big(F(\{\eta' < |x| \leq \eta'(1 + \theta^{1/3})/2\}) + F(\{|x| > \eta'\})\, \mathbb{P}(|Y'_s| > \eta'\theta^{1/3})\Big)$$

$$\leq \theta\Big(1 - (1 + \theta^{1/3})^{-\beta} + \frac{4}{\eta'^2\theta^{2/3}}\, \mathbb{E}((Y'_s)^2)\Big) \leq K\theta^{4/3}, \qquad (3.264)$$

where we have used (3.263) for the last inequality.

Now, we have

$$\mathbb{P}(D \cap B^c) = \mathbb{P}(D) - \mathbb{P}(B) + \mathbb{P}(B \cap D^c).$$

Observe also that $D \cap B' = D' \cap B'$, and D' and B' are independent, hence

$$\mathbb{P}(D') = \frac{\mathbb{P}(D' \cap B')}{\mathbb{P}(B')} = \frac{\mathbb{P}(D \cap B')}{\mathbb{P}(B')} \leq \frac{\mathbb{P}(D \cap B^c)}{\mathbb{P}(B')} \leq K\mathbb{P}(D \cap B^c)$$

because $\mathbb{P}(B') = e^{-\theta} \geq e^{-1/2}$. The last two displays, plus (3.261), (3.262) and (3.264) give us $\mathbb{P}(D') \leq K\theta^{4/3}$, hence the result.   □

Now we turn to semimartingales. We have (3.246) and there exists a predictable subset Φ of $\Omega \times (0, \infty) \times \mathbb{R}$ such that

$$\begin{array}{l} F_t''(\omega, .) \text{ is supported by the set } \{x : (\omega, t, x) \in \Phi\} \\ F_t'(\omega, .) \text{ is supported by the set } \{x : (\omega, t, x) \notin \Phi\}. \end{array} \tag{3.265}$$

Next we will derive a decomposition of X a bit similar to (3.146), but here we have a control on the Lévy measure of X itself, through (SL), so it is more convenient to truncate at the value taken by ΔX_t rather than by the function γ. Recall that the jumps of X are bounded, so we can write X in the form (3.71), with still b_t bounded. For any $\eta \in (0, 1]$ we set

$$b(\eta)_t = b_t - \int_{\{|x|>\eta\}} F_t'(dx)x - \int F_t''(dx)x$$

By (3.258) and (SL) the process $b(\eta)_t$ is well defined and satisfies $|b(\eta)_t| \leq K/\eta$. Then by (3.71) we can write $X = X(\eta) + X'(\eta)$, where $X'(\eta) = \widehat{X}(\eta) + \widehat{X}'(\eta) + \widehat{X}''(\eta)$ and

$$X(\eta) = (x1_{\{|x|>\eta\}}) \star \mu, \qquad \widehat{X}(\eta)_t = X_0 + \int_0^t b(\eta)_s ds + \int_0^t \sigma_s dW_s$$

$$\widehat{X}'(\eta) = (x1_{\{|x|\leq\eta\}} 1_{\Phi^c}) \star (\mu - \nu), \; \widehat{X}''(\eta) = (x1_{\{|x|\leq\eta\}} 1_{\Phi}) \star \mu.$$

Lemma 3.60 *Assume (SL). We have for all $p \geq 2$:*

$$\left. \begin{array}{l} \mathbb{E}_{i-1}^n(|\Delta_i^n \widehat{X}(\eta)|^p) \leq K_p (\Delta_n^{p/2} + \eta^{-p}\Delta_n^p) \\ \mathbb{E}_{i-1}^n(|\Delta_i^n \widehat{X}'(\eta)|^2) \leq K \Delta_n \eta^{2-\beta} \\ \mathbb{E}_{i-1}^n(|\Delta_i^n \widehat{X}''(\eta)|^{\beta'}) \leq K \Delta_n. \end{array} \right\} \tag{3.266}$$

Proof. The first estimate is obvious (see after (3.160)), whereas the second one is obtained from the second line of (3.258). Since $\beta' < 1$, we have $|\sum_j x_j|^{\beta'} \leq \sum_j |x_j|^{\beta'}$ for any sequence (x_j), hence

$$\mathbb{E}_{i-1}^n(|\Delta_i^n \widehat{X}''(\eta)|^{\beta'}) \leq \mathbb{E}_{i-1}^n \left(\Delta_i^n \left((|x|^{\beta'} 1_{\{|x|\leq\eta\}} 1_\Phi 1_{(t,\infty)}) \star \mu \right) \right)$$

$$= \mathbb{E}_{i-1}^n \left(\int_{(i-1)\Delta_n}^{i\Delta_n} dr \int_{\{|x|\leq\eta\}} |x|^{\beta'} F_r'(dx) \right) \leq K\Delta_n.$$

□

Next, we give a general result on counting processes. Let N be a counting process (that is, right continuous with $N_0 = 0$, piecewise constant, with jumps equal to 1) adapted to $(\mathcal{F}_t)$ and with predictable compensator of the form $G_t = \int_0^t g_s ds$.

Lemma 3.61 With N and G as above, and if further $g_t \leq u$ for some constant $u > 0$, we have

$$\left| \mathbb{P}_{i-1}^n(\Delta_i^n N = 1) - \mathbb{E}_{i-1}^n(\Delta_i^n G) \right| + \mathbb{P}_{i-1}^n(\Delta_i^n N \geq 2) \leq (u\Delta_n)^2. \quad (3.267)$$

Proof. Introduce the successive jump times $T_1, T_2, \cdots$ of N after time $(i-1)\Delta_n$, the sets $D = \{\Delta_i^n N = 1\}$ and $D' = \{\Delta_i^n N \geq 2\}$ and the variable $G_i'^n = \mathbb{E}_{i-1}^n(\Delta_i^n G)$. Then

$$\begin{aligned} \mathbb{P}_{i-1}^n(D) &= \mathbb{E}_{i-1}^n(N_{(i\Delta_n)\wedge T_1} - N_{(i-1)\Delta_n}) \\ &= \mathbb{E}_{i-1}^n\left(\int_{(i-1)\Delta_n}^{(i\Delta_n)\wedge T_1} g_r dr \right) \leq G_i'^n \leq u\Delta_n, \end{aligned}$$

$$G_i'^n - \mathbb{P}_{i-1}^n(D) = \mathbb{E}_{i-1}^n\left(\int_{(i\Delta_n)\wedge T_1}^{i\Delta_n} g_r dr \right) \leq u\Delta_n \mathbb{P}_{i-1}^n(D) \leq (u\Delta_n)^2$$

This gives us the first estimate. Next,

$$\begin{aligned} \mathbb{P}_{i-1}^n(D') &= \mathbb{P}_{i-1}^n(T_2 \leq i\Delta_n) \\ &= \mathbb{E}_{i-1}^n\left(1_{\{T_1 < i\Delta_n\}} \mathbb{P}_{i-1}^n(T_2 \leq i\Delta_n \mid \mathcal{F}_{T_1}) \right) \\ &= \mathbb{E}_{i-1}^n\left(1_{\{T_1 < i\Delta_n\}} \mathbb{E}\left(\int_{T_1}^{(i\Delta_n)\wedge T_2} g_r dr \mid \mathcal{F}_{T_1} \right) \right) \\ &\leq u\Delta_n \mathbb{P}_{i-1}^n(D) \leq (u\Delta_n)^2, \end{aligned}$$

hence the second estimate.   □

Lemma 3.62 With the notation $N(\eta)_t = \sum_{s \leq t} 1_{\{|\Delta X_s| > \eta\}}$, for all $\eta \in (0, 1]$, $\zeta \in (0, \frac{1}{2})$ and $p \geq 2$ we have

$$\mathbb{P}_{i-1}^n(\Delta_i^n N(\eta) \geq 1, |\Delta_i^n X'(\eta)| > \eta\zeta) \quad\quad\quad (3.268)$$

$$\leq K_p\left(\frac{\Delta_n^{p/2}}{\zeta^p \eta^p} + \frac{\Delta_n^p}{\zeta^p \eta^{2p}} + \frac{\Delta_n^2}{\zeta^2 \eta^{2\beta}} + \frac{\Delta_n}{\eta^{\beta'} \zeta^{\beta'}} \right).$$

Proof. (3.266) and Bienaymé–Tchebycheff inequality yield

$$\mathbb{P}_{i-1}^n\left(|\Delta_i^n \widehat{X}(\eta)| > \frac{\eta\zeta}{4} \right) \leq K_p\left(\frac{\Delta_n^{p/2}}{\eta^p \zeta^p} + \frac{\Delta_n^p}{\eta^{2p} \zeta^p} \right),$$

$$\mathbb{P}_{i-1}^n\left(|\Delta_i^n \widehat{X}''(\eta)| > \frac{\eta\zeta}{4} \right) \leq K \frac{\Delta_n}{\eta^{\beta'} \zeta^{\beta'}}.$$

Since $X'(\eta) = \widehat{X}(\eta) + \widehat{X}'(\eta) + \widetilde{X}'(\eta)$ it remains to prove

$$\mathbb{P}^n_{i-1}\left(\Delta^n_i N(\eta) \geq 1, \; |\Delta^n_i \widehat{X}'(\eta)| > \frac{\eta\zeta}{2}\right) \leq K\frac{\Delta^2_n}{\zeta^2 \eta^{2\beta}}. \qquad (3.269)$$

Define $N_s = N(\eta)_{(i-1)\Delta_n+s} - N(\eta)_{(i-1)\Delta_n}$ and $Y_s = \widehat{X}'(\eta)_{(i-1)\Delta_n+s} - \widehat{X}'(\eta)_{(i-1)\Delta_n}$ (for simplicity). By Bienaymé–Tchebycheff inequality again the left side of (3.269) is not bigger than $4\mathbb{E}(N_{\Delta_n} Y^2_{\Delta_n})/\eta^2\zeta^2$. Now, N is a counting process and Y is a purely discontinuous square-integrable martingale, and they have no common jumps, so Itô's formula yields

$$N_s Y^2_s = 2\int_0^s N_{r-}Y_{r-}dY_r + \int_0^s Y^2_{r-}dN_r + \sum_{r\leq s} N_{r-}(\Delta Y_r)^2.$$

Moreover, the compensator N is as in the previous lemma, with $g_s \leq K\eta^{-\beta}$, and the predictable quadratic variation of Y is $G'_s = \int_0^s g'_r dr$ with $g'_r \leq K\eta^{2-\beta}$ (see Lemma 3.60). Then taking expectations in the above display, and since the first term of the right side above is a martingale, we get

$$\mathbb{E}^n_{i-1}(N_{\Delta_n} Y^2_{\Delta_n}) = \mathbb{E}^n_{i-1}\left(\int_0^{\Delta_n} Y^2_r dG_r + \int_0^{\Delta_n} N_r dG'_r\right)$$

$$\leq K\eta^{-\beta}\int_0^{\Delta_n} \mathbb{E}^n_{i-1}\left(Y^2_r + \eta^2 N_r\right) dr$$

$$= K\eta^{-\beta}\int_0^{\Delta_n} \mathbb{E}^n_{i-1}\left(G'_r + \eta^2 G_r\right) dr \leq K\eta^{2(1-\beta)}\Delta^2_n.$$

(3.269) is then obvious. $\qquad\qquad\qquad\qquad\qquad\qquad\qquad\qquad\qquad \square$

The following lemma is key to the whole proof. We use the notation $u_n = \alpha\Delta^\varpi_n$.

Lemma 3.63 *Let $\alpha > 0$, $\varpi \in (0, \frac{1}{2})$) and $\eta \in (0, \frac{1}{2} - \varpi)$, and set*

$$\rho = \eta \wedge (\varpi(\beta - \beta') - \beta'\eta) \wedge (\varpi\gamma) \wedge (1 - \varpi\beta - 2\eta) \qquad (3.270)$$

There is a constant K depending on (α, ϖ, η), and also on the characteristics of X, such that

$$\left|\mathbb{P}^n_{i-1}(|\Delta^n_i X| > u_n) - \mathbb{E}^n_{i-1}\left(\int_{(i-1)\Delta_n}^{i\Delta_n} \overline{F}_r(u_n)dr\right)\right| \qquad (3.271)$$

$$\leq K\Delta^{1-\varpi\beta+\rho}_n$$

$$\mathbb{P}^n_{i-1}(u_n < |\Delta^n_i X| \leq u_n(1 + \Delta^\eta_n)) \leq K\Delta^{1-\varpi\beta+\rho}_n \qquad (3.272)$$

$$\mathbb{P}^n_{i-1}(|\Delta^n_i X| > u_n) \leq K\Delta^{1-\varpi\beta}_n. \qquad (3.273)$$

Proof. 1) Observe that $\rho > 0$, and it is clearly enough to prove the results when Δ_n is smaller than some number $\xi \in (0, 1)$ to be chosen later, and independent of i and n.

We can apply (3.266) and Bienaymé–Tchebycheff inequality to obtain

$$\mathbb{P}^n_{i-1}(|\Delta^n_i \widehat{X}(u_n)| > u_n \Delta^\eta_n/2) \leq K_p \Delta^{p(1-2\varpi-2\eta)/2}_n$$

$$\mathbb{P}^n_{i-1}(|\Delta^n_i \widehat{X}''(u_n)| > u_n \Delta^\eta_n/2) \leq K \Delta^{1-\beta'(\varpi+2\eta)}_n.$$

Then by choosing p large enough and by (3.270), we see that $Y^n = \widehat{X}(u_n) + \widehat{X}''(u_n)$ satisfies

$$\mathbb{P}^n_{i-1}(|\Delta^n_i Y^n| > u_n \Delta^\eta_n) \leq K \Delta^{1-\varpi\beta+\rho}_n. \tag{3.274}$$

2) By (SL) we have $F'_r(dx) \leq (L'/|x|^{1+\beta})dx$ in restriction to $[-1, 1]$, for some constant L'. We fix n. For each $\omega \in \Omega$ we endow the canonical (Skorokhod) space $(\Omega', \mathcal{F}', (\mathcal{F}'_t))$ of all càdlàg functions on $\mathbb{R}_+$ starting from 0 with the (unique) probability measure Q_ω under which the canonical process X' is a semimartingale with characteristics $(0, 0, \nu'_\omega)$, where

$$\nu'_\omega(\omega', dr, dx) = dr \, 1_{\{|x| \leq u_n\}} \left(\frac{L'}{|x|^{1+\beta}} \, dx - F'_r(\omega, dx) \right). \tag{3.275}$$

This measure does not depend on ω', hence under Q_ω the process X' has independent increments; $\nu'_\omega(\omega', dr, dx)$ depends measurably on ω, hence $Q_\omega(d\omega')$ is a transition probability from $(\Omega, \mathcal{F})$ into $(\Omega', \mathcal{F}')$. Then we extend X, X' and other quantities defined on Ω or Ω' in the usual way (without changing the symbols) to the product $\widetilde{\Omega} = \Omega \times \Omega'$ endowed with the product σ-field $\widetilde{\mathcal{F}}$, the product filtration $(\widetilde{\mathcal{F}}_t)$, and the probability measure $\widetilde{\mathbb{P}}(d\omega, d\omega') = \mathbb{P}(d\omega) \, Q_\omega(d\omega')$.

Because of (3.258) and (3.275), and as in Lemma 3.60, $\mathbb{E}_{Q_\omega}(|\Delta^n_i X'|^2 \mid \widetilde{\mathcal{F}}_{(i-1)\Delta_n}) \leq K \Delta_n u^{2-\beta}_n$, so for some constant C depending on α and β but not on n and ω we have

$$Q_\omega(|\Delta^n_i X'| > u_n \Delta^\eta_n \mid \widetilde{\mathcal{F}}_{(i-1)\Delta_n}) \leq C\Delta^{1-\varpi\beta-2\eta}_n \leq C\Delta^\rho_n. \tag{3.276}$$

3) By well known results on extensions of spaces (see e.g. Jacod and Shiryaev (2003), Section II.7; note that the present extension of the original space is a "very good extension"), X' is a semimartingale on the extension with characteristics $(0, 0, \nu')$, where $\nu'((\omega, \omega'), dr, dx) = \nu'_\omega(dr, dx)$, and any semimartingale on the original space is a semimartingale on the extension, with the same characteristics. Moreover X and X' have almost surely no common jump, so the sum $Y'(u_n) = \widehat{X}'(u_n) + X'$ is a semimartingale with characteristics $(0, 0, \nu')$, where

$$\nu'(dr, dx) = dr \, 1_{\{|x| \leq u_n\}} F'_r(dx) + \nu_\omega(dr, dx) = 1_{\{|x| \leq u_n\}} \frac{L'}{|x|^{1+\beta}} \, dr \, dx,$$

where the last equality comes from (3.275). It follows that $Y'(u_n)$ is a Lévy process with Lévy measure given above, or in other words, it is a version of the process $Y'(u_n)$ of (3.259) with $A = 2L'/\beta$. Hence, recalling (3.258), we deduce from (3.260) and from the Lévy property of $Y'(u_n)$ that, as soon as $\Delta_n^\eta \leq 1/4$, and if $A \in \mathcal{F}_{(i-1)\Delta_n}$:

$$\widetilde{\mathbb{P}}(A \cap \{|\Delta_i^n Y'(u_n)| > u_n(1 - 2\Delta_n^\eta)\}) \leq K\Delta_n^{4/3 - 4\varpi\beta/3}. \qquad (3.277)$$

Next, let ξ be such that $C\xi^\rho \leq 1/2$. With A as above, and if $\Delta_n \leq \xi$, we can write

$$\widetilde{\mathbb{P}}\big(A \cap \{|\Delta_i^n Y'(u_n)| > u_n(1 - 2\Delta_n^\eta)\}\big)$$
$$\geq \widetilde{\mathbb{P}}\Big(A \cap \{|\Delta_i^n \widehat{X}'(u_n)| > u_n(1 - \Delta_n^\eta)\} \cap \{|\Delta_i^n X'| \leq u_n\Delta_n^\eta\}\Big)$$
$$= \widetilde{\mathbb{E}}\Big(1_{\{A \cap \{|\Delta_i^n \widehat{X}'(u_n)| > u_n(1 - \Delta_n^\eta)\}\}} Q.\Big(|X'_{t+s} - X'_t| \leq u_n\Delta_n^\eta\Big)\Big)$$
$$\geq \frac{1}{2} \mathbb{P}\Big(A \cap \{|\Delta_i^n \widehat{X}'(u_n)| > u_n(1 - \Delta_n^\eta)\}\Big)$$

where the last inequality comes from (3.276). Then by (3.277) and the facts that A is arbitrary in $\mathcal{F}_{(i-1)\Delta_n}$ and that $\rho \leq \frac{1 - \varpi\beta}{3}$ we deduce

$$\mathbb{P}_{i-1}^n\Big(|\Delta_i^n \widehat{X}'(u_n)| > u_n(1 - \Delta_n^\eta)\Big) \leq K\Delta_n^{4/3 - 4\varpi\beta/3} \leq K\Delta_n^{1 - \varpi\beta + \rho}.$$

In turn, combining this with (3.274), we readily obtain

$$\mathbb{P}_{i-1}^n\Big(|\Delta_i^n X'(u_n)| > u_n\Big) \leq K\,\Delta_n^{1 - \varpi\beta - \rho}. \qquad (3.278)$$

4) Now we write $u'_n = u_n(1 + \Delta_n^\eta)$ and also

$$\theta_i^n = \mathbb{E}_{i-1}^n\Big(\int_{(i-1)\Delta_n}^{i\Delta_n} \overline{F}_r(u_n)dr\Big), \qquad \theta_i'^n = \mathbb{E}\Big(\int_{(i-1)\Delta_n}^{i\Delta_n} \overline{F}_r(u'_n)dr\Big),$$

and introduce the following two counting process

$$N_t^n = \sum_{s \leq t} 1_{\{|\Delta X_s| > u_n\}}, \qquad N_t'^n = \sum_{s \leq t} 1_{\{|\Delta X_s| > u'_n\}}.$$

Their predictable compensators are $\int_0^t \overline{F}_r(u_n)dr$ and $\int_0^t \overline{F}_r(u'_n)dr$, whereas both $\overline{F}_r(u_n)$ and $\overline{F}_r(u'_n)$ are smaller than $K/\Delta_n^{\varpi\beta}$. Hence (3.267) gives

$$|\mathbb{P}_{i-1}^n(\Delta_i^n N^n = 1) - \theta_i^n| + \mathbb{P}_{i-1}^n(\Delta_i^n N^n \geq 2) \leq K\Delta_n^{2(1 - \varpi\beta)},$$
$$|\mathbb{P}_{i-1}^n(\Delta_i^n N'^n = 1) - \theta_i'^n| \leq K\Delta_n^{2(1 - \varpi\beta)}. \qquad (3.279)$$

Since $N^n - N'^n$ is non-decreasing, we have

$$\mathbb{P}_{i-1}^n(\Delta_i^n N^n = 1, \Delta_i^n N'^n = 0) = \mathbb{P}_{i-1}^n(\Delta_i^n N^n = 1)$$
$$- \mathbb{P}_{i-1}^n(\Delta_i^n N'^n = 1) + \mathbb{P}_{i-1}^n(\Delta_i^n N^n \geq 2, \Delta_i^n N'^n = 1).$$

Then (3.279) yields

$$|\mathbb{P}^n_{i-1}(\Delta^n_i N^n = 1, \; \Delta^n_i N'^n = 0) - (\theta^n_i - \theta'^n_i)| \; \leq \; K\Delta^{2(1-\varpi\beta)}_n. \quad (3.280)$$

Moreover, (3.258) clearly implies $\theta^n_i - \theta'^n_i \leq K\Delta^{1-\varpi\beta}_n(\Delta^\eta_n + \Delta^{\varpi(\gamma\wedge(\beta-\beta'))}_n) \leq K\Delta^{1-\varpi\beta+\rho}_n$. We then deduce from (3.280) that

$$\mathbb{P}^n_{i-1}(\Delta^n_i N^n = 1, \; \Delta^n_i N'^n = 0) \; \leq \; K\Delta^{1-\varpi\beta+\rho}_n. \quad (3.281)$$

5) If $\Delta^n_i N^n = \Delta^n_i N'^n = 1$ and $|\Delta^n_i X| \leq u_n$, then necessarily $|\Delta^n_i X(u'_n)| > u_n\Delta^\eta_n$. Hence

$$\mathbb{P}^n_{i-1}(\Delta^n_i N^n = 1, \; |\Delta^n_i X| \leq u_n) \leq \mathbb{P}^n_{i-1}(\Delta^n_i N^n = 1, \; \Delta^n_i N'^n = 0)$$
$$+\mathbb{P}^n_{i-1}(\Delta^n_i N^n = 1, \; |\Delta^n_i X'(u_n)| > u_n\Delta^\eta_n).$$

Then if we apply (3.268) with p large enough and $\eta = u_n$ and $\zeta = \Delta^\eta_n$, and (3.281), we deduce

$$\mathbb{P}^n_{i-1}(\Delta^n_i N^n = 1, \; |\Delta^n_i X| \leq u_n) \; \leq \; K\Delta^{1-\varpi\beta+\rho}_n. \quad (3.282)$$

Finally $\Delta^n_i X = \Delta^n_i X'(u_n)$ on the set $\{\Delta^n_i N^n = 0\}$, so

$$\begin{aligned}
\mathbb{P}^n_{i-1}(|\Delta^n_i X| > u_n) \;=\; & \mathbb{P}^n_{i-1}(\Delta^n_i N^n = 1) \\
& - \mathbb{P}^n_{i-1}(\Delta^n_i N^n = 1, \; |\Delta^n_i X| \leq u_n) \\
& + \mathbb{P}^n_{i-1}(\Delta^n_i N^n = 0, \; |\Delta^n_i X'(u_n)| > u_n) \\
& + \mathbb{P}^n_{i-1}(\Delta^n_i N^n \geq 2, \; |\Delta^n_i X| > u_n).
\end{aligned}$$

Then if we combine (3.278), (3.279) and (3.282), if $\Delta_n \leq \xi h$ we readily obtain (3.271). We also trivially deduce (3.273) from (3.258) and (3.271),

6) Finally, a close look at the previous argument shows that (3.271) also holds with $\alpha\Delta^\varpi_n(1+\Delta^\eta_n)$ and θ'^n_i in place of $\alpha\Delta^\varpi_n$ and θ^n_i. Therefore (3.272) follows, upon using the property $\theta^n_i - \theta'^n_i \leq K\Delta^{1-\varpi\beta+\rho}_n$ proved above. $\square$

Lemma 3.64 *Under the assumption and with the notation of Lemma 3.63, and if M is a bounded continuous martingale, we have (with K depending also on M):*

$$\left|\mathbb{E}^n_{i-1}\left(\Delta^n_i M \, 1_{\{|\Delta^n_i X| > u_n\}}\right)\right| \quad (3.283)$$
$$\leq \; K\Delta^{1-\varpi\beta+\rho}_n + K\Delta^{1-(\varpi+\eta)\beta}_n \, \mathbb{E}^n_{i-1}(|\Delta^n_i M|).$$

Proof. 1) There exist C^2 functions f_n such that (with K independent of n):

$$1_{\{|x|>u_n(1+2\Delta^\eta_n/3)\}} \leq f_n(x) \leq 1_{\{|x|>u_n(1+\Delta^\eta_n/3)\}}$$
$$|f'_n(x)| \leq \frac{K}{\Delta^{\varpi+\eta}_n}, \qquad |f''_n(x)| \leq \frac{K}{\Delta^{2(\varpi+\eta)}_n}. \quad (3.284)$$

With $\widehat{X}' = X - B - X^c$, and since M is bounded, we have

$$\left| \mathbb{E}^n_{i-1}(\Delta^n_i M \, 1_{\{|\Delta^n_i X| > u_n\}}) - \mathbb{E}^n_{i-1}(\Delta^n_i M \, f_n(\Delta^n_i \widehat{X}')) \right| \qquad (3.285)$$
$$\leq K\mathbb{P}^n_{i-1}(u_n < |\Delta^n_i X| \leq u_n(1 + \Delta^n_n))$$
$$+ K\mathbb{E}^n_{i-1}(|f_n(\Delta^n_i X) - f_n(\Delta^n_i \widehat{X}')|).$$

Now we have

$$|f_n(x + y) - f_n(x)| \leq 1_{\{|y| > u_n \Delta^n_n/3)\}} + \frac{K}{\Delta^{\varpi + \eta}_n}|y| 1_{\{u_n < |x+y| \leq u_n(1 + \Delta^n_n)\}}.$$

If we apply this with $x = \Delta^n_i \widehat{X}'$ and $y = \Delta^n_i(B + X^c)$, plus (3.266) for p large enough and Bienaymé–Tchebycheff inequality and $1 - 2\varpi - 2\eta > 0$, plus (3.272) and (3.266) again and Hölder's inequality, we obtain that the right side of (3.285) is smaller than $K\Delta^{1-\varpi\beta+\rho}_n$. Therefore it remains to prove that

$$\left| \mathbb{E}^n_{i-1}(\Delta^n_i M \, f_n(\Delta^n_i \widehat{X}')) \right| \leq \qquad (3.286)$$
$$K\Delta^{1-\varpi\beta+\rho)}_n + K\Delta^{1-(\varpi+\eta)\beta}_n \, \mathbb{E}^n_{i-1}(|\Delta^n_i M|).$$

2) For simplicity we write $Y_t = \widehat{X}'_{(i-1)\Delta_n + t} - \widehat{X}'_{(i-1)\Delta_n}$ and $Z_r = M_{(i-1)\Delta_n + t} - M_{(i-1)\Delta_n}$. Since Z is a bounded continuous martingale and Y a semimartingale with vanishing continuous martingale part, and $f_n(Y)$ is bounded, we deduce from Itô's formula that the product $Z_t f_n(Y_t)$ is the sum of a martingale plus the process $\int_0^t \Gamma^n_u du$, where

$$\Gamma^n_t = Z_t \int F_{(i-1)\Delta_n + t}(dx) \, g_n(Y_t, x),$$

$$g_n(y, x) = f_n(y + x) - f_n(y) - f'_n(y)x 1_{\{|x| \leq 1\}}.$$

An easy computation allows to deduce of (3.284) that

$$|g_n(y, x)| \leq 1_{\{|x| > u_n \Delta^n_n/3\}} + K 1_{\{u_n < |y| \leq u_n(1+\Delta^n_n)\}} \cdot$$
$$\left(\frac{x^2}{\Delta^{2\varpi + 2\eta}_n} 1_{\{|x| \leq u_n \Delta^n_n\}} + \frac{|x| \wedge 1}{\Delta^{\varpi + \eta}_n} 1_{\{|x| > u_n \Delta^n_n\}} \right).$$

Now, we apply (3.258) to get for any $\varepsilon > 0$:

$$|\Gamma^n_t| \leq K |Z_t| \Delta^{-(\varpi + \eta)\beta}_n + K_\varepsilon |Z_t| \Delta^{-(\beta + \varepsilon)(\varpi + \eta)}_n \, 1_{\{u_n < |Y_t| \leq u_n(1+\Delta^n_n)\}}.$$

Since $\eta < 1/2 - \varpi$ we have $\beta(\varpi + \eta) < 1$ and thus $(\beta + \varepsilon)(\varpi + \eta) = 1$ for a suitable $\varepsilon > 0$. Moreover, $\mathbb{E}(|Z_u|) \leq \mathbb{E}(|Z_s|)$ if $u \leq s$ because Z is a martingale. Therefore, since Z is bounded we obtain

$$\left| \mathbb{E}_{i-1}^n (\Delta_i^n M \, f_n(\Delta_i^n \widehat{X}')) \right| = \left| \mathbb{E}_{i-1}^n \left(\int_0^{\Delta_n} \Gamma_t^n \, dt \right) \right| \leq \int_0^{\Delta_n} \mathbb{E}(|\Gamma_t^n|) \, dt$$

$$\leq K \Delta_n^{1-(\varpi+\eta)\beta} \, \mathbb{E}_{i-1}^n (|\Delta_i^n M|)$$

$$+ K \Delta_n^{-1} \int_0^{\Delta_n} \mathbb{P}_{i-1}^n (u_n < |Y_t| \leq u_n(1 + \Delta_n^\eta)) \, dt.$$

By (3.272) for the process $\widehat{X}'$ instead of X, we readily deduce (3.286). □

3.12.5 Some auxiliary limit theorems

Below, recall the process $\overline{A}$ of (3.250). We still assume (SL) and write $u_n = \alpha \Delta_n^\varpi$.

Lemma 3.65 *Let $\rho' < \frac{1}{2} \wedge (\varpi\gamma) \wedge (\varpi(\beta - \beta'))$. Then for all $t > 0$ we have*

$$\Delta_n^{-\rho'} \left(\sum_{i=1}^{[t/\Delta_n]} \Delta_n^{\varpi\beta} \, \mathbb{E}_{i-1}^n \left(\int_{(i-1)\Delta_n}^{i\Delta_n} \overline{F}_t(u_n) dt \right) - \frac{\overline{A}_t}{\alpha^\beta} \right) \xrightarrow{\mathbb{P}} 0. \qquad (3.287)$$

Proof. Let $\theta_i^n = \int_{(i-1)\Delta_n}^{i\Delta_n} \overline{F}_t(u_n) dt$ and $\eta_i^n = \int_{(i-1)\Delta_n}^{i\Delta_n} A_t dt$. We deduce from (3.258) that

$$\left| \Delta_n^{\varpi\beta} \theta_i^n - \frac{1}{\alpha^\beta} \eta_i^n \right| \leq K \Delta_n^{1+\varpi(\beta-(\beta-\gamma)\vee\beta')} \leq K (\Delta_n^{1+\varpi\gamma} + \Delta_n^{1+\varpi(\beta-\beta')}).$$

Then obviously

$$\mathbb{E} \left(\Delta_n^{-\rho'} \sum_{i=1}^{[t/\Delta_n]} \mathbb{E}_{i-1}^n \left(\left| \Delta_n^{\varpi\beta} \theta_i^n - \frac{1}{\alpha^\beta} \eta_i^n \right| \right) \right) \to 0,$$

and since A_t is bounded we have $\left| \overline{A}_t - \sum_{i=1}^{[t/\Delta_n]} \eta_i^n \right| \leq K t \Delta_n$, whereas $\rho' < 1$. It thus remains to prove that

$$\Delta_n^{-\rho'} \left(\sum_{i=1}^{[t/\Delta_n]} (\eta_i^n - \mathbb{E}_{i-1}^n(\eta_i^n)) \right) \xrightarrow{\mathbb{P}} 0. \qquad (3.288)$$

Since $\zeta_i^n = \Delta_n^{-\rho'} (\eta_i^n - \mathbb{E}_{i-1}^n(\eta_i^n))$ is a martingale increment, for (3.288) it is enough to check that $a_n(t) = \mathbb{E} \left(\sum_{i=1}^{[t/\Delta_n]} (\zeta_i^n)^2 \right)$ goes to 0. However, since A_t is bounded, we have $|\zeta_i^n|^2 \leq K \Delta_n^{2-2\rho'}$, so $a_n(t) \leq K t \Delta_n^{1-2\rho'} \to 0$ because $\rho' < 1/2$. □

Lemma 3.66 *a) Let* $\chi < (\varpi\gamma) \wedge \frac{1-\varpi}{3} \wedge \frac{\varpi(\beta-\beta')}{1+\beta'} \wedge \frac{1-2\varpi}{2}$. *Then for all* $t > 0$ *we have*

$$\Delta_n^{-\chi}\left(\Delta_n^{\varpi\beta}\sum_{i=1}^{[t/\Delta_n]}\mathbb{P}_{i-1}^n(|\Delta_i^n X| > u_n) - \frac{\overline{A}_t}{\alpha^\beta}\right) \xrightarrow{\mathbb{P}} 0, \qquad (3.289)$$

and in particular

$$\Delta_n^{\varpi\beta}\sum_{i=1}^{[t/\Delta_n]}\mathbb{P}_{i-1}^n(|\Delta_i^n X| > u_n) \xrightarrow{\mathbb{P}} \frac{\overline{A}_t}{\alpha^\beta} \qquad (3.290)$$

b) If further $\beta' < \frac{\beta}{2+\beta}$ *and* $\gamma > \frac{\beta}{2}$ *and* $\varpi < \frac{1}{2+\beta} \wedge \frac{1}{3\beta}$, *and if* M *is a bounded continuous martingale, we also have*

$$\Delta_n^{-\varpi\beta/2}\left(\Delta_n^{\varpi\beta}\sum_{i=1}^{[t/\Delta_n]}\mathbb{P}_{i-1}^n(|\Delta_i^n X| > u_n) - \frac{\overline{A}_s}{\alpha^\beta}\right) \xrightarrow{\mathbb{P}} 0. \qquad (3.291)$$

$$\Delta_n^{\varpi\beta/2}\sum_{i=1}^{[t/\Delta_n]}\left|\mathbb{E}_{i-1}^n\left(\Delta_i^n M 1_{\{|\Delta_i^n X|>u_n\}}\right)\right| \xrightarrow{\mathbb{P}} 0. \qquad (3.292)$$

Proof. a) In Lemma 3.63 we can take $\eta = \frac{1-\varpi\beta}{3} \wedge \frac{\varpi(\beta-\beta')}{1+\beta'} \wedge \frac{1-2\varpi-\varepsilon}{2}$ for some $\varepsilon > 0$, and ρ is given by (3.270). Upon taking ε small enough, we then have $\chi < \rho$, and also $\chi \le \rho'$ for a ρ' satisfying the conditions of Lemma 3.65. Then (3.289) readily follows from (3.271) and (3.287).

b) Our conditions on γ, β' and ϖ imply (after some calculations) that one may take $\chi = \varpi\beta/2$ satisfying the condition in (a), so (3.291) follows from (3.289).

It remains to prove (3.292). By (3.283), the left side of (3.292) is smaller than

$$K t \Delta_n^{\rho-\varpi\beta/2} + K\Delta_n^{1-\eta\beta-\varpi\beta/2}\sum_{i=1}^{[t/\Delta_n]}\mathbb{E}_{i-1}^n(|\Delta_i^n M|).$$

By the Cauchy-Schwarz inequality this is smaller than

$$K(t + \sqrt{t})\left(\Delta_n^{\rho-\varpi\beta/2} + \Delta_n^{1/2-\eta\beta-\varpi\beta/2}\left(\sum_{i=1}^{[t/\Delta_n]}\mathbb{E}_{i-1}^n(|\Delta_i^n M|^2)\right)^{1/2}\right).$$

A well known property of martingales yields

$$\mathbb{E}\left(\sum_{i=1}^{[t/\Delta_n]}\mathbb{E}_{i-1}^n(|\Delta_i^n M|^2)\right) = \mathbb{E}\left((M_{\Delta_n[t/\Delta_n]} - M_0)^2\right),$$

which is bounded (in n). Therefore we deduce (3.292), provided we have $\rho > \varpi\beta/2$ and also $1 - 2\eta\beta > \varpi\beta$. The first condition has already been checked, but the second one may fail with our previous choice of η. However since $\varpi < 1/3\beta$ we have $\varpi\beta/2 < (1-\varpi\beta)/2\beta$, and we can find η' strictly between these two numbers. Then we replace ρ and η by $\bar{\rho} = \rho \wedge \eta$ and $\bar{\eta} = \eta \wedge \eta'$, which still satisfy (3.270), and now the required conditions are fulfilled by $\bar{\rho}$ and $\bar{\eta}$. This ends the proof. $\qquad\square$

Proposition 3.67 *Assume (SL). For each $t > 0$ we have*

$$\Delta_n^{\varpi\beta} U(\alpha\Delta_n^{\varpi}, \Delta_n)_t \overset{\mathbb{P}}{\longrightarrow} \frac{\overline{A}_t}{\alpha^\beta}. \tag{3.293}$$

Proof. Set

$$\zeta_i^n = \Delta_n^{\varpi\beta/2} \left(1_{\{|\Delta_i^n X| > \alpha\Delta_n^\varpi\}} - \mathbb{P}_{i-1}^n(|\Delta_i^n X| > \alpha\Delta_n^\varpi)\right). \tag{3.294}$$

By virtue of (3.290), it suffices to prove that the sequence $\sum_{i=1}^{[t/\Delta_n]} \zeta_i^n$ is tight. Since the ζ_i^n's are martingale increments, it is enough to show that the sequence $a_n(t) = \sum_{i=1}^{[t/\Delta_n]} \mathbb{E}((\zeta_i^n)^2)$ is bounded. But (3.273) yields $\mathbb{E}((\zeta_i^n)^2) \leq K\Delta_n$, which in turn yields $a_n(t) \leq Kt$. $\qquad\square$

Proposition 3.68 *Assume (SL). Let $\alpha' > \alpha$. If we have $\beta' < \frac{\beta}{2+\beta}$ and $\gamma > \frac{\beta}{2}$ and $\varpi < \frac{1}{2+\beta} \wedge \frac{1}{3\beta}$, the pair of processes*

$$\Delta_n^{-\varpi\beta/2} \left(\Delta_n^{\varpi\beta} U(\alpha\Delta_n^\varpi, \Delta_n)_t - \frac{\overline{A}_t}{\alpha^\beta}, \ \Delta_n^{\varpi\beta} U(\alpha'\Delta_n^\varpi, \Delta_n)_t - \frac{\overline{A}_t}{\alpha'^\beta}\right) \tag{3.295}$$

converges stably in law to a process $(\overline{W}, \overline{W}')$ defined on an extension of $(\Omega, \mathcal{F}, (\mathcal{F}_t)_{t\geq 0}, \mathbb{P})$, and with conditionally on $\mathcal{F}$ is a continuous Gaussian martingale with

$$\widetilde{\mathbb{E}}(\overline{W}_t^2 \mid \mathcal{F}) = \frac{\overline{A}_t}{\alpha^\beta}, \quad \widetilde{\mathbb{E}}(\overline{W}_t'^2 \mid \mathcal{F}) = \frac{\overline{A}_t}{\alpha'^\beta}, \quad \widetilde{\mathbb{E}}(\overline{W}_t \overline{W}_t' \mid \mathcal{F}) = \frac{\overline{A}_t}{\alpha'^\beta}. \tag{3.296}$$

Proof. Define ζ_i^n by (3.294), and associate $\zeta_i'^n$ with α' in the same way. In view of (3.291) the result amounts to proving the stable convergence in law of the pair of processes $\left(\sum_{i=1}^{[t/\Delta_n]} \zeta_i^n, \sum_{i=1}^{[t/\Delta_n]} \zeta_i'^n\right)$ to $(\overline{W}, \overline{W}')$. The variables $\zeta^n i$ and $\zeta_i'^n$ are martingale increments and are smaller than $K\Delta_n^{\varpi\beta/2}$, so in view of Lemma 3.7 it is enough to prove the following

$$\left.\begin{aligned}
\sum_{i=1}^{[t/\Delta_n]} \mathbb{E}_{i-1}^n((\zeta_i^n)^2) &\overset{\mathbb{P}}{\longrightarrow} \frac{\overline{A}_t}{\alpha^\beta}, \quad & \sum_{i=1}^{[t/\Delta_n]} \mathbb{E}_{i-1}^n((\zeta_i'^n)^2) &\overset{\mathbb{P}}{\longrightarrow} \frac{\overline{A}_t}{\alpha'^\beta}, \\
\sum_{i=1}^{[t/\Delta_n]} \mathbb{E}_{i-1}^n(\zeta_i^n \zeta_1'^n) &\overset{\mathbb{P}}{\longrightarrow} \frac{\overline{A}_t}{\alpha'^\beta}.
\end{aligned}\right\} \tag{3.297}$$

$$\sum_{i=1}^{[t/\Delta_n]} \mathbb{E}_{i-1}^n (\zeta_i^n \, \Delta_i^n M) \xrightarrow{\;\mathbb{P}\;} 0, \qquad \sum_{i=1}^{[t/\Delta_n]} \mathbb{E}_{i-1}^n (\zeta_i'^n \, \Delta_i^n M) \xrightarrow{\;\mathbb{P}\;} 0, \qquad (3.298)$$

where M is any bounded martingale.

Since $\alpha < \alpha'$, we have

$$\mathbb{E}_{i-1}^n (\zeta_i^n \, \zeta_i'^n) = \Delta_n^{\varpi\beta} \Big(\mathbb{P}_{i-1}^n (|\Delta_i^n X| > \alpha' \Delta_n^\varpi)$$
$$- \mathbb{P}_{i-1}^n (|\Delta_i^n X| > \alpha \Delta_n^\varpi) \, \mathbb{P}_{i-1}^n (|\Delta_i^n X| > \alpha' \Delta_n^\varpi) \Big),$$

whereas $\mathbb{P}_{i-1}^n (|\Delta_i^n X| > \alpha \Delta_n^\varpi) \le K \Delta_n^{1-\varpi\beta}$ by (3.273). Therefore we deduce the last part of (3.297) from (3.293), and the first two parts are proved in the same way.

Now we turn to (3.298). Since $\mathbb{E}_{i-1}^n (\Delta_i^n M) = 0$, this follows from (3.292), which has been proved when M is continuous. Now, since any bounded martingale is the sum of a continuous martingale and a purely discontinuous martingale with bounded jumps, and up to a localization, it remains to prove (3.298) when M is a bounded purely discontinuous martingale.

In this case, we consider the discretized process $M_t^n = M_{\Delta_n [t/\Delta_n]}$, and we set $Z^n = \sum_{i=1}^{[t/\Delta_n]} \zeta_i^n$. We know by (3.297) that the sequence (of discrete-time martingales) Z^n is tight, whereas the convergence $M^n \to M$ (pathwise, in the Skorokhod sense) is a known fact. Since further any limiting process of Z^n is continuous, the pair (Z^n, M^n) is tight. From any subsequence of indices we pick a further subsequence, say (n_k), such that (Z^{n_k}, M^{n_k}) is tight, with the limit (Z, M). Another well known fact is that the quadratic covariation $[M^{n_k}, Z^{n_k}]$ converges to $[M, Z]$, and since M is purely discontinuous and Z is continuous it follows that $[M, Z] = 0$. Then by Lenglart inequality (since the jumps of the discrete processes $[M^n, Z^n]$ are bounded by a constant), the predictable compensators of $[M^n, Z^n]$ also go to 0 in probability. Now, those compensators are exactly $\sum_{i=1}^{[t/\Delta_n]} \mathbb{E}_{i-1}^n (\zeta_i^n \Delta_i^n M)$, which thus goes to 0 in probability along the subsequence n_k; it readily follows that the first part of (3.298) holds, and the second part is similarly analyzed. $\qquad\square$

3.12.6 Proof of Theorem 3.54

At this point, the proof is nearly trivial. As said before, it is no restriction to assume (SL). Then in view of Proposition 3.67 the consistency result (a) is obvious.

As for (b), we apply Proposition 3.68, to obtain that

$$U(\alpha \Delta_n^\varpi, \Delta_n)_t = \frac{\overline{A}_t}{\Delta_n^{\varpi\beta} \alpha^\beta} + \Delta_n^{\varpi\beta/2} V_n,$$

$$U(\alpha' \Delta_n^\varpi, \Delta_n)_t = \frac{\overline{A}_t}{\Delta_n^{\varpi\beta} \alpha'^\beta} + \Delta_n^{\varpi\beta/2} V_n',$$

where the pair (V_n, V_n') converge stably in law to a variable (V, V') which is $\mathcal{F}$-conditionally Gaussian centered with covariance matrix

$$\begin{pmatrix} \overline{A}_t/\alpha^\beta & \overline{A}_t/\alpha'^\beta \\ \overline{A}_t/\alpha'^\beta & \overline{A}_t/\alpha'^\beta \end{pmatrix}.$$

Then a simple computation shows that the variable

$$\frac{1}{\Delta_n^{\varpi\beta/2}} \left(\widehat{\beta}_n(t, \varpi, \alpha, \alpha') - \beta \right)$$

is equivalent (in probability) to

$$\frac{\alpha^\beta V_n - \alpha'^\beta V_n'}{\overline{A}_t \log(\alpha'/\alpha)},$$

on the set $\{\overline{A}_t > 0\}$, and the result readily follows .   □

References

Aït–Sahalia, Y. (2004). Disentangling diffusion from jumps. *J. Financial Econometrics, 74,* 487–528.

Aït–Sahalia, Y., & Jacod, J. (2008). Fisher's information for discretely sampled Lévy processes. *Econometrica, 76,* 727–761.

Aït–Sahalia, Y., & Jacod, J. (2009a). Estimating the degree of activity of jumps in high frequency financial data. *Ann. Statist., 37,* 2202–2244.

Aït–Sahalia, Y., & Jacod, J. (2009b). Testing for jumps in a discretely observed process. *Ann. Statist., 37,* 184–222.

Aldous, D., & Eagleson, G. K. (1978). On mixing and stability of limit theorems. *Ann. Probab., 6,* 325–331.

Barndorff–Nielsen, O. E., Graversen, S. E., Jacod, J., Podolskij, M., & Shephard, N. (2006). A central limit theorem for realised power and bipower variations of continuous semimartingales. In Y. Kabanov, R. Liptser, & J. Stoyanov (Eds.), *From Stochastic Calculus to Mathematical Finance* (pp. 33–68). Berlin: Springer.

Barndorff–Nielsen, O. E., & Shephard, N. (2003). Realised power variation and stochastic volatility. *Bernoulli,, 9,* 243–265. Correction published on pages 1109–1111.

Barndorff–Nielsen, O. E., & Shephard, N. (2004). Power and bipower variation with stochastic volatility and jumps. *J. Financial Econometrics, 2,* 1–48.

Barndorff–Nielsen, O. E., Shephard, N., & Winkel, M. (2006). Limit theorems for multipower variation in the presence of jumps. *Stochastic Processes Appl., 116,* 796–826.

Billingsley, P. (1999). *Convergence of Probability Measures* (Second ed.). New York: John Wiley & Sons Inc.

Carr, P., & Wu, L. (2003). What type of process underlies options? A simple robust test. *Finance, 58,* 2581–2610.

Huang, X., & Tauchen, G. (2006). The relative contribution of jumps to total price variance. *J. Financial Econometrics, 4,* 456–499.

Jacod, J. (2008). Asymptotic properties of realized power variations and related functionals of semimartingales. *Stochastic Processes Appl., 118,* 517–559.

Jacod, J., & Protter, P. (1998). Asymptotic error distributions for the Euler method for stochastic differential equations. *Ann. Probab.*, *26*, 267–307.

Jacod, J., & Shiryaev, A. N. (2003). *Limit Theorems for Stochastic Processes* (Second ed.). Berlin: Springer–Verlag.

Jacod, J., & Todorov, V. (2009). Testing for common arrivals for discretely observed multidimensional processes. *Ann. Statist.*, *37*, 1792–1838.

Jiang, G. J., & Oomen, R. C. (2008). A new test for jumps in asset prices. *Journal of Econometrics*, *144*, 352–370.

Lee, S., & Mykland, P. (2008). Jumps in financial markets: a new nonparametric test and jump clustering. *Review of Financial Studies*, *21*, 2535–2563.

Lépingle, D. (1976). La variation d'ordre p des semimartingales. *Zeit. Wahrsch. Verw. Gebiete*, *36*, 285–316.

Mancini, C. (2001). Disentangling the jumps of the diffusion in a geometric jumping Brownian motion. *Giornale dell'Instituto Italiano degli Attuari*, *LXIV*, 19–47.

Mancini, C. (2009). Non-parametric threshold estimation for models with stochastic diffusion coefficient and jumps. *Scand. J. Stat.*, *36*(2), 270–296.

Rényi, A. (1963). On stable sequences of events. *Sankhyā Ser. A*, *25*, 293–302.

Tukey, J. W. (1939). On the distribution of the fractional part of a statistical variable. *Math. Sb.*, *4*, 561–562.

Zolotarev, V. M. (1986). *One-dimensional Stable Distributions*. Providence, RI: American Mathematical Society.

CHAPTER 4

Importance sampling techniques for estimation of diffusion models

Omiros Papaspiliopoulos and Gareth Roberts

Department of Economics, Universitat Pompeu Fabra
Ramon Trias Fargas 25-27, Barcelona 08005, Spain
and
Department of Statistics, University of Warwick
Coventry, CV4 7AL, UK

4.1 Overview of the chapter

This article develops a class of Monte Carlo (MC) methods for simulating conditioned diffusion sample paths, with special emphasis on importance sampling schemes. We restrict attention to a particular type of conditioned diffusion, the so-called diffusion bridge processes. The diffusion bridge is the process obtained by conditioning a diffusion to start and finish at specific values at two consecutive times $t_0 < t_1$.

Diffusion bridge simulation is a highly non-trivial problem. At an even more elementary level, unconditional simulation of diffusions, that is, without fixing the value of the process at t_1, is difficult. This is a simulation from the transition distribution of the diffusion which is typically intractable. This intractability stems from the implicit specification of the diffusion as a solution of a stochastic differential equation (SDE). Although the unconditional simulation can be carried out by various approximate schemes based on discretizations of the SDE, it is not feasible to devise similar schemes for diffusion bridges in general. This has motivated active research in the last 15 years or so for the development of an MC methodology for diffusion bridges.

The research in this direction has been fuelled by the fundamental role that diffusion bridge simulation plays in the statistical inference for diffusion processes. Any statistical analysis which requires the transition density of the process is halted whenever the latter is not explicitly available, which is typically the case. Hence it is challenging to fit diffusion models employed in applications to the incomplete data typically available. An interesting possibility is to approximate the intractable transition density using an appropriate MC scheme and carry out the analysis using the approximation. It is of course desirable that the MC scheme is such that the approximation error in the analysis decreases to 0 as the MC effort increases. It turns out that basically all such MC schemes require diffusion bridge simulation.

We have been vague about the nature of "statistical analysis" mentioned above, since a range of statistical problems can be tackled using the diffusion bridge simulation methodology we develop here: parameter estimation of discretely observed diffusions, on-line inference for diffusions observed with error, off-line posterior estimation of partially observed diffusions, etc. Additionally, a range of computational statistics tools can be combined with the simulation methodology to give answers to the aforementioned problems: the EM algorithm, simulated likelihood, Sequential Monte Carlo, Markov chain Monte Carlo, etc. Given the wide range of applications where diffusions are employed, it is not surprising that important methodological contributions in bridge simulation are published in scientific journals in statistics, applied probability, econometrics, computational physics, signal processing, etc. Naturally, there is a certain lack of communication across disciplines, and one of the aims of this chapter is to unify some fundamental techniques.

In this article we concentrate on two specific methodological components of the wide research agenda described above. Firstly, we derive importance sampling schemes for diffusion bridge simulation. We refer to diffusion bridge simulation as an *imputation* problem, since we wish to recover an unobserved path given its end points. Secondly, we demonstrate how the samples can provide estimators of the diffusion transition density. Such estimators can be directly used in a simulated likelihood framework to yield approximations to the maximum likelihood estimator for the parameters of a discretely observed diffusion. We refer to estimation of the transition density as an *estimation* problem.

A fundamental complication in this context is that the diffusion bridge is an infinite dimensional random variable. One strategy to tackle this issue is to first approximate the stochastic process with a finite-dimensional vector, a so-called skeleton of the bridge obtained at a collection of n intermediate time points in $[t_0, t_1]$. This step adds a further approximation error in the analysis, which to be eliminated n has to be chosen large enough. Subsequently, one has to devise a MC sampling scheme for the corresponding n-dimensional distribution. Let us for convenience call this paradigm the projection-simulation strategy. The problem with this approach is that typically reducing the approximation bias (increasing n) leads to an increase of the MC variance.

An alternative strategy is to design an appropriate MC scheme which operates on the infinite dimensional space, hence in principle it returns diffusion bridges. In our specific framework, we are interested in importance sampling for diffusion bridges. There is a certain mathematical hurdle in this direction since it requires changes of measure in infinite dimensional spaces. For practical implementation we might have to approximate the output of the simulation algorithm using a skeleton of the bridge based on n intermediate points. Again, the approximation bias is eliminated as $n \to \infty$. Let us refer to this paradigm as the simulation-projection strategy.

There are two main advantages of this strategy over the projection-simulation. Firstly, it often results in a much better bias/variance tradeoff. Roughly speaking, the fact that by construction a valid MC scheme exists in the limit $n \to \infty$ avoids the curse of dimensionality from which they often suffer the projection-simulation strategies. Secondly, in some contexts the approximation step is unnecessary, and unbiased MC methods can be devised. The retrospective simulation technique is instrumental in this context, and from an operational point of view the output of the algorithm consists of a skeleton of the diffusion bridge unveiled at a collection of random times.

The organisation of the chapter is as follows. In Section 4.2 essential introductory material is presented on diffusions and Monte Carlo methods. Sections 4.3 and 4.4 consider the fundamental problem of Importance Sampling and related methods for diffusion bridges. Finally, Section 4.5 discusses exact and unbiased methods for diffusion simulation.

4.2 Background

4.2.1 Diffusion processes

Diffusion processes are extensively used for modelling continuous-time phenomena in many scientific areas; an indicative list includes economics (Merton (1971); Bergstrom (1990)), finance (Cox, Ingersoll, and Ross (1985); Sundaresan (2000); Chan, Karolyi, Longstaff, and Sanders (1992)), biology (McAdams and Arkin (1997); Wilkinson (2006)), genetics (Kimura and Ohta (1971); Tan (2002)), chemistry and physics (Kampen (1981)), dynamical systems (Arnold (1998); Givon, Kupferman, and Stuart (2004)) and engineering (Bar-Shalom, Kirubarajan, and Li (2002)). Their appeal lies in the fact that the model is built by specifying the instantaneous mean and variance of the process through a stochastic differential equation (SDE). Specifically, a d-dimensional diffusion process $V \in R^d$ is a strong Markov process defined as the solution of an SDE of the type:

$$dV_s = b(s, V_s)\, ds + \sigma(s, V_s)\, dB_s, \quad s \in [0, T], V_0 = v_0; \qquad (4.1)$$

B is an m-dimensional standard Brownian motion, $b(\cdot, \cdot) : R_+ \times R^d \to R^d$ is
called the *drift*, $\sigma(\cdot, \cdot) : R_+ \times R^d \to R^{d \times m}$ is called the *diffusion coefficient*.
We will treat the initial point v_0 as fixed by the design, although it is straight-
forward to model it with a distribution on R^d. In applications the drift and the
diffusion matrix are only known up to some parameters, which have to be esti-
mated. It is convenient to introduce also $\Gamma = \sigma\sigma^*$, where σ^* denotes the matrix
transpose. We assume that all coefficients are sufficiently regular so that (4.1)
has a unique weak non-explosive solution, and (crucially) that the Cameron-
Martin-Girsanov theorm holds. Details of these conditions can be found for
example in Rogers and Williams (2000).

The popularity of SDEs in time-series modelling is due to various reasons:
they provide a flexible framework for modelling both stationary processes with
quite general invariant distributions and non-stationary processes, see for example Sørensen (2012); in many applications they arise as limits of discrete-
time and/or discrete-space Markov processes; they are a natural stochastic
counterpart to deterministic modelling using Ordinary Differential Equations
(ODEs); the Markov property is particularly convenient from a computational
perspective, allowing fast statistical inference for long time-series; smooth processes can also be modelled by allowing the drift and diffusion coefficient to
depend on the past of the process.

The SDE (4.1) describes the *microscopic* behaviour of the process, i.e. its dynamics in infinitesimal time increments. One the other hand, the *exact macroscopic* dynamics of the diffusion process are governed by its transition density:

$$p_{s,t}(v, w) = \Pr\left[V_t \in dw \mid V_s = v\right] / dw, \quad t > s, \, w, v \in R^d. \quad (4.2)$$

There are very few examples of SDEs with tractable transition densities. One
generic class of such processes is the so-called linear SDEs, where the drift
is linear in the state variable and the diffusion matrix is constant with respect
to the state variable; see the corresponding subsection of this section. This
class incorporates the Ornstein-Uhlenbeck process, a special case of which is
a model due to Vasicek (1977) for the term structure of interest rates. Also in
the context of interest rate modelling Cox et al. (1985) proposed a non-linear
SDE, which, however, has a known transition density.

Although typically intractable, the transition density has various representations which suggest different approaches for its approximation. We could identify two main representations. First, it is given as a solution of the Fokker-
Planck partial differential equation (PDE) with appropriate initial and boundary conditions. There are various methods to solve the PDE numerically, see
for example Hurn, Jeisman, and Lindsay (2007) for a recent article which in-
vestigates this possibility in the context of estimation of diffusions. Second, it
can be expressed in various ways as an expectation, and these expressions lend

themselves to Monte Carlo approximation. It is this second approach which is pursued in this article and linked with diffusion bridge simulation.

Numerical approximation

The core of Monte Carlo methodology for SDEs is the simulation of a skeleton of the process $\{V_{t_0}, V_{t_1}, \ldots, V_{t_n}\}$. In fact, there are two types of simulations which can be considered for SDEs. Simulating the strong solution essentially corresponds to jointly constructing V, the solution of (4.1), and B, the Brownian motion which drives it. On the other hand, simulating the weak solution only asks for simulating V according to the probability law implied by (4.1).

Note that due to the strong Markov property, *exact simulation* of a skeleton entails sequential simulation from the transition density (4.2), which however is typically intractable. Exact simulation of strong solutions is clearly an even harder problem. Nevertheless, a vast collection of *approximate simulation* schemes are available based on discretizations of the SDE (4.1). The simplest approximation scheme is the Euler-Maruyama approximation (Maruyama (1955)):

$$V_{t+\Delta t} \approx V_t + b(t, V_t)\Delta t + \sigma(t, V_t) \cdot (B_{t+\Delta t} - B_t). \qquad (4.3)$$

In this form the discretisation tries to approximate the diffusion in a *strong* sense. A weak Euler method is not constrained to have Gaussian innovations, but it will have innovations with the correct mean and variance (up to a certain order of Δt). Let V_T^Δ denote an approximate solution based on a Δ-discretization of $[0, T]$. We say that a strong approximation scheme is of order γ if

$$\mathrm{E}[\|V_T - V_T^\Delta\|] \leq K\Delta^\gamma$$

and correspondingly a weak approximation scheme is of order γ if

$$|\mathrm{E}[g(V_T)] - \mathrm{E}[g(V_T^\Delta)]| \leq K\Delta^\gamma$$

for suitable test functions g. Under suitable regularity conditions on the coefficients of the SDE, a strong Euler scheme is of order $1/2$ in the strong sense, whereas in general Euler schemes are of order 1 in the weak sense. Many higher order schemes exist. Some are based on the Itô-Taylor expansion, and there are also implicit and split-step methods (which are particularly important in the construction of MCMC methods using diffusion dynamics). For a detailed exposition of numerical approximation of SDEs we refer to Kloeden and Platen (1995).

Diffusion bridges

We now consider the dynamics of the process V not only conditioned on its initial point, $V_0 = u$, but also on its ending point, $V_T = v$. The conditioned

process, which we will also denote by V (the distinction from the unconditioned process will be clear from the context), is still Markov, and the theory of
h-transforms, see for example Chapter IV.39 of Rogers and Williams (2000),
allows us to derive its SDE:

$$dV_s = \tilde{b}(s, V_s)\, ds + \sigma(s, V_s)\, dB_s, \quad s \in [0, T], V_0 = u;$$
$$\tilde{b}(s, x) = b(s, x) + [\sigma\sigma^*](s, x)\, \nabla_x \log p_{s,T}(x, v). \tag{4.4}$$

There are three main remarks on this representation. First, note that the local
characteristics of the unconditioned and conditioned processes are the same, in
the sense that they share the same diffusion coefficient. Second, the drift of the
conditioned process includes an extra term which forces the process to hit v at
time T. Third, although insightful for the bridge dynamics, (4.4) is typically
intractable since the drift is expressed in terms of the transition density.

Therefore, the diffusion bridge solves an SDE whose drift is intractable, hence
even approximate simulation using the schemes described above is infeasible.
Application of that technology would require first the approximation of the
transition density, and consequently a discretization of (4.4). This is clearly
impractical, and it is difficult to quantify the overall error of the approach.
Instead, we will consider alternative Monte Carlo schemes for simulating diffusion bridges.

Data and likelihood

In some contexts we can observe directly a *path* of the modelled process $V = (V_s, s \in [0, T])$. More realistically we might be able to observe a *skeleton* of
the process $\{V_{t_0}, V_{t_1}, \ldots, V_{t_n}\}$, but where the frequency of the data can be
chosen arbitrarily high. A typical example is molecular dynamics modelling,
where the data is simulated according to a complex deterministic model, see for
example Pavliotis, Pokern, and Stuart (2012). A rich mathematical framework
is available for statistical analyses in this high frequency regime, see for example Prakasa Rao (1999). Two main components of this theory is the quadratic
variation identity and the Cameron-Martin-Girsanov change of measure. According to the former, the local characteristics of the SDE can be completely
identified given an observed path. In particular, for any $t \in [0, T]$,

$$\lim_{\Delta \to 0} \sum_{t_j \leq t} (V_{t_{j+1}} - V_{t_j})(V_{t_{j+1}} - V_{t_j})^* = \int_0^t [\sigma\sigma^*](s, V_s)\, ds \tag{4.5}$$

in probability for any partition $0 = t_0 \leq t_1 \leq \cdots \leq t_n = t$, whose mesh
is Δ; see Jacod (2012). This implies that from high frequency data we can
consistently estimate the diffusion coefficient. A further implication is that the
probability laws which correspond to SDEs with different diffusion coefficients
are mutually singular. On the contrary, under weak conditions the laws which

correspond to SDEs with the same diffusion coefficient but different drifts are equivalent and a simple expression for the Radon-Nikodym derivative is available. This is the context of the Cameron-Martin-Girsanov theorem for Itô processes, see for example Theorem 8.6.6 of Øksendal (1998).

In the context of (4.1) consider functionals u and α of the dimensions of b and assume that u solves the equation:

$$\sigma(s, x)h(s, x) = b(s, x) - \alpha(s, x).$$

Additionally, let $\mathbb{P}_b$ and $\mathbb{P}_\alpha$ be the probability laws implied by the (4.1) with drift b and α respectively. Then, under certain conditions $\mathbb{P}_b$ and $\mathbb{P}_\alpha$ are equivalent with density (*continuous time likelihood*) on $\mathcal{F}_t = \sigma(V_s, s \leq t), t \leq T$, given by

$$\left.\frac{d\mathbb{P}_b}{d\mathbb{P}_\alpha}\right|_t = \exp\left\{\int_0^t h(s, V_s)^* dB_s - \frac{1}{2}\int_0^t [h^*h](s, V_s)ds\right\}. \qquad (4.6)$$

In this expression, B is the $\mathbb{P}_\alpha$ Brownian motion, and although this is the usual probabilistic statement of the Cameron-Martin-Girsanov theorem, it is not a natural expression to be used in statistical inference, and alternatives are necessary. For example, note that when σ can be inverted, the expression can be considerably simplified. Recall that $\Gamma = \sigma\sigma^*$, then the density becomes

$$\exp\left\{\int_0^t [(b-\alpha)^*\Gamma^{-1}](s, V_s)dV_s - \frac{1}{2}\int_0^t [(b-\alpha)^*\Gamma^{-1}(b+\alpha)](s, V_s)ds\right\}. \qquad (4.7)$$

Appendix 2 contains a simple presentation of change of measure for Gaussian multivariate distributions, which might be useful for the intuition behind the Girsanov theorem. For statistical inference about the drift (4.1) the Girsanov theorem is used with $\alpha = 0$. Any unknown parameters in the drift can be estimated by using (4.7) as a likelihood function. In practice, the integrals in the density are approximated by sums, leading to an error which can be controlled provided the data are available at arbitrarily high frequency.

Nevertheless, in the majority of applications, V can only be partially observed. The simplest case is that of a *discretely observed diffusion*, where we observe a skeleton of the process $\{V_{t_0}, V_{t_1}, \ldots, V_{t_n}\}$, but without any control on the frequency of the data. As a result, the approach described above is not feasible since it might lead to large biases, see for example Dacunha-Castelle and Florens-Zmirou (1986). From a different persective, we deal with data from a Markov process, hence the joint Lebesgue density of a sample (*discrete time likelihood*) is simply given by the product of the transition densities

$$\prod_{i=0}^{n-1} p_{t_i, t_{i+1}}(V_{t_i}, V_{t_{i+1}}). \qquad (4.8)$$

Unknown parameters in the drift and diffusion coefficient can be estimated

working with this discrete-time likelihood. Theoretical properties of such estimators are now well known in particular under ergodicity assumptions, see for
example Kessler (1997); Gobet (2002); Kutoyants (2004). Unfortunately the
discrete-time likelihood is not practically useful in all those cases where the
transition density of the diffusion is analytically unavailable.

More complicated data structures are very common. In many applications V
consists of many components which might not be synchroneously observed,
or there the observation might be subject to measurement error, or there might
be components completely latent. However, likelihood estimation of discretely
observed diffusions will serve as a motivating problem throughout the subsequent methodological sections.

Linear SDEs

A great part of the diffusion bridge simulation methodology is based on the
tractability of linear SDEs and uses this class as a building block. Hence, it
is useful to include a short description of this class. This is a large family of
SDEs characterised by state-independent diffusion coefficient and drift which
is linear in the state variable. In the most general form we have

$$dV_s = (D(s) V_s + G(s))\, ds + E(s)\, dB_s\,; \qquad (4.9)$$

hence $b(s, x) = D(s)x + G(s)$, and $\sigma(s, x) = E(s)$, where D, G, E are
matrix-valued functions of appropriate dimensions, which are allowed to depend only on time.

The SDE in (4.9) can be solved explicitly when $d = m = 1$. We define $P(s)$
to be the solution of the linear ODE

$$\frac{dP}{ds} = D(s)P(s)\,, \quad P(0) = 1\,. \qquad (4.10)$$

Then the SDE is solved by

$$V_t = P(t) \int_0^t P(s)^{-1}(G(s)ds + E(s)dB_s) + P(t)v_0\,. \qquad (4.11)$$

It follows that V is a *Gaussian process* with mean $m_t := \mathrm{E}(V_t)$ and covariance
matrix $C_t := \mathrm{Cov}(V_t, V_t)$, which solve the systems of ODEs

$$\frac{dm_t}{dt} = D(t)m_t + G(t)\,, \quad m_0 = v_0$$

$$\frac{dC_t}{dt} = D(t)C_t + C_t D(t)^* + \Gamma(t)\,, \quad C_0 = 0.$$

In various contexts these ODEs can be solved analytically. When $d = m = 1$,
$D(t)^* = D(t)$. When $d > 1$, it is not generally possible to solve explicitly
(4.9) at that level of generality. The above argument can be carried through to

the $d > 1$ case when $D(s)D(t) = D(t)D(s)$ for all t, s, in which case the matrix/vector versions of the previous equations remain valid.

The transition density of linear SDEs is Gaussian with mean and variance which can be identified in the cases discussed above. A consequence of this is a further appealing feature of linear SDEs, that the corresponding bridge processes have tractable dynamics. This can be seen directly from the h-transform, since the gradient of the log-density is a linear function of the state, hence from (4.4) we have that the bridge process is also a linear SDE. This can be proved also from first principles working with the finite-dimensional distributions of the conditioned process, see for example Theorem 2 of Delyon and Hu (2006). In the simplest setup where $b = 0$ and σ is the identity matrix, the linear SDE is the Brownian motion and the bridge process conditioned upon $V_T = v$ is known as *Brownian bridge*, which solves the time-inhomogeneous SDE

$$\mathrm{d}V_s = \frac{v - V_s}{T - s}\mathrm{d}s + \mathrm{d}B_s \,. \tag{4.12}$$

Although the coefficient of V in the drift is time-inhomogenous, it satisfies the commutability condition given above, and the SDE can be solved to yield for $0 < t_1 < t_2 < T$,

$$V_{t_2} \mid V_{t_1} \sim N\left(V_{t_1} + \frac{t_2 - t_1}{T - t_1}(v - V_{t_1}), \frac{(t_2 - t_1)(T - t_2)}{T - t_1}\right) \,. \tag{4.13}$$

4.2.2 Importance sampling and identities

Importance sampling (IS) is a classic Monte Carlo technique for obtaining samples from a probability measure $\mathbb{P}$ using samples from another probability measure $\mathbb{Q}$, see for example Chapter 2.5 of Liu (2008) for an introduction. Mathematically it is based on the concept of *change of measure*. Suppose that $\mathbb{P}$ is *absolutely continuous* with respect to $\mathbb{Q}$ with Radon-Nikodym density $f(x) = \mathbb{P}(\mathrm{d}x)/\mathbb{Q}(\mathrm{d}x)$. Then, in its simplest form IS consists of constructing a set of *weighted particles* (x_i, w_i), $i = 1, \ldots, N$, where $x_i \sim \mathbb{Q}$, and $w_i = f(x_i)$. This set gives a Monte Carlo approximation of $\mathbb{P}$, in the sense that for suitably integrable functions g, we have that

$$\frac{\sum_{i=1}^{N} g(x_i)w_i}{N} \tag{4.14}$$

is an unbiased and consistent estimator of

$$\mathrm{E}_{\mathbb{P}}[g] := \int g(x)\mathbb{P}(\mathrm{d}x) \,.$$

However, IS can be cast in much more general terms, an extension particularly attractive in the context of stochastic processes. First, note that in most applications f is known only up to a normalising constant, $f(x) = cf_u(x)$, where only f_u can be evaluated and

$$c = \mathrm{E}_\mathbb{Q}[f_u] \,. \tag{4.15}$$

The notion of a *properly weighted sample*, see for example Section 2.5.4 of Liu (2008) refers to a set of weighted particles (x_i, w_i), where $x_i \sim \mathbb{Q}$ and w_i is an *unbiased estimator* of $f_u(x_i)$, that is

$$\mathrm{E}_\mathbb{Q}[w_i \mid x_i] = f_u(x_i) \,.$$

In this setup we have the fundamental equality for any integrable g

$$\mathrm{E}_\mathbb{Q}[gw] = \mathrm{E}_\mathbb{P}[g]\,\mathrm{E}_\mathbb{Q}[w] \,. \tag{4.16}$$

Rearranging the expression we find that a *consistent* estimator of $\mathrm{E}_\mathbb{P}[g]$ is given by

$$\frac{\sum_{i=1}^N g(x_i)w_i}{\sum_{i=1}^N w_i} \,. \tag{4.17}$$

When w_i is an unbiased estimator of $f(x_i)$ we have the option of using (4.14), thus yielding an unbiased estimator. However, (4.17) is a feasible estimator when c is unknown.

Although the first moment of w (under $\mathbb{Q}$) exists by construction, the same is not true for its second moment. Hence it is a minimal requirement of a "good" proposal distribution $\mathbb{Q}$ that $\mathrm{E}_\mathbb{Q}[w^2] < \infty$. In this case, and using the Delta method for ratio of averages it can be shown that (4.17) is often preferable to (4.14) in a mean square error sense because the denominator acts effectively as a control variable. Therefore it might be preferable even when c is known. The same analysis leads to an interesting approximation for the variance of (4.17). Assume for simplicity that $\mathrm{E}_\mathbb{Q}[w] = 1$. Then exploiting the fact that

$$\mathrm{E}_\mathbb{P}[w] = \mathrm{var}_\mathbb{Q}[w] + 1 \,,$$

using a further Taylor expansion we obtain the following approximation for the variance of (4.17):

$$\frac{1}{N}\mathrm{var}_\mathbb{P}[g](1 + \mathrm{var}_\mathbb{Q}[w]) \,. \tag{4.18}$$

Although this might be a poor approximation when the residual terms are significant, the expression motivates the notion of the *effective sample size (ESS)*, $1/(\mathrm{var}_\mathbb{Q}[w]+1)$. This corresponds to an approximation of the ratio of variances of a Monte Carlo estimator of $\mathrm{E}_\mathbb{P}[g]$ based on independent samples from $\mathbb{P}$, and the IS estimator (4.17). The most appealing feature of the above approximation is that it does not depend on the function g, hence ESS can be used as a rough indication of the effectiveness of the IS approximation of $\mathbb{P}$. $N\times$ESS can be interpreted as the equivalent number of independent samples from $\mathbb{P}$. For more details see Section 2.5.3 of Liu (2008) and references therein.

The general framework where w is an unbiased estimator of the Radon-Niko-dym derivative between $\mathbb{P}$ and $\mathbb{Q}$, opens various possibilities: constructing new Monte Carlo schemes, see e.g Partial Rejection Control, see Section 2.6 of Liu (2008), devising schemes whose computational complexity scales well with the number of particles N, e.g. the auxiliary particle filter of Pitt and Shephard (1999), or applying IS in cases where even the computation of f_u is infeasible, see e.g. the random weight IS for diffusions of Beskos, Papaspiliopoulos, Roberts, and Fearnhead (2006); Fearnhead, Papaspiliopoulos, and Roberts (2008) which is also covered in detail in this article.

IS includes exact simulation as a special case when $\mathbb{Q} = \mathbb{P}$. Another special case is *rejection sampling* (RS), which assumes further that $f_u(x)$ is bounded in x by some calculable $K < \infty$. Then, if we accept each draw x_i with probability $f_u(x_i)/K$, the resulting sample (of random size) consists of independent draws from $\mathbb{P}$. This is a special case of the generalised IS where w_i is a binary 0-1 random variable taking the value 1 with probability $f_u(x_i)/K$.

The IS output can be used in estimating various normalising constants and density values involved in the costruction. It follows directly from the previous exposition that $c = \mathrm{E}_{\mathbb{Q}}[w]$. Moreover, note that

$$f(x) = \mathrm{E}_{\mathbb{Q}}[w \mid x]/\mathrm{E}_{\mathbb{Q}}[w] \,. \tag{4.19}$$

4.3 IS estimators based on bridge processes

Let V be a multivariate diffusion (4.1) observed at two consecutive time points $V_0 = u$, $V_T = v$, and consider the following two problems: a) (*imputation*) the design of efficient IS scheme for the corresponding diffusion bridge, and b) (*estimation*) the MC estimation of the corresponding transition density (4.2). In this section we consider these problems for a specific class of diffusions: those for which the diffusion coefficient is indepedent of V. In this context the methodology is much simpler and important developments have been made since the mid-80s. Furthermore, under additional structure exact simulation of diffusion bridges is feasible (see Section 4.5).

Before detailing the approach let us briefly discuss the restriction imposed by the assumption that the diffusion coefficient is independent of V. For scalar diffusions when $\sigma(\cdot, \cdot)$ is appropriately differentiable, by Itô's rule the transformation $V_s \to \eta(s, V_s) =: X_s$, where

$$\eta(s, u) = \int^u \frac{1}{\sigma(s, z)} \, \mathrm{d}z, \tag{4.20}$$

is any anti-derivative of $\sigma^{-1}(s, \cdot)$, yields a diffusion X with diffusion coefficient 1. A particle approximation of the diffusion bridge of X directly implies one for V and the transition densities of the two processes are linked by a change of variables formula. Therefore, the methodology of this section effectively covers all scalar diffusions, as well as a wide variety of multivariate processes used in applications. At the end of this section we discuss the limitations of the methodology based on bridge processes.

Let $\mathbb{P}_b$ be the law of the diffusion V on $[0, T]$ with $V_0 = u$ (abusing slighlty the notation set up in Section 4.2.1). Similarly, let $\mathbb{P}_0$ denote the law of the driftless process $dV_s = \sigma(s)dB_s$. Crucially, in this setting the driftless process is a linear SDE and $\mathbb{P}_0$ is a Gaussian measure. Additionally, let $p_{0,T}(u, v)$ and $\mathcal{G}_{0,T}(u, v)$ denote the transition densities of the two processes. Let $\mathbb{P}_b^*$ and $\mathbb{P}_0^*$ denote the laws of the corresponding diffusion bridges conditioned on $V_T = v$. As we discussed in Section 4.2.1, the conditioned driftless process is also a linear SDE.

We present a heuristic argument for deriving the density $d\mathbb{P}_b^*/d\mathbb{P}_0^*$. Consider the decomposition of the laws $\mathbb{P}_b$ and $\mathbb{P}_0$ into the marginal distributions at time T and the diffusion bridge laws conditioned on V_T. Then by a marginal-conditional decomposition we have that for a path V with $V_0 = u$,

$$\frac{d\mathbb{P}_b}{d\mathbb{P}_0}(V) \, 1[V_T = v] = \frac{p_{0,T}(u, v)}{\mathcal{G}_{0,T}(u, v)} \frac{d\mathbb{P}_b^*}{d\mathbb{P}_0^*}(V). \tag{4.21}$$

The term on the left-hand side is given by the Cameron-Martin-Girsanov theorem (see Section 4.2.1). Hence, by a rearrangement we get the density between the diffusion bridge laws:

$$\frac{d\mathbb{P}_b^*}{d\mathbb{P}_0^*}(V) = \frac{\mathcal{G}_{0,T}(u, v)}{p_{0,T}(u, v)} \exp\left\{ \int_0^T h(s, V_s)^* dB_s - \frac{1}{2} \int_0^T [h^*h](s, V_s) ds \right\}, \tag{4.22}$$

where h solves $\sigma h = b$ (see Section 4.2.1), and B is Brownian motion.

Additional structure on b and σ can lead to further simplifications of (4.22). We emphasize the setting where σ is the identity matrix, the diffusion is time-homogenous and of *gradient-type*, i.e. there exists a field H such that $b(v) = \nabla_v H(v)$. When the function $\rho(v) \propto \exp\{H(v)/2\}$ is integrable, the diffusion is a reversible Markov process with ρ as the invariant density. In this setting, we can use Itô's rule to perform integration by parts in the exponent of (4.22) to eliminate the stochastic integral, and obtain

$$\frac{d\mathbb{P}_b^*}{d\mathbb{P}_0^*}(V)$$

$$= \frac{\mathcal{G}_{0,T}(u, v)}{p_{0,T}(u, v)} \exp\left\{ H(v) - H(u) - \frac{1}{2} \int_0^T \left(\|b(V_s)\|^2 + \nabla^2 H(V_s) \right) ds \right\}. \tag{4.23}$$

(4.22) forms the basis for a particle approximation of the law of $\mathbb{P}_b^*$ using proposals from $\mathbb{P}_0^*$. An idealized algorithm proceeds by first generating a linear SDE according to $\mathbb{P}_0^*$, and subsequently, by assigning weight according to (4.22). Note in particular that B in (4.22) is the Brownian motion driving the proposed linear bridge. The weights are known only up to a normalizing constant due to the presence of $p_{0,T}(u, v)$. However, as we saw in Section 4.2.2 this poses no serious complication in the applciation of IS. Note that $\mathcal{G}_{0,T}(u, v)$ is a Gaussian density which can be computed and be included explicitly in the weights, although this is not necessary for the IS.

Practically, we will have to simulate the proposed bridge at a finite collection of M times in $[0, T]$ and approximate the integrals in the weights by sums. This is an instance of the simulation-projection strategy outlined in Section 4.1. It introduces a bias in the MC approximations which is eliminated as $M \to \infty$. It is a subtle and largely unresolved issue how to distribute a fixed computational effort between M and N in order to minimize the MC variance of estimates of expectations of a class of test functions. In Section 4.5 we will see that in the more specific case of (4.23) the approximations can be avoided altogether and construct a properly weighted sample using unbiased estimators of the weights.

It follows directly from the general development of Section 4.2.2 that the diffusion transition density can be consistently estimated using a particle approximation of $\mathbb{P}_b^*$. From (4.15) it follows the key identity

$$p_{0,T}(u, v)$$

$$= \mathcal{G}_{0,T}(u, v) \mathrm{E}_{\mathbb{P}_0^*} \left[\exp \left\{ \int_0^T h(s, V_s)^* \mathrm{d}B_s - \frac{1}{2} \int_0^T [h^*h](s, V_s) \mathrm{d}s \right\} \right].$$
(4.24)

In the case where b is of gradient form we can correspondingly write

$$p_{0,T}(u, v) = \mathcal{G}_{0,T}(u, v) \exp\{H(v) - H(u)\} \mathrm{E}_{\mathbb{P}_0^*} \left[\exp \left\{ - \int_0^T \phi(V_s) \mathrm{d}s \right\} \right],$$
(4.25)

where $\phi(z) = (\|b(z)\|^2 + \nabla^2 H(z))/2$. Hence, the transition density is estimated by the average of the IS weights. It is at this stage where the explicit computation of the Gaussian density in the denominator of (4.22) becomes indispensable: if it were unknown we could only estimate the ratio of the two transition densities, but not $p_{0,T}(u, v)$.

Historical development

The expressions (4.22) and (4.24) have been derived several times in the literature with different motives. Remarkably, there is almost no cross-referencing among the papers which have derived the expressions. To our best knowledge, the expressions appear for the first time for scalar diffusions in the proof

of Theorem 1 of Rogers (1985). The context of the Theorem is to establish
smoothness of the transition density. Again for scalar diffusions the expressions appear in the proofs of Lemma 1 of Dacunha-Castelle and Florens-Zmirou
(1986). The context of that paper is a quantification of the error in parameter
estimates obtained using approximations of the transition density. Since both
papers deal with scalar diffusions, they apply the integration by parts to get
the simplified expression (4.23). More recently, Durham and Gallant (2002)
working in a projection-simulation paradigm, derive effectively an IS for $\mathbb{P}_b^*$
and an estimator of $p_{0,T}(u, v)$, which in the case of constant diffusion coefficient are discretizations of (4.22) and (4.24) (see also Section 4.4 below). The
context here is MC estimation of diffusion models. Since the authors work in a
time-discretized framework from the beginning, the possibility to perform integration by parts when possible, is not at all considered. Nicolau (2002) uses
the Dacunha-Castelle and Florens-Zmirou (1986) expression for the transition
density as a basis for MC estimation using approximation of the weights based
on M intermediate points. Beskos, Papaspiliopoulos, Roberts, and Fearnhead
(2006) used (4.23) as a starting point for the exact simulation of diffusions and
(4.24) as a basis for unbiased estimation of the transition density (see also Section 4.5). Finally, Delyon and Hu (2006) state (4.22) as Theorem 2 and prove
it for multivariate processes.

Limitations of the methodology

The outline of the methodology we have described in this section for IS approximation of $\mathbb{P}_b^*$ is to find a probability measure $\mathbb{P}_0$ which is absolutely continuous
with respect to the unconditional measure $\mathbb{P}_b$, and *probabilistically condition*
the former on the same event that the latter is conditioned upon. Hence, the
proposed random variables are indeed *bridge processes*. The same development can be carried out even when σ depends on the state variable. In the more
general setup $\mathbb{P}_0$ is the law of $dV_s = \sigma(s, V_s)dB_s$, which is now a non-linear
diffusion. Hence the corresponding bridge process will be typically intractable
(see Section 4.2.1) and the IS practically infeasible. Therefore, the methodology of this section applies only to diffusions which can be transformed to have
state-independent diffusion coefficient.

For multivariate diffusions with V-dependent volatility the generalized version
of transformation (4.20) involves the solution of an appropriate vector differential equation which is often intractable or insolvable, see for example Aït-
Sahalia (2008). A very popular class of models which have state-dependent
volatility are stochastic volatility models employed in financial econometrics.
The methodology of space transformation can be generalised to include time-
change methodology to allow models like stochastic volatility models to be
addressed, see Kalogeropoulos, Roberts, and Dellaportas (2010).

Summarising, constructing valid and easy to simulate proposals by *conditioning* is difficult when we deal with multivariate processes with state-dependent diffusion coefficient. Instead, the next section considers a different possibility where the proposals are generated by a process which is explicitly constructed to hit the desired end-point at time T. We call such processes *guided*.

4.4 IS estimators based on guided processes

In this section we consider the same two problems described in the beginning of Section 4.3 but for processes with state-dependent diffusion coefficient. We consider IS approximation of the law of the target diffusion bridge, $\mathbb{P}_b^*$, using appropriate diffusion processes as proposals. The design of such proposals is guided by the SDE of the target bridge, given in (4.4), and the Cameron-Martin-Girsanov theorem. Hence, the diffusion coefficient of any valid proposal has to be precisely $\sigma(s, v)$, and the drift has to be such that it forces the process to hit the value v at time T almost surely. The following processes are natural candidates under these considerations:

$$[G1] \quad dV_s = -\frac{V_s - v}{T - s} ds + \sigma(s, V_s) dB_s \tag{4.26}$$

$$[G2] \quad dV_s = -\frac{V_s - v}{T - s} ds + b(s, V_s) ds + \sigma(s, V_s) dB_s . \tag{4.27}$$

Note that the drift of [G1] ("G" stands for "Guided") is precisely the one of the Brownian bridge (4.12); the one of [G2] mimics the structure of the drift of the target bridge process (4.4) but substitutes the intractable term in the drift by the Brownian bridge drift. Let $\mathbb{Q}_{G1}$ and $\mathbb{Q}_{G2}$ denote the laws of [G1] and [G2] correspondingly. We will use [G] to refer to a generic guided process.

The mathematical argument which yields the IS is formally presented in Section 4 of Delyon and Hu (2006). The construction requires for tractability that σ is invertible so that we can work with (4.7) and introduce explicitly V (instead of the driving B) in the weights. To simplify the exposition, we present the argument when $d = 1$; the formulae extend naturally to the multidimensional case, under the same assumptions.

We define for any z and $s \leq T$,

$$A(s, z) = (\sigma(s, z))^{-2} .$$

Up to any time $t < T$, $\mathbb{Q}_{G2}|_t$ (resp. $\mathbb{Q}_{G1}|_t$) is absolutely continuous with respect to $\mathbb{P}_b^*|_t$, and we can apply the Cameron-Martin-Girsanov theorem (4.7). The resulting likelihood ratio, although of expected value 1, it converges almost surely to 0 as $t \to T$. To identify the leading term which drives the weights to 0 (denoted ψ_t below) we apply an integration by parts to the stochastic integral in the exponent. Let us define the following functionals

$$\psi_t = \exp\left\{-\frac{1}{2(T-t)}(V_t - v)^2 A(t, V_t)\right\}$$

$$C_t = \frac{1}{(T-t)^{1/2}}$$

$$\log(\phi_t^{G2}) = -\int_0^t \frac{(V_s - v)}{T - s} A(s, V_s) b(s, V_s) ds$$

$$-\int_0^t \frac{(V_s - v)^2}{2(T - s)} \diamond dA(s, V_s)$$

$$\log(\phi_t^{G1}) = \int_0^t b(s, V_s) A(s, V_s) dV_s - \frac{1}{2}\int_0^t b(s, V_s)^2 A(s, V_s) ds$$

$$-\int_0^t \frac{(V_s - v)^2}{2(T - s)} \diamond dA(s, V_s)$$

where the $\diamond$-stochastic integral is understood as the limit of approximating sums where the integrand is evaluated at the right-hand time-points of each sub-interval (as opposed to the left-hand in the definition of the Itô stochastic integral). Then, we have that

$$\left.\frac{d\mathbb{Q}_{G2}}{d\mathbb{P}_b}\right|_t = \sqrt{T}\exp\left\{\frac{(u - v)^2 A(0, u)}{2T}\right\} C_t \psi_t / \phi_t^{G2}$$

$$\left.\frac{d\mathbb{Q}_{G1}}{d\mathbb{P}_b}\right|_t = \sqrt{T}\exp\left\{\frac{(u - v)^2 A(0, u)}{2T}\right\} C_t \psi_t / \phi_t^{G1}.$$

Therefore, for any measurable (with respect to the filtration of V up to t) non-negative function f_t, we have that

$$E_{\mathbb{P}_b}[f_t(V)\psi_t] = C_t^{-1} T^{-1/2} \exp\left\{-\frac{(u - v)^2 A(0, u)}{2T}\right\} E_{\mathbb{Q}_{G2}}[f_t(V)\phi_t^{G2}]$$

$$= C_t^{-1} T^{-1/2} \exp\left\{-\frac{(u - v)^2 A(0, u)}{2T}\right\} E_{\mathbb{Q}_{G1}}[f_t(V)\phi_t^{G1}]$$

$$E_{\mathbb{P}_b}[\psi_t] = C_t^{-1} T^{-1/2} \exp\left\{-\frac{(u - v)^2 A(0, u)}{2T}\right\} E_{\mathbb{Q}_{G2}}[\phi_t^{G2}]$$

$$= C_t^{-1} T^{-1/2} \exp\left\{-\frac{(u - v)^2 A(0, u)}{2T}\right\} E_{\mathbb{Q}_{G1}}[\phi_t^{G1}],$$

where the expression for $E_{\mathbb{P}_b}[\psi_t]$ is obtained from the first expression with $f_t = 1$.

Hence, we derive the key equality, that for any positive measurable f_t,

$$\frac{E_{\mathbb{P}_b}[f_t(V)\psi_t]}{E_{\mathbb{P}_b}[\psi_t]} = \frac{E_{\mathbb{Q}_{G2}}[f_t(V)\phi_t^{G2}]}{E_{\mathbb{Q}_{G2}}[\phi_t^{G2}]} = \frac{E_{\mathbb{Q}_{G1}}[f_t(V)\phi_t^{G1}]}{E_{\mathbb{Q}_{G1}}[\phi_t^{G1}]}.$$

The final part of the argument consists of taking the limit $t \to T$ on each part of the previous equality (this requires a careful non-trivial technical argument, see proof of Theorem 5 and related Lemmas of Delyon and Hu (2006)). The limit on the left hand side converges to the regular conditional expectation $E_{\mathbb{P}_b^*}[f_T(V)] = E_{\mathbb{P}_b}[f_T(V) \mid V_T = v]$; intuitively this can be verified by the form of ψ_t given above. The other two terms converge to $E_{\mathbb{Q}_{G2}}[f_T(V)\phi_T^{G2}]/E_{\mathbb{Q}_{G2}}[\phi_T^{G2}]$ and $E_{\mathbb{Q}_{G1}}[f_T(V)\phi_T^{G1}]/E_{\mathbb{Q}_{G1}}[\phi_T^{G1}]$, respectively. Therefore, we have that

$$\frac{d\mathbb{P}_b^*}{d\mathbb{Q}_{G2}}(V) = \frac{\phi_T^{G2}}{E_{\mathbb{Q}_{G2}}[\phi_T^{G2}]} \tag{4.28}$$

$$\frac{d\mathbb{P}_b^*}{d\mathbb{Q}_{G1}}(V) = \frac{\phi_T^{G1}}{E_{\mathbb{Q}_{G1}}[\phi_T^{G1}]}, \tag{4.29}$$

where the denominators on the right-hand side in each expression are normalising constants. These two expressions are all is needed for the IS approximation of the diffusion bridge. Practically, as in Section 4.3, we will have to simulate the proposed bridge at a finite collection of M times in $[0, T]$ and approximate the integrals in the weights by sums, which introduces a bias in the MC approximations which is eliminated as $M \to \infty$.

We now address the problem of deriving a transition density identity, as we did in (4.24). To our best knowledge, this is the first time that such an expression appears in the literature. Note, however, that our argument is informal and certain technical conditions (outside the scope of this article) will have to be imposed for a formal derivation. Working with guided processes, this derivation is much less immediate than in Section 4.3.

Since ψ_t is a function of V_t only, we have that

$$E_{\mathbb{P}_b}[C_t\psi_t]$$
$$= \int \frac{1}{\sqrt{T-t}} \exp\left\{-\frac{1}{2(T-t)}(w-v)^2 A(t,w)\right\} p_{0,t}(u,w)dw$$
$$= \int \exp\left\{-\frac{1}{2}z^2 A(t, z\sqrt{T-t}+v)\right\} p_{0,t}(u, z\sqrt{T-t}+v)dz$$
$$\to_{t \to T} p_{0,T}(u,v) \int \exp\left\{-\frac{1}{2}z^2 A(T,v)\right\} dz = \frac{\sqrt{2\pi}\, p_{0,T}(u,v)}{\sqrt{A(T,v)}},$$

where taking the limit we have used dominated convergence (which clearly requires certain assumptions). We can use this expression, together with the identities which link $E_{\mathbb{P}_b}[C_t\psi_t]$ with $E_{\mathbb{Q}_{Gi}}[\phi_t^{Gi}]$, $i = 1, 2$ given above, and the fact that the latter converge as $t \to T$ to $E_{\mathbb{Q}_{Gi}}[\phi_T^{Gi}]$, $i = 1, 2$ (which is shown in Delyon and Hu (2006)), to establish the fundamental identity

$$p_{0,T}(u,v) = \sqrt{\frac{A(T,v)}{2\pi T}} \exp\left\{ -\frac{(u-v)^2 A(0,u)}{2T} \right\} E_{\mathbb{Q}_{Gi}}[\phi_T^{Gi}], \, i = 1,2 \, .$$

(4.30)

Therefore, given the IS output the transition density can be estimated.

Connections to the literature and to Section 4.3

The first major contribution to IS in this context was made in the seminal article of Durham and Gallant (2002) in the context of estimating the transition density of non-linear diffusions for statistical inference. They took, however, a projection-simulation approach, they first discretized the unobserved paths and then considered discrete-time processes as proposals. They suggest two different discrete-time processes as proposals, the so-called "Brownian bridge" proposal and the "modified Brownian bridge." They both are inspired by (and intend to be a type of discretization of) (4.26). Indeed, their "Brownian bridge" proposal is precisely a first-order Euler approximation of (4.26). The Euler approximation is not very appealing here since it is unable to capture the inhomogeneity in the variance of the transition distribution of (4.26). To see this more clearly, consider the simplified case where $\sigma = 1$ and contrast (4.12) with (4.13) when $t_2 - t_1$ is small. The Euler approximation suggests constant variance for the time-increments of the process, which is a very poor approximation when $t \approx T$. To mitigate against this lack of heteroscedasticity, Durham and Gallant (2002) use Bayes' theorem together with heuristic approximations to find a better approximation to the transition density of (4.26). The process with the new dynamics is termed "modified Brownian bridge," and it corresponds to the exact solution of the Brownian bridge SDE when $\sigma = 1$. Generally, due to various approximations at various stages the connection between IS for paths and estimation of the transition density is not particularly clear in their paper.

It is important to observe that the samplers and identities in this section become precisely those of Section 4.3 when σ is constant. An interesting deviation, is the use of (4.27) when σ is constant, which does not correspond to the setup of Section 4.3, and has the effect of making non-linear the proposal process (hence exact skeletons cannot be simulated in this case) but removes the stochastic integral from the weights (which typically has as a variance reduction effect).

As a final remark, note that when possible it is advisable to transform the diffusion to have unit volatility. Apart from facilitating the methodology of Section 4.3 the tansformation has been empirically shown to be a good variance reduction technique even for schemes which do not require it, see for example the discussion in Durham and Gallant (2002).

4.5 Unbiased Monte Carlo for diffusions

In many cases (including most all one-dimensional diffusions with sufficiently smooth coefficients) the need to use a fine discretisation for the diffusion sample path (and the associated approximation error) can be completely removed. This section will very briefly describe some of the basic ideas behind this approach, though for detailed account the reader is referred to Beskos, Papaspiliopoulos, Roberts, and Fearnhead (2006); Fearnhead et al. (2008). The methodology here is closely related to allied exact simulation algorithms for diffusions as described in Beskos and Roberts (2005); Beskos, Papaspiliopoulos, and Roberts (2006, 2008).

For simplicity we shall focus on the problem of estimating (unbiasedly) the diffusion transition density. The use of this approach in Monte Carlo maximum likelihood and related likelihood inference methodology is described in detail in Beskos, Papaspiliopoulos, Roberts, and Fearnhead (2006). We shall assume that the diffusion can be reduced to unit diffusion coefficient, and that the drift b can be expressed in gradient form $b = \nabla H$. We can therefore express the transition density according to (4.25).

Here we describe the so-called *generalised Poisson estimator* for estimating $p_{0,T}(u,v)$. A simple Taylor expansion of (4.25) for an arbitrary constant c gives

$p_{0,T}(u,v)$

$$= \mathcal{G}_{0,T}(u,v) \exp\{H(v) - H(u)\} e^{-cT} \mathrm{E}_{\mathbb{P}_0^*}\left[\sum_{\kappa=0}^{\infty} \frac{\left(\int_0^T (c - \phi(V_s))ds\right)^{\kappa}}{\kappa!}\right].$$

$$(4.31)$$

For an arbitrary function $g(\cdot)$, $\mathrm{E}(\int_0^T g(X_s)ds)$ is readily estimated unbiasedly by $Tg(U)$ where $U \sim U(0,T)$. This idea is easily generalised to consider $\mathrm{E}((\int_0^T g(X_s)ds)^{\kappa})$ for arbitrary positive integer κ. In this case the unbiased estimator is just $T^{\kappa} \prod_{i=1}^{\kappa} g(X_{U_i})$ where $\{U_i\}$ denote an independent collection of $U(0,T)$ variables.

Therefore, letting $\{q_i\}$ denote positive probabilities for all non-negative integers i, an unbiased estimator for $p_{0,T}(u,v)$ is given for arbitrary constant c by

$$\hat{p}_{0,T}(u,v) = \frac{\mathcal{G}_{0,T}(u,v) \exp\{H(v) - H(u) - cT\} T^I \prod_{i=1}^{I}(c - \phi(V_{U_i}))}{q_I I!}$$

$$(4.32)$$

where $I \sim q$. The choice of the importance proposal q is critical in determining the efficiency of the estimator $\hat{p}_{0,T}(u,v)$, and this is discussed in more detail in Beskos and Roberts (2005); Beskos, Papaspiliopoulos, and Roberts (2006); Beskos et al. (2008).

For the purposes of parameter estimation, it is critical to be able to obtain density estimates simultaneously for a collection of plausible parameter values. This is one important advantage of the form of the estimator in (4.32) since the estimator can be applied simultaneously to provide unbiased estimators of densities at a continuum of parameter values in such a way that the estimated likelihood surface is itself continuous (and of course unbiased for each parameter choice). As well as being of practical use, these properties are indispensible for proving consistency of Monte Carlo MLEs for large sample sizes Beskos, Papaspiliopoulos, and Roberts (2009).

Acknowledgements

The first author would like to acknowledge financial support from the Spanish Goverment via grant MTM2008-06660 and the "Ramon y Cajal" fellowship. We would like also to think Jaya Bishwal and Giorgos Sermaidis for helpful suggestions. The second author gratefully acknowledges the support of the Centre for Research in Statistical Methodology, funded by EPSRC.

4.6 Appendix 1: Typical problems of the projection-simulation paradigm in MC for diffusions

In this article we have advocated a simulation-projection paradigm, that is, designing Monte Carlo methods on the path space which are then, if necessary, discretized for practical implementation. Apart from the transparency of the resulting methods, the motivation for adopting this paradigm is also due to typical problems faced by the projection-simulation alternative. In this Appendix we mention two typical problematic cases. The first concerns the estimation of the transition density by Monte Carlo, and the the simultaneous parameters estimation and imputation of unobserved paths using the Gibbs sampler. A common characteristic in both is that decrease in approximation bias comes with an increase in Monte Carlo variance. For the sake of presentation we only consider scalar homogeneous diffusions.

The problem of estimating the transition density by Monte Carlo and use of the approximation for likelihood inference for unknown parameters was first considered by Pedersen (1995). Using the Chapman-Kolmogorov equation and

Euler approximation he obtained

$$
\begin{aligned}
p_{0,T}(u,v) \\
= \quad & \mathbb{E}_{\mathbb{P}_b}\left[p_{t,T}(V_t, v)\right] \\
\approx \quad & \mathbb{E}_{\mathbb{P}_b}\left[C_t \psi_t \frac{A(t, V_t)^{1/2}}{\sqrt{2\pi}}\right. \\
& \left. \times \exp\left\{-\frac{1}{2}A(t, V_t)(b(t, V_t)^2(T - t) + 2(v - V_t)b(t, V_t))\right\}\right]
\end{aligned}
$$

with the definitions as in Section 4.4. This suggest an IS approximation where we generate (unconditionally) paths up to time $t < T$ and associate weights to each path given by

$$
\psi_t A(t, V_t)^{1/2} \exp\left\{-\frac{1}{2}A(t, V_t)(b(t, V_t)^2(T - t) + 2(v - V_t)b(t, V_t))\right\}.
$$

Due to the Euler approximation on $[t, T]$ the weights have a bias which is eliminated as $t \to T$. On the other hand, the leading term in the weights for $t \approx T$ is ψ_t, thus the variance of the weights tends to infinity as $t \to T$. (There is of course additional potential bias in simulating V_t using a discretization method; this however, can be eliminated with increasing Monte Carlo effort without inflating the variance of the weights). The approach we expose in Sections 4.3 and 4.4 is designed to overcome this problem.

Bayesian inference for unknown parameters in the drift and the volatility of the SDE and the simultaneous imputation of unobserved paths for discretely observed diffusions, was originally considered by Elerian, Chib, and Shephard (2001); Eraker (2001); Roberts and Stramer (2001) (and remains a topic of active research). The first two articles work in a projection-simulation framework, hence the unobserved path between each pair of observations (i.e each diffusion bridge) is approximated by a skeleton of, M say, points. The joint distribution of the augmented dataset can be approximated using for example the Euler scheme (which gets increasingly accurate as M increases). This is effectively equivalent to using a Riemann approximation to the continuous-time likelihood (4.7). Therefore, we deal with a missing data problem where, given additional data (the imputed values in-between the observations) the likelihood is available, although here the missing data (for each pair of observations) are in principle infinite-dimensional and are approximated by an M-dimensional vector. Hence the computations are subject to a model-approximation bias which is eliminated in the limit $M \to \infty$. The *Gibbs sampler* is a popular computational tool for parameter estimation and simultaneous imputation of missing data in such a context. It consists of iterative simulation of missing data given the observed data and current values of prameters, and the simulation of the parameters according to their posterior distribution conditionally on the augmented dataset.

There are two main challenges in designing a Gibbs sampler for discretely observed diffusions: how to efficiently simulate the M intermediate points given
the endpoints for each pair of observations, and how to reduce the dependence
between the missing data and the parameters. As far as the first problem is
concerned, note that it is directly related to the diffusion bridge simulation,
and it is best understood thinking of the simulation in the infinite-dimensional
space. For diffusions which can be transformed to have unit volatility (as in
Section 4.3) Roberts and Stramer (2001) describe a Markov chain Monte Carlo
(MCMC) scheme which uses global moves on the path space, an approach very
closely related to that described in Section 4.3. More recently, global moves
MCMC using the processes discussed in Section 4.4 as proposals has been con-
sidered by Golightly and Wilkinson (2008); Chib, Shephard, and Pitt (2004).
For local moves MCMC designed on the path space, see for example Beskos,
Roberts, and Stuart (2009).

However, it is the second challenge we wish to emphasize in this section, i.e.
the dependence between the imputed data and the parameters. Strong posterior
dependence between missing data and parameters is known to be the principal
reason for slow convergence of the Gibbs sampler and results in high variance of the estimates based on its output, see for example Papaspiliopoulos,
Roberts, and Sköld (2007). The dependence between imputed data and parameters in this application can only be understood by considering a Gibbs
sampler on the infinite-dimensional space, i.e the product space of parameters and diffusion bridges. This approach was adopted in Roberts and Stramer
(2001) where it was noticed that due to the quadratic variation identity (4.5)
there is complete dependence between the missing paths and any parameters
involved in the volatility. Hence, an idealized algorithm ($M = \infty$) would be
completely reducible, whereas in practical applications where M is finite we
observe that decreasing the bias (increasing M) causes an increase of the mixing time of the algorithm and a corresponding increase in the variance of estimates based on its output. A solution to this problem is given in Roberts and
Stramer (2001) by appropriate transformations in the path space which break
down this dependence. It turns out that the strong dependence between parameters and unobserved processes is very common in many hierarchical models
and a generic methodology for reducing it, which includes the one considered
in Roberts and Stramer (2001), is known as *non-centred parametrisations*, see
Papaspiliopoulos, Roberts, and Sköld (2003); Papaspiliopoulos et al. (2007).

4.7 Appendix 2: Gaussian change of measure

The concept of change of measure is very central to the approaches we have
treated in this article. The aim of this section is to give a simplified presentation
of the change of measure between two Gaussian laws, and to the various ways

this result might be put in use. It is easy to see the correspondence between the expressions we obtain here and those of Section 4.2.1, but the greatly simplified context of this section has the educational value of pointing out some of the main elements of the construction, which can be understood without knowledge of stochastic calculus.

Let $(\Omega, \mathcal{F})$ be a measure space with elements $\omega \in \Omega$, $B : \Omega \to R^m$ a random variable on that space, let σ be a $d \times m$ matrix, $\Gamma = \sigma\sigma^*$, a, b, be $d \times 1$ vectors, and define a random variable V via the equation

$$V(\omega) = b + \sigma B(\omega)\,.$$

Let $\mathbb{R}_b$ be the probability measure on $(\Omega, \mathcal{F})$ such that B is a standard Gaussian vector. Therefore, under this measure V is a Gaussian vector with mean b (hence the indexing of the measure by b). Assume now that we can find a $m \times 1$ vector h which solves the equation

$$\sigma h = (b - a)\,, \tag{4.33}$$

and define $\hat{B}(\omega) = B(\omega) + h$. Thus, we have the alternative representation

$$V(\omega) = a + \sigma\hat{B}(\omega)\,,$$

which follows directly from the definitions of V and h. Let $\mathbb{R}_a$ be the measure defined by its density with respect to $\mathbb{R}_b$,

$$\frac{d\mathbb{R}_a}{d\mathbb{R}_b}(\omega) = \exp\left\{-h^*B(\omega) - h^*h/2\right\}\,, \tag{4.34}$$

which is well-defined, since the right-hand side has finite expectation with respect to $\mathbb{R}_b$. Notice that under this new measure, $\hat{B}$ is a standard Gaussian vector. To see this, notice that for any Borel set $A \subset R^m$,

$$
\begin{aligned}
\mathbb{R}_a[\hat{B} \in A] &= \int_{\{\omega:\hat{B}(\omega)\in A\}} \exp\left\{-u^*B(\omega) - u^*u/2\right\} d\mathbb{R}[\omega] \\
&= \int_{\{y:y+u\in A\}} \exp\left\{-u^*y - u^*u/2 - y^*y/2\right\}(2\pi)^{-m/2}dy \\
&= \int_A e^{-v^*v/2}(2\pi)^{-m/2}dv\,,
\end{aligned}
$$

where the last equality follows from a change of variables.

Notice that directly from (4.34) we have

$$\frac{d\mathbb{R}_b}{d\mathbb{R}_a}(\omega) = \exp\{h^*B(\omega) + h^*h/2\} = \exp\{h^*\hat{B}(\omega) - h^*h/2\}\,. \tag{4.35}$$

Let E_b and E_a denote expectations with respect to $\mathbb{R}_b$ and $\mathbb{R}_a$ respectively.

Thus, for any measurable $\mathbb{R}_b$-integrable function f defined on R^d,

$$
\begin{aligned}
\mathrm{E}_b[f(V)] &= \mathrm{E}_a\left[f(V)\exp\{h^*B + h^*h/2\}\right] \\
&= \mathrm{E}_a\left[f(V)\exp\{h^*\hat{B} - h^*h/2\}\right].
\end{aligned}
$$

Let X be another random variable, defined as $X(\omega) = a + \sigma B(\omega)$. Since under $\mathbb{R}_a$, the pair $(V, \hat{B})$ has the same law as the pair (X, B) under $\mathbb{R}_b$, we have that

$$
\mathrm{E}_b[f(V)] = \mathrm{E}_b[f(X)\exp\{h^*B - h^*h/2\}].
$$

If further σ is invertible we get

$$
\mathrm{E}_b[f(V)] = \mathrm{E}_b\left[f(X)\exp\left\{(b-a)^*\Gamma^{-1}X - \frac{1}{2}(b-a)^*\Gamma^{-1}(b+a)\right\}\right].
\tag{4.36}
$$

Let $\mathbb{P}_b$ and $\mathbb{P}_a$ be the law of V implied by $\mathbb{R}_b$ and $\mathbb{R}_a$ respectively. Then, assuming that σ is invertible and taking $\alpha = 0$, we can obtain from the previous expression the likelihood ratio between the hypotheses that V has mean b against that it has mean 0, but a Gaussian distribution with covariance Γ in both cases. Therefore, we get the likelihood function for estimating b on the basis of observed data V, while treating Γ as known:

$$
L(b) = \frac{d\mathbb{P}_b}{d\mathbb{P}_0}(V) = \exp\left\{b^*\Gamma^{-1}V - \frac{1}{2}b^*\Gamma^{-1}b\right\}.
\tag{4.37}
$$

It is interesting to consider the cases where (4.33) has many or no solutions. We will do so by looking at two characteristic examples. We first consider the case where (4.33) has multiple solutions and take $d = 1$, $m = 2$, $\sigma = (1,1)$, in which case (4.33) has infinite solutions. Notice that in this case there are more sources of randomness than observed variables. To simplify matters (and without loss of generality) we take $a = 0$. Then, for any $\phi \in R, u = (\phi, b-\phi)^*$ solves (4.33), and the measure $\mathbb{R}_0^\phi$ defined by (4.34), makes $\hat{B}$ a standard Gaussian vector. Then, writing $B = (B_1, B_2)$, the importance weights in (4.36) become

$$
\exp\{(b - \phi)B_1 + \phi B_2 - \phi^2 - b^2/2 + \phi b\}.
\tag{4.38}
$$

Direct calculation verifies that the change of measure in (4.36) holds for any ϕ; it is instructive to do directly the calculations using a change of variables and check that the right-hand side of (4.36) does not depend on ϕ. Additionally, using the moment generating function of the Gaussian distribution, one can verify that the expected value of the importance weights (4.38) under $\mathbb{R}_0^\phi$ is 1. However, the second moment of the importance weights is $\exp\{2(\phi - b/2)^2 + b^2/2\}$, which is minimized for $\phi = b/2$. Additionally, we can re-express (4.38) in terms of V as

$$
\exp\{(b - \phi)V + (2\phi - b)\hat{B}_1 - \phi^2 - b^2/2 + \phi b\}.
$$

In a statistical application only V will be observed whereas $\hat{B}_1$ will be unobserved, therefore we cannot use the expression directly to estimate b. Notice
that for $\phi = b/2$ the $\hat{B}_1$ terms cancels out from the density.

We now consider the case where (4.33) has no solution. An example of that
is produced under the setting $d = 2$, $m = 1$, $\sigma = (0,1)^*$. Writing $V =$
$(V_1, V_2)^*$ and $a = (a_1, a_2)$, $b = (b_1, b_2)^*$, notice the example implies that
$V_1 = b_1$. Therefore, it is expected that $\mathbb{R}_b$ will be mutually singular with any
measure which implies that $V_1 = a_1$, if $b_1 \neq a_1$. However, notice that (4.33)
can be solved by $u = b_2 - a_2$ provided that $a_1 = b_1$. Then, (4.34)-(4.36) hold.
Moreover, defining $\mathbb{P}_{(b_1, b_2)}$ and $\mathbb{P}_{(b_1, 0)}$ analogously as before, and noticing that
$B = V_2 - b_2$, we have the following likelihood ratio which can be used for the
estimation of b_2:

$$L(b_2) = \frac{d\mathbb{P}_{(b_1, b_2)}}{d\mathbb{P}_{(b_1, 0)}}(V) = \exp\left\{ b_2 V - \frac{1}{2} b_2^2 \right\}.$$

References

Aït-Sahalia, Y. (2008). Closed-form likelihood expansions for multivariate diffusions. *Ann. Statist.*, *36*, 906–937.

Arnold, L. (1998). *Random Dynamical Systems.* Berlin: Springer.

Bar-Shalom, Y., Kirubarajan, T., & Li, X.-R. (2002). *Estimation with Applications to Tracking and Navigation.* New York: John Wiley & Sons, Inc.

Bergstrom, A. R. (1990). *Continuous Time Econometric Modelling.* Oxford: Oxford University Press.

Beskos, A., Papaspiliopoulos, O., & Roberts, G. (2006). Retrospective exact simulation of diffusion sample paths with applications. *Bernoulli*, *12*(6), 1077–1098.

Beskos, A., Papaspiliopoulos, O., & Roberts, G. (2008). A factorisation of diffusion measure and finite sample path constructions. *Method. Comp. Appl. Probab.*, *10*(1), 85–104.

Beskos, A., Papaspiliopoulos, O., & Roberts, G. (2009). Monte Carlo maximum likelihood estimation for discretely observed diffusion processes. *Ann. Statist.*, *37*(1), 223–245.

Beskos, A., Papaspiliopoulos, O., Roberts, G. O., & Fearnhead, P. (2006). Exact and efficient likelihood–based inference for discretely observed diffusions (with Discussion). *J. Roy. Statist. Soc. Ser. B*, *68*(3), 333-82.

Beskos, A., & Roberts, G. O. (2005). Exact simulation of diffusions. *Ann. Appl. Probab.*, *15*(4), 2422-2444.

Beskos, A., Roberts, G. O., & Stuart, A. M. (2009). Optimal scalings for local metropolis-hastings chains on non-product targets in high dimensions. *Ann. Appl. Probab.*, *19*(3), 863–898.

Chan, K., Karolyi, A. G., Longstaff, F. A., & Sanders, A. B. (1992). An empirical comparison of alternative models of the short-term interest rate. *J. Finance*, *47*(3), 1209–1227.

Chib, S., Shephard, N., & Pitt, M. (2004). *Likelihood based inference for diffusion driven models.* (Available from http://ideas.repec.org/p/sbs/wpsefe/2004fe17.html)

Cox, J. C., Ingersoll, J. E., Jr., & Ross, S. A. (1985). A theory of the term structure of interest rates. *Econometrica*, *53*(2), 385–407.

Dacunha-Castelle, D., & Florens-Zmirou, D. (1986). Estimation of the coefficients of a diffusion from discrete observations. *Stochastics, 19*(4),
263–284.

Delyon, B., & Hu, Y. (2006). Simulation of conditioned diffusion and application to parameter estimation. *Stochastic Process. Appl., 116*(11),
1660–1675.

Durham, G. B., & Gallant, A. R. (2002). Numerical techniques for maximum
likelihood estimation of continuous-time diffusion processes. *J. Bus.
Econom. Statist., 20*(3), 297–338. (With comments and a reply by the
authors)

Elerian, O., Chib, S., & Shephard, N. (2001). Likelihood inference for discretely observed nonlinear diffusions. *Econometrica, 69*(4), 959–993.

Eraker, B. (2001). MCMC analysis of diffusion models with application to
finance. *J. Bus. Econom. Statist., 19*(2), 177–191.

Fearnhead, P., Papaspiliopoulos, O., & Roberts, G. O. (2008). Particle filters
for partially observed diffusions. *J. Roy. Statist. Soc. Ser. B, 70*, 755–
777.

Givon, D., Kupferman, R., & Stuart, A. (2004). Extracting macroscopic
dynamics: model problems and algorithms. *Nonlinearity, 17*(6), R55–
R127.

Gobet, E. (2002). LAN property for ergodic diffusions with discrete observations. *Ann. Inst. H. Poincaré Probab. Statist., 38*(5), 711–737.

Golightly, A., & Wilkinson, D. J. (2008). Bayesian inference for nonlinear
multivariate diffusion models observed with error. *Comput. Statist. Data
Anal., 52*(3), 1674–1693.

Hurn, A., Jeisman, J. I., & Lindsay, K. (2007). Seeing the wood for the trees:
A critical evaluation of methods to estimate the parameters of stochastic
differential equations. *J. Financial Econometrics, 5*(3), 390–455.

Jacod, J. (2012). *Statistics and high frequency data.* (Chapter 3 in this book)

Kalogeropoulos, K., Roberts, G. O., & Dellaportas, P. (2010). Inference for
stochastic volatility models using time change transformations. *Ann.
Statist.*(38), 784–807.

Kampen, N. G. van. (1981). *Stochastic Processes in Physics and Chemistry.*
Amsterdam: North-Holland Publishing Co. (Lecture Notes in Mathematics, 888)

Kessler, M. (1997). Estimation of an ergodic diffusion from discrete observations. *Scand. J. Statist., 24*(2), 211–229.

Kimura, M., & Ohta, T. (1971). *Theoretical Aspects of Population Genetics.*
Princeton: Princeton University Press.

Kloeden, P., & Platen, E. (1995). *Numerical Solution of Stochastic Differential
Equations.* Springer-Verlag.

Kutoyants, Y. A. (2004). *Statistical Inference for Ergodic Diffusion Processes.*
London: Springer-Verlag London Ltd.

Liu, J. S. (2008). *Monte Carlo Strategies in Scientific Computing*. New York: Springer.

Maruyama, G. (1955). Continuous Markov processes and stochastic equations. *Rend. Circ. Mat. Palermo (2)*, *4*, 48 90.

McAdams, H., & Arkin, A. (1997). Stochastic mechanisms in gene expression. *Proc. Natl. Acad. Sci. USA*, *94*, 814–819.

Merton, R. C. (1971). Optimum consumption and portfolio rules in a continuous-time model. *J. Econom. Theory*, *3*(4), 373–413.

Nicolau, J. (2002). A new technique for simulating the likelihood of stochastic differential equations. *Econom. J.*, *5*(1), 91–103.

Øksendal, B. K. (1998). *Stochastic Differential Equations: An Introduction with Applications*. Springer-Verlag.

Papaspiliopoulos, O., Roberts, G. O., & Sköld, M. (2003). Non-centered parameterizations for hierarchical models and data augmentation. In *Bayesian Statistics, 7 (Tenerife, 2002)* (pp. 307–326). New York: Oxford Univ. Press. (With a discussion by Alan E. Gelfand, Ole F. Christensen and Darren J. Wilkinson, and a reply by the authors)

Papaspiliopoulos, O., Roberts, G. O., & Sköld, M. (2007). A general framework for the parametrization of hierarchical models. *Statist. Sci.*, *22*(1), 59–73.

Pavliotis, G. A., Pokern, Y., & Stuart, A. M. (2012). *Parameter estimation for multiscale diffusions: an overview.* (Chapter 7 in this book)

Pedersen, A. R. (1995). Consistency and asymptotic normality of an approximate maximum likelihood estimator for discretely observed diffusion processes. *Bernoulli*, *1*(3), 257–279.

Pitt, M. K., & Shephard, N. (1999). Filtering via simulation: auxiliary particle filters. *J. Amer. Statist. Assoc.*, *94*(446), 590–599.

Prakasa Rao, B. L. S. (1999). *Statistical Inference for Diffusion Type Processes* (Vol. 8). London: Edward Arnold.

Roberts, G. O., & Stramer, O. (2001). On inference for partially observed nonlinear diffusion models using the Metropolis-Hastings algorithm. *Biometrika*, *88*(3), 603–621.

Rogers, L. C. G. (1985). Smooth transition densities for one-dimensional diffusions. *Bull. London Math. Soc.*, *17*(2), 157–161.

Rogers, L. C. G., & Williams, D. (2000). *Diffusions, Markov Processes, and Martingales. Vol. 1*. Cambridge: Cambridge University Press. (Foundations, Reprint of the second (1994) edition)

Sørensen, M. (2012). *Estimating functions for diffusion type processes.* (Chapter 1 in this book)

Sundaresan, S. M. (2000). Continuous-time methods in finance: A review and an assessment. *J. Finance*, *55*, 1569–1622.

Tan, W.-Y. (2002). *Stochastic Models with Applications to Genetics, Cancers, AIDS and Other Biomedical Systems* (Vol. 4). River Edge: World Scientific Publishing Co. Inc.

Vasicek, O. (1977). An equilibrium characterization of the term structure. *J. Financial Economics, 5,* 177–188.

Wilkinson, D. J. (2006). *Stochastic Modelling for Systems Biology.* Boca Raton: Chapman & Hall/CRC.

CHAPTER 5

Non-parametric estimation of the coefficients of ergodic diffusion processes based on high-frequency data

Fabienne Comte, Valentine Genon-Catalot and Yves Rozenholc

UFR de Mathématiques et Informatique, Université Paris Descartes – Paris 5
45 rue des Saints-Pères, 75270 Paris cedex 06, France

5.1 Introduction

The content of this chapter is directly inspired by Comte, Genon-Catalot, and Rozenholc (2006; 2007). We consider non-parametric estimation of the drift and diffusion coefficients of a one-dimensional diffusion process. The main assumption on the diffusion model is that it is ergodic and geometrically β-mixing. The sample path is assumed to be discretely observed with a small regular sampling interval Δ. The estimation method that we develop is based on a penalized mean square approach. This point of view is fully investigated for regression models in Comte and Rozenholc (2002, 2004). We adapt it to discretized diffusion models.

5.2 Model and assumptions

Let $(X_t)_{t \geq 0}$ be a one-dimensional diffusion process with dynamics described by the stochastic differential equation:

$$dX_t = b(X_t)dt + \sigma(X_t)dW_t, \quad t \geq 0, \quad X_0 = \eta, \tag{5.1}$$

where (W_t) is a standard Brownian motion and η is a random variable independent of (W_t). Consider the following assumptions:

[A1] $-\infty \le l < r \le +\infty$, b and σ belong to $C^1((l,r))$, $\sigma(x) > 0$ for all $x \in (l,r)$.

[A2] For $x_0, x \in (l,r)$, let $s(x) = \exp\left(-2\int_{x_0}^{x} b(u)/\sigma^2(u)du\right)$ denote the scale density and $m(x) = 1/[\sigma^2(x)s(x)]$ the speed density. We assume

$$\int_l s(x)dx = +\infty = \int^r s(x)dx, \quad \int_l^r m(x)dx = M < \infty.$$

[A3] The initial random variable η has distribution

$$\pi(x)dx = m(x)/M\mathbf{1}_{(l,r)}(x)dx.$$

Under [A1] – [A2], equation (5.1) admits a unique strong solution with state space the open interval (l,r) of the real line. Moreover, it is positive recurrent on this interval and admits as unique stationary distribution the normalized speed density π. With the additional assumption [A3], the process (X_t) is strictly stationary, with marginal distribution $\pi(x)dx$, ergodic and β-mixing, i.e. $\lim_{t\to+\infty}\beta_X(t) = 0$ where $\beta_X(t)$ denotes the β-mixing coefficient of (X_t). For stationary Markov processes such as (X_t), the β-mixing coefficient has the following explicit expression

$$\beta_X(t) = \int_l^r \pi(x)dx\|P_t(x,dx') - \pi(x')dx'\|_{TV}. \quad (5.2)$$

The norm $\|.\|_{TV}$ is the total variation norm and P_t denotes the transition probability (see e.g. Genon-Catalot, Jeantheau, and Larédo (2000) for a review). The statistical study relies on a stronger mixing condition which is satisfied in most standard models.

[A4] There exist constants $K > 0$ and $\theta > 0$ such that:

$$\beta_X(t) \le Ke^{-\theta t}. \quad (5.3)$$

In some cases, assumption [A4] can be checked directly using formula (5.2) (see Proposition 5.14 below). Otherwise, simple sufficient conditions are available (see e.g. Proposition 1 in Pardoux and Veretennikov (2001)). Lastly, we strengthen assumptions on b and σ to deal altogether with finite or infinite boundaries and keep a general, simple and clear framework. We also need a moment assumption for π and that σ be bounded.

[A5] Let $I = [l,r] \cap \mathbb{R}$.

  (i) Assume that $b \in C^1(I)$, b' bounded on I, $\sigma^2 \in C^2(I)$, $(\sigma^2)''$ bounded on I.

  (ii) For all $x \in I$, $\sigma^2(x) \le \sigma_1^2$.

[A6] $E(\eta^8) < \infty$.

The following property will be useful.

Lemma 5.1 *Under Assumptions [A1] – [A3] and [A5] – [A6], for all t, s such that $|t - s| \leq 1$, for $1 \leq i \leq 4$, $\mathrm{E}((X_t - X_s)^{2i}) \leq c|t - s|^i$.*

Proof. Using the strict stationarity, we only need to study $\mathrm{E}((X_t - X_0)^{2i})$ for $t \leq 1$. By the Minkowski inequality,

$$(X_t - X_0)^{2i} \leq 2^{2i-1}[(\int_0^t b(X_s)ds)^{2i} + (\int_0^t \sigma(X_s)dW_s)^{2i}].$$

For the drift term, we use the Hölder inequality, [A5] and the strict stationarity to get:

$$\mathrm{E}\left[(\int_0^t b(X_s)ds)^{2i}\right] \leq t^{2i-1}\int_0^t \mathrm{E}(b^{2i}(X_s))ds \leq t^{2i}C(1 + \mathrm{E}(\eta^{2i})),$$

with C a constant. For the diffusion term, we use the Burkholder–Davis–Gundy inequality and obtain:

$$\mathrm{E}\left[(\int_0^t \sigma(X_s)dW_s)^{2i}\right] \leq C\,\mathrm{E}(\int_0^t \sigma^2(X_s)ds)^i \leq C\,t^i\sigma_1^{2i}$$

with C a constant. This gives the result. $\quad\square$

5.3 Observations and asymptotic framework

We assume that the sample path X_t is observed at $n + 1$ discrete instants with sampling interval Δ. The asymptotic framework that we consider is the context of high-frequency data: the sampling interval $\Delta = \Delta_n$ tends to 0 as n tends to infinity. Moreover, we assume that the total length $n\Delta_n$ of the time interval where observations are taken tends to infinity. This is a classical framework for ergodic diffusion models: it allows us to estimate simultaneously the drift b and the diffusion coefficient σ and to enlighten us about the different rates of estimation of these two coefficients. For the penalized non-parametric method developed here, the asymptotic framework has to fulfill the following strengthened condition.

[A7] As n tends to infinity, $\Delta = \Delta_n \to 0$ and $n\Delta_n/\ln^2 n \to +\infty$.

To simplify notations, we will drop the subscript and simply write Δ for the sampling interval.

5.4 Estimation method

5.4.1 General description

We aim at estimating functions b and σ^2 of model (5.1) on a compact subset A of the state space (l, r). For simplicity, and without loss of generality, we assume from now on that

$$A = [0, 1],$$

and we set

$$b_A = b\mathbf{1}_A, \quad \sigma_A = \sigma\mathbf{1}_A. \tag{5.4}$$

The estimation method is inspired by what is done for regression models (see e.g. Comte and Rozenholc (2002, 2004)). Suppose we have observations (x_i, y_i), $i = 1, \ldots, n$ such that

$$y_i = f(x_i) + \text{noise}, \tag{5.5}$$

where f is unknown. We consider a regression contrast of the form

$$t \to \gamma_n(t) = \frac{1}{n} \sum_{i=1}^{n} (y_i - t(x_i))^2.$$

The aim is to build estimators for f by minimizing the least-square criterion $\gamma_n(t)$. For that purpose, we consider a collection of finite dimensional linear subspaces of $\mathbb{L}^2([0, 1])$ and compute for each space an associated least-squares estimator. Afterwards, a data-driven procedure chooses from the resulting collection of estimators the "best" one, in a sense to be precised, through a penalization device. For adapting the method to discretized diffusion processes, we have to find a regression equation analogous to (5.5), i.e. find for $f = b, \sigma^2$ the appropriate couple (x_i, y_i) and the adequate regression equation. Of course, starting with a diffusion model, we do not find an exact regression equation but only a regression-type equation, one for the drift and another one for the diffusion coefficient. Hence, estimators for b and for σ^2 are built from two distinct constructions and the method does not require one to estimate the stationary density π.

5.4.2 Spaces of approximation

Let us describe now some possible collections of spaces of approximation. We focus on two specific collections, the collection of dyadic regular piecewise polynomial spaces, denoted hereafter by [DP], and the collection of general piecewise polynomials, denoted by [GP], which is more complex. As for numerical implementation, algorithms for both collections are available and have been implemented on several examples.

In what follows, several norms for $[0, 1]$-supported functions are needed and the following notations will be used:

$$\|t\| = (\textstyle\int_0^1 t^2(x)dx)^{1/2}, \quad \|t\|_\pi = (\textstyle\int_0^1 t^2(x)\pi(x)dx)^{1/2}, \\ \|t\|_\infty = \sup_{x \in [0,1]} |t(x)|. \tag{5.6}$$

By our assumptions, the stationary density π is bounded from below and above on every compact subset of (l, r). Hence, let π_0, π_1 denote two positive real numbers such that,

$$\forall x \in [0, 1], \quad 0 < \pi_0 \leq \pi(x) \leq \pi_1. \tag{5.7}$$

Thus, for $[0, 1]$-supported functions, the norms $\|.\|$ and $\|.\|_\pi$ are equivalent.

Dyadic regular piecewise polynomials

Let $r \geq 0$, $p \geq 0$ be integers. On each subinterval $I_j = [(j-1)/2^p, j/2^p)$, $j = 1, \ldots, 2^p$, consider $r + 1$ polynomials of degree ℓ, $\varphi_{j,\ell}(x)$, $\ell = 0, 1, \ldots r$ and set $\varphi_{j,\ell}(x) = 0$ outside I_j. The space S_m, $m = (p, r)$, is defined as generated by the $D_m = 2^p(r+1)$ functions $(\varphi_{j,\ell})$. A function t in S_m may be written as

$$t(x) = \sum_{j=1}^{2^p} \sum_{\ell=0}^{r} t_{j,\ell} \varphi_{j,\ell}(x).$$

The collection of spaces $(S_m, m \in \mathcal{M}_n)$ is such that, for r_{max} a fixed integer,

$$\mathcal{M}_n = \{m = (p, r), p \in \mathbb{N}, r \in \{0, 1, \ldots, r_{max}\}, 2^p(r_{max} + 1) \leq N_n\}. \tag{5.8}$$

In other words, $D_m \leq N_n$ with $N_n \leq n$. The maximal dimension N_n will be subject to additional constraints. The role of N_n is to bound all dimensions D_m, even when m is random. In practice, it corresponds to the maximal number of coefficients to estimate. Thus it must not be too large.

More concretely, consider the orthogonal collection in $\mathbb{L}^2([-1, 1])$ of Legendre polynomials $(Q_\ell, \ell \geq 0)$, where the degree of Q_ℓ is equal to ℓ, generating $\mathbb{L}^2([-1, 1])$ (see Abramowitz and Stegun (1972), p.774). They satisfy $|Q_\ell(x)| \leq 1, \forall x \in [-1, 1]$, $Q_\ell(1) = 1$ and $\int_{-1}^{1} Q_\ell^2(u)du = 2/(2\ell + 1)$. Then we set $P_\ell(x) = (2\ell + 1)^{1/2}Q_\ell(2x - 1)$, to get an orthonormal basis of $\mathbb{L}^2([0, 1])$. Finally,

$$\varphi_{j,\ell}(x) = 2^{p/2} P_\ell(2^p x - j + 1)\mathbf{1}_{I_j}(x), \quad j = 1, \ldots, 2^p, \ \ell = 0, 1, \ldots, r.$$

The space S_m has dimension $D_m = 2^p(r+1)$ and its orthonormal basis described above satisfies

$$\left\| \sum_{j=1}^{2^p} \sum_{\ell=0}^{r} \varphi_{j,\ell}^2 \right\|_\infty \leq D_m(r+1).$$

Hence, using notations (5.6), for all $t \in S_m$, $\|t\|_\infty \leq (r+1)^{1/2} D_m^{1/2}\|t\|$.

Collection [DP] is simple in the sense that one dimension D_m is associated with a single space S_m. In particular, since $N_n \leq n$, the following holds:

$$\Sigma = \sum_{m \in \mathcal{M}_n} \exp(-D_m) = \sum_{r=0}^{r_{max}} \sum_{p:2^p(r+1)\leq N_n} \exp(-2^p(r+1)) < \infty. \quad (5.9)$$

Finally, let us sum up the two key properties that are fulfilled by this collection. The collection $(S_m)_{m \in \mathcal{M}_n}$ is composed of finite dimensional linear sub-spaces of $\mathbb{L}^2([0,1])$, indexed by a set $\mathcal{M}_n$ depending on n. The space S_m has dimension $D_m \leq N_n \leq n, \forall m \in \mathcal{M}_n$, where N_n designates a maximal dimension, and S_m is equipped with an orthonormal basis $(\varphi_\lambda)_{\lambda \in \Lambda_m}$ with $|\Lambda_m| = D_m$. The following holds:

($\mathcal{H}_1$) Norm connection: There exists $\Phi_0 > 0$, such that,

$$\forall m \in \mathcal{M}_n, \forall t \in S_m, \|t\|_\infty \leq \Phi_0 D_m^{1/2} \|t\|. \quad (5.10)$$

($\mathcal{H}_2$) Nesting condition: There exists a space denoted by $\mathcal{S}_n$, belonging to the collection, with $\forall m \in \mathcal{M}_n, S_m \subset \mathcal{S}_n$, with dimension denoted by N_n.

In Birgé and Massart (1998, p.337, Lemma 1), it is proved that property (5.10) is equivalent to:

$$\text{There exists } \Phi_0 > 0, \| \sum_{\lambda \in \Lambda_m} \varphi_\lambda^2 \|_\infty \leq \Phi_0^2 D_m. \quad (5.11)$$

There are other collections of spaces satisfying the above two properties and for which our proofs apply (for instance, the trigonometric spaces, or the dyadic wavelet generated spaces).

General piecewise polynomials

A more general family can be described, the collection of general piecewise polynomials spaces denoted by [GP]. We first build the largest space $\mathcal{S}_n$ of the collection whose dimension is denoted as above by N_n ($N_n \leq n$). For this, we fix an integer r_{max} and let d_{max} be an integer such that $d_{max}(r_{max}+1) = N_n$. The space $\mathcal{S}_n$ is linearly spanned by piecewise polynomials of degree r_{max} on the regular subdivision of $[0,1]$ with step $1/d_{max}$. Any other space S_m of the collection is described by a multi-index $m = (d, j_1, \ldots, j_{d-1}, r_1, \ldots, r_d)$ where d is the number of intervals of the partition, $j_0 := 0 < j_1 < \cdots < j_{d-1} < j_d := 1$ are integers such that $j_i \in \{1, \ldots, d_{max}-1\}$ for $i = 1, \ldots d-1$. The latter integers define the knots j_i/d_{max} of the subdivision. Lastly $r_i \leq r_{max}$ is the degree of the polynomial on the interval $[j_{i-1}/d_{max}, j_i/d_{max})$, for $i = 1, \ldots, d$. A function t in S_m can thus be described as

$$t(x) = \sum_{i=1}^{d} P_i(x) \mathbf{1}_{[j_{i-1}/d_{max}, j_i/d_{max})}(x),$$

with P_i a polynomial of degree r_i. The dimension of S_m is still denoted by D_m and is equal to $\sum_{i=1}^{d}(r_i + 1)$ for all the $\binom{d_{max}-1}{d-1}$ choices of the knots $(j_1, \ldots, j_{d-1})$. Note that the P_i's can still be decomposed by using the Legendre basis rescaled on the intervals $[j_{i-1}/d_{max}, j_i/d_{max})$. The collection [GP] of models $(S_m)_{m \in \mathcal{M}_n}$ is described by the set of indexes

$$\mathcal{M}_n = \{m = (d, j_1, \ldots, j_{d-1}, r_1, \ldots, r_d), \ 1 \leq d \leq d_{max},$$
$$j_i \in \{1, \ldots, d_{max} - 1\}, r_i \in \{0, \ldots, r_{max}\}\}.$$

It is important to note that now, for all $m \in \mathcal{M}_n$, for all $t \in S_m$, since $S_m \subset S_n$,

$$\|t\|_\infty \leq \sqrt{(r_{max} + 1)N_n}\|t\|. \tag{5.12}$$

Hence, the norm connection property still holds but only on the maximal space and no more on each individual space of the collection as for collection [DP]. This comes from the fact that regular partitions are involved in all spaces of [DP] and only in the maximal space of [GP]. Obviously, collection [GP] has higher complexity than [DP]. The complexity of a collection is usually evaluated through a set of weights (L_m) that must satisfy $\sum_{m \in \mathcal{M}_n} e^{-L_m D_m} < \infty$. For [DP], $L_m = 1$ suits (see (5.9)). For [GP], we have to look at

$$\sum_{m \in \mathcal{M}_n} e^{-L_m D_m} =$$

$$\sum_{d=1}^{d_{max}} \sum_{1 \leq j_1 < \cdots < j_{d-1} < d_{max}} \sum_{0 \leq r_1, \ldots, r_d \leq r_{max}} e^{-L_m \sum_{i=1}^{d}(r_i+1)}.$$

From the equality above, we deduce that the choice

$$L_m D_m = D_m + \ln\left(\begin{matrix} d_{max} - 1 \\ d - 1 \end{matrix}\right) + d \ln(r_{max} + 1) \tag{5.13}$$

can suit. Actually, this relation guides the choice of the penalty function used in the practical implementation. To see more clearly what orders of magnitude are involved, let us choose $L_m = L_n$ for all $m \in \mathcal{M}_n$. Then, we have a further bound for the series:

$$\sum_{m \in \mathcal{M}_n} e^{-L_m D_m} \leq \sum_{d=1}^{d_{max}} \left(\begin{matrix} d_{max} - 1 \\ d - 1 \end{matrix}\right)(r_{max} + 1)^d e^{-dL_n}$$

$$\leq \sum_{d=0}^{d_{max}-1} \left(\begin{matrix} d_{max} - 1 \\ d \end{matrix}\right)[(r_{max} + 1)e^{-L_n}]^{d+1}$$

$$\leq (r_{max} + 1)\left[1 + (r_{max} + 1)e^{-L_n}\right]^{d_{max}-1}$$

$$\leq (r_{max} + 1)\exp(d_{max}(r_{max} + 1)e^{-L_n})$$

$$\leq (r_{max} + 1)\exp(N_n e^{-L_n}).$$

Thus $L_m = L_n = \ln(N_n)$ ensures that the series is bounded. (For more details on these collections, see *e.g.* Comte and Rozenholc (2004) or Baraud, Comte, and Viennet (2001b).

5.5 Drift estimation

5.5.1 Drift estimators: Statements of the results

The regression-type equation for the drift is as follows:

$$Y_{k\Delta} := \frac{X_{(k+1)\Delta} - X_{k\Delta}}{\Delta} = b(X_{k\Delta}) + Z_{k\Delta} + R_{k\Delta} \qquad (5.14)$$

where

$$Z_{k\Delta} = \frac{1}{\Delta} \int_{k\Delta}^{(k+1)\Delta} \sigma(X_s) dW_s, \quad R_{k\Delta} = \frac{1}{\Delta} \int_{k\Delta}^{(k+1)\Delta} (b(X_s) - b(X_{k\Delta})) ds. \qquad (5.15)$$

The couple $(X_{k\Delta}, Y_{k\Delta})$ stands for the data (x_k, y_k). The term $Z_{k\Delta}$ is a martingale increment (with respect to the filtration $\mathcal{F}_{k\Delta} = \sigma(X_s, s \le k\Delta)$) and plays the role of the noise term. The term $R_{k\Delta}$ is a remainder due to the discretization. Now, we consider a collection which may be either [DP] or [GP]. For S_m a space of the collection and for $t \in S_m$, we set

$$\gamma_n(t) = \frac{1}{n} \sum_{k=1}^{n} [Y_{k\Delta} - t(X_{k\Delta})]^2. \qquad (5.16)$$

We define an estimator $\hat{b}_m$ of b_A belonging to S_m as any solution of:

$$\hat{b}_m = \arg \min_{t \in S_m} \gamma_n(t). \qquad (5.17)$$

Note that, with this definition, only the random $\mathbb{R}^n$-vector $(\hat{b}_m(X_\Delta), \ldots, \hat{b}_m(X_{n\Delta}))'$ is uniquely defined. Indeed, let Π_m denote the orthogonal projection (with respect to the inner product of $\mathbb{R}^n$) onto the subspace $\{(t(X_\Delta), \ldots, t(X_{n\Delta}))', t \in S_m\}$ of $\mathbb{R}^n$. Then $(\hat{b}_m(X_\Delta), \ldots, \hat{b}_m(X_{n\Delta}))' = \Pi_m Y$ where $Y = (Y_\Delta, \ldots, Y_{n\Delta})'$. Any function t in S_m such that $t(X_{k\Delta}) = \hat{b}_m(X_{k\Delta})$, $k = 1, \ldots, n$, is a solution of (5.17).

This is the reason why we adopt a specific definition of the risk of an estimator. Consider the following empirical norm for a function t:

$$\|t\|_n^2 = \frac{1}{n} \sum_{k=1}^{n} t^2(X_{k\Delta}). \qquad (5.18)$$

The risk of $\hat{b}_m$ is defined as the expectation of the empirical norm:

$$\mathbb{E}(\|\hat{b}_m - b_A\|_n^2).$$

Note that, for a deterministic function t, $\mathbb{E}(\|t\|_n^2) = \|t\|_\pi^2 = \int t^2(x)\pi(x)dx$.

The following proposition provides an upper bound for the risk of an estimator $\hat{b}_m$ with fixed m. Let b_m denote the orthogonal projection of b on S_m.

Proposition 5.2 *Assume that* [A1] – [A7] *hold. Consider a space S_m in collection* [DP] *or* [GP] *with maximal dimension satisfying $N_n = o(n\Delta/\ln^2(n))$. Then the estimator $\hat{b}_m$ of b is such that (see (5.4))*

$$\mathbb{E}(\|\hat{b}_m - b_A\|_n^2) \leq 7\pi_1\|b_m - b_A\|^2 + K\frac{\sigma_1^2 D_m}{n\Delta} + K'\Delta + \frac{K''}{n\Delta}, \quad (5.19)$$

where K, K' and K'' are some positive constants.

As usual, there appears to be a squared bias term $\|b_m - b_A\|^2$ and a variance term of order $D_m/(n\Delta)$, plus additional terms due to the discretization. It is standard for diffusion models in high-frequency data to assume that $\Delta = o(1/(n\Delta))$ so that the two last terms in (5.19) are negligible with respect to the variance term. It remains to select the dimension D_m that leads to the best compromise between the squared bias term and the variance term.

Consider the case [DP]. To compare the result of Proposition 5.2 with the optimal non-parametric rates exhibited by Hoffmann (1999), let us assume that b_A belongs to a ball of some Besov space, $b_A \in \mathcal{B}_{\alpha,2,\infty}([0,1])$, and that $r+1 \geq \alpha$. Then, for $\|b_A\|_{\alpha,2,\infty} \leq L$, it is known that $\|b_A - b_m\|^2 \leq C(\alpha, L)D_m^{-2\alpha}$ (see DeVore and Lorentz (1993, p.359) or Lemma 12 in Barron, Birgé, and Massart (1999)). Thus, choosing $D_m = (n\Delta)^{1/(2\alpha+1)}$, we obtain

$$\mathbb{E}(\|\hat{b}_m - b_A\|_n^2) \leq C(\alpha, L)(n\Delta)^{-2\alpha/(2\alpha+1)} + K'\Delta + \frac{K''}{n\Delta}.$$

The first term $(n\Delta)^{-2\alpha/(2\alpha+1)}$ is exactly the optimal non-parametric rate (see Hoffmann (1999)). Under the condition $\Delta = o(1/(n\Delta))$, the last two terms in (5.19) are negligible with respect to $(n\Delta)^{-2\alpha/(2\alpha+1)}$.

As a second step, we must ensure an automatic selection of D_m, which does not use any knowledge on b, and in particular which does not require knowing its regularity α. This selection is done by defining

$$\hat{m} = \arg\min_{m \in \mathcal{M}_n} \left[\gamma_n(\hat{b}_m) + \text{pen}(m)\right], \quad (5.20)$$

with pen(m) a penalty to be properly chosen. We denote by $\hat{b}_{\hat{m}}$ the resulting estimator and we need to determine pen(.) such that, ideally,

$$\mathbb{E}(\|\hat{b}_{\hat{m}} - b_A\|_n^2) \leq C \inf_{m \in \mathcal{M}_n} \left(\|b_A - b_m\|^2 + \frac{\sigma_1^2 D_m}{n\Delta}\right) + K'\Delta + \frac{K''}{n\Delta},$$

with C a constant, which should not be too large. We almost reach this aim.

Theorem 5.3 *Assume that [A1] – [A7] hold and consider the nested collection of models [DP] with $L_m = 1$ or the collection [GP] with L_m given by (5.13), both with maximal dimension $N_n = o(n\Delta/\ln^2(n))$. Let*

$$\text{pen}(m) \geq \kappa\sigma_1^2 \frac{(1 + L_m)D_m}{n\Delta}, \tag{5.21}$$

where κ is a universal constant. Then the estimator $\hat{b}_{\hat{m}}$ of b with $\hat{m}$ defined in (5.20) is such that

$$\mathbb{E}(\|\hat{b}_{\hat{m}} - b_A\|_n^2) \leq C \inf_{m \in \mathcal{M}_n} \left(\|b_m - b_A\|^2 + \text{pen}(m)\right) + K'\Delta + \frac{K''}{n\Delta}. \tag{5.22}$$

Inequality (5.22) shows that the adaptive estimator automatically realizes the bias-variance compromise. Nevertheless, some comments need to be made. It is possible to choose the equality in (5.21) but this is not what is done in practice. It is better to have additional terms to avoid underpenalization. The constant κ in the penalty is numerical and must be calibrated for the problem. Another important point is that σ_1^2 is unknown. In practice, it is replaced by a rough estimator.

5.5.2 Proof of Proposition 5.2

Let us set (see (5.15)), for any function $t(.)$,

$$\nu_n(t) = \frac{1}{n}\sum_{k=1}^{n} t(X_{k\Delta})Z_{k\Delta}, \quad R_n(t) = \frac{1}{n}\sum_{k=1}^{n} t(X_{k\Delta})R_{k\Delta}. \tag{5.23}$$

Using (5.14) – (5.16) – (5.18), we have:

$$\gamma_n(t) - \gamma_n(b) = \|t - b\|_n^2 + 2\nu_n(b - t) + 2R_n(b - t).$$

Recall that b_m denotes the orthogonal projection of b on S_m. By definition of $\hat{b}_m$, $\gamma_n(\hat{b}_m) \leq \gamma_n(b_m)$. So, $\gamma_n(\hat{b}_m) - \gamma_n(b) \leq \gamma_n(b_m) - \gamma_n(b)$. This implies

$$\|\hat{b}_m - b\|_n^2 \leq \|b_m - b\|_n^2 + 2\nu_n(\hat{b}_m - b_m) + 2R_n(\hat{b}_m - b_m).$$

The functions $\hat{b}_m$ and b_m being A-supported, we can cancel the terms $\|b\mathbb{1}_{A^c}\|_n^2$ that appear in both sides of the inequality. This yields

$$\|\hat{b}_m - b_A\|_n^2 \leq \|b_m - b_A\|_n^2 + 2\nu_n(\hat{b}_m - b_m) + 2R_n(\hat{b}_m - b_m).$$

Recall notations (5.6) and let

$$B_m^{\pi} = \{t \in S_m, \|t\|_{\pi} = 1\}.$$

We use the standard inequality $2xy \leq \theta^2 x^2 + y^2/\theta^2$ which holds for all $\theta \neq 0$ with $\theta^2 = 8$:

$$2\nu_n(\hat{b}_m - b_m) \leq 2\|\hat{b}_m - b_m\|_{\pi} \sup_{t \in B_m^{\pi}} |\nu_n(t)| \leq \frac{1}{8}\|\hat{b}_m - b_m\|_{\pi}^2 + 8 \sup_{t \in B_m^{\pi}} [\nu_n(t)]^2.$$

Similarly,

$$2R_n(\hat{b}_m - b_m) \leq 2\|\hat{b}_m - b_m\|_n \left(\frac{1}{n} \sum_{k=1}^{n} R_{k\Delta}^2 \right)^{1/2}$$

$$\leq \frac{1}{8}\|\hat{b}_m - b_m\|_n^2 + 8\frac{1}{n} \sum_{k=1}^{n} R_{k\Delta}^2.$$

Because the $\mathbb{L}_\pi^2$-norm, $\|.\|_\pi$, and the empirical norm (5.18) are not equivalent, we must introduce a set on which they are, and afterwards prove that this set has small probability. Let us define (see (5.8))

$$\Omega_n = \left\{ \omega / \left| \frac{\|t\|_n^2}{\|t\|_\pi^2} - 1 \right| \leq \frac{1}{2}, \ \forall t \in \bigcup_{m,m' \in \mathcal{M}_n} (S_m + S_{m'})/\{0\} \right\}. \quad (5.24)$$

On Ω_n, $\|\hat{b}_m - b_m\|_\pi^2 \leq 2\|\hat{b}_m - b_m\|_n^2$, and $\|\hat{b}_m - b_m\|_n^2 \leq 2(\|\hat{b}_m - b_A\|_n^2 + \|b_m - b_A\|_n^2)$. Hence, some elementary computations yield:

$$\frac{1}{4}\|\hat{b}_m - b_A\|_n^2 \mathbf{1}_{\Omega_n} \leq \frac{7}{4}\|b_m - b_A\|_n^2 + 8 \sup_{t \in B_m^\pi} [\nu_n(t)]^2 + \frac{8}{n} \sum_{k=1}^{n} R_{k\Delta}^2.$$

Now, using [A5] and Lemma 5.1, we get

$$\mathrm{E}(R_{k\Delta}^2) \leq \frac{1}{\Delta}\mathrm{E} \int_{k\Delta}^{(k+1)\Delta} (b(X_s) - b(X_{k\Delta}))^2 ds \leq c'\Delta.$$

Consequently,

$$\mathbb{E}(\|\hat{b}_m - b_A\|_n^2 \mathbf{1}_{\Omega_n}) \leq 7\|b_m - b_A\|_\pi^2 + 32\,\mathbb{E} \left(\sup_{t \in B_m^\pi} [\nu_n(t)]^2 \right) + 32c'\Delta. \quad (5.25)$$

Consider a basis of S_m, say $\{\psi_\lambda, \lambda \in J_m\}$, which is orthonormal with respect to L_π^2, and with $|J_m| = D_m$. For $t \in B_m^\pi$,

$$[\nu_n(t)]^2 \leq \sum_{\lambda \in J_m} [\nu_n^2(\psi_\lambda)].$$

Using the martingale property of (5.23) and the bound of $\sigma^2(.)$, we get:

$$\mathbb{E}[\nu_n^2(\psi_\lambda)] = \frac{1}{n^2\Delta^2} \sum_{k=1}^{n} \mathbb{E} \left\{ \psi_\lambda^2(X_{k\Delta}) \int_{k\Delta}^{(k+1)\Delta} \sigma^2(X_s) ds \right\}$$

$$\leq \frac{\sigma_1^2}{n\Delta} \int \psi_\lambda^2(x)\pi(x)dx = \frac{\sigma_1^2}{n\Delta}.$$

Therefore,

$$\mathbb{E} \left(\sup_{t \in B_m^\pi} [\nu_n(t)]^2 \right) \leq \frac{\sigma_1^2}{n\Delta} D_m.$$

Gathering bounds, and using the upper bound π_1 defined in (5.7), we get

$$\mathbb{E}(\|\hat{b}_m - b_A\|_n^2 \mathbf{1}_{\Omega_n}) \leq 7\pi_1 \|b_m - b_A\|^2 + 32\frac{\sigma_1^2 D_m}{n\Delta} + 32c'\Delta.$$

Now, it remains to deal with Ω_n^c. Since $\|\hat{b}_m - b_A\|_n^2 \leq \|\hat{b}_m - b\|_n^2$, it is enough to check that $\mathbb{E}(\|\hat{b}_m - b\|_n^2 \mathbf{1}_{\Omega_n^c}) \leq c/n$. Write $Y_{k\Delta} = b(X_{k\Delta}) + \varepsilon_{k\Delta}$ with $\varepsilon_{k\Delta} = Z_{k\Delta} + R_{k\Delta}$. Recall that Π_m denotes the orthogonal projection (with respect to the inner product of $\mathbb{R}^n$) onto the subspace $\{(t(X_\Delta), \ldots, t(X_{n\Delta}))', t \in S_m\}$ of $\mathbb{R}^n$. We have $(\hat{b}_m(X_\Delta), \ldots, \hat{b}_m(X_{n\Delta}))' = \Pi_m Y$ where $Y = (Y_\Delta, \ldots, Y_{n\Delta})'$. Using the same notation for the function t and the vector $(t(X_\Delta), \ldots, t(X_{n\Delta}))'$, we see that

$$\|b - \hat{b}_m\|_n^2 = \|b - \Pi_m b\|_n^2 + \|\Pi_m \varepsilon\|_n^2 \leq \|b\|_n^2 + n^{-1} \sum_{k=1}^n \varepsilon_{k\Delta}^2.$$

Therefore,

$$\begin{aligned}
\mathbb{E}\left(\|b - \hat{b}_m\|_n^2 \mathbf{1}_{\Omega_n^c}\right) &\leq \mathbb{E}\left(\|b\|_n^2 \mathbf{1}_{\Omega_n^c}\right) + \frac{1}{n}\sum_{k=1}^n \mathbb{E}\left(\varepsilon_{k\Delta}^2 \mathbf{1}_{\Omega_n^c}\right) \\
&\leq \left(\mathbb{E}^{1/2}(b^4(X_0)) + \mathbb{E}^{1/2}(\varepsilon_\Delta^4)\right) \mathbb{P}^{1/2}(\Omega_n^c).
\end{aligned}$$

Using [A5], we have $\mathbb{E}(b^4(X_0)) \leq c(1 + \mathbb{E}(X_0^4)) = K$. With the Burholder-Davis-Gundy inequality, we find

$$\mathbb{E}(\varepsilon_\Delta^4) \leq 2^3 \left\{\frac{1}{\Delta}\int_0^\Delta \mathbb{E}[(b(X_s) - b(X_\Delta))^4]ds + \frac{36}{\Delta^3}\mathbb{E}\left(\int_0^\Delta \sigma^4(X_s)ds\right)\right\}.$$

Under [A1] – [A3], [A5] – [A6] and using Lemma 5.1, we obtain $\mathbb{E}(\varepsilon_\Delta^4) \leq C(1 + \sigma_1^4/\Delta^2) := C'/\Delta^2$. The next lemma enables us to complete the proof.

Lemma 5.4 *Let Ω_n be defined by (5.24). Then, if $N_n \leq O(n\Delta_n/\ln^2(n))$*

$$\mathbb{P}(\Omega_n^c) \leq \frac{c}{n^4}. \tag{5.26}$$

The proof of this technical lemma is given in Comte et al. (2007) and relies on inequalities proved in Baraud, Comte, and Viennet (2001a). We stress the fact that it is for this lemma that we need the exponential β-mixing assumption for (X_t). It is also for this lemma that we have constraints on the maximal dimension N_n.

Now, we gather all terms and use (5.26) to get (5.19). $\square$

5.5.3 Proof of Theorem 5.3

The proof relies on the following Bernstein-type inequality:

Lemma 5.5 *Under the assumptions of Theorem 5.3, for any positive numbers ϵ and v, we have (see (5.23)):*

$$\mathbb{P}\left(\nu_n(t) \geq \epsilon, \|t\|_n^2 \leq v^2\right) \leq \exp\left(-\frac{n\Delta\epsilon^2}{2\sigma_1^2 v^2}\right).$$

Proof of Lemma 5.5. Consider the process:

$$H_u^n = H_u = \sum_{k=1}^{n} 1_{[k\Delta,(k+1)\Delta[}(u) t(X_{k\Delta})\sigma(X_u)$$

which satisfies $H_u^2 \leq \sigma_1^2\|t\|_\infty^2$ for all $u \geq 0$. Then, denoting by $M_s = \int_0^s H_u dW_u$, we get that

$$M_{(n+1)\Delta} = \sum_{k=1}^{n} t(X_{k\Delta}) \int_{k\Delta}^{(k+1)\Delta} \sigma(X_s)dW_s = n\Delta\nu_n(t)$$

and that

$$\langle M \rangle_{(n+1)\Delta} = \sum_{k=1}^{n} t^2(X_{k\Delta}) \int_{k\Delta}^{(k+1)\Delta} \sigma^2(X_s)ds.$$

Moreover, $\langle M \rangle_s = \int_0^s H_u^2 du \leq n\sigma_1^2\Delta\|t\|_n^2, \forall s \geq 0$, so that (M_s) and $\exp(\lambda M_s - \lambda^2\langle M \rangle_s/2)$ are martingales with respect to the filtration $\mathcal{F}_s = \sigma(X_u, u \leq s)$. Therefore, for all $s \geq 0, c > 0, d > 0, \lambda > 0$,

$$\mathbb{P}(M_s \geq c, \langle M \rangle_s \leq d) \leq \mathbb{P}\left(e^{\lambda M_s - \frac{\lambda^2}{2}\langle M \rangle_s} \geq e^{\lambda c - \frac{\lambda^2}{2}d}\right) \leq e^{-(\lambda c - \frac{\lambda^2}{2}d)}.$$

Therefore,

$$\mathbb{P}(M_s \geq c, \langle M \rangle_s \leq d) \leq \inf_{\lambda > 0} e^{-(\lambda c - \frac{\lambda^2}{2}d)} = e^{-\frac{c^2}{2d}}.$$

Finally,

$$\mathbb{P}\left(\nu_n(t) \geq \epsilon, \|t\|_n^2 \leq v^2\right) = \mathbb{P}(M_{(n+1)\Delta} \geq n\Delta\epsilon, \langle M \rangle_{(n+1)\Delta} \leq nv^2\sigma_1^2\Delta)$$
$$\leq \exp\left(-\frac{(n\Delta\epsilon)^2}{2nv^2\sigma_1^2\Delta}\right) = \exp\left(-\frac{n\epsilon^2\Delta}{2v^2\sigma_1^2}\right). \quad \square$$

Now we turn to the proof of Theorem 5.3. As in the proof of Proposition 5.2, we have to split $\|\hat{b}_{\hat{m}} - b_A\|_n^2 = \|\hat{b}_{\hat{m}} - b_A\|_n^2 1_{\Omega_n} + \|\hat{b}_{\hat{m}} - b_A\|_n^2 1_{\Omega_n^c}$. For the study on Ω_n^c, the end of the proof of Proposition 5.2 can be used.

It remains to look at what happens on Ω_n. Let us introduce the notation

$$G_m(m') = \sup_{t \in S_m + S_{m'}, \|t\|_\pi = 1} |\nu_n(t)|.$$

From the definition of $\hat{b}_{\hat{m}}$, we have, $\forall m \in \mathcal{M}_n$, $\gamma_n(\hat{b}_{\hat{m}}) + \mathrm{pen}(\hat{m}) \le \gamma_n(b_m) + \mathrm{pen}(m)$. We proceed as in the proof of Proposition 5.2 with the additional penalty terms (see (5.25)) and obtain

$$
\begin{aligned}
\mathbb{E}(\|\hat{b}_{\hat{m}} - b_A\|_n^2 \mathbf{1}_{\Omega_n}) \le\ & 7\pi_1\|b_m - b_A\|^2 + 4\mathrm{pen}(m) \\
& + 32\mathbb{E}\left([G_m(\hat{m})]^2 \mathbf{1}_{\Omega_n}\right) - 4\mathbb{E}(\mathrm{pen}(\hat{m})) + 32c'\Delta.
\end{aligned}
$$

The main problem here is to control the supremum of $\nu_n(t)$ on a random ball (which depends on the random $\hat{m}$). This will be done using Lemma 5.5 and Proposition 5.6 below, proceeding first as follows. We plug in a function $p(m, m')$, which will in turn fix the penalty:

$$
\begin{aligned}
[G_m(\hat{m})]^2 \mathbf{1}_{\Omega_n} \le\ & \{([G_m(\hat{m})]^2 - p(m, \hat{m}))\mathbf{1}_{\Omega_n}\}_+ + p(m, \hat{m}) \\
\le\ & \sum_{m' \in \mathcal{M}_n} \{([G_m(m')]^2 - p(m, m'))\mathbf{1}_{\Omega_n}\}_+ + p(m, \hat{m}).
\end{aligned}
$$

The penalty $\mathrm{pen}(.)$ is chosen such that $8p(m, m') \le \mathrm{pen}(m) + \mathrm{pen}(m')$. More precisely, the next proposition determines the choice of $p(m, m')$.

Proposition 5.6 *Under the assumptions of Theorem 5.3, there exists a numerical constant κ_1 such that, for $p(m, m') = \kappa_1 \sigma_1^2(D_m + (1 + L_{m'})D_{m'})/(n\Delta)$, we have*

$$
\mathbb{E}\{([G_m(m')]^2 - p(m, m'))\mathbf{1}_{\Omega_n}\}_+ \le c\sigma_1^2 \frac{e^{-L_{m'}D_{m'}}}{n\Delta}.
$$

Proof of Proposition 5.6. The result of Proposition 5.6 follows from Lemma 5.5 applying the $\mathbb{L}_\pi^2$-chaining technique used in Baraud et al. (2001b) (see Proposition 6.1, p.42, and Section 7, pp. 44–47, Lemma 7.1, with $s^2 = \sigma_1^2/\Delta$). $\square$

Recall that the weights L_m are such that $\Sigma = \sum_{m' \in \mathcal{M}_n} e^{-L_{m'}D_{m'}} < \infty$. Thus, the result of Theorem 5.3 follows from Proposition 5.6 with $\mathrm{pen}(m) \ge \kappa \sigma_1^2(1 + L_m)D_m/(n\Delta)$ and $\kappa = 8\kappa_1$. $\square$

5.5.4 *Bound for the $\mathbb{L}^2$-risk*

In Theorem 5.3, the risk of $\hat{b}_{\hat{m}}$ is not measured as a standard $\mathbb{L}^2$-risk. In this paragragh, we prove that a simple truncation of $\hat{b}_{\hat{m}}$ allows to study an integrated loss over a compact set instead of our empirical loss.

Proposition 5.7 *Let*

$$
\tilde{b}^* = \begin{cases} \tilde{b} & \text{if } \|\tilde{b}\| \le k_n \\ 0 & \text{else,} \end{cases}
$$

where $\tilde{b} = \hat{b}_{\hat{m}}$ and $k_n = O(n)$. Then,

$$
\mathbb{E}(\|\tilde{b}^* - b_A\|^2) \le C \inf_{m \in \mathcal{M}_n} \left(\|b_m - b_A\|^2 + \mathrm{pen}(m)\right) + K'\Delta + \frac{K''}{n\Delta}.
$$

Proof. Recall that $\|\tilde{b}^* - b_A\|^2 \leq (1/\pi_0)\|\tilde{b}^* - b_A\|_\pi^2$. Then, we decompose the $\mathbb{L}_\pi^2$-norm into:

$$
\begin{aligned}
\|\tilde{b}^* - b_A\|_\pi^2 \\
= \|\tilde{b}^* - b_A\|_\pi^2 \mathbf{1}_{\|\tilde{b}\| \leq k_n} \mathbf{1}_{\Omega_n} + \|\tilde{b}^* - b_A\|_\pi^2 \mathbf{1}_{\|\tilde{b}\| > k_n} \mathbf{1}_{\Omega_n} + \|\tilde{b}^* - b_A\|_\pi^2 \mathbf{1}_{\Omega_n^c} \\
= T_1 + T_2 + T_3.
\end{aligned}
$$

First, it is easy to see that

$$
\mathbb{E}(T_3) \leq 2(\pi_1 k_n^2 + \|b_A\|_\pi^2)\mathbb{P}(\Omega_n^c),
$$

and with $k_n = O(n)$, as we know that $\mathbb{P}(\Omega_n^c) \leq c/n^4$, we get a negligible term of order $1/n^2$.

Next, T_1 can be studied as above, except that some constants are increased:

$$
\begin{aligned}
T_1 \leq \|\tilde{b}^* - b_A\|_\pi^2 \mathbf{1}_{\Omega_n} &\leq 2(\|\tilde{b} - b_m\|_n^2 + \|b_m - b_A\|_\pi^2)\mathbf{1}_{\Omega_n} \\
&\leq 4\|\tilde{b} - b_m\|_n^2 \mathbf{1}_{\Omega_n} + 2\|b_m - b_A\|_\pi^2 \mathbf{1}_{\Omega_n} \\
&\leq 8\|\tilde{b} - b_A\|_n^2 \mathbf{1}_{\Omega_n} + 8\|b_m - b_A\|_n^2 \mathbf{1}_{\Omega_n} \\
&\quad + 2\|b_m - b_A\|_\pi^2 \mathbf{1}_{\Omega_n}
\end{aligned}
$$

and we can use the bound obtained in Theorem 5.3 to get that, for all $m \in \mathcal{M}_n$,

$$
\mathbb{E}(T_1) \leq C(\|b_m - b_A\|^2 + \operatorname{pen}(m)) + K\Delta + \frac{K'}{n\Delta}.
$$

Lastly, $T_2 = \|b_A\|_\pi^2 \mathbf{1}_{\|\tilde{b}\| > k_n} \mathbf{1}_{\Omega_n}$. On Ω_n,

$$
\|\tilde{b}\|^2 \leq \frac{1}{\pi_0}\|\tilde{b}\|_\pi^2 \leq \frac{3}{2\pi_0}\|\tilde{b}\|_n^2 \leq \frac{3}{\pi_0}(\|b_A - \tilde{b}\|_n^2 + \|b_A\|_n^2)
$$

and the study of this term leads to the bound

$$
\|b_A - \tilde{b}\|_n^2 \leq \|b_A\|_n^2 + \frac{1}{n}\sum_{k=1}^n \varepsilon_{k\Delta}^2
$$

with $\mathbb{E}(\varepsilon_{k\Delta}^2) \leq c/\Delta$. It follows that, with Markov's inequality,

$$
\begin{aligned}
\mathbb{E}(T_2) &\leq \|b_A\|_\pi^2 \mathbb{P}(\{\|\tilde{b}\| \geq k_n\} \cap \Omega_n) \\
&\leq \|b_A\|_\pi^2 \left(\mathbb{P}(6\|b\|_n^2 \geq \pi_0 k_n^2) + \mathbb{P}(\frac{3}{n}\sum_{k=1}^n \varepsilon_{k\Delta}^2 \geq \pi_0 k_n^2) \right) \\
&\leq \|b_A\|_\pi^2 \left(\frac{4\|b_A\|_\pi^2}{\pi_0 k_n^2} + \frac{c}{\Delta}\frac{3}{\pi_0 k_n^2} \right) = o(\frac{1}{n}),
\end{aligned}
$$

since $k_n = O(n)$.

Gathering all terms gives that Theorem 5.3 extends to $\mathbb{E}(\|\tilde{b}^* - b_A\|^2)$. $\quad\square$

5.6 Diffusion coefficient estimation

5.6.1 Diffusion coefficient estimator: Statement of the results

For diffusion coefficient estimation under our asymptotic framework, it is now well known that rates of convergence are faster than for drift estimation. This is the reason why the regression-type equation has to be more precise than for b. We set

$$U_{k\Delta} = \frac{(X_{(k+1)\Delta} - X_{k\Delta})^2}{\Delta}.$$

The couple of data is now $(U_{k\Delta}, X_{k\Delta})$. The regression-type equation is as follows:

$$U_{k\Delta} = \sigma^2(X_{k\Delta}) + V_{k\Delta} + \tau_{k\Delta}, \tag{5.27}$$

where $V_{k\Delta} = V_{k\Delta}^{(1)} + V_{k\Delta}^{(2)} + V_{k\Delta}^{(3)}$ with

$$V_{k\Delta}^{(1)} = \frac{1}{\Delta}\left[\left\{\int_{k\Delta}^{(k+1)\Delta} \sigma(X_s)dW_s\right\}^2 - \int_{k\Delta}^{(k+1)\Delta} \sigma^2(X_s)ds\right],$$

$$V_{k\Delta}^{(2)} = \frac{1}{\Delta}\int_{k\Delta}^{(k+1)\Delta} ((k+1)\Delta - s)(\sigma^2)'(X_s)\sigma(X_s)dW_s,$$

$$V_{k\Delta}^{(3)} = 2b(X_{k\Delta})\int_{k\Delta}^{(k+1)\Delta} \sigma(X_s)dW_s,$$

$\tau_{k\Delta} = \tau_{k\Delta}^{(1)} + \tau_{k\Delta}^{(2)} + \tau_{k\Delta}^{(3)}$ with

$$\tau_{k\Delta}^{(1)} = \frac{1}{\Delta}\left(\int_{k\Delta}^{(k+1)\Delta} b(X_s)ds\right)^2,$$

$$\tau_{k\Delta}^{(2)} = \frac{2}{\Delta}\int_{k\Delta}^{(k+1)\Delta} (b(X_s) - b(X_{k\Delta}))ds \int_{k\Delta}^{(k+1)\Delta} \sigma(X_s)dW_s,$$

$$\tau_{k\Delta}^{(3)} = \frac{1}{\Delta}\int_{k\Delta}^{(k+1)\Delta} [(k+1)\Delta - s]\psi(X_s)ds,$$

and

$$\psi = \frac{\sigma^2}{2}(\sigma^2)'' + b(\sigma^2)' = L\sigma^2, \tag{5.28}$$

where $Lf = \frac{\sigma^2}{2}f'' + bf'$ is the infinitesimal generator of (5.1). The above relations are obtained by applying the Itô and the Fubini formulae. The term $V_{k\Delta}$ is a sum of martingale increments whose variances have different orders. The term $V_{k\Delta}^{(1)}$ plays the role of the main noise. The term $\tau_{k\Delta}$ is a remainder due to the discretization and to the presence of the drift. The scheme is similar to what is done for the drift. To estimate σ^2 on $A = [0, 1]$, we define

$$\hat{\sigma}_m^2 = \arg\min_{t \in S_m} \breve{\gamma}_n(t), \text{ with } \breve{\gamma}_n(t) = \frac{1}{n}\sum_{k-1}^{n}[U_{k\Delta} - t(X_{k\Delta})]^2. \quad (5.29)$$

And, we obtain the following result.

Proposition 5.8 *Assume that [A1]-[A7] hold and consider a model S_m in collection [DP] or [GP] with maximal dimension $N_n = o(n\Delta/\ln^2(n))$. Then the estimator $\hat{\sigma}_m^2$ of σ^2 defined by (5.29) is such that*

$$\mathbb{E}(\|\hat{\sigma}_m^2 - \sigma_A^2\|_n^2) \leq 7\pi_1 \|\sigma_m^2 - \sigma_A^2\|^2 + K\frac{\sigma_1^4 D_m}{n} + K'\Delta^2 + \frac{K''}{n}, \quad (5.30)$$

where K, K', K'' are some positive constants.

Let us make some comments on the rates of convergence for estimators built with [DP]. If σ_A^2 belongs to a ball of some Besov space, say $\sigma_A^2 \in \mathcal{B}_{\alpha,2,\infty}([0, 1])$, and $\|\sigma_A^2\|_{\alpha,2,\infty} \leq L$, with $r + 1 \geq \alpha$, then $\|\sigma_A^2 - \sigma_m^2\|^2 \leq C(\alpha, L)D_m^{-2\alpha}$. Therefore, if we choose $D_m = n^{1/(2\alpha+1)}$, we obtain

$$\mathbb{E}(\|\hat{\sigma}_m^2 - \sigma_A^2\|_n^2) \leq C(\alpha, L)n^{-2\alpha/(2\alpha+1)} + K'\Delta^2 + \frac{K''}{n}.$$

The first term $n^{-2\alpha/(2\alpha+1)}$ is the optimal non-parametric rate proved by Hoffmann (1999). Moreover, under the standard condition $\Delta^2 = o(1/n)$, the last two terms are $O(1/n)$, i.e. negligible with respect to $n^{-2\alpha/(2\alpha+1)}$.

As previously, the second step is to ensure an automatic selection of D_m, which does not use any knowledge on σ^2. This selection is done by

$$\hat{m} = \arg\min_{m \in \mathcal{M}_n} \left[\breve{\gamma}_n(\hat{\sigma}_m^2) + \widetilde{\text{pen}}(m)\right]. \quad (5.31)$$

We denote by $\hat{\sigma}_{\hat{m}}^2$ the resulting estimator and we need to determine the penalty $\widetilde{\text{pen}}$ as for b. For simplicity, we use the same notation $\hat{m}$ in (5.31) as in (5.20) although they are different. We can prove the following theorem.

Theorem 5.9 *Assume that [A1]-[A7] hold. Consider collection [DP] with $L_m = 1$ or [GP] with L_m given by (5.13) both with maximal dimension $N_n \leq n\Delta/\ln^2(n)$. Let*

$$\widetilde{\text{pen}}(m) \geq \tilde{\kappa}\sigma_1^4\frac{(1 + L_m)D_m}{n},$$

where $\tilde{\kappa}$ is a universal constant. Then, the estimator $\hat{\sigma}_{\hat{m}}^2$ of σ^2 with $\hat{m}$ defined by (5.31) is such that

$$\mathbb{E}(\|\hat{\sigma}_{\hat{m}}^2 - \sigma_A^2\|_n^2) \leq C \inf_{m \in \mathcal{M}_n} \left(\|\sigma_m^2 - \sigma_A^2\|^2 + \widetilde{\text{pen}}(m)\right) + K'\Delta^2 + \frac{K''}{n}.$$

Analogous comments as those given for the drift can be made.

5.6.2 Proof of Proposition 5.8

Let us set

$$\breve{\nu}_n(t) = \breve{\nu}_n^{(1)}(t) + \breve{\nu}_n^{(2)}(t) + \breve{\nu}_n^{(3)}(t) \tag{5.32}$$

with

$$\breve{\nu}_n^{(i)}(t) = \frac{1}{n}\sum_{k=1}^{n} t(X_{k\Delta})V_{k\Delta}^{(i)}, \tag{5.33}$$

and

$$\breve{\tau}_n(t) = \frac{1}{n}\sum_{k=1}^{n} t(X_{k\Delta})\tau_{k\Delta}.$$

We begin with some lemmas. The first one concerns the remainder term.

Lemma 5.10 *We have (see (5.27))*

$$\mathbb{E}(\frac{1}{n}\sum_{k=1}^{n}\tau_{k\Delta}^2) \le K\Delta^2. \tag{5.34}$$

Proof of Lemma 5.10. We prove that $\mathbb{E}[(\tau_{k\Delta}^{(i)})^2] \le K_i\Delta^2$ for $i = 1, 2, 3$. Using [A5] and Lemma 5.1,

$$\begin{aligned}
\mathbb{E}[(\tau_{k\Delta}^{(1)})^2] &\le \mathbb{E}\left(\int_{k\Delta}^{(k+1)\Delta} b^2(X_s)ds\right)^2 \le \Delta\mathbb{E}\left(\int_{k\Delta}^{(k+1)\Delta} b^4(X_s)ds\right) \\
&\le \Delta^2\mathbb{E}(b^4(X_0)) \le c\Delta^2,
\end{aligned}$$

$$\begin{aligned}
\mathbb{E}[(\tau_{k\Delta}^{(2)})^2] &\le \frac{1}{\Delta^2}\left(\mathbb{E}\left(\int_{k\Delta}^{(k+1)\Delta} (b(X_s)-b(X_{k\Delta}))ds\right)^4 \right. \\
&\quad \left. \times \mathbb{E}\left(\int_{k\Delta}^{(k+1)\Delta} \sigma(X_s)dW_s\right)^4\right)^{1/2}
\end{aligned}$$

Using [A5], Lemma 5.1 and the Burkholder–Davis–Gundy inequality, we get

$$\mathbb{E}[(\tau_{k\Delta}^{(2)})^2] \le c'\Delta^2.$$

Lastly, [A5] implies that $|\psi(x)| \le K(1+x^2)$ (see (5.28)), hence

$$\begin{aligned}
\mathbb{E}[(\tau_{k\Delta}^{(3)})^2] &\le \frac{1}{\Delta}\mathbb{E}\left(\int_{k\Delta}^{(k+1)\Delta} ((k+1)\Delta - s)^2\psi^2(X_s)ds\right) \\
&\le \mathbb{E}(\psi^2(X_0))\frac{\Delta^2}{3} \le c''\Delta^2.
\end{aligned}$$

Therefore (5.34) is proved. $\square$

Now, we deal with the noise terms and show that $i = 1$ gives the main term. In the statement below, K, K' denote constants which may vary from line to line.

Lemma 5.11 1. For S_m in collection [DP] or [GP],

$$\mathbb{E}\left(\sup_{t \in S_m, \|t\|_\pi = 1} (\breve{\nu}_n^{(1)}(t))^2\right) \leq K \frac{D_m}{n} \sigma_1^4.$$

2. Recall that $\mathcal{S}_n$ denotes the maximal space for both collections. For $i = 2, 3$,

$$\mathbb{E}\left(\sup_{t \in \mathcal{S}_n, \|t\|_\pi = 1} (\breve{\nu}_n^{(i)}(t))^2\right) \leq K \frac{\Delta N_n}{n} \leq K' \Delta^2. \tag{5.35}$$

Proof of Lemma 5.11. To study $\breve{\nu}_n^{(1)}(t)$, we consider, as for the drift case, an orthonormal basis $(\psi_\lambda, \lambda \in J_m)$ of S_m with respect to $\mathbb{L}_\pi^2$. So,

$$\mathbb{E}\left(\sup_{t \in S_m, \|t\|_\pi = 1} (\breve{\nu}_n^{(1)}(t))^2\right) \leq \sum_{\lambda \in J_m} \mathbb{E}((\breve{\nu}_n^{(1)}(\psi_\lambda))^2).$$

Then, we use the fact that $V_{k\Delta}^{(1)}$ is a martingale increment and obtain:

$$\mathbb{E}((\breve{\nu}_n^{(1)}(\psi_\lambda))^2) = \frac{1}{n^2} \sum_{k=1}^n \mathbb{E}(\psi_\lambda^2(X_{k\Delta}) \mathbb{E}([V_{k\Delta}^{(1)}]^2 | \mathcal{F}_{k\Delta})).$$

Then,

$$\mathbb{E}((V_{k\Delta}^{(1)})^2 | \mathcal{F}_{k\Delta}) \leq \frac{2}{\Delta^2} \left[\mathbb{E}\left(\left(\int_{k\Delta}^{(k+1)\Delta} \sigma(X_s) dW_s\right)^4 | \mathcal{F}_{k\Delta}\right) \right.$$
$$\left. + \mathbb{E}\left(\left(\int_{k\Delta}^{(k+1)\Delta} \sigma^2(X_s) ds\right)^2 | \mathcal{F}_{k\Delta}\right) \right].$$

Using the Burkholder–Davis–Gundy inequality, we obtain:

$$\mathbb{E}((V_{k\Delta}^{(1)})^2 | \mathcal{F}_{k\Delta}) \leq C \sigma_1^4.$$

This gives the first part.

For the second part, note that the maximal space $\mathcal{S}_n$ is equipped with an orthonormal basis $(\varphi_\lambda, \lambda \in \mathcal{L}_n)$ with respect to $\mathbb{L}^2$ which satisfies for both collections (see $(5.11) - (5.12)$)

$$\left\| \sum_{\lambda \in \mathcal{L}_n} \varphi_\lambda^2 \right\|_\infty \leq \Phi_0^2 N_n,$$

with $\Phi_0^2 = r_{max} + 1$. For $i = 2, 3$,

$$\mathbb{E}\left(\sup_{t \in \mathcal{S}_n, \|t\| \leq 1} (\breve{\nu}_n^{(i)}(t))^2\right) \leq \sum_{\lambda \in \mathcal{L}_n} \mathbb{E}((\breve{\nu}_n^{(i)}(\varphi_\lambda))^2).$$

Since the martingale increments $(V_{k\Delta}^{(i)})$ are uncorrelated, we have:

$$E((\breve{\nu}_n^{(i)}(\varphi_\lambda))^2) = \frac{1}{n^2}\sum_{k=1}^{n} E\left(\varphi_\lambda^2(X_{k\Delta})(V_{k\Delta}^{(i)})^2\right).$$

Therefore, interchanging sums in λ and k, we get:

$$\sum_{\lambda \in \mathcal{L}_n} E((\breve{\nu}_n^{(i)}(\varphi_\lambda))^2) \leq \frac{\Phi_0^2 N_n}{n}\frac{1}{n}\sum_{k=1}^{n} E((V_{k\Delta}^{(i)})^2). \qquad (5.36)$$

Now,

$$\begin{aligned}
E((V_{k\Delta}^{(2)})^2) &= \frac{1}{\Delta^2}E[((\sigma^2)'(X_0)\sigma(X_0))^2]\int_{k\Delta}^{(k+1)\Delta}((k+1)\Delta - s)^2 ds \\
&\leq C\Delta(1 + (E(X_0))^4)
\end{aligned}$$

and

$$\begin{aligned}
E((V_{k\Delta}^{(3)})^2) &= 4E(b^2(X_{k\Delta})\int_{k\Delta}^{(k+1)\Delta}\sigma^2(X_s)ds) \\
&\leq 4\left(E(b^4(X_{k\Delta}))E[(\int_{k\Delta}^{(k+1)\Delta}\sigma^2(X_s)ds)^2]\right)^{1/2} \\
&\leq 4\left(E(b^4(X_0))E(\sigma^4(X_0))\right)^{1/2}\Delta \leq C\Delta(1 + E(X_0^4)).
\end{aligned}$$

Since $\|t\|^2 \leq \|t\|_\pi^2/\pi_0$, we join the above bounds and (5.36) and obtain the first inequality in (5.35). Since $N_n \leq n\Delta/\ln^2 n$, $N_n\Delta/n \leq \Delta^2/\ln^2 n$. This gives the second inequality. $\square$

Now, we can prove Proposition 5.8. As for the drift, the starting point is:

$$\breve{\gamma}_n(t) - \breve{\gamma}_n(\sigma^2) = \|\sigma^2 - t\|_n^2 + 2\breve{\nu}_n(\sigma^2 - t) + 2\breve{\tau}_n(\sigma^2 - t).$$

Introducing the orthogonal projection σ_m^2 of σ^2 on S_m, we have:

$$\breve{\gamma}_n(\hat{\sigma}_m^2) - \breve{\gamma}_n(\sigma^2) \leq \breve{\gamma}_n(\sigma_m^2) - \breve{\gamma}_n(\sigma^2).$$

After some computations analogous to those done for the drift study, we are led to the following inequality which holds on Ω_n (see (5.24)):

$$\frac{1}{4}\|\hat{\sigma}_m^2 - \sigma_A^2\|_n^2 \leq \frac{7}{4}\|\sigma_m^2 - \sigma_A^2\|_n^2 + 8\sup_{t \in B_m^\pi(0,1)}\breve{\nu}_n^2(t) + \frac{8}{n}\sum_{k=1}^{n}\tau_{k\Delta}^2,$$

where $B_m^\pi(0,1) = \{t \in S_m, \|t\|_\pi = 1\}$. Now we apply Lemma 5.10 and Lemma 5.11. This yields the first three terms of the right-hand-side of (5.30). The study on Ω_n^c is the same as for b with the regression model $U_{k\Delta} = \sigma^2(X_{k\Delta}) + \xi_{k\Delta}$, where $\xi_{k\Delta} = V_{k\Delta} + \tau_{k\Delta}$. By standard inequalities, $\mathbb{E}(\xi_\Delta^4) \leq K\{\Delta^4\mathbb{E}(b^8(X_0)) + \mathbb{E}(\sigma^8(X_0))\}$. Hence, $\mathbb{E}(\xi_\Delta^4)$ is bounded. Moreover, using Lemma 5.4, $\mathbb{P}(\Omega_n^c) \leq c/n^2$. $\square$

5.6.3 Proof of Theorem 5.9

This proof follows the same lines as the proof of Theorem 5.3. We start with a Bernstein-type inequality.

Lemma 5.12 *Under the assumptions of Theorem 5.9,*

$$\mathbb{P}\left(\breve{\nu}_n^{(1)}(t)) \geq \epsilon, \|t\|_n^2 \leq v^2\right) \leq \exp\left(-Cn\frac{\epsilon^2/2}{2\sigma_1^4 v^2 + \epsilon\|t\|_\infty \sigma_1^2 v}\right)$$

and

$$\mathbb{P}\left(\breve{\nu}_n^{(1)}(t) \geq v\sigma_1^2(2x)^{1/2} + \sigma_1^2\|t\|_\infty x, \|t\|_n^2 \leq v^2\right) \leq \exp(-Cnx). \quad (5.37)$$

The proof that the first inequality implies the second one above is rather tricky and proved in Birgé and Massart (1998). Consequently, we just prove the first one.

Proof of Lemma 5.12. First we note that:

$$\begin{aligned}
\mathbb{E}\left(e^{ut(X_{n\Delta})V_{n\Delta}^{(1)}}|\mathcal{F}_{n\Delta}\right) &= 1 + \sum_{p=2}^{+\infty}\frac{u^p}{p!}\mathbb{E}\left\{(t(X_{n\Delta})V_{n\Delta}^{(1)})^p|\mathcal{F}_{n\Delta}\right\} \\
&\leq 1 + \sum_{p=2}^{+\infty}\frac{u^p}{p!}|t(X_{n\Delta})|^p\mathbb{E}\left(|V_{n\Delta}^{(1)}|^p|\mathcal{F}_{n\Delta}\right).
\end{aligned}$$

Next we apply successively the Minkowski inequality and the Burkholder–Davis–Gundy inequality with best constant (Proposition 4.2 of Barlow and Yor (1982)). For a continuous martingale (M_t), with $M_0 = 0$, for $k \geq 2$, $M_t^* = \sup_{s \leq t}|M_s|$ satisfies $\|M^*\|_k \leq ck^{1/2}\|\langle M\rangle^{1/2}\|_k$, with c a universal constant. And we obtain:

$$\begin{aligned}
\mathbb{E}(|V_{n\Delta}^{(1)}|^p|\mathcal{F}_{n\Delta}) &\leq \frac{2^{p-1}}{\Delta^p}\left\{\mathbb{E}\left(\left|\int_{n\Delta}^{(n+1)\Delta}\sigma(X_s)dW_s\right|^{2p}|\mathcal{F}_{n\Delta}\right)\right. \\
&\quad +\left.\mathbb{E}\left(\left|\int_{n\Delta}^{(n+1)\Delta}\sigma^2(X_s)ds\right|^p|\mathcal{F}_{n\Delta}\right)\right\} \\
&\leq \frac{2^{p-1}}{\Delta^p}(c^{2p}(2p)^p\Delta^p\sigma_1^{2p} + \Delta^p\sigma_1^{2p}) \leq (2\sigma_1 c)^{2p}p^p.
\end{aligned}$$

Therefore,

$$\mathbb{E}\left(e^{ut(X_{n\Delta})V_{n\Delta}^{(1)}}|\mathcal{F}_{n\Delta}\right) \leq 1 + \sum_{p=2}^{\infty}\frac{p^p}{p!}(4u\sigma_1^2 c^2)^p|t(X_{n\Delta})|^p.$$

Using $p^p/p! \leq e^{p-1}$, we find

$$\mathbb{E}\left(e^{ut(X_{n\Delta})V^{(1)}_{n\Delta}}|\mathcal{F}_{n\Delta}\right) \leq 1 + e^{-1}\sum_{p=2}^{\infty}(4u\sigma_1^2c^2e)^p|t(X_{n\Delta})|^p$$

$$\leq 1 + e^{-1}\frac{(4u\sigma_1^2c^2e)^2t^2(X_{n\Delta})}{1-(4u\sigma_1^2c^2e\|t\|_{\infty})}.$$

Now, let us set

$$a = e(4\sigma_1^2c^2)^2 \text{ and } b = 4\sigma_1^2c^2e\|t\|_{\infty}.$$

Since for $x \geq 0$, $1 + x \leq e^x$, we get, for all u such that $bu < 1$,

$$\mathbb{E}\left(e^{ut(X_{n\Delta})V^{(1)}_{n\Delta}}|\mathcal{F}_{n\Delta}\right) \leq 1 + \frac{au^2t^2(X_{n\Delta})}{1-bu} \leq \exp\left(\frac{au^2t^2(X_{n\Delta})}{1-bu}\right).$$

This can also be written:

$$\mathbb{E}\left(\exp\left(ut(X_{n\Delta})V^{(1)}_{n\Delta} - \frac{au^2t^2(X_{n\Delta})}{1-bu}\right)|\mathcal{F}_{n\Delta}\right) \leq 1.$$

Therefore, iterating conditional expectations yields

$$\mathbb{E}\left[\exp\left\{\sum_{k=1}^{n}\left(ut(X_{k\Delta})V^{(1)}_{k\Delta} - \frac{au^2t^2(X_{k\Delta})}{1-bu}\right)\right\}\right] \leq 1.$$

Then, we deduce that

$$\mathbb{P}\left(\sum_{k=1}^{n}t(X_{k\Delta})V^{(1)}_{k\Delta} \geq n\epsilon, \|t\|_n^2 \leq v^2\right)$$

$$\leq e^{-nu\epsilon}\mathbb{E}\left\{1_{\|t\|_n^2 \leq v^2}\exp\left(u\sum_{k=1}^{n}t(X_{k\Delta})V^{(1)}_{k\Delta}\right)\right\}$$

$$\leq e^{-nu\epsilon}\mathbb{E}\left[1_{\|t\|_n^2 \leq v^2}\exp\left\{\sum_{k=1}^{n}(ut(X_{k\Delta})V^{(1)}_{k\Delta} - \frac{au^2t^2(X_{k\Delta})}{1-bu})\right\}\right.$$

$$\left. \times e^{(au^2)/(1-bu)\sum_{k=1}^{n}t^2(X_{k\Delta})}\right]$$

$$\leq e^{-nu\epsilon}e^{(nau^2v^2)/(1-bu)}\mathbb{E}\left[\exp\left\{\sum_{k=1}^{n}(ut(X_{k\Delta})V^{(1)}_{k\Delta} - \frac{au^2t^2(X_{k\Delta})}{1-bu})\right\}\right]$$

$$\leq e^{-nu\epsilon}e^{(nau^2v^2)/(1-bu)}.$$

The inequality holds for any u such that $bu < 1$. In particular, $u = \epsilon/(2av^2 + \epsilon b)$ gives $-u\epsilon + av^2u^2/(1-bu) = -(1/2)(\epsilon^2/(2av^2 + \epsilon b))$ and therefore

$$\mathbb{P}\left(\sum_{k=1}^{n}t(X_{k\Delta})V^{(1)}_{k\Delta} \geq n\epsilon, \|t\|_n^2 \leq v^2\right) \leq \exp\left(-n\frac{\epsilon^2/2}{2av^2 + \epsilon b}\right). \quad \square$$

We now finish the proof of Theorem 5.9. As for $\hat{b}_{\hat{m}}$, we introduce the additional penalty terms and obtain that the risk satisfies

$$\mathbb{E}(\|\hat{\sigma}_{\hat{m}}^2 - \sigma_A^2\|_n^2 \mathbf{1}_{\Omega_n})$$

$$\leq \quad 7\pi_1 \|\sigma_m^2 - \sigma_A^2\|^2 + 4\widetilde{\mathrm{pen}}(m) + 32\mathbb{E}\left(\sup_{t \in B_{m,\hat{m}}^\pi(0,1)} (\check{\nu}_n(t))^2 \mathbf{1}_{\Omega_n}\right)$$

$$-4\mathbb{E}(\widetilde{\mathrm{pen}}(\hat{m})) + K'\Delta^2 \tag{5.38}$$

where $B_{m,m'}^\pi(0,1) = \{t \in S_m + S_{m'}, \|t\|_\pi = 1\}$. We use that

$$(\check{\nu}_n(t))^2 \leq 2[(\check{\nu}_n^{(1)}(t))^2 + (\check{\nu}_n^{(2)}(t) + \check{\nu}_n^{(3)}(t))^2].$$

By Lemma 5.11, since $B_{m,m'}^\pi(0,1) \subset \{t \in \mathcal{S}_n, \|t\|_\pi = 1\}$,

$$\mathbb{E}\left(\sup_{t \in B_{m,\hat{m}}^\pi(0,1)} (\check{\nu}_n^{(2)}(t) + \check{\nu}_n^{(3)}(t))^2\right) \leq K\Delta^2.$$

There remains the main term to study

$$\check{G}_m(m') = \sup_{t \in B_{m,m'}^\pi(0,1)} |\check{\nu}_n^{(1)}(t)|. \tag{5.39}$$

As for the drift, we write

$$\mathbb{E}(\check{G}_m^2(\hat{m})) \quad \leq \quad \mathbb{E}[(\check{G}_m^2(\hat{m}) - \tilde{p}(m, \hat{m}))\mathbf{1}_{\Omega_n}]_+ + \mathbb{E}(\tilde{p}(m, \hat{m}))$$

$$\leq \quad \sum_{m' \in \mathcal{M}_n} \mathbb{E}[(\check{G}_m^2(m') - \tilde{p}(m, m'))\mathbf{1}_{\Omega_n}]_+ + \mathbb{E}(\tilde{p}(m, \hat{m})).$$

Now we have the following statement.

Proposition 5.13 *Under the assumptions of Theorem 5.9, for*

$$\tilde{p}(m, m') = \kappa\sigma_1^4 \frac{D_m + D_{m'}(1 + L_{m'})}{n} + K\Delta^2,$$

where κ is a numerical constant, we have

$$\mathbb{E}[(\check{G}_m^2(m') - \tilde{p}(m, m'))\mathbf{1}_{\Omega_n}]_+ \leq c\sigma_1^4 \frac{e^{-D_{m'}L_{m'}}}{n}.$$

The result of Proposition 5.13 is obtained from inequality (5.37) of Lemma 5.12 by a $L_\pi^2 - L^\infty$ chaining technique. A description of this method, in a more general setting, is given in Propositions 2–4, pp. 282–287, in Comte (2001), Theorem 5 in Birgé and Massart (1998) and Proposition 7, Theorem 8 and Theorem 9 in Barron et al. (1999). For the sake of completeness and since the context is slightly different, we detail the proof in the Appendix, Section 5.9. Note that there is a difference between Propositions 5.6 and 5.13 which comes from the additional term $\|t\|_\infty$ appearing in Lemma 5.12.

Choosing $\widetilde{\text{pen}}(m) \geq \tilde{\kappa}\sigma_1^4 D_m(1+L_m)/n$ with $\tilde{\kappa} = 16\kappa$, we deduce from (5.38), Proposition 5.13 and $D_m \leq N_n \leq n\Delta/\ln^2(n)$ that,

$$\mathbb{E}(\|\hat{\sigma}_{\hat{m}}^2 - \sigma_A^2\|_n^2) \leq 7\pi_1\|\sigma_m^2 - \sigma_A^2\|^2 + 8\widetilde{\text{pen}}(m) + c\sigma_1^4 \sum_{m' \in \mathcal{M}_n} \frac{e^{-D_{m'}L_{m'}}}{n}$$

$$+ K'\Delta^2 + \mathbb{E}(\|\hat{\sigma}_{\hat{m}}^2 - \sigma_A^2\|_n^2 \mathbf{1}_{\Omega_n^c}).$$

The bound for $\mathbb{E}(\|\hat{\sigma}_{\hat{m}}^2 - \sigma^2\|_n^2 \mathbf{1}_{\Omega_n^c})$ is the same as the one given in the end of the proof of Proposition 5.8. It is less than c/n. The result of Theorem 5.9 follows. $\square$

5.7 Examples and practical implementation

In this section, we consider classical examples of diffusions for which an exact simulation of sample paths is possible and for which the estimation method has been implemented with [GP]. For exact simulation of sample paths, when it is not directly possible, we have in view the retrospective exact simulation algorithms proposed by Beskos, Papaspiliopoulos, and Roberts (2006) and Beskos and Roberts (2005). Models of Families 1 and 2 below can be simulated by the algorithm EA1. Among the assumptions, requiring that σ be bounded is rather stringent and not always satisfied in our examples. The other assumptions hold. More details may be found in Comte et al. (2006, 2007)

5.7.1 Examples of diffusions

Family 1

First, we consider (5.1) with

$$b(x) = -\theta x, \quad \sigma(x) = c(1+x^2)^{1/2}.$$

Standard computations of the scale and speed densities show that the model is positive recurrent for $\theta + c^2/2 > 0$. In this case, its stationary distribution has density

$$\pi(x) \propto \frac{1}{(1+x^2)^{1+\theta/c^2}}.$$

If $X_0 = \eta$ has distribution $\pi(x)dx$, then, setting $\nu = 1 + 2\theta/c^2$, $\nu^{1/2}\eta$ has Student distribution $t(\nu)$. This distribution satisfies the moment condition [A6] for $2\theta/c^2 > 7$. See Figure 5.1 for the estimation of b and σ^2 in this case.

Then, we consider $F_1(x) = \int_0^x 1/(c(1+x^2)^{1/2}dx = \arg\sinh(x)/c$. By the Itô formula, $\xi_t = F_1(X_t)$ is solution of a stochastic differential equation with $\sigma(\xi) = 1$ and

$$b(\xi) = -(\theta/c + c/2)\tanh(c\xi).$$

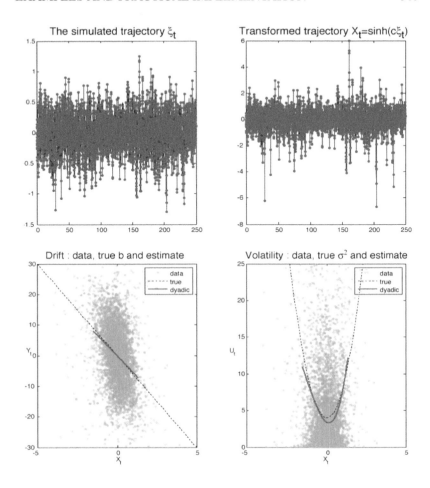

Figure 5.1 *First example:* $dX_t = -\theta X_t dt+, c\sqrt{1+X_t^2}dW_t$, $n = 5000$, $\Delta = 1/20$, $\theta = 2$, $c = 1$, *dotted line: true function, full line: estimated function.*

Assumptions [A1] – [A3] and [A5] hold for (ξ_t) with $\xi_0 = F_1(X_0)$. More-over, (ξ_t) satisfies the conditions of Proposition 1 in Pardoux and Veretennikov (2001) implying that (ξ_t) is exponentially β-mixing and has moments of any order. Hence, [A4] and [A6] hold. See Figure 5.2 for the estimation of b and σ^2 in this case.

Since $X_t = F_1^{-1}(\xi_t)$, this process is also β-mixing. It satisfies all assumptions except that $\sigma^2(x)$ is not bounded from above.

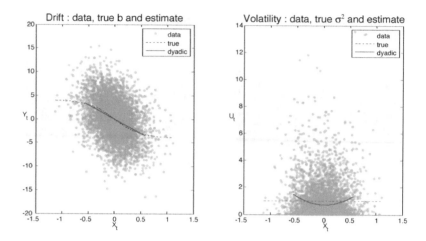

Figure 5.2 *Second example:* $d\xi_t = -(\theta/c + c/2)\tanh(c\xi_t) + dW_t$, $n = 5000$, $\Delta = 1/20$, $\theta = 6$, $c = 2$, *dotted line: true function, full line: estimated function.*

Family 2

For the second family of models, we consider a process (ξ_t) with diffusion coefficient $\sigma(\xi) = 1$ and drift

$$b(\xi) = -\theta \frac{\xi}{(1 + c^2\xi^2)^{1/2}}, \qquad (5.40)$$

(see Barndorff-Nielsen (1978)). The model is positive recurrent on $\mathbb{R}$ for $\theta > 0$. Its stationary distribution is a hyperbolic distribution given by

$$\pi(\xi)d\xi \propto \exp(-2\frac{\theta}{c^2}(1 + c^2\xi^2)^{1/2}).$$

Assumptions [A1] – [A3], [A5] – [A6] hold for this model. For [A4], we apply Proposition 1 of Pardoux and Veretennikov (2001).

Next, we consider $X_t = F_2(\xi_t) = \arg\sinh(c\xi_t)$ which satisfies a stochastic differential equation with coefficients:

$$b(x) = -\left(\theta + \frac{c^2}{2\cosh(x)}\right)\frac{\sinh(x)}{\cosh^2(x)}, \qquad \sigma(x) = \frac{c}{\cosh(x)}.$$

The process (X_t) is exponentially β-mixing as (ξ_t). The diffusion coefficient $\sigma(x)$ has an upper bound. See Figure 5.3 for the estimation of b and σ^2 in this case.

To obtain a different shape for the diffusion coefficient, showing two bumps,

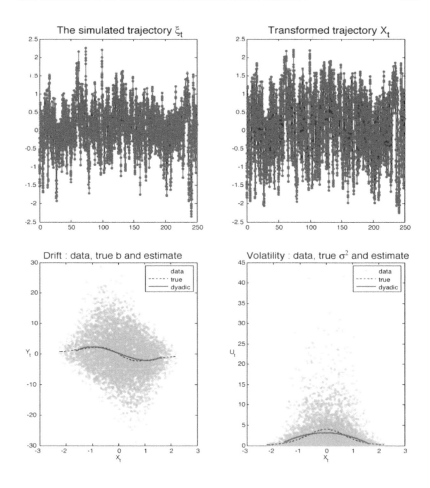

Figure 5.3 *Third example,* $dX_t = -[\theta + c^2/(2\cosh(X_t))](\sinh(X_t)/\cosh^2(X_t))dt$ $+(c/\cosh(X_t))dW_t$, $n = 5000$, $\Delta = 1/20$, $\theta = 3$, $c = 2$, *dotted line: true function, full line: estimated function.*

we consider $X_t = G(\xi_t) = \arg\sinh(\xi_t - 5) + \arg\sinh(\xi_t + 5)$. The function $G(.)$ is invertible and its inverse has the following explicit expression,

$$G^{-1}(x) = \frac{1}{2^{1/2}\sinh(x)}\left[49\sinh^2(x) + 100 + \cosh(x)(\sinh^2(x) - 100)\right]^{1/2}.$$

The diffusion coefficient of (X_t) is given by

$$\sigma(x) = \frac{1}{(1 + (G^{-1}(x) - 5)^2)^{1/2}} + \frac{1}{(1 + (G^{-1}(x) + 5)^2)^{1/2}}.$$

The drift is given by $G'(G^{-1}(x))b(G^{-1}(x)) + \frac{1}{2}G''(G^{-1}(x))$ with b given in (5.40). See Figure 5.4 for the estimation of b and σ^2 in this case.

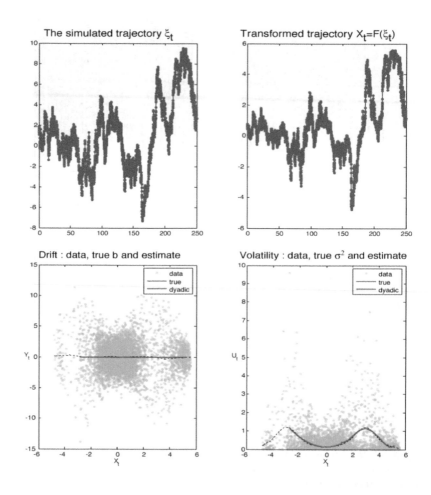

Figure 5.4 *Fourth example, the two-bumps diffusion coefficient* $X_t = G(\xi_t)$, $d\xi_t = -\theta\xi_t/\sqrt{1+c^2\xi_t^2}dt + dW_t$, $G(x) = \arg\sinh(x-5) + \arg\sinh(x+5)$, $n = 5000$, $\Delta = 1/20$, $\theta = 1$, $c = 10$, *dotted line: true function, full line: estimated function.*

Family 3

Consider Y_t a stationary Ornstein-Uhlenbeck process given by $dY_t = -\theta Y_t dt + cdW_t$ with $\theta > 0$ and $Y_0 \rightsquigarrow \mathcal{N}(0, c^2/(2\theta))$. The β-mixing coefficient of (Y_t)

can be evaluated using the exact formula (5.2). This gives a direct proof of (5.3).

Proposition 5.14 *The β-mixing coefficient of (Y_t) satisfies*

$$\beta_Y(t) \leq \frac{\exp(-\theta t)}{2(1 - \exp(-\theta t))}.$$

Proof. We use the expansion of the transition density $p_t(x, y)$ of (Y_t) in terms of the sequence of eigenvalues and eigenfunctions of the infinitesimal generator $Lf(y) = \frac{\sigma^2}{2} f''(y) - \theta y f'(y)$. For this, we refer e.g. to Karlin and Taylor (1981, p.333). To simplify notations, we assume that $\sigma^2/2\theta = 1$ so that the stationary distribution of (Y_t) is $\pi(y)dy = \mathcal{N}(0, 1)$. Let us now consider the n-th Hermite polynomial given, for $n = 0, 1, \ldots$, by:

$$H_n(x) = \frac{(-1)^n}{\sqrt{n!}} \exp(x^2/2) \frac{d^n}{dx^n}[\exp(-x^2/2)].$$

As defined above, this sequence is an orthonormal basis of $L^2(\pi)$ and satisfies, for all $n \geq 0$, $LH_n = -n\theta H_n$, i.e., H_n is the eigenfunction associated with the eigenvalue $-n\theta$ of L. This gives the following expansion:

$$p_t(x, y) = \pi(y) \sum_{n=0}^{+\infty} \exp(-n\theta t) H_n(x) H_n(y).$$

Since $H_0(x) = 1$ and the H_n have $L^2(\pi)$-norm equal to 1, we get

$$\|p_t(x, y)dy - \pi(y)dy\|_{TV}$$
$$= \frac{1}{2} \int_{\mathbb{R}} |p_t(x, y) - \pi(y)| dy$$
$$\leq \frac{1}{2} \sum_{n=1}^{+\infty} \exp(-n\theta t) |H_n(x)| \int_{\mathbb{R}} |H_n(y)| \pi(y) dy$$
$$\leq \frac{1}{2} \sum_{n=1}^{+\infty} \exp(-n\theta t) |H_n(x)|.$$

Integrating w.r.t. $\pi(x)dx$ and repeating the same tool, we obtain:

$$\beta_Y(t) \leq \frac{1}{2} \sum_{n=1}^{+\infty} \exp(-n\theta t) = \frac{\exp(-\theta t)}{2(1 - \exp(-\theta t))}.$$

□

The interest of this proof is that it can be mimicked for all models for which the infinitesimal generator has a discrete spectrum with explicit eigenfunctions and eigenvalues.

Now, we consider $X_t = \tanh(Y_t)$. By the Itô formula, we get that (X_t) has coefficients

$$b(x) = -(1 - x^2) \left[c^2 x + \frac{\theta}{2} \ln \left(\frac{1+x}{1-x} \right) \right], \quad \sigma(x) = c(1 - x^2).$$

Assumptions [A1] – [A6] are satisfied for (X_t).

Finally, we consider

$$dX_t = \left[\frac{dc^2}{4} - \theta X_t \right] dt + c\sqrt{X_t} dW_t.$$

With $d \geq 2$ integer, (X_t) has the distribution of $\sum_{i=1}^{d} Y_{i,t}^2$ where $(Y_{i,t})$ are i.i.d. Ornstein-Uhlenbeck processes as above. The process (X_t) satisfies all assumptions except that its diffusion coefficient is not bounded.

5.7.2 Calibrating the penalties

It is not easy to calibrate the penalties. The method is studied in full details in Comte and Rozenholc (2004). Implementation with [DP] is done on the above examples in Comte et al. (2007) and for [GP] in Comte et al. (2006). We only give here a brief description.

For collection [GP], the drift penalty $(i = 1)$ and the diffusion penalty $(i = 2)$ are given by

$$2\frac{\hat{s}_i^2}{n} \left(d - 1 + \ln \left(\frac{d_{max} - 1}{d - 1} \right) + \ln^{2.5}(d) + \sum_{j=1}^{d} (r_j + \ln^{2.5}(r_j + 1)) \right).$$

Moreover, $d_{max} = [n\Delta/\ln^{1.5}(n)]$, $r_{max} = 5$. The constants κ and $\tilde{\kappa}$ in both drift and diffusion penalties have been set equal to 2. The term $\hat{s}_1^2$ replaces σ_1^2/Δ for the estimation of b and $\hat{s}_2^2$ replaces σ_1^4 for the estimation of σ^2. Let us first explain how $\hat{s}_2^2$ is obtained. We run once the estimation algorithm of σ^2 with a preliminary penalty where $\hat{s}_2^2$ is taken equal to $2 \max_m (\breve{\gamma}_n(\hat{\sigma}_m^2))$. This gives a preliminary estimator $\tilde{\sigma}_0^2$. Now, we take $\hat{s}_2$ equal to twice the 99.5%-quantile of $\tilde{\sigma}_0^2$. The use of the quantile here is to avoid extreme values. We get $\tilde{\sigma}^2$. We use this estimate and set $\hat{s}_1^2 = \max_{1 \leq k \leq n}(\tilde{\sigma}^2(X_{k\Delta}))/\Delta$ for the penalty of b. In all the examples, parameters have been chosen in the admissible range of ergodicity. The sample size $n = 5000$ and the step $\Delta = 1/20$ are in accordance with the asymptotic context (great n's and small Δ's).

5.8 Bibliographical remarks

Non-parametric estimation of the coefficients of diffusion processes has been widely investigated in the last decades. There are two first reference papers which are only devoted to drift estimation. One is by Banon (1978), who uses a spectral approach and a continuous time observation of the sample path. The other one is by Tuan (1981) who constructs and studies kernel estimators of the drift based on a continuous time observation of the sample path and also on a discrete observation of the sample path for an ergodic diffusion process.

More recently, several authors have considered drift estimation based on a continuous time observation of the sample path for ergodic models. Asymptotic results are given as the length of the observation time interval tends to infinity (Prakasa Rao (2010), Spokoiny (2000), Kutoyants (2004) or Dalalyan (2005)).

Discrete sampling of observations has also been investigated, with different asymptotic frameworks, implying different statistical strategies. It is now classical to distinguish between low-frequency and high-frequency data. In the former case, observations are taken at regularly spaced instants with fixed sampling interval Δ and the asymptotic framework is that the number of observations tends to infinity. Then, only ergodic models are usually considered. Parametric estimation in this context has been studied by Bibby and Sørensen (1995), Kessler and Sørensen (1999), see also Bibby, Jacobsen, and Sørensen (2009). A non-parametric approach using spectral methods is investigated in Gobet, Hoffmann, and Reiß (2004), where non-standard non-parametric rates are exhibited.

In high-frequency data, the sampling interval $\Delta = \Delta_n$ between two successive observations is assumed to tend to zero as the number of observations n tends to infinity. Taking $\Delta_n = 1/n$, so that the length of the observation time interval $n\Delta_n = 1$ is fixed, can only lead to estimating the diffusion coefficient consistently with no need of ergodicity assumptions. This is done by Hoffmann (1999) who generalizes results by Jacod (2000), Florens-Zmirou (1993) and Genon-Catalot, Larédo, and Picard (1992).

Now, estimating both drift and diffusion coefficients requires that the sampling interval Δ_n tends to zero while $n\Delta_n$ tends to infinity. For ergodic diffusion models, Hoffmann (1999) proposes non-parametric estimators using projections on wavelet bases together with adaptive procedures. He exhibits minimax rates and shows that his estimators automatically reach these optimal rates up to logarithmic factors. Hence, Hoffmann's paper gives the benchmark for studying non-parametric estimation in this framework and assumptions. Nevertheless, Hoffmann's estimators are based on computations of some random times which make them difficult to implement.

Finally, let us mention that Bandi and Phillips (2003) also consider the same asymptotic framework but with nonstationary diffusion processes: they study kernel estimators using local time estimations and random normalization.

5.9 Appendix. Proof of Proposition 5.13

The proof relies on the following Lemma (Lemma 9 in Barron et al. (1999)):

Lemma 5.15 *Let μ be a positive measure on $[0, 1]$. Let $(\psi_\lambda)_{\lambda \in \Lambda}$ be a finite orthonormal system in $\mathbb{L}_2 \cap \mathbb{L}_\infty(\mu)$ with $|\Lambda| = D$ and $\bar{S}$ be the linear span of $\{\psi_\lambda\}$. Let*

$$\bar{r} = \frac{1}{\sqrt{D}} \sup_{\beta \neq 0} \frac{\|\sum_{\lambda \in \Lambda} \beta_\lambda \psi_\lambda\|_\infty}{|\beta|_\infty}.$$

For any positive δ, one can find a countable set $T \subset \bar{S}$ and a mapping p from $\bar{S}$ to T with the following properties:

- *for any ball $\mathcal{B}$ with radius $\sigma \geq 5\delta$,*

$$|T \cap \mathcal{B}| \leq (B'\sigma/\delta)^D \text{ with } B' < 5.$$

- $\|u - p(u)\|_\mu \leq \delta$ *for all u in $\bar{S}$, and*

$$\sup_{u \in p^{-1}(t)} \|u - t\|_\infty \leq \bar{r}\delta, \text{ for all } t \text{ in } T.$$

To use this lemma, the main difficulty is often to evaluate $\bar{r}$ in the different contexts. In our problem, the measure μ is π. We consider a collection of models $(S_m)_{m \in \mathcal{M}_n}$ which can be [DP] or [GP]. Recall that $B^\pi_{m,m'}(0,1) = \{t \in S_m + S_{m'}, \|t\|_\pi = 1\}$. We have to compute $\bar{r} = \bar{r}_{m,m'}$ corresponding to $\bar{S} = S_m + S_{m'}$. We denote by $D(m, m') = \dim(S_m + S_{m'})$.

Collection [DP]– $S_m + S_{m'} = S_{\max(m,m')}$, $D(m, m') = \max(D_m, D_{m'})$, an orthonormal $\mathbb{L}_2(\pi)$-basis $(\psi_\lambda)_{\lambda \in \Lambda(m,m')}$ can be built by orthonormalisation, on each sub-interval, of $(\varphi_\lambda)_{\lambda \in \Lambda(m,m')}$. Then

$$
\begin{aligned}
\sup_{\beta \neq 0} & \frac{\|\sum_{\lambda \in \Lambda(m,m')} \beta_\lambda \psi_\lambda\|_\infty}{|\beta|_\infty} \\
&\leq \| \sum_{\lambda \in \Lambda(m,m')} |\psi_\lambda|\|_\infty \leq (r_{max} + 1) \sup_{\lambda \in \Lambda(m,m')} \|\psi_\lambda\|_\infty \\
&\leq (r_{max} + 1)^{3/2} \sqrt{D(m, m')} \sup_{\lambda \in \Lambda(m,m')} \|\psi_\lambda\| \\
&\leq (r_{max} + 1)^{3/2} \sqrt{D(m, m')} \sup_{\lambda \in \Lambda(m,m')} \|\psi_\lambda\|_\pi / \sqrt{\pi_0} \\
&\leq (r_{max} + 1)^{3/2} \sqrt{D(m, m')/\pi_0}.
\end{aligned}
$$

Thus here $\bar{r}_{m,m'} \leq (r_{max} + 1)^{3/2}/\sqrt{\pi_0}$.

Collection [GP]– Here we have $\bar{r}_{m,m'} \leq [(r_{max} + 1)\sqrt{N_n}]/\sqrt{D(m, m')\pi_0}$.

We now prove Proposition 5.13. We apply Lemma 5.15 to the linear space $S_m + S_{m'}$ of dimension $D(m, m')$ and norm connection measured by $\bar{r}_{m,m'}$ bounded above. We consider δ_k-nets, $T_k = T_{\delta_k} \cap B_{m,m'}^{\pi}(0, 1)$, with $\delta_k = \delta_0 2^{-k}$ with $\delta_0 \leq 1/5$, to be chosen later and we set

$$H_k = \ln(|T_k|) \leq D(m, m') \ln(5/\delta_k) = D(m, m')[k \ln(2) + \ln(5/\delta_0)].$$
(5.41)

Given some point $u \in B_{m,m'}^{\pi}(0, 1)$, we can find a sequence $\{u_k\}_{k \geq 0}$ with $u_k \in T_k$ such that $\|u - u_k\|_{\pi}^2 \leq \delta_k^2$ and $\|u - u_k\|_{\infty} \leq \bar{r}_{m,m'}\delta_k$. Thus we have the following decomposition that holds for any $u \in B_{m,m'}^{\pi}(0, 1)$,

$$u = u_0 + \sum_{k=1}^{\infty}(u_k - u_{k-1}).$$

Clearly $\|u_0\|_{\pi} \leq 1$, $\|u_0\|_{\infty} \leq \bar{r}_{(m,m')}$ and for all $k \geq 1$, $\|u_k - u_{k-1}\|_{\pi}^2 \leq 2(\delta_k^2 + \delta_{k-1}^2) = 5\delta_{k-1}^2/2$ and $\|u_k - u_{k-1}\|_{\infty} \leq 3\bar{r}_{(m,m')}\delta_{k-1}/2$. In the sequel we denote by $\mathbb{P}_n(.)$ the measure $\mathbb{P}(. \cap \Omega_n)$, see (5.24), (actually only the inequality $\|t\|_n^2 \leq \frac{3}{2}\|t\|_{\pi}^2$ holding for any $t \in S_m + S_{m'}$ is required).

Let $(\eta_k)_{k \geq 0}$ be a sequence of positive numbers that will be chosen later on and η such that $\eta_0 + \sum_{k \geq 1}\eta_k \leq \eta$. Recall that $\breve{\nu}_n^{(1)}$ is defined by $(5.27) - (5.32) - (5.33)$. We have

$$\mathbb{P}_n\left[\sup_{u \in B_{m,m'}^{\pi}(0,1)} \breve{\nu}_n^{(1)}(u) > \eta\right]$$

$$= \mathbb{P}_n\left[\exists(u_k)_{k \in \mathbb{N}} \in \prod_{k \in \mathbb{N}} T_k \; / \right.$$

$$\left. \breve{\nu}_n^{(1)}(u_0) + \sum_{k=1}^{+\infty} \breve{\nu}_n^{(1)}(u_k - u_{k-1}) > \eta_0 + \sum_{k \geq 1}\eta_k \right]$$

$$\leq \mathbb{P}_1 + \mathbb{P}_2$$

where

$$\mathbb{P}_1 = \sum_{u_0 \in T_0} \mathbb{P}_n(\breve{\nu}_n^{(1)}(u_0) > \eta_0),$$

$$\mathbb{P}_2 = \sum_{k=1}^{\infty} \sum_{\substack{u_{k-1} \in T_{k-1} \\ u_k \in T_k}} \mathbb{P}_n(\breve{\nu}_n^{(1)}(u_k - u_{k-1}) > \eta_k).$$

Then using inequality (5.37) of Lemma 5.12 and (5.41), we straightforwardly infer that $\mathbb{P}_1 \leq \exp(H_0 - Cnx_0)$ and $\mathbb{P}_2 \leq \sum_{k \geq 1} \exp(H_{k-1} + H_k - Cnx_k)$

if we choose

$$\begin{cases} \eta_0 = \sigma_1^2(\sqrt{3x_0} + \bar{r}_{(m,m')}x_0) \\ \eta_k = (\sigma_1^2/\sqrt{2})\delta_{k-1}(\sqrt{15x_k} + 3\bar{r}_{(m,m')}x_k). \end{cases}$$

Fix $\tau > 0$ and choose x_0 such that

$$Cnx_0 = H_0 + L_{m'}D_{m'} + \tau$$

and for $k \geq 1$, x_k such that

$$Cnx_k = H_{k-1} + H_k + kD_{m'} + L_{m'}D_{m'} + \tau.$$

If $D_{m'} \geq 1$, we infer that

$$\mathbb{P}_n \left(\sup_{t \in B^\pi_{m,m'}(0,1)} \breve{\nu}_n^{(1)}(t) > \eta_0 + \sum_{k \geq 1} \eta_k \right)$$
$$\leq e^{-L_{m'}D_{m'}-\tau} \left(1 + \sum_{k=1}^{\infty} e^{-kD_{m'}} \right) \leq 1.6e^{-L_{m'}D_{m'}-\tau}.$$

Now, it remains to compute $\sum_{k \geq 0} \eta_k$. We note that $\sum_{k=0}^{\infty} \delta_k = \sum_{k=0}^{\infty} k\delta_k = 2\delta_0$. This implies

$$x_0 + \sum_{k=1}^{\infty} \delta_{k-1} x_k$$
$$\leq \left[\ln(5/\delta_0) + \delta_0 \sum_{k=1}^{\infty} 2^{-(k-1)} [(2k-1)\ln(2) + 2\ln(5/\delta_0) + k] \right]$$
$$\times \frac{D(m,m')}{nC}$$
$$+ \left(1 + \delta_0 \sum_{k \geq 1} 2^{-(k-1)} \right) \left(\frac{L_{m'}D_{m'}}{nC} + \frac{\tau}{nC} \right)$$
$$\leq a(\delta_0)\frac{D(m,m')}{n} + \left(\frac{1+2\delta_0}{C}\right)\left(\frac{L_{m'}D_{m'}}{n} + \frac{\tau}{n}\right), \qquad (5.42)$$

where $Ca(\delta_0) = \ln(5/\delta_0) + \delta_0(4\ln(5/\delta_0) + 6\ln(2) + 4)$. This leads to

$$
\left(\sum_{k=0}^{\infty} \eta_k\right)^2
$$

$$
\leq \frac{\sigma_1^4}{2}\left[\sqrt{2}\left(\sqrt{3x_0} + +\bar{r}_{m,m'}x_0\right) + \sum_{k=1}^{\infty}\delta_{k-1}\left(\sqrt{15x_k} + 3\bar{r}_{m,m'}x_k\right)\right]^2
$$

$$
\leq \frac{\sigma_1^4}{2}\left[\left(\sqrt{6x_0} + \sum_{k=1}^{\infty}\delta_{k-1}\sqrt{15x_k}\right) + \bar{r}_{m,m'}\left(\sqrt{2}x_0 + 3\sum_{k=1}^{\infty}\delta_{k-1}x_k\right)\right]^2
$$

$$
\leq 15\sigma_1^4\left[\left(\sqrt{x_0} + \sum_{k=1}^{\infty}\delta_{k-1}\sqrt{x_k}\right)^2 + \bar{r}_{m,m'}^2\left(x_0 + \sum_{k=1}^{\infty}\delta_{k-1}x_k\right)^2\right]
$$

$$
\leq 15\sigma_1^4\left[\left(1 + \sum_{k=1}^{\infty}\delta_{k-1}\right)\left(x_0 + \sum_{k=1}^{\infty}\delta_{k-1}x_k\right)\right.
$$

$$
\left. + \bar{r}_{m,m'}^2\left(x_0 + \sum_{k=1}^{\infty}\delta_{k-1}x_k\right)^2\right].
$$

Now, fix $\delta_0 \leq 1/5$ (say, $\delta_0 = 1/10$) and use the bound (5.42). The bound for $(\sum_{k=0}^{+\infty}\eta_k)^2$ is less than a quantity proportional to

$$
\sigma_1^4\left[\frac{D(m,m')}{n} + \frac{L_{m'}D_{m'}}{n} + \bar{r}_{m,m'}^2\left(\frac{D(m,m')}{n} + \frac{L_{m'}D_{m'}}{n}\right)^2\right.
$$

$$
\left. + \frac{\tau}{n} + \bar{r}_{m,m'}^2\frac{\tau^2}{n^2}\right].
$$

Now in the case of collection [DP], we have $L_m = 1$, $\bar{r}_{m,m'}$ is bounded uniformly with respect to m and m' and $(D(m,m')/n)^2 \leq (N_n/n)^2 \leq \Delta^2/\ln^4(n)$ with $N_n \leq n\Delta/\ln^2(n)$. Thus the bound for $(\sum \eta_k)^2$ reduces to

$$
C'\sigma_1^4\left[\frac{D(m,m')}{n} + (1 + r_{max})^3\Delta^2/\pi_0 + \frac{\tau}{n} + \bar{r}_{m,m'}^2\frac{\tau^2}{n^2}\right].
$$

Next, for collection [GP], we use that $L_m \leq c \ln(n)$, $\bar{r}_{m,m'}^2 \leq (r_{max} + 1)^3 N_n/(\pi_0 D(m,m'))$ and $N_n \leq n\Delta/\ln^2(n)$ to obtain the bound

$$
\begin{aligned}
\bar{r}_{m,m'}^2 &\left(\frac{D(m,m')}{n} + \frac{L_{m'} D_{m'}}{n} \right)^2 \\
&\leq (r_{max}+1)^3 \frac{N_n}{\pi_0 D(m,m')} \frac{D(m,m')^2}{n^2} (1+\ln(n))^2 \\
&\leq (r_{max}+1)^3 \frac{N_n D(m,m')}{\pi_0 n^2} (1+\ln(n))^2 \\
&\leq (r_{max}+1)^3 \frac{N_n^2}{\pi_0 n^2} (1+\ln(n))^2 \leq 2(r_{max}+1)^3 \Delta^2/\pi_0.
\end{aligned}
$$

Thus, the bound for $\left(\sum \eta_k \right)^2$ is proportional to

$$
\sigma_1^4 \left[\frac{D(m,m')}{n} + \frac{L_{m'} D_{m'}}{n} + 2(r_{max}+1)^3 \Delta^2/\pi_0 + \frac{\tau}{n} + \bar{r}_{m,m'}^2 \frac{\tau^2}{n^2} \right].
$$

This term defines $\tilde{p}(m,m')$ as given in Proposition 5.13.

We obtain, for $K = (r_{max}+1)^3/\pi_0$,

$$
\begin{aligned}
\mathbb{P}_n &\left[\sup_{u \in B_{m,m'}^\pi(0,1)} [\breve{\nu}_n^{(1)}(u)]^2 > \kappa \sigma_1^4 \left(\frac{D_m + D_{m'}(1+L_{m'})}{n} \right. \right. \\
&\left. \left. + K\Delta^2 + 2\left(\frac{\tau}{n} \vee 2\bar{r}_{m,m'}^2 \frac{\tau^2}{n^2} \right) \right) \right] \\
&\leq \mathbb{P}_n \left[\sup_{u \in B_{m,m'}^\pi(0,1)} [\breve{\nu}_n^{(1)}(u)]^2 > \eta^2 \right] \\
&\leq 2\mathbb{P}_n \left[\sup_{u \in B_{m,m'}^\pi(0,1)} \breve{\nu}_n^{(1)}(u) > \eta \right] \\
&\leq 3.2 e^{-L_{m'} D_{m'} - \tau}
\end{aligned}
$$

so that, reminding that $\breve{G}_m(m')$ is defined by (5.39),

$$
\mathbb{E}\left[\left(\breve{G}_m^2(m') - \kappa\sigma_1^4 \frac{D_m + D_{m'}(1 + L_{m'})}{n} + K\Delta^2\right)_+ 1_{\Omega_n}\right]
$$

$$
\leq \int_0^\infty \mathbb{P}_n\left(\breve{G}_m^2(m') > \kappa\sigma_1^4 \frac{D_m + D_{m'}(1 + L_{m'})}{n} + K\Delta^2 + \tau\right) d\tau
$$

$$
\leq e^{-L_{m'} D_{m'}} \left(\int_{2\kappa\sigma_1^4/\bar{r}_{(m,m')}^2}^\infty e^{-n\tau/(2\kappa\sigma_1^4)} d\tau \right.
$$

$$
\left. + \int_0^{2\kappa\sigma_1^4/\bar{r}_{m,m'}^2} e^{-n\sqrt{\tau}/(2\sqrt{\kappa}\bar{r}_{m,m'}\sigma_1^2)} d\tau\right)
$$

$$
\leq e^{-L_{m'} D_{m'}} \frac{2\kappa\sigma_1^4}{n} \left(\int_0^\infty e^{-v} dv + \frac{2\bar{r}_{m,m'}^2}{n} \int_0^\infty e^{-\sqrt{v}} dv\right)
$$

$$
\leq e^{-L_{m'} D_{m'}} \frac{2\kappa\sigma_1^4}{n} (1 + \frac{4\bar{r}_{m,m'}^2}{n}) \leq \kappa' e^{-L_{m'} D_{m'}} \frac{\sigma_1^4}{n}
$$

which ends the proof. $\square$

References

Abramowitz, M., & Stegun, A. (1972). *Handbook of mathematical functions with formulas, graphs, and mathematical tables.* Wiley, New York.

Bandi, F. M., & Phillips, P. C. B. (2003). Fully nonparametric estimation of scalar diffusion models. *Econometrica, 71,* 241–283.

Banon, G. (1978). Nonparametric identification for diffusion processes. *SIAM J. Control Optim., 16,* 380–395.

Baraud, Y., Comte, F., & Viennet, G. (2001a). Adaptive estimation in autoregression or β-mixing regression via model selection. *Ann. Statist., 29,* 839–875.

Baraud, Y., Comte, F., & Viennet, G. (2001b). Model selection for (auto)-regression with dependent data. *ESAIM Probab. Statist., 5,* 33–49.

Barlow, M. T., & Yor, M. (1982). Semimartingale inequalities via the Garsia-Rodemich-Rumsey lemma, and applications to local times. *J. Funct. Anal., 49,* 198–229.

Barndorff-Nielsen, O. E. (1978). Hyperbolic distributions and distributions on hyperbolae. *Scand. J. Statist., 5,* 151–157.

Barron, A. R., Birgé, L., & Massart, P. (1999). Risk bounds for model selection via penalization. *Probab. Theory Related Fields, 113,* 301–413.

Beskos, A., Papaspiliopoulos, O., & Roberts, G. O. (2006). Retrospective exact simulation of diffusion sample paths with applications. *Bernoulli, 12,* 1077–1098.

Beskos, A., & Roberts, G. O. (2005). Exact simulation of diffusions. *Ann. Appl. Probab., 15,* 2422–2444.

Bibby, B. M., Jacobsen, M., & Sørensen, M. (2009). Estimating functions for discretely sampled diffusion-type models. In Y. Aït-Sahalia & L. Hansen (Eds.), *Handbook of Financial Econometrics* (pp. 203–268). North Holland, Oxford.

Bibby, B. M., & Sørensen, M. (1995). Martingale estimation functions for discretely observed diffusion processes. *Bernoulli, 1,* 17–39.

Birgé, L., & Massart, P. (1998). Minimum contrast estimators on sieves: exponential bounds and rates of convergence. *Bernoulli, 4,* 329–375.

Comte, F. (2001). Adaptive estimation of the spectrum of a stationary Gaussian sequence. *Bernoulli, 7,* 267–298.

Comte, F., Genon-Catalot, V., & Rozenholc, Y. (2006). *Nonparametric estimation of a discretely observed integrated diffusion model.* (Tech. Rep.). MAP 5, Mathématiques Appliquées - Paris 5, UMR CNRS 8145.

Comte, F., Genon-Catalot, V., & Rozenholc, Y. (2007). Penalized nonparametric mean square estimation of the coefficients of diffusion processes. *Bernoulli, 13*, 514–543.

Comte, F., & Rozenholc, Y. (2002). Adaptive estimation of mean and volatility functions in (auto-)regressive models. *Stochastic Process. Appl., 97*, 111–145.

Comte, F., & Rozenholc, Y. (2004). A new algorithm for fixed design regression and denoising. *Ann. Inst. Statist. Math., 56*, 449–473.

Dalalyan, A. (2005). Sharp adaptive estimation of the drift function for ergodic diffusions. *Ann. Statist., 33*, 2507–2528.

DeVore, R. A., & Lorentz, G. G. (1993). *Constructive Approximation.* Berlin: Springer.

Florens-Zmirou, D. (1993). On estimating the diffusion coefficient from discrete observations. *J. Appl. Probab., 30*, 790–804.

Genon-Catalot, V., Jeantheau, T., & Larédo, C. (2000). Stochastic volatility models as hidden markov models and statistical applications. *Bernoulli, 6*, 1051–1079.

Genon-Catalot, V., Larédo, C., & Picard, D. (1992). Nonparametric estimation of the diffusion coefficient by wavelet methods. *Scand. J. Statist., 19*, 319–335.

Gobet, E., Hoffmann, M., & Reiß, M. (2004). Nonparametric estimation of scalar diffusions based on low frequency data. *Ann. Statist., 32*, 2223–2253.

Hoffmann, M. (1999). Adaptive estimation in diffusion processes. *Stochastic Process. Appl., 79*, 135–163.

Jacod, J. (2000). Non-parametric kernel estimation of the coefficient of a diffusion. *Scand. J. Statist., 27*, 83–96.

Karlin, S., & Taylor, H. M. (1981). *A Second Course in Stochastic Processes.* New York: Academic Press.

Kessler, M., & Sørensen, M. (1999). Estimating equations based on eigenfunctions for a discretely observed diffusion process. *Bernoulli, 5*, 299–314.

Kutoyants, Y. A. (2004). *Statistical Inference for Ergodic Diffusion Processes.* London: Springer.

Pardoux, E., & Veretennikov, A. Y. (2001). On the Poisson equation and diffusion approximation. I. *Ann. Probab., 29*, 1061–1085.

Prakasa Rao, B. L. S. (2010). *Statistical Inference for Fractional Diffusion Processes.* Chichester: Wiley.

Spokoiny, V. G. (2000). Adaptive drift estimation for nonparametric diffusion model. *Ann. Statist., 28*, 815–836.

Tuan, P. D. (1981). Nonparametric estimation of the drift coefficient in the diffusion equation. *Math. Operationsforsch. Statist., Ser. Statistics*, *12*, 61–73.

CHAPTER 6

Ornstein–Uhlenbeck related models driven by Lévy processes

Peter J. Brockwell and Alexander Lindner

Department of Statistics, Colorado State University
Fort Collins, CO 80523–1877, USA
and
Institut für Mathematische Stochastik, Technische Universität Braunschweig
Pockelsstraße 14, 38106 Braunschweig, Germany

6.1 Introduction

Recently, there has been increasing interest in continuous-time stochastic models with jumps, a class of models which has applications in the fields of finance, insurance mathematics and storage theory, to name just a few. In this chapter we shall collect known results about a prominent class of these continuous-time models with jumps, namely the class of Lévy-driven Ornstein–Uhlenbeck processes, and their generalisations. In Section 6.2, basic facts about Lévy processes, needed in the sequel, are reviewed. Then, in Section 6.3.1 the Lévy-driven Ornstein–Uhlenbeck process, defined as a solution of the stochastic differential equation

$$dV_t = -\lambda V_t dt + dL_t,$$

where L is a driving Lévy process, is introduced. An application to storage theory is mentioned, followed by the volatility model of Barndorff–Nielsen and Shephard (2001a, 2001b). Then, in Sections 6.3.2 and 6.3.3, two generalisations of Ornstein–Uhlenbeck processes are considered, both of which are based on the fact that an Ornstein–Uhlenbeck process can be seen as a continuous-time analogue of an AR(1) process with i.i.d. noise. In Section 6.3.2 we consider CARMA processes, which are continuous-time analogues of discrete time ARMA processes, and in Section 6.3.3 we consider generalised Ornstein–Uhlenbeck processes, which are continuous time analogues of AR(1) processes with i.i.d. random coefficients. Special emphasis is given to conditions for stationarity of these processes and existence of moments. Then, in Section 6.3.4, we introduce the COGARCH(1,1) process, which is a continuous-time analogue of the ARCH process of Engle (1982) and the GARCH(1,1) process of Bollerslev (1986). It is an example of a generalised Ornstein–Uhlenbeck process. An extension to COGARCH(q,p) processes is also given.

Finally, in Section 6.4 we present some estimation methods for these processes, mostly based on high frequency data. In Section 6.4.1, Jongbloed, van der Meulen and van der Vaart's (2005) non-parametric estimator for the underlying Lévy measure of a subordinator driven Ornstein–Uhlenbeck process is presented, together with an estimator for the underlying parameter λ and a method for reconstruction of the sample paths of the driving Lévy process. In Section 6.4.2, a least squares estimator for the parameters of a CARMA process is considered, and Section 6.4.3 presents a generalised method of moments estimator for the COGARCH(1,1) process.

Throughout the paper, $\mathbb{N} = \{1, 2, \ldots\}$ is the set of strictly positive integers, $\mathbb{N}_0 = \mathbb{N} \cup \{0\}$, $\Re(z)$ is the real part of the complex number z, and $\mathbf{b}'$ is the transpose of a vector $\mathbf{b} \in \mathbb{C}^n$, where $\mathbb{C}^n$ is understood to consist of column vectors. For a real number x, $\lfloor x \rfloor$ denotes the largest integer not exceeding x. The symbol "$\overset{d}{=}$" will be used to denote equality in distribution, and "a.s." as well as "i.i.d." are abbreviations for "almost surely" and "independent and identically distributed."

6.2 Lévy processes

In this section we collect some basic aspects of Lévy processes which will be needed in the following. We refer to the books by Applebaum (2004), Bertoin (1996), Kyprianou (2006) and Sato (1999) for the cited results and for further information about Lévy processes.

Definition 6.1 A Lévy process *with values in* $\mathbb{R}^d$ $(d \in \mathbb{N})$ *defined on a probability space* $(\Omega, \mathcal{F}, P)$ *is a stochastic process* $M = (M_t)_{t \geq 0}$, $M_t : \Omega \to \mathbb{R}^d$, *which satisfies the following properties:*

(i) *it has independent increments, i.e. the random variables* M_{t_0}, $M_{t_1} - M_{t_0}$, $M_{t_2} - M_{t_1}, \ldots, M_{t_n} - M_{t_{n-1}}$ *are independent for every* $n \in \mathbb{N}$ *and* $0 \leq t_0 < t_1 < \ldots < t_n$,

(ii) *it has stationary increments, i.e.* $M_{s+t} - M_s \overset{d}{=} M_t$ *for every* $s, t \geq 0$,

(iii) *it starts almost surely at 0, i.e.* $M_0 = 0$ *a.s.*,

(iv) *its paths are almost surely càdlàg functions, i.e. there is a set* $\Omega_0 \in \mathcal{F}$ *such that* $P(\Omega_0) = 1$ *and for every* $\omega \in \Omega_0$ *the path* $[0, \infty) \to \mathbb{R}^d$, $t \mapsto M_t(\omega)$ *is right-continuous in* $t \geq 0$ *and has finite left limits in* $t > 0$.

It can be shown that any Lévy process is continuous in probability, i.e. for every $t \geq 0$ and $\varepsilon > 0$

$$\lim_{s \to t} P(|M_s - M_t| > \varepsilon) = 0.$$

Elementary examples of Lévy processes include linear deterministic processes of the form $[0, \infty) \to \mathbb{R}^d$, $t \mapsto tb$, where $b \in \mathbb{R}^d$, every Brownian motion, and every compound Poisson process with parameter $c > 0$ and jump size distribution P_Y, i.e. all processes of the form

$$M_t = \sum_{j=1}^{N_t} Y_j,$$

where $(Y_j)_{j \in \mathbb{N}}$ is an i.i.d. sequence of random variables with values in $\mathbb{R}^d \setminus \{0\}$ and distribution P_Y, and $(N_t)_{t \geq 0}$ is a Poisson process with parameter $c > 0$, independent of $(Y_j)_{j \in \mathbb{N}}$.

Lévy processes are conveniently described in terms of their characteristic function. Let $M = (M_t)_{t \geq 0}$ be a Lévy process. Then by the Lévy-Khintchine formula, there exists a unique triplet (A_M, ν_M, γ_M) consisting of a symmetric non-negative $d \times d$-matrix A_M, a measure ν_M on $\mathbb{R}^d$ satisfying

$$\nu_M(\{0\}) = 0 \quad \text{and} \quad \int_{\mathbb{R}^d} \min\{|x|^2, 1\} \, \nu_M(dx) < \infty, \qquad (6.1)$$

and a constant $\gamma_M \in \mathbb{R}^d$ such that

$$E e^{i \langle M_t, z \rangle} = e^{t \psi_M(z)}, \quad z \in \mathbb{R}^d, \quad t \geq 0, \qquad (6.2)$$

where

$$\psi_M(z) = i \langle \gamma_M, z \rangle - \frac{1}{2} \langle z, A_M z \rangle + \int_{\mathbb{R}^d} (e^{i \langle z, x \rangle} - 1 - i \langle z, x \rangle \mathbf{1}_{\{|x| \leq 1\}}) \, \nu_M(dx). \qquad (6.3)$$

Conversely, if $\gamma_M \in \mathbb{R}^d$, A_M is a symmetric non-negative definite $d \times d$-matrix and ν_M is a measure satisfying (6.1), then there exists a Lévy process M, unique up to identity in law, such that (6.2) and (6.3) both hold. The triplet (A_M, ν_M, γ_M) is called the *characteristic triplet* of the Lévy process M and ψ_M given by (6.3) the *characteristic exponent* of M. Every measure ν_M satisfying (6.1) is called a *Lévy measure*, more precisely the *Lévy measure of M*. A distribution μ on $\mathbb{R}^d$ is *infinitely divisible* if for every $n \in \mathbb{N}$ there exist independent and identically distributed random variables $W_1, \ldots, W_n$ such that $W_1 + \ldots + W_n \overset{d}{=} \mu$. A distribution μ is infinitely divisible if and only if there exists a Lévy process $(M_t)_{t \geq 0}$ with $M_1 \overset{d}{=} \mu$.

Example 6.2 The characteristic triplet of the deterministic process $t \mapsto tb$ is

$(0, 0, b)$, the characteristic triplet of a Brownian motion with drift b and covariance matrix Σ is $(\Sigma, 0, b)$. For a compound Poisson process $M = (M_t)_{t \geq 0}$ with parameter $c > 0$ and jump size distribution P_Y, the characteristic triplet is given by $(0, cP_Y, c \int_{|x| \leq 1} x \, P_Y(dx))$. A further example of a Lévy process in $\mathbb{R}^1$ is the *Gamma process* with parameters $c, \lambda > 0$, in which case the distribution of M_t has Lebesgue density $x \mapsto (\Gamma(ct))^{-1} \lambda^{ct} x^{ct-1} e^{-\lambda x} 1_{(0, \infty)}(x)$. The characteristic triplet of the Gamma process is then $(0, \nu_M, \int_0^1 ce^{-\lambda x} dx)$ with Lévy measure ν_M given by

$$\nu_M(dx) = \frac{ce^{-\lambda x}}{x} 1_{(0, \infty)}(x) \, dx.$$

Observe that if $M^{(1)}$ and $M^{(2)}$ are two independent Lévy processes in $\mathbb{R}^d$ with characteristic triplets $(A_{M^{(1)}}, \nu_{M^{(1)}}, \gamma_{M^{(1)}})$ and $(A_{M^{(2)}}, \nu_{M^{(2)}}, \gamma_{M^{(2)}})$, respectively, then $M^{(1)} + M^{(2)}$ is a Lévy process with characteristic triplet $(A_{M^{(1)}} + A_{M^{(2)}}, \nu_{M^{(1)}} + \nu_{M^{(2)}}, \gamma_{M^{(1)}} + \gamma_{M^{(2)}})$.

For a càdlàg process $X = (X_t)_{t \geq 0}$ in $\mathbb{R}^d$, the *jump* ΔX_t of X at time t is given by

$$\Delta X_t := X_t - X_{t-},$$

where X_{t-} denotes the left-limit at $t > 0$ with the convention that $X_{0-} := 0$. Due to the càdlàg property of X, the process $(\Delta X_t)_{t \geq 0}$ is almost surely well defined, and further it is seen that for each fixed $K > 0$ and $t > 0$ the number of jumps of X on $[0, t]$ exceeding K in magnitude must be almost surely finite. For a Lévy process M, the probabilistic structure of the jumps is described in terms of the Lévy measure. More precisely, for any Borel set Λ in $\mathcal{B}_d$ not containing 0, the Lévy measure of Λ is the expected number of jumps of M with size in Λ during the time interval $(0, 1]$, i.e.

$$\nu_M(\Lambda) = E \sum_{0 < s \leq 1} 1_\Lambda(\Delta M_s), \quad \Lambda \in \mathcal{B}_d, \quad 0 \notin \Lambda.$$

This is a consequence of the Lévy-Itô decomposition, which further states that every Lévy process M is a sum of three independent Lévy processes, namely a Brownian motion $M^{(1)}$ with drift, a compound Poisson process $M^{(2)}$ whose jumps are greater than 1 in magnitude, and a pure jump square integrable martingale $M^{(3)}$ whose jumps are less than or equal to 1 in magnitude. The characteristic triplets of $M^{(1)}$, $M^{(2)}$ and $M^{(3)}$ are $(A_M, 0, b_{M^{(1)}})$, $(0, \nu_{M | \{x \in \mathbb{R}^d : |x| > 1\}}, 0)$ and $(0, \nu_{M | \{x \in \mathbb{R}^d : 0 < |x| \leq 1\}}, b_{M^{(3)}})$, respectively. In particular, the matrix A_M appearing in the characteristic triplet of M can be identified as the covariance matrix of the Brownian motion component $M^{(1)}$ at time 1, which is why A_M is also called the Gaussian covariance matrix of a Lévy process. A further consequence of the Lévy-Itô decomposition is that $\sum_{0 < s \leq t} |\Delta M_s|^2 < \infty$ a.s. for every $t > 0$ and every Lévy process M, while

M has a.s. paths of finite variation on compacts if and only if both

$$A_M = 0 \quad \text{and} \quad \int_{\{|x| \leq 1\}} |x| \, \nu_M(dx) < \infty.$$

In the case of finite variation on compacts, the characteristic exponent ψ_M of M can be expressed in the form

$$\psi_M(z) = i\langle \gamma_M^0, z \rangle + \int_{\mathbb{R}^d} (e^{i\langle z, x \rangle} - 1) \, \nu_M(dx),$$

and γ_M^0 is called the *drift* of M. If M has paths of finite variation on compacts, then the drift γ_M^0 and the constant γ_M appearing in the characteristic triplet of M are related by

$$\gamma_M^0 = \gamma - \int_{|x| \leq 1} x \, \nu_M(dx).$$

A Lévy process is called a *subordinator* if it has increasing sample paths (a.s.), a property which is characterized by

$$A_M = 0, \quad \int_{\{|x| \leq 1\}} |x| \, \nu_M(dx) < \infty, \quad \nu_M(\mathbb{R}^d \backslash [0, \infty)^d) = 0 \quad \text{and} \quad \gamma_M^0 \geq 0.$$

Every Poisson process in $\mathbb{R}^d$ is of finite variation with drift 0, and a subordinator if the jump size distribution is concentrated on $[0, \infty)^d$. The gamma process is a subordinator with infinite Lévy measure and drift 0.

From the point of view of stochastic integration it is necessary to identify appropriate filtrations. So suppose that $M = (M_t)_{t \geq 0}$ is a d-dimensional Lévy process, defined on a probability space $(\Omega, \mathcal{F}, P)$. Denote by $\mathcal{N}$ the set of all subsets of P-null-sets in $\mathcal{F}$. Then it can be shown that the completed natural filtration $\mathbb{F} = (\mathcal{F}_t)_{t \geq 0}$ of M, given by

$$\mathcal{F}_t = \sigma(\mathcal{N}, M_s : 0 \leq s \leq t), \quad t \geq 0,$$

is right-continuous (i.e. $\mathcal{F}_t = \bigcap_{s > t} \mathcal{F}_s$ for all $t \geq 0$), and hence the completed natural filtration satisfies the usual hypothesis (in the completed probability space of $(\Omega, \mathcal{F}, P)$). It is clear that M remains a Lévy process under the completed probability measure, and further that the independent increments property implies that $\mathcal{F}_t$ is independent of $(M_s - M_t)_{s \geq t}$. Each of the three summands $M^{(1)}$, $M^{(2)}$ and $M^{(3)}$ in the Lévy-Itô decomposition remain adapted to $\mathbb{F}$, and in particular $M^{(3)}$ remains a square integrable martingale with respect to $\mathbb{F}$. It follows that each d-dimensional Lévy process is a d-dimensional vector of semimartingales with respect to its completed natural filtration, and when working with stochastic integrals for Lévy processes we shall assume that we are working with this setup. For further information regarding stochastic integration theory for Lévy processes refer to Applebaum (2004) or Protter (2005).

6.3 Ornstein–Uhlenbeck related models

While Ornstein–Uhlenbeck processes and their generalisations have found many applications in finance, insurance and other research areas, the original motivation for the Ornstein–Uhlenbeck process stems from physics. Einstein (1905) described the movement of a free particle in gas or liquid by a Brownian motion, and this was extended by Uhlenbeck and Ornstein (1930) to allow for the effects of friction. They modeled the velocity v_t of such a particle as the solution of

$$m\, dv_t = -\lambda v_t dt + dB_t, \qquad (6.4)$$

where m denotes the mass of the particle, $\lambda > 0$ a friction coefficient and $B = (B_t)_{t\in\mathbb{R}}$ a Brownian motion (times a constant). Equation (6.4) is also known as the *Langevin equation*, and its solution satisfies

$$v_t = e^{-\lambda(t-s)/m} v_s + m^{-1} \int_s^t e^{\lambda(u-t)/m}\, dB_u, \quad s \le t \in \mathbb{R}.$$

The solution is called an *Ornstein–Uhlenbeck process*. A strictly stationary solution, i.e. a solution with shift invariant finite dimensional distributions, is given by

$$v_t = m^{-1} \int_{-\infty}^t e^{\lambda(u-t)/m}\, dB_u.$$

In this section we consider the Lévy-driven Ornstein–Uhlenbeck models, extensions of them, and discuss some applications.

6.3.1 The Lévy-driven Ornstein–Uhlenbeck process

Let $L = (L_t)_{t\ge 0}$ be a real valued Lévy process and $\lambda \in \mathbb{R}$ a fixed constant. Then a *Lévy-driven Ornstein–Uhlenbeck process* $V = (V_t)_{t\ge 0}$ is defined by the stochastic differential equation

$$dV_t = -\lambda V_t\, dt + dL_t, \quad t \ge 0, \qquad (6.5)$$

or in integral form as

$$V_t = V_0 - \lambda \int_0^t V_u\, du + L_t, \quad t \ge 0. \qquad (6.6)$$

The solution to this equation is given by

$$V_t = e^{-\lambda t}\left(V_0 + \int_0^t e^{\lambda u}\, dL_u\right), \quad t \ge 0, \qquad (6.7)$$

where V_0 is a starting random variable, which is usually assumed to be independent of the driving Lévy process $(L_t)_{t\ge 0}$. Often, the driving Lévy process is extended to a Lévy process on the whole real line, by defining $L_t := -L'_{-t-}$

for $t < 0$, where $(L'_t)_{t\geq 0}$ is an independent copy of L. Equations (6.6) and (6.7) then extend to

$$V_t - V_s - \lambda \int_s^t V_u \, du + L_t - L_s = e^{-\lambda(t-s)} \left(V_0 + \int_s^t e^{\lambda(u-s)} \, dL_s \right), \quad (6.8)$$

for all $s \leq t \in \mathbb{R}$. If $\lambda > 0$ then a stationary solution of (6.8) exists if and only if

$$E \log^+ |L_1| = E \log \max(1, |L_1|) < \infty,$$

in which case it is unique and given by

$$V_t = e^{-\lambda t} \int_{-\infty}^t e^{\lambda u} \, dL_u, \quad t \in \mathbb{R},$$

and the marginal distribution is self-decomposable, see e.g. Wolfe (1982), Sato and Yamazato (1983, 1984), or Jacod (1985). A distribution μ on $\mathbb{R}$ is called *self-decomposable* if for every $b \in (0,1)$ there exist random variables X, X' and W_b such X and X' both have distribution μ, X' is independent of W_b and

$$X \overset{d}{=} bX' + W_b.$$

In fact, all self-decomposable distributions can be represented as the law of $\int_{-\infty}^0 e^u dL_u$, when L ranges over all Lévy processes with finite log-moment (e.g. Sato (1999, Chapter 17)). Self-decomposability of the stationary distribution is used intensively by the non-parametric estimator of Jongbloed et al. (2005), presented in Section 6.4.1.

We now give two examples where Lévy-driven Ornstein–Uhlenbeck processes have been applied, the first one coming from storage theory, the second one from finance.

Example 6.3 Denote by V_t the water content of a dam at time t, with initial content V_0. Then a reasonable model for the cumulated inputs into the dam until time t is given by a Lévy process $(L_t)_{t\geq 0}$ with increasing sample paths, i.e. a subordinator. Now let $r : [0, \infty) \to [0, \infty)$ be a function which is strictly positive and càglàd on $(0, \infty)$, which describes the release rule for the dam. A model for the water content at time t is then given by

$$dV_t = -r(V_t) \, dt + dL_t, \quad t \geq 0,$$

see e.g. Çinlar and Pinsky (1972) or Brockwell, Resnick, and Tweedie (1982). The Lévy-driven Ornstein–Uhlenbeck process is obtained in the particular case when r is linear, i.e. $r(t) = \lambda t$, corresponding to a constant release rate λ.

The following example is the famous volatility model of Barndorff–Nielsen and Shephard (2001a, 2001b).

Example 6.4 Let λ be strictly greater than 0 and let $L = (L_t)_{t \geq 0}$ be a subordinator. Then the squared volatility process $(\sigma_t^2)_{t \geq 0}$ in the volatility model of Barndorff–Nielsen and Shephard is defined as the solution of the equation

$$d\sigma_t^2 = -\lambda \sigma_t^2 \, dt + dL_{\lambda t},$$

i.e. it is a subordinator driven Ornstein–Uhlenbeck process with driving Lévy process $(t \mapsto L_{\lambda t})$ and coefficient λ. Then σ_t^2 is positive if $\sigma_0^2 \geq 0$ or if we are dealing with the stationary version. The logarithmic price process $(G_t)_{t \geq 0}$ is then modeled by

$$dG_t = (\mu + b\sigma_t^2) \, dt + \sigma_t \, dW_t,$$

where μ and b are constants, $\sigma_t = \sqrt{\sigma_t^2}$ and $(W_t)_{t \geq 0}$ is a standard Brownian motion, independent of σ_0^2 and $(L_t)_{t \geq 0}$. When $EL_1^2 < \infty$, the autocovariance function of the stationary squared volatility process is given by the exponentially decaying function,

$$\text{Cov}(\sigma_t^2, \sigma_{t+h}^2) = 2^{-1} \text{Var}(L_1) e^{-\lambda h}, \quad t, h \geq 0,$$

(cf. Barndorff–Nielsen and Shephard (2001b, p. 172)). If additionally $\mu = b = 0$, then non-overlapping increments of G of length $r > 0$ are uncorrelated, i.e.

$$\text{Cov}(G_t - G_{t-r}, G_{t+h} - G_{t+h-r}) = 0, \quad t, h \geq r,$$

while, if $EL_1^4 < \infty$, the squared increments are correlated with the autocorrelation function,

$$\text{Cov}((G_t - G_{t-r})^2, (G_{t+h} - G_{t+h+r})^2) = C_r e^{-\lambda h}$$

for strictly positive integer multiples h of $r > 0$, where $C_r > 0$ is some constant (cf. Barndorff–Nielsen and Shephard (2001b, Section 4)). The process $((G_{rh} - G_{r(h-1)})^2)_{h \in \mathbb{N}}$ thus has the autocovariance structure of an ARMA (1,1) process. The fact that the increments of the log-price process are uncorrelated while its squares are not is one of the important *stylized features* of financial time series. The tail behaviour of the stationary volatility process depends on the tail behaviour of the driving Lévy process. In particular, it can be seen that σ_t^2 has Pareto tails, i.e. that $P(\sigma_t^2 > x)$ behaves asymptotically as a constant times $x^{-\alpha}$ for some $\alpha > 0$ as $x \to \infty$, if and only if L_1 has Pareto tails with the same index α (e.g. Fasen, Klüppelberg, and Lindner (2006); the converse follows from the monotone density theorem for regularly varying functions, see e.g. Theorem 1.7.2 in Bingham, Goldie, and Teugels (1989)).

The Lévy-driven Ornstein–Uhlenbeck process can be seen as a natural continuous-time generalisation of a an autoregressive process of order 1 (AR(1)) with i.i.d. noise. Indeed, if $(V_t)_{t \in \mathbb{R}}$ is a stationary Ornstein–Uhlenbeck process driven by $(L_t)_{t \geq 0}$ and $\lambda > 0$, then it follows from (6.8) that for any grid size

$r > 0$,

$$V_{rn} - e^{-\lambda r}V_{r(n-1)} = \int_{(n-1)r}^{nr} e^{\lambda(u-nr)}\, dL_u, \quad r \in \mathbb{Z}, \tag{6.9}$$

and the noise $(\int_{(n-1)r}^{nr} e^{\lambda(u-nr)}\, dL_u)_{n \in \mathbb{Z}}$ is i.i.d.

6.3.2 Continuous-time ARMA processes

If $(L_t)_{t \in \mathbb{R}}$ is a two-sided Lévy process with $v^2 := \mathrm{Var}(L(1)) < \infty$, then the strictly stationary solution of (6.5) with $\lambda > 0$ is

$$V_t = \int_{-\infty}^{t} e^{-\lambda(t-u)}\, dL_u, \tag{6.10}$$

with autocovariance function,

$$\mathrm{Cov}(V_{t+h}, V_t) = \frac{v^2}{2\lambda}e^{-\lambda|h|}, \quad h \in \mathbb{R},$$

and, for any spacing $r > 0$, the sampled process $(V_{nr})_{n \in \mathbb{Z}}$ is a discrete-time AR(1) process defined by (6.9). The exponential shape of the autocovariance function of $(V_t)_{t \in \mathbb{R}}$, however, severely limits its application to the realistic modelling of observed continuous-time data.

In discrete time the class of AR(1) processes has a natural extension to the class of ARMA(p, q) processes, the latter being defined as the stationary solution of a pth-order linear difference equation of the form,

$$\Phi(B)V_n = \Theta(B)Z_n, \quad n \in \mathbb{Z}, \tag{6.11}$$

where $\Phi(z)$ and $\Theta(z)$ are polynomials of the form,

$$\Phi(z) = 1 - \phi_1 z - \cdots - \phi_p z^p$$

and

$$\Theta(z) = 1 + \theta_1 z + \cdots + \theta_q z^q,$$

$(Z_n)_{n \in \mathbb{Z}}$ is a sequence of i.i.d. random variables, B denotes the backward shift operator, $B^j Y_t := Y_{t-j}$, $\phi_p \neq 0$ and $\theta_q \neq 0$. The extra generality achieved by moving from the AR(1) process to the class of ARMA(p, q) processes extends the class of attainable autocovariance functions drastically. In fact, given a set of sample autocovariances up to any arbitrarily high lag K, it is possible to find a value of p and an ARMA($p, 0$) process whose autocovariance function matches the sample autocovariances exactly from lags 0 to K (see e.g. Brockwell and Davis (1991)).

The analogous generalization of the stationary Ornstein–Uhlenbeck process is the CARMA(p, q) (continuous-time ARMA(p, q)) process, $0 \leq q < p$, defined (in a formal sense only) to be a strictly stationary solution of the pth-order stochastic differential equation,

$$a(D)V_t = b(D)DL_t,\qquad(6.12)$$

where $a(z)$ and $b(z)$ are polynomials of the form,

$$a(z) = z^p + a_1 z^{p-1} + \cdots + a_p$$

and

$$b(z) = b_0 + b_1 z + \cdots + b_{p-1} z^{p-1},$$

where $a_1, \ldots, a_p, b_0, \ldots, b_{p-1}$ are complex-valued coefficients such that $a_p \neq 0$, $b_q = 1$ and $b_j = 0$ for $j > q$. The operator D denotes differentiation with respect to t, but since the derivatives on the right-hand side of equation (6.12) do not exist in the usual sense, we make sense of the equation by reexpressing it in the state-space form,

$$Y_t = \mathbf{b}'\mathbf{X}_t, \quad t \in \mathbb{R},\qquad(6.13)$$

where $\mathbf{X} = (\mathbf{X}_t)_{t \in \mathbb{R}}$ is a $\mathbb{C}^p$-valued process satisfying the stochastic differential equation,

$$d\mathbf{X}_t = A\mathbf{X}_t\, dt + \mathbf{e}\, dL_t,\qquad(6.14)$$

or equivalently

$$\mathbf{X}_t = e^{A(t-s)}\mathbf{X}_s + \int_s^t e^{A(t-u)}\mathbf{e}\, dL_u, \quad \forall s \leq t \in \mathbb{R},\qquad(6.15)$$

with

$$A = \begin{bmatrix} 0 & 1 & 0 & \cdots & 0 \\ 0 & 0 & 1 & \cdots & 0 \\ \vdots & \vdots & \vdots & \ddots & \vdots \\ 0 & 0 & 0 & \cdots & 1 \\ -a_p & -a_{p-1} & -a_{p-2} & \cdots & -a_1 \end{bmatrix} \in \mathbb{C}^{p\times p},\qquad(6.16)$$

$$\mathbf{e} = [0, 0, \cdots, 0, 1]' \in \mathbb{C}^p \quad \text{and} \quad \mathbf{b} = [b_0, b_1, \cdots, b_{p-2}, b_{p-1}]' \in \mathbb{C}^p. \quad(6.17)$$

(For $p = 1$ the matrix A is to be understood as $-a_1$.) The state-space formulation (6.13), (6.14) enables us now to make a precise definition of the CARMA process.

Definition 6.5 *For $0 \leq q < p$, the CARMA(p, q) process with background driving Lévy process (or BDLP) L, autoregressive polynomial $a(z) = z^p + a_1 z^{p-1} + \cdots + a_p$, and moving-average polynomial $b(z) = b_0 + b_1 z + \cdots + b_{p-1} z^{p-1}$, where $a_p \neq 0$, $b_q = 1$ and $b_j = 0$ for $j > q$, is defined to be a strictly stationary solution Y of equations (6.13) and (6.14).*

The first questions arising from the definition concern the conditions on L, $a(z)$

and $b(z)$ under which a strictly stationary solution exists and is unique. Sufficient conditions have been known for some time (see e.g. Brockwell (2001)) but necessary and sufficient conditions were derived more recently by Brockwell and Lindner (2009). These are stated in the following theorem.

Theorem 6.6 *Suppose that $p \geq 1$, that $b(z)$ is not identically zero and that the Lévy process L is not deterministic. Then the CARMA equations (6.13) and (6.14) have a strictly stationary solution Y on $\mathbb{R}$ if and only if $E \log^+ |L_1| < \infty$ and all singularities of the meromorphic function $z \mapsto b(z)/a(z)$ on the imaginary axis are removable, i.e. if $a(\cdot)$ has a zero λ_1 of multiplicity $\mu(\lambda_1)$ on the imaginary axis, then $b(\cdot)$ has also a zero at λ_1 of multiplicity greater than or equal to $\mu(\lambda_1)$. In this case, the solution is unique and is given by*

$$Y_t = \int_{-\infty}^{\infty} g(t - u) \, dL_u, \quad t \in \mathbb{R}, \tag{6.18}$$

where

$$g(t) = \sum_{\lambda: \Re(\lambda) < 0} \sum_{k=0}^{\mu(\lambda)-1} c_{\lambda k} t^k e^{\lambda t} \mathbf{1}_{(0,\infty)}(t) - \sum_{\lambda: \Re(\lambda) > 0} \sum_{k=0}^{\mu(\lambda)-1} c_{\lambda k} t^k e^{\lambda t} \mathbf{1}_{(-\infty,0)}(t). \tag{6.19}$$

The sums are over the distinct zeroes λ of $a(z)$, $\mu(\lambda)$ denotes the multiplicity of the zero λ and $\sum_{k=0}^{\mu(\lambda)-1} c_{\lambda k} t^k e^{\lambda t}$ is the residue of $z \mapsto e^{zt} b(z)/a(z)$ at λ (which reduces to $b(\lambda)e^{\lambda t}/a'(\lambda)$ if $\mu(\lambda) = 1$).

Outline of Proof. Sufficiency of the conditions is established by defining $\mathbf{l}(t)$, $\mathbf{r}(t)$ and $\mathbf{n}(t)$ to be the residues of the column vector $e^{zt} a^{-1}(z)[1, z \cdots, z^{p-1}]'$ at the zeroes of $a(\cdot)$ with strictly negative, strictly positive and zero real parts respectively, and

$$\mathbf{X}_t := \int_{-\infty}^{t} \mathbf{l}(t - u) dL_u - \int_{t}^{\infty} \mathbf{r}(t - u) dL_u + \int_{0}^{t} \mathbf{n}(t - u) dL_u.$$

Then it can be verified directly that $\mathbf{X}$ satisfies (6.14) and that if $Y_t := \mathbf{b}' \mathbf{X}_t$, Y is strictly stationary as required.

Necessity of the conditions is more difficult to establish. We suppose that $(Y_t)_{t \in \mathbb{R}}$ is a strictly stationary solution of (6.13) and (6.14). Then $(Y_n)_{n \in \mathbb{Z}}$ is also strictly stationary and satisfies the equations,

$$\Phi(B)Y_n = Z_n^1 + Z_{n-1}^2 + \cdots + Z_{n-p+1}^p, \tag{6.20}$$

where Φ is the polynomial,

$$\Phi(z) := \prod_{j=1}^{p} (1 - e^{\lambda_j} z),$$

$\lambda_1, \ldots, \lambda_p$ are the p zeroes of a, and the right-hand side of (6.20) is a sum of independent, non-identically distributed random variables expressible in terms of L. If λ_1 is any zero of $a(\cdot)$ which is not a zero of $b(\cdot)$, then it can be shown, by considering the strictly stationary process,

$$W_n := \tilde{\Phi}(B)Y_n, \quad n \in \mathbb{Z},$$

where $\tilde{\Phi}(z) := \Phi(z)/(1 - e^{\lambda_1}z)$, and making use of the Borel-Cantelli Lemma, that $\Re(\lambda_1) \neq 0$ and $E \log^+ |L_1| < \infty$. In the case when $a(\cdot)$ and $b(\cdot)$ have common zeroes, Theorem 4.1 of Brockwell and Lindner (2009) permits cancellation of the corresponding factors, thereby establishing that the process Y can be reexpressed as a lower-order CARMA process with no common factors in the reduced autoregressive and moving-average polynomials. $\quad\square$

Remark 6.7 *If the conditions of the theorem are satisfied and if the zeroes of $a(z)$ all have strictly negative real parts, then the strictly stationary solution Y is a causal function of L, i.e. for each t, Y_t is independent of the increments $\{L_u - L_t, u > t\}$.*

Remark 6.8 *The equations (6.18) and (6.19) show that the strictly stationary solution Y can be decomposed into a linear combination of continuous-time autoregressions $Y^{(\lambda k)}$, all with the same driving Lévy process L, namely*

$$Y_t = \sum_{\lambda:\Re(\lambda)<0} \sum_{k=0}^{\mu(\lambda)-1} k! \, c_{\lambda k} Y_t^{(\lambda k)} - \sum_{\lambda:\Re(\lambda)>0} \sum_{k=0}^{\mu(\lambda)-1} k! \, c_{\lambda k} Y_t^{(\lambda k)},$$

where $Y^{(\lambda k)}$ is the CAR$(k+1)$ process with autoregressive polynomial $(z - \lambda)^{k+1}$ and BDLP L. In particular, if $\mu(\lambda) = 1$ for each λ then Y is the linear combination of CAR(1) processes,

$$Y_t = \sum_{\lambda:\Re(\lambda)<0} \frac{b(\lambda)}{a'(\lambda)} \int_{-\infty}^{t} e^{\lambda(t-u)} dL_u - \sum_{\lambda:\Re(\lambda)>0} \frac{b(\lambda)}{a'(\lambda)} \int_{t}^{\infty} e^{\lambda(t-u)} dL_u.$$

Remark 6.9 *In the case when $EL(1)^2 < \infty$ the CARMA processes play a role analogous to that of discrete-time ARMA processes, i.e. they constitute a family of parametric models capable of approximating a very wide range of autocovariance functions and marginal distributions of second-order strictly stationary processes. Suppose that $v^2 := \mathrm{Var}(L_1) < \infty$ and $EL_1 = m$. Then from (6.19) we find that*

$$h(\omega) := \int_{-\infty}^{\infty} e^{-i\omega t} g(t) dt = \frac{b(i\omega)}{a(i\omega)}. \tag{6.21}$$

Taking expectations in (6.18) and setting $\omega = 0$ in (6.21) then gives

$$EY(t) = mb_0/a_p.$$

Assuming that the coefficients of the polynomials $a(z)$ and $b(z)$, and hence the strictly stationary solution Y, are all real-valued, the autocovariance function of Y is

$$\gamma(h) = v^2 \int_{-\infty}^{\infty} g(t+h)g(t)dt, \quad h \in \mathbb{R},$$

and its spectral density is therefore

$$f(\omega) = \frac{1}{2\pi} \int_{-\infty}^{\infty} e^{-i\omega h} \gamma(h)dh = \frac{v^2}{2\pi}|h(\omega)|^2 = \frac{v^2}{2\pi}\left|\frac{b(i\omega)}{a(i\omega)}\right|^2, \quad \omega \in \mathbb{R}.$$

This spectral density is the same as that of the causal CARMA(p,q) process with moving-average polynomial b and autoregressive polynomial $a^*(z) = \prod_{j=1}^{p}(z - \lambda_j^*)$ where the zeroes λ_j^* of $a^*(z)$ coincide with those of $a(z)$ having negative real parts and are equal to $-\overline{\lambda}_j$ for those which do not. Rewriting the autocovariance function γ as

$$\gamma(h) = \frac{v^2}{2\pi} \int_{-\infty}^{\infty} e^{i\omega h} \frac{b(i\omega)b(-i\omega)}{a^*(i\omega)a^*(-i\omega)}d\omega,$$

changing the variable of integration to $z = i\omega$ and using contour integration, shows that $\gamma(h)$ is the sum of residues of $v^2 e^{z|h|}[b(z)b(-z)]/[a^*(z)a^*(-z)]$ at the zeroes of $a^*(z)$. If the zeroes of $a^*(z)$ are distinct, this reduces to

$$\gamma(h) = v^2 \sum_{j=1}^{p} \frac{e^{\lambda_j^*|h|}b(\lambda_j^*)b(-\lambda_j^*)}{a^{*\prime}(\lambda_j^*)a^*(-\lambda_j^*)},$$

where $a^{*\prime}(\lambda_j^*)$ denotes the derivative of a^* at λ_j^*.

Example 6.10 In the stochastic volatility model of Barndorff–Nielsen and Shephard (2001a, 2001b) considered in Example 6.4, we can expand the achievable autocorrelation functions for the stochastic volatility process and the squared log returns $(G_t - G_{t-r})^2$ if we replace the Ornstein–Uhlenbeck model for volatility with a non-negative CARMA process. Suppose, for example, that V is a CARMA(2,1) process with autoregressive polynomial $a(z) = (z - \lambda_1)(z - \lambda_2)$, moving average polynomial $b(z) = b_0 + z$ and BDLP L such that $EL_1^2 < \infty$. Suppose also that λ_1 and λ_2 both have strictly negative real parts and that $\lambda_1 \neq \lambda_2$. In order to ensure that V is a suitable model for stochastic volatility we must ensure that V takes only non-negative values. This can be achieved if the BDLP is a subordinator and if the kernel function g is non-negative (like the exponential kernel of the Ornstein–Uhlenbeck process). Necessary and sufficient conditions for the kernel of V to be non-negative are that λ_1 and λ_2 be real and that $b_0 \geq \min(|\lambda_i|)$. (See Brockwell and Davis (2001). For conditions implying non-negativity of more general CARMA kernels see Tsai and Chan (2005).) Under the above assumptions we can use Theorem 6.6

to write the volatility process as

$$V_t = \sum_{i=1}^{2} \frac{b(\lambda_i)}{a'(\lambda_i)} \int_{-\infty}^{t} e^{\lambda_i(t-u)} dL_u,$$

where L is the background driving subordinator. By Remark 6.9, we have $EV_t = mb_0/(\lambda_1 \lambda_2)$ and

$$\mathrm{Cov}(V_{t+h}, V_t) = \sum_{i=1}^{2} \alpha_i e^{\lambda_i h}, \quad h \geq 0,$$

where $\alpha_i = v^2[b(\lambda_i)b(-\lambda_i)]/[a(-\lambda_i)a'(\lambda_i)]$. As in Example 6.4, the logarithm G_t of the asset price at time t is modelled by the equation

$$dG_t = (\mu + bV_t)\, dt + \sqrt{V_t}\, dW_t,$$

where μ and b are constants and $(W_t)_{t\geq 0}$ is a standard Brownian motion, independent of $(L_t)_{t\geq 0}$. If $\mu = b = 0$, then, as in Example 6.9, non-overlapping increments of G of length $r > 0$ are uncorrelated, i.e.

$$\mathrm{Cov}(G_t - G_{t-r}, G_{t+h} - G_{t+h-r}) = 0, \quad t, h \geq r,$$

while the squared increments are dependent. If $r > 0$, h is any strictly positive integer multiple of r and $EL_1^4 < \infty$ then the covariance of the squared increments, calculated as in Barndorff–Nielsen and Shephard (2001b, Section 4.2), is

$$\mathrm{Cov}((G_t - G_{t-r})^2, (G_{t+h} - G_{t+h-r})^2) = \sum_{i=1}^{2} c_{i,r} e^{\lambda_i h}$$

for some constants $c_{1,r}, c_{2,r}$ which are both different from zero provided a and b have no common zero. This result implies that the sequence of squared log returns over successive intervals of length r, i.e. $((G_{rh} - G_{r(h-1)})^2)_{h\in\mathbb{N}}$, has the autocovariance structure of an ARMA(2,2) process with constrained coefficients.

6.3.3 Generalized Ornstein–Uhlenbeck processes

Another way to extend the class of Ornstein–Uhlenbeck processes is to replace the deterministic function $t \mapsto \lambda t$ in (6.6) by a second Lévy process $(\xi_t)_{t\geq 0}$. This leads to the so-called generalized Ornstein–Uhlenbeck processes: let $(\xi, \eta) = (\xi_t, \eta_t)_{t\geq 0}$ be a bivariate Lévy process and consider a starting random variable V_0, which is assumed to be independent of (ξ, η). Then the process defined by

$$V_t = e^{-\xi_t}\left(V_0 + \int_0^t e^{\xi_{s-}}\, d\eta_s\right), \quad t \geq 0, \qquad (6.22)$$

is called a *generalized Ornstein–Uhlenbeck process*, driven by (ξ, η). This terminology is due to de Haan and Karandikar (1989) and Carmona, Petit, and Yor (1997, 2001), who studied various properties of these processes. It is clear that if $(\zeta_t, \eta_t) = (\lambda t, L_t)$, $t \geq 0$, then a generalized Ornstein–Uhlenbeck process reduces to the Lévy driven Ornstein–Uhlenbeck process as defined in Section 6.3.1, and if further η_t is Brownian motion, then we get the classical Ornstein–Uhlenbeck process.

Before we give further examples and properties, let us have a look at the stochastic differential equation which is satisfied by a generalized Ornstein–Uhlenbeck process. The following result can be found as Exercise V.27 in the book of Protter (2005), who refers to an unpublished note of Yoeurp and Yor (1977). An extension of it can be found in Jaschke (2003), see also Behme, Lindner, and Maller (2011). As usual, in the following $\mathcal{E}(U)$ denotes the stochastic exponential of a semimartingale U (e.g. Protter (2005, Theorem II.37)), which is the unique semimartingale $(Z_t)_{t \geq 0}$ satisfying $Z_0 = 1$ and $dZ_t = Z_{t-} dU_t$, and $[\cdot, \cdot]$ denotes the quadratic variation of two semimartingales.

Proposition 6.11 *Let $(\xi_t, \eta_t)_{t \geq 0}$ be a bivariate Lévy process, independent of the starting random variable V_0. Let $(\mathcal{F}_t)_{t \geq 0}$ be the smallest filtration satisfying the usual hypotheses such that (ξ, η) is adapted and V_0 is $\mathcal{F}_0$-measurable. Define the bivariate process $(U_t, L_t)_{t \geq 0}$ by*

$$e^{-\xi_t} = \mathcal{E}(U)_t \quad and \quad L_t = \eta_t + [U, \eta]_t, \quad t \geq 0, \tag{6.23}$$

i.e.

$$\begin{pmatrix} U_t \\ L_t \end{pmatrix} = \begin{pmatrix} -\xi_t + \sum_{0 < s \leq t} \left(e^{-\Delta \xi_s} - 1 + \Delta \xi_s \right) + t \sigma_\xi^2/2 \\ \eta_t + \sum_{0 < s \leq t} (e^{-\Delta \xi_s} - 1) \Delta \eta_s - t \sigma_{\xi, \eta} \end{pmatrix}, \quad t \geq 0, \tag{6.24}$$

where $\begin{pmatrix} \sigma_\xi^2 & \sigma_{\xi, \eta} \\ \sigma_{\xi, \eta} & \sigma_\eta^2 \end{pmatrix}$ denotes the Gaussian covariance matrix $A_{\xi, \eta}$ appearing in the characteristic triplet of (ξ, η). Then (U, L) is a bivariate Lévy process, and the generalized Ornstein–Uhlenbeck process driven by (ξ, η) is the unique solution of the stochastic differential equation

$$dV_t = V_{t-} dU_t + dL_t, \quad t \geq 0, \tag{6.25}$$

with starting random variable V_0.

Outline of Proof. That (U, L) as given by (6.24) is a bivariate Lévy process is clear, since the required properties carry over directly from (ξ, η) to (U, L). That (6.24) is the same as (6.23) can be seen from the Doléans–Dade formula for the stochastic exponential and the properties of the quadratic variation. Finally, to show that the generalized Ornstein–Uhlenbeck process satisfies (6.25), write $A_t = \mathcal{E}(U)_t$ and $B_t = V_0 + \int_{(0,t]} \mathcal{E}(U)_{s-}^{-1} d\eta_s$. Then $V_t = A_t B_t$ and

A, B, V are semimartingales with respect to $(\mathcal{F}_t)_{t\geq 0}$. Partial integration then gives

$$
\begin{aligned}
& V_t - V_0 \\
= {} & \int_{(0,t]} A_{s-} dB_s + \int_{(0,t]} B_{s-} dA_s + [A, B]_t \\
= {} & \int_{(0,t]} d\eta_s + \int_{(0,t]} V_{s-}\mathcal{E}(U)_{s-}^{-1} d\mathcal{E}(U)_s + \int_{(0,t]} [\mathcal{E}(U)_{s-}]^{-1} d[\mathcal{E}(U), \eta]_s \\
= {} & \int_{(0,t]} dL_s + \int_{(0,t]} V_{s-} dU_s,
\end{aligned}
$$

where the last line follows from $d\mathcal{E}(U)_t = \mathcal{E}(U)_{t-} dU_t$ and from

$$
[\mathcal{E}(U), \eta]_t = \left[1 + \int_0^\bullet \mathcal{E}(U)_{s-} dU_s, \int_0^\bullet 1 d\eta_s \right]_t = \int_0^t \mathcal{E}(U)_{s-} d[U, \eta]_s
$$

together with the definition of L. Hence the generalized Ornstein–Uhlenbeck process satisfies (6.25), and the uniqueness of the solution follows from standard results for stochastic differential equations, see e.g. Theorem V.7 in Protter (2005). □

Remark 6.12 *If ξ and η are independent, then so are U and η, and $[U, \eta] = 0$, so that $L = \eta$ in (6.23). For a general bivariate Lévy process (ξ, η) however, $L \neq \eta$.*

We have already seen that generalized Ornstein–Uhlenbeck processes arise as Lévy-driven Ornstein–Uhlenbeck processes in the stochastic volatility model of Barndorff–Nielsen and Shephard (2001a, 2001b). In Section 6.3.4 we shall see that generalized Ornstein–Uhlenbeck processes also arise as volatility processes of continuous-time GARCH(1,1) processes, when ξ is deterministic and η random in contrast to the situation of the Lévy-driven Ornstein–Uhlenbeck process. The next example is the risk model of Paulsen (1993), who modelled a risk process by a generalized Ornstein–Uhlenbeck process with independent non-deterministic ξ and η.

Example 6.13 In the classical Cramér–Lundberg model disturbed by a Brownian motion, the capital of an insurance company at time t is given by

$$
V_t = V_0 + pt - \sum_{i=1}^{N_{\eta,t}} S_{\eta,i} + B_{\eta,t}, \quad t \geq 0. \tag{6.26}
$$

Here, V_0 is the initial capital, p the premium intensity, $N_{\eta,t}$ denotes the number of claims up to time t modelled by a Poisson process, $S_{\eta,i} \geq 0$ denotes the size of the ith claim, and $(S_{\eta,i})_{i\in\mathbb{N}}$ is assumed to be an i.i.d. sequence, independent of $(N_{\eta,t})_{t\geq 0}$, so that the claim process $t \mapsto \sum_{i=1}^{N_{\eta,i}} S_{\eta,i}$ is a compound Poisson process. The process $(B_{\eta,t})_{t\geq 0}$ is a Brownian motion with variance σ_η^2 and independent of the claim process, and describes the disturbance term to the classical Cramér–Lundberg model. Writing

$$\eta_t := pt - \sum_{i=1}^{N_{\eta,t}} S_{\eta,i} + B_{\eta,t}, \quad t \geq 0,$$

$(\eta_t)_{t\geq 0}$ is a Lévy process and we can write

$$V_t = V_0 + \eta_t \tag{6.27}$$

for (6.26). Now suppose that the insurance company invests its capital in a market. A popular model for the cumulated investment return of the market is given by the Lévy process,

$$U_t = rt + B_{U,t} + \sum_{i=1}^{N_{U,t}} S_{U,i}, \quad t \geq 0,$$

where r describes the interest rate, $B_{U,t}$ is a Brownian motion component with variance σ_U^2, and $t \mapsto \sum_{i=1}^{N_{U,t}} S_{U,i}$ is a compound Poisson process, independent of $B_{U,t}$. An amount Y_t in the market at time t evolves in accordance with the stochastic differential equation,

$$dY_t = Y_{t-} dU_t, \quad t \geq 0. \tag{6.28}$$

The combination of both models (6.27) and (6.28), i.e. for the capital of an insurance company that invests its capital in a market described by U, is then given by

$$dV_t = V_{t-} dU_t + d\eta_t, \quad t \geq 0.$$

In this model, proposed by Paulsen (1993), U and η are assumed to be independent. Under the natural assumption that $S_{U,1} > -1$ almost surely, which prevents V_t from becoming negative due to a jump in the investment process U, the stochastic exponential $\mathcal{E}(U)_t$ is positive for all t and $(\xi_t := -\log \mathcal{E}(U)_t)_{t\geq 0}$ is a Lévy process. It follows from Proposition 6.11 that $(V_t)_{t\geq 0}$ is a generalized Ornstein–Uhlenbeck process, driven by (ξ, η).

In Section 6.3.1 we interpreted the Lévy-driven Ornstein–Uhlenbeck process as a natural continuous-time analogue of discrete AR(1) processes with i.i.d. noise. Similarly, generalized Ornstein–Uhlenbeck processes can be seen as continuous-time analogues of AR(1) processes with random coefficients. This result is due to de Haan and Karandikar (1989) and stated in the next theorem.

Theorem 6.14 *Let $(V_t)_{t\geq 0}$ be a generalized Ornstein–Uhlenbeck process, driven by (ξ, η). For $0 \leq s \leq t$ define*

$$A_t^s := e^{-(\xi_t - \xi_s)} \quad \text{and} \quad B_t^s := e^{-(\xi_t - \xi_s)} \int_{(s,t]} e^{\xi_u - \xi_s} \, d\eta_u. \tag{6.29}$$

Then

(i) $V_t = A_t^s V_s + B_t^s$ *a.s., for all* $t \geq s \geq 0$,

(ii) $(V_s)_{s \leq a}$ *and* $(A_s^t, B_s^t)_{t \geq s \geq a}$ *are independent for all* $a \geq 0$,

(iii) $A_t^s = A_u^s A_t^u$ *and* $B_t^s = A_u^t B_u^s + B_u^t$ *a.s., for all* $0 \leq s \leq u \leq t$,

(iv) $(A_t^s, B_t^s)_{a \leq s \leq t \leq b}$ *and* $(A_t^s, B_t^s)_{c \leq s \leq t \leq d}$ *are independent for all* $a \leq b \leq c \leq d$,

(v) the distribution of $(A_{t+h}^{s+h}, B_{t+h}^{s+h})_{0 \leq s \leq t}$ *does not depend on* $h \geq 0$, *and*

(vi) $A_s^t > 0$ *a.s. for all* $t \geq 0$ *and* (A_t^0, B_t^0) *converges in probability to (1,0) as* $t \downarrow 0$.

Conversely, if $(A_t^s, B_t^s)_{0 \leq s \leq t}$ *are random processes such that conditions (iii) – (vi) hold, then* (A_t^s, B_t^s) *admit versions which are càdlàg in* t *for every* s, *and with these versions chosen, there exists a bivariate Lévy process* (ξ, η) *such that* A_s^t *and* B_s^t *are given by (6.29). Further, if* V_0 *is chosen independent of* $(A_t^s, B_t^s)_{t \geq s \geq 0}$, *then the process defined by* $V_t = A_t^0 V_0 + B_t^0$ *is a generalized Ornstein–Uhlenbeck process driven by* (ξ, η) *and hence satisfies (i) and (ii).*

Proof. That a generalized Ornstein–Uhlenbeck process satisfies (i) - (vi) is an easy consequence of (6.22) and the fact that (ξ, η) is a bivariate Lévy process. For the converse, we refer to de Haan and Karandikar (1989). □

Statements (i), (ii), (iv) and (v) of the previous theorem imply in particular that a generalized Ornstein–Uhlenbeck process sampled at equidistant times of length r satisfies

$$V_{rn} = A_{rn}^{r(n-1)} V_{r(n-1)} + B_{rn}^{r(n-1)}, \quad n \in \mathbb{N},$$

where

$$(A_{rn}^{r(n-1)}, B_{rn}^{r(n-1)})_{n \in \mathbb{N}}$$
$$= (e^{-(\xi_{rn} - \xi_{r(n-1)})}, e^{-(\xi_{rn} - \xi_{r(n-1)})} \int_{(r-1)n}^{rn} e^{\xi_{u^-} - \xi_{r(n-1)}} \, d\eta_u)_{n \in \mathbb{N}}$$

is an i.i.d. sequence, with $V_{r(n-1)}$ being independent of $(A_{rn}^{r(n-1)}, B_{rn}^{r(n-1)})$ for every $n \in \mathbb{N}$. Hence, a generalized Ornstein–Uhlenbeck process sampled at integer times satisfies an AR(1) process with i.i.d. noise and i.i.d. coefficients.

It is interesting to know for which bivariate Lévy processes (ξ, η) there exists a starting random variable V_0, independent of (ξ, η), such that the corresponding generalized Ornstein–Uhlenbeck process becomes strictly stationary. This is addressed in the following theorem, which is from Lindner and Maller (2005).

Theorem 6.15 *Let* (ξ, η) *be a bivariate Lévy process such that neither* ξ *nor* η *is the zero-process. Then a random variable* V_0, *independent of* (ξ, η), *can be chosen such that* $(V_t)_{t \geq 0}$ *is strictly stationary if and only if*

(i) there is $k \in \mathbb{R} \setminus \{0\}$ *such that* $e^\xi = \mathcal{E}(\eta/k)$,

or

(ii) $V_\infty := \lim_{z\to\infty} \int_0^t e^{-\xi_{s-}} dL_s$ exists as an almost sure limit as a finite random variable.

If any of the conditions (i) or (ii) is satisfied, then the marginal stationary distribution is unique and given for $t \ge 0$ by $V_t = k$ in case (i) and by $V_t \stackrel{d}{=} V_\infty$ in case (ii).

Outline of Proof. Fix $t > 0$ and define

$$(\widehat{\xi}_s, \widehat{\eta}_s) := (\xi_t - \xi_{(t-s)-}, \eta_t - \eta_{(t-s)-}), \quad 0 \le s \le t.$$

Then it is well known from the duality lemma for Lévy processes (Lemma II.2 in Bertoin (1996)) that $(\widehat{\xi}_s, \widehat{\eta}_s)_{0\le s\le t}$ has the same distribution as $(\xi_s, \eta_s)_{0\le s\le t}$. Further,

$$e^{-\xi_t} \int_0^t e^{\xi_{s-}} d\eta_s = \int_0^t e^{-\xi_t} e^{\xi_{s-}} d\eta_s = -\int_0^t e^{-(\xi_t - \xi_{s-})} d(\eta_t - \eta_s). \quad (6.30)$$

Here the first equality needs explanation. Since $s \mapsto \xi_t - \xi_s$ for fixed t is not adapted to the natural filtration, it is not clear that the second integral makes sense. The trick is to enlarge the underlying filtration to a filtration $(\mathcal{G}_s)_{0\le s\le t}$ such that ξ_t is $\mathcal{G}_0$ measurable, and then use the fact that, as a Lévy process, η remains a semimartingale with respect to $(\mathcal{G}_s)_{0\le s\le t}$ (e.g. Protter (2005, Theorem VI.3)). Then a time reversal argument (e.g. Protter (2005, Theorem VI.22)) can be applied to show that the right-hand side of (6.30) is equal to $\int_0^t e^{-\widehat{\xi}_{s-}} d\widehat{\eta}_s + [e^{-\widehat{\xi}}, \widehat{\eta}]_t$, which by the duality lemma is equal in distribution to

$$\int_0^t e^{-\xi_{s-}} d\eta_s + [e^{-\xi}, \eta]_t = \int_0^t e^{-\xi_{s-}} d\eta_s + \int_0^t e^{-\xi_{s-}} d[U, \eta] = \int_0^t e^{-\xi_{s-}} dL_s.$$

Similarly, one gets for each fixed $t > 0$ that

$$V_t \stackrel{d}{=} e^{-\xi_t} V_0 + \int_0^t e^{-\xi_{s-}} dL_s. \quad (6.31)$$

The key idea is then to use a Theorem of Erickson and Maller (2005) which, apart from providing a characterisation for almost sure convergence of $\int_0^\infty e^{-\xi_{s-}} dL_s$, shows that if $\int_0^\infty e^{-\xi_{s-}} dL_s$ does not converge almost surely, then either $|\int_0^t e^{-\xi_{s-}} dL_s|$ converges in probability to $+\infty$ as $t \to \infty$, or there is a constant $k \in \mathbb{R}\setminus\{0\}$ such that for every $t > 0$, $\int_0^t e^{-\xi_{s-}} dL_s = k(1 - e^{-\xi_t})$ a.s. This, together with (6.31) and the assumed independence of V_0 and (ξ, η), can then be used to show that the stated conditions are necessary for a stationary solution to exist, and it is easy to see that they are also sufficient. Further details can be found in Lindner and Maller (2005). $\square$

Remark 6.16 *The definition of U in (6.23) requires the Lévy process U to* *have no jump of size ≤ -1, since $\mathcal{E}(U)$ has to be strictly positive to satisfy (6.23). Nevertheless, the stochastic differential equation (6.25) makes sense for a general bivariate Lévy process (U, L), and Behme et al. (2011) have extended Theorem 6.15 to this more general situation and also dropped the assumption of V_0 and (U, L) being independent, thereby giving rise also to non-causal solutions.*

The next theorem gives a complete characterisation in terms of the characteristic triplet of the underlying Lévy processes for condition (ii) in Theorem 6.15 to hold. Part (i) of the theorem below is due to Erickson and Maller (2005), part (ii) to Doney and Maller (2002) and part (iii) to Lindner and Maller (2005). We omit the proof.

Theorem 6.17 *Let (ξ, η) be a bivariate Lévy process with Lévy measure $\nu_{\xi,\eta}$ and let L be constructed as in (6.23). Denote the Lévy measures of ξ and L by ν_ξ and ν_L, respectively. The following statements then hold.*
(i) The integral $\int_0^t e^{-\xi_{s-}} dL_s$ exists as an almost sure limit as $t \to \infty$ if and only if

$$\lim_{t \to \infty} \xi_t = +\infty \ a.s. \quad \text{and} \quad \int_{\mathbb{R} \setminus [-e,e]} \frac{\log |y|}{R_\xi(\log |y|)} \nu_L(dy) < \infty,$$

where

$$R_\xi(y) := 1 + \int_1^y \nu_\xi((z, \infty)) \, dz, \quad y > 1.$$

(ii) $\lim_{t \to \infty} \xi_t = +\infty$ a.s. if and only either $0 < E\xi_1 \leq E|\xi_1| < \infty$, or

$$\lim_{t \to \infty} R_\xi(t) = +\infty \quad \text{and} \quad \int_{-\infty}^1 \frac{|x|}{R_\xi(|x|)} \nu_\xi(dx) < \infty.$$

(iii) The Lévy measure ν_L of L is $\nu_L = (T\nu_{\xi,\eta})_{|\mathbb{R} \setminus \{0\}}$, the image measure of $\nu_{\xi,\eta}$ under the mapping $T : \mathbb{R}^2 \to \mathbb{R}$, $(x, y) \mapsto e^{-x} y$, when restricted to $\mathbb{R} \setminus \{0\}$.

An extension of the previous Theorem to the case of integrals of the form $\int_0^\infty \mathcal{E}(U)_{t-} dL_t$ for a general bivariate Lévy process (U, L) can be found in Behme et al. (2011).

If ξ_1 has positive and finite expectation, then a more convenient criterion for the existence of stationary solutions is given in the theorem below. The sufficiency of the criterion has been shown by de Haan and Karandikar (1989), while in Lindner and Maller (2005) it has been shown that the stated condition is equivalent to condition (ii) of Theorem 6.15.

Corollary 6.18 *Let (ξ, η) be a bivariate Lévy process such that $0 < E\xi_1 \leq$*

$E|\xi_1| < \infty$. *Then a necessary and sufficient condition for the existence of a stationary solution of the generalized Ornstein–Uhlenbeck process driven by (ξ, η) is that $E\log^+ |\eta_1| < \infty$.*

Outline of Proof. If ξ has finite and positive expectation, then it follows from Theorem 3.1 (a,b) in Lindner and Maller (2005) that $\int_0^\infty e^{-\xi_{s-}} dL_s$ converges a.s. if and only if $\int_0^\infty e^{-\xi_{s-}} d\eta_s$ does. Since $\lim_{t\to\infty} R_\xi(t) < \infty$, Theorem 6.17 (i) then shows that this is equivalent to $\int_{|x|\geq 1} \log |x|\, \nu_\eta(dx) < \infty$, which in turn is equivalent to $E\log^+ |\eta_1| < \infty$ (e.g. Sato (1999, Theorem 25.3)). Hence condition (ii) of Theorem 6.15 is equivalent to $E\log^+ |\eta_1| < \infty$. To complete the proof, it suffices to observe that in the degenerate case (i) of Theorem 6.15, i.e. when $e^{\xi_t} = \mathcal{E}(\eta/k)_t$, it also follows that $\int_0^\infty e^{-\xi_{t-}} dL_t$ converges almost surely (since additionally $\lim_{t\to\infty} \xi_t = +\infty$ by assumption), as has been shown in Theorem 2.1 (a) of Behme et al. (2011). $\square$

Observe that Corollary 6.18 contains the condition for stationary versions of Lévy-driven Ornstein–Uhlenbeck processes, considered in Section 6.3.1, as a special case.

The autocorrelation structure of a generalized Ornstein–Uhlenbeck process is described in the following theorem. It is always of exponential form.

Theorem 6.19 *Let (ξ, η) be a bivariate Lévy process such that*

$$Ee^{-2p\xi_1} < \infty \quad \text{and} \quad E|\eta_1|^{2q} < \infty, \qquad (6.32)$$

where $p, q > 1$ such that $p^{-1} + q^{-1} = 1$. Suppose further that $Ee^{-2\xi_1} < 1$. Then condition (ii) of Theorem 6.15 is satisfied, and the stationary version $(V_t)_{t\geq 0}$ of the generalized Ornstein–Uhlenbeck process driven by (ξ, η) satisfies $EV_0^2 < \infty$ and

$$\operatorname{Cov}(V_t, V_{t+h}) = \operatorname{Var}(V_0)e^{-h|\Psi_\xi(1)|}, \quad t, h \geq 0,$$

where

$$\Psi_\xi(\kappa) := \log Ee^{-\kappa\xi_1} \quad \text{for} \quad \kappa > 0.$$

If ξ and η are independent, then condition (6.32) can be replaced by

$$Ee^{-2\xi_1} < 1 \quad \text{and} \quad E\eta_1^2 < \infty.$$

Proof. See Prop. 4.1, Theorem 4.3 and Remark 8.1 in Lindner and Maller (2005). $\square$

Remark 6.20 *Sufficient conditions for the above theorem to hold can also be given in terms of U and L as defined in (6.23). Behme (2011) shows that if $EL_1^2 < \infty$ and $2E(U_1) + \operatorname{Var}(U_1) < 0$, then the stationary version satisfies $EV_t^2 < \infty$ and*

$$\operatorname{Cov}(V_t, V_{t+h}) = \operatorname{Var}(V_0)e^{-h|E(U_1)|}, \quad t, h \geq 0,$$

in particular no Hölder type condition as in (6.32) is needed. An expression for the variance of V_0 in terms of U and L can also be found in Behme (2011).

The result that the Lévy-driven Ornstein–Uhlenbeck process has Pareto tails if and only if the driving Lévy process has Pareto tails does not apply to the generalized Ornstein–Uhlenbeck process. This is due to the fact that the stationary distribution satisfies the random recurrence equation $V_\infty \overset{d}{=} A_1^0 V_\infty + B_1^0$, with A_1^0 and B_1^0 as defined in (6.29) and with V_∞ independent of (A_1^0, B_1^0) on the right hand side of this equation. This allows the application of results of Kesten (1973) and Goldie (1991) to ensure Pareto tails of the stationary solution for a large class of driving Lévy processes; see Theorem 4.5 of Lindner and Maller (2005), or also Behme (2011).

On the other hand, while it is known that the stationary distribution of Lévy-driven Ornstein–Uhlenbeck processes is always self-decomposable, in particular infinitely divisible and absolutely continuous unless degenerate to a constant, this is not the case for generalized Ornstein–Uhlenbeck processes. While $\int_0^\infty e^{-\xi_{s-}} dL_s$ will be self-decomposable if ξ_t drifts to $-\infty$ a.s. and has no positive jumps, as shown by Samorodnitsky (2004) (see e.g. Klüppelberg, Lindner, and Maller (2006)), in general, the stationary solution may be neither infinitely divisible nor absolutely continuous. Simple examples when continuity properties and infinite divisibility depend critically on underlying parameters of the Lévy process are given in Lindner and Sato (2009). However, as recently observed by Watanabe (2009), applying results of Alsmeyer, Iksanov, and Rösler (2009), the stationary distribution of a generalized Ornstein–Uhlenbeck process satisfies a pure type theorem, in the sense that it is either degenerate to a constant, singular continuous or absolutely continuous.

We finish this section with an example showing that different driving bivariate Lévy processes may lead to the same stationary generalized Ornstein–Uhlenbeck process. The underlying Lévy process is therefore not necessarily identifiable without further assumptions:

Example 6.21 Let $\eta^{(1)}$ and $\eta^{(2)}$ be Lévy processes on $\mathbb{R}$ such that the Lévy measures $\nu_{\eta^{(1)}}$ and $\nu_{\eta^{(2)}}$ are concentrated on $(-1, \infty)$. Then the stochastic exponentials of $\eta^{(1)}$ and $\eta^{(2)}$ are strictly positive, and

$$\xi_t^{(j)} := \log \mathcal{E}(\eta^{(j)})_t, \quad t \geq 0, \quad j = 1, 2,$$

are Lévy processes. By Theorem 6.15, the strictly stationary solutions satisfy

$$V_t^{(j)} = e^{-\xi_t^{(j)}} \left(1 + \int_0^t e^{\xi_{s-}^{(j)}} d\eta_s^{(j)}\right) = 1, \quad t \geq 0, \quad j = 1, 2,$$

and hence $(\xi^{(1)}, \eta^{(1)})$ and $(\xi^{(2)}, \eta^{(2)})$ lead to the same stationary generalized Ornstein–Uhlenbeck process. Thus in general the driving Lévy process cannot be identified from the generalized Ornstein–Uhlenbeck process.

The previous example implies that one has to restrict to subclasses of generalized Ornstein–Uhlenbeck processes if one wants to estimate them, and some methods for the Lévy-driven Ornstein–Uhlenbeck process and for the COGARCII(1,1) process will be given in Section 6.4.3.

6.3.4 The COGARCH process

The ARCH and GARCH processes of Engle (1982) and Bollerslev (1986) specify the squared volatility of a a financial time series as an affine linear combination of the squared past observations and the squared past volatility. To be more specific, let $(\varepsilon_n)_{n \in \mathbb{N}_0}$ be an i.i.d. sequence of random variables, and let $\beta > 0$, $\lambda > 0$ and $\delta \geq 0$. A GARCH(1,1) process $(Y_n)_{n \in \mathbb{N}_0}$ with *volatility process* $(\sigma_n)_{n \in \mathbb{N}_0}$ is then a solution of the equations

$$Y_n = \sigma_n \varepsilon_n, \quad n \subset \mathbb{N}_0, \tag{6.33}$$

$$\sigma_n^2 = \beta + \lambda Y_{n-1}^2 + \delta \sigma_{n-1}^2, \quad n \in \mathbb{N}, \tag{6.34}$$

and such that σ_n is independent of $(\varepsilon_{n+h})_{h \in \mathbb{N}_0}$ and non-negative for every $n \in \mathbb{N}_0$. For $\delta = 0$, this gives the ARCH(1) process.

Since in financial mathematics a continuous-time setting is often used, it is of interest to have continuous-time counterparts to GARCH processes. The most prominent continuous-time GARCH process is the diffusion limit of Nelson (1990). By considering GARCH processes on fine grids $h\mathbb{N}_0$ and rescaling the parameters appropriately as $h \downarrow 0$, Nelson obtained the following diffusion limit $(G_t, \sigma_t)_{t \geq 0}$ given by

$$dG_t = \sigma_t \, dB_t^{(1)}, \quad t \geq 0, \tag{6.35}$$

$$d\sigma_t^2 = (\omega - \theta \sigma_t^2) \, dt + \alpha \sigma_t^2 \, dB_t^{(2)}, \quad t \geq 0, \tag{6.36}$$

where $B^{(1)}$ and $B^{(2)}$ are two independent Brownian motions and $\theta \in \mathbb{R}$, $\omega \geq 0$ and $\alpha > 0$ are parameters. Observe that (6.36) specifies a generalized Ornstein–Uhlenbeck process for σ_t^2 driven by $(\xi_t, \eta_t) = (-\alpha B_t^{(2)} + (\theta + \alpha^2/2)t, \omega t)$, as follows easily from Proposition 6.11. A drawback of the diffusion limit of Nelson as a continuous-time GARCH model is that in (6.35) and (6.36) there are two independent sources of randomness, namely $B^{(1)}$ and $B^{(2)}$, while the GARCH(1,1) process (6.33), (6.34) is driven by a single noise process $(\varepsilon_n)_{n \in \mathbb{N}_0}$. Another drawback of the diffusion limit as a model for financial time series is the absence of jumps which are frequently observed in financial data. Finally, as shown by Wang (2002), statistical inference for GARCH modelling and for Nelson's diffusion limit are not asymptotically equivalent in the sense of Le Cam (1986).

Motivated by these limitations, Klüppelberg, Lindner, and Maller (2004) constructed a continuous-time GARCH process driven by a single Lévy process. In the following we will sketch the construction and the properties of the process, which is called COGARCH(1,1), where "CO" stands for continuous-time.

Recall that for the GARCH(1,1) process, we obtain by iterating (6.34)

$$\sigma_n^2 = \beta + \lambda Y_{n-1}^2 + \delta \sigma_{n-1}^2 = \beta + (\delta + \lambda \varepsilon_{n-1}^2)\sigma_{n-1}^2$$

$$\vdots$$

$$= \beta \sum_{i=0}^{n-1} \prod_{j=i+1}^{n-1} (\delta + \lambda \varepsilon_j^2) + \sigma_0^2 \prod_{j=0}^{n-1} (\delta + \lambda \varepsilon_j^2)$$

$$= \left(\beta \int_0^n \exp\left\{ -\sum_{j=0}^{\lfloor s \rfloor} \log(\delta + \lambda \varepsilon_j^2) \right\} ds + \sigma_0^2 \right)$$

$$\times \exp\left\{ \sum_{j=0}^{n-1} \log(\delta + \lambda \varepsilon_j^2) \right\}, \tag{6.37}$$

$$X_n = \sigma_n \left(\sum_{j=0}^{n} \varepsilon_j - \sum_{j=0}^{n-1} \varepsilon_j \right). \tag{6.38}$$

Observe that $n \mapsto \sum_{j=0}^{n} \varepsilon_j$ and $n \mapsto \sum_{j=0}^{n-1} \log(\delta + \lambda \varepsilon_j^2)$ are random walks, and that the second one can be constructed from the first one. The key idea in the construction of Klüppelberg et al. (2004) was to replace these random walks by their continuous-time analogues, i.e. by Lévy processes. More precisely, for a given driving Lévy process $M = (M_t)_{t \geq 0}$, ε_j will be replaced by the jump ΔM_t of M, and the random walk

$$n \mapsto -\sum_{j=0}^{n} \log(\delta + \lambda \varepsilon_j^2) = -(n+1)\log \delta - \sum_{j=0}^{n} \log(1 + \lambda \delta^{-1} \varepsilon_j^2)$$

by

$$\xi_t := -t \log \delta - \sum_{0 < s \leq t} \log(1 + \lambda \delta^{-1}(\Delta M_s)^2), \quad t \geq 0. \tag{6.39}$$

Observe that this sum converges since $\sum_{0<s\leq t}(\Delta M_s)^2 < \infty$ a.s. for any Lévy process M, and that ξ defined this way is again a Lévy process. Comparing with (6.37) and (6.38), this leads to the following definition:

Definition 6.22 *Let $M = (M_t)_{t \geq 0}$ be a Lévy process with nonzero Lévy measure ν_M and let $\beta, \delta > 0$ and $\lambda \geq 0$ be fixed constants. Then the CO-GARCH(1,1) process $(G_t)_{t \geq 0}$ with volatility process $(\sigma_t)_{t \geq 0}$ is defined by*

$$G_0 = 0, \quad dG_t = \sigma_{t-} dM_t, \quad t \geq 0,$$

where $\sigma_t = \sqrt{\sigma_t^2}$,

$$\sigma_t^2 = \left(\beta \int_0^t e^{\xi_{s-}} \, ds + \sigma_0^2 \right) e^{-\xi_t}, \quad t \geq 0,$$

and ξ is defined by (6.39).

From this definition we see that the squared volatility process is none other than a generalized Ornstein–Uhlenbeck process, driven by the bivariate Lévy process $(\xi_t, \beta t)_{t \geq 0}$. In particular, with $U_t = t \log \delta + \lambda \delta^{-1} \sum_{0 < s \leq t} (\Delta M_s)^2$ and $L_t = \beta t$ it satisfies the stochastic differential equation (6.25), i.e.

$$d\sigma_t^2 = \sigma_{t-}^2 d(t \log \delta + \lambda \delta^{-1} [M, M]_t^{(d)}) + \beta dt, \quad t \geq 0,$$

where $[M, M]_t^{(d)} = \sum_{0 < s \leq t} (\Delta M_s)^2$ denotes the discrete part of the quadratic variation of M. It is easy to see that $-\xi_t - t \log \delta$ is increasing, hence a subordinator, and that ξ is of bounded variation with drift $-\log \delta$. Further, since $\Delta \xi_t = -\log(1 + \lambda \delta^{-1} (\Delta M_t)^2)$, it follows that the Lévy measure ν_ξ of ξ is the image measure of ν_M under the transformation

$$T : \mathbb{R} \setminus \{0\} \to (-\infty, 0), \quad x \mapsto -\log(1 + \lambda \delta^{-1} x^2). \quad (6.40)$$

Having determined the characteristic triplet of ξ in terms of the driving Lévy process M, it is easy to obtain conditions for a stationary COGARCH(1,1) volatility to exist. Thus, by Theorem 6.15, a stationary version exists if and only if $\int_0^\infty e^{-\xi_{s-}} \, ds < \infty$ a.s., which by Theorem 6.17 is equivalent to $E\xi_1 > 0$. Further, since

$$E\xi_1 = -\log \delta + \int_{-\infty}^0 x \, \nu_\xi(dx) = -\log \delta - \int_{\mathbb{R}} \log(1 + \lambda \delta^{-1} y^2) \, \nu_M(dx),$$

where the first equality follows from Example 25.12 in Sato (1999) and the second from (6.40), we see that the COGARCH(1,1) process admits a strictly stationary volatility if and only if

$$\int_{\mathbb{R}} \log(1 + \lambda \delta^{-1} x^2) \, \nu_M(dx) < -\log \delta, \quad (6.41)$$

which in particular requires M to have finite log-moment and $\delta < 1$. The stationarity condition was obtained in Klüppelberg et al. (2004).

Since ξ has no positive jumps but drifts to infinity, the stationary version σ_t^2 is self-decomposable as pointed out in Section 6.3.3, hence absolutely continuous and infinitely divisible. The second moment structure of σ_t^2 can be obtained from Theorem 6.19. More generally we have the following result.

Proposition 6.23 *Let $(G_t)_{t \geq 0}$ be a COGARCH(1,1) process with parameters β, δ and λ, driven by M, such that (6.41) holds. Let $(\sigma_t^2)_{t \geq 0}$ be the stationary* *volatility process. For $\kappa > 0$, denote*

$$\Psi_\xi(\kappa) \quad := \quad \log E e^{-\kappa\xi_1}$$

$$= \quad \kappa \log \delta + \int_{\mathbb{R}} \left((1 + \lambda \delta^{-1} y^2)^\kappa - 1 \right) \nu_M(dy) \in (-\infty, \infty].$$

Then for $k \in \mathbb{N}$, $E\sigma_0^{2k} < \infty$ if and only if $EM_1^{2k} < \infty$ and $\Psi_\xi(k) < 0$, in which case $\Psi_\xi(l) < 0$ for all $l \in \{1, \dots, k\}$ and

$$E\sigma_0^{2k} = k!\beta^k \prod_{l=1}^{k} (-\Psi_\xi(l))^{-1}.$$

Further, if $EM_1^4 < \infty$ and $\Psi_\xi(2) < 0$, then

$$\mathrm{Cov}(\sigma_t^2, \sigma_{t+h}^2) = \beta^2 \left(2\Psi_\xi^{-1}(1)\Psi_\xi^{-1}(2) - \Psi_\xi^{-2}(1) \right) e^{-h|\Psi_\xi(1)|}, \quad t, h \geq 0.$$

Proof. The sufficiency of the condition for $E\sigma_0^4 < \infty$ and the covariance formula follows from Theorem 6.19. For the remaining assertions, we refer to Proposition 4.2 and Corollary 4.1 in Klüppelberg et al. (2004). □

As for the volatility model of Barndorff–Nielsen and Shephard, under certain assumptions, non-overlapping increments of the COGARCH(1,1) process G are uncorrelated, while the autocovariance function of $((G_{rh} - G_{r(h-1)})^2)_{h \in \mathbb{N}}$ is that of an ARMA(1,1) process for any $r > 0$. This will be stated more precisely in Section 6.4.3, when it will be used to derive a generalized method of moment estimator for the COGARCH parameters.

We conclude this section by presenting the construction of COGARCH(q, p) processes, as defined by Brockwell, Chadraa, and Lindner (2006). Recall that for $p, q \geq 1$ a GARCH(q, p) process $(X_n, \sigma_n)_{n \in \mathbb{N}_0}$ satisfies the relations

$$X_n = \sigma_n \varepsilon_n, \quad n \in \mathbb{N}_0$$

$$\sigma_n^2 = \beta + \sum_{i=1}^{q} b_{i-1} X_{n-i}^2 + \sum_{i=1}^{p} \alpha_i \sigma_{t-i}^2, \quad n \geq \max\{q, p\},$$

with an i.i.d. sequence $(\varepsilon_n)_{n \in \mathbb{N}_0}$, appropriate starting random variables $\sigma_0^2, \dots,$ $\sigma_{\max\{q-1, p-1\}}^2$ and coefficients $\beta > 0$, $b_0, \dots, b_{q-1} \geq 0$, $a_1, \dots, a_p \geq 0$ such that $a_p > 0$ and $b_{q-1} > 0$. In particular, $(\sigma_n^2)_{n \in \mathbb{N}_0}$ can be seen as a "self-exciting" ARMA$(p, q - 1)$ process driven by $(\sigma_{n-1}^2 \varepsilon_{n-1}^2)$ together with the "mean correction" β. This motivates the definition of the squared volatility process $(V_t)_{t \geq 0}$ of a continuous-time GARCH(q, p) process as a "self-exciting mean corrected" CARMA$(p, q - 1)$ process driven by an appropriate noise term. In discrete time, the driving noise is defined through the increments of the process $(\sum_{i=0}^{n-1} \sigma_i^2 \varepsilon_i^2)_{n \in \mathbb{N}}$. Replacing σ_i^2 by $V_{t-} = \sigma_{t-}^2$ in continuous-time and ε_i by the jump ΔM_t of a Lévy process $(M_t)_{t\geq 0}$ suggests the use of

$$\sum_{0<s\leq t} V_{s-}(\Delta M_s)^2 = \int_0^t V_{s-} d[M, M]_s^{(d)}, \quad t \geq 0,$$

as driving noise for the CARMA equations. If the squared volatility process $(V_t)_{t\geq 0}$ specified this way is then non-negative, the COGARCH process should be defined by $dG_t = \sqrt{V_{t-}}\, dM_t$. This is made precise in the following definition:

Definition 6.24 *Let $M = (M_t)_{t\geq 0}$ be a Lévy process with nonzero Lévy measure. With $p, q \in \mathbb{N}$ such that $q \leq p$, $a_1, \ldots, a_p, b_0, \ldots, b_{p-1} \in \mathbb{R}$, $\beta > 0$, $a_p \neq 0$, $b_{q-1} \neq 0$ and $b_q = \ldots = b_{p-1} = 0$, define the $p\times p$-matrix A and the vectors $\mathbf{b}, \mathbf{e} \in \mathbb{C}^p$ as in (6.16) and (6.17), respectively. Define the squared volatility process $(V_t)_{t\geq 0}$ with parameters A, $\mathbf{b}$ and driving Lévy process M by*

$$V_t = \beta + \mathbf{b}'\mathbf{X}_t, \quad t \geq 0,$$

where the state process $\mathbf{X} = (\mathbf{X}_t)_{t\geq 0}$ is the unique càdlàg solution of the stochastic differential equation

$$d\mathbf{X}_t = A\mathbf{X}_{t-}\, dt + \mathbf{e}V_{t-}\, d[M, M]_t^{(d)} = A\mathbf{X}_{t-}\, dt + \mathbf{e}(\beta + \mathbf{b}'\mathbf{X}_{t-})\, d[M, M]_t^{(d)},$$

with initial value $\mathbf{X}_0$, independent of $(M_t)_{t\geq 0}$. If the process $(V_t)_{t\geq 0}$ is strictly stationary and non-negative almost surely, then $G = (G_t)_{t\geq 0}$, defined by

$$G_0 = 0, \quad dG_t = \sqrt{V_{t-}}\, dM_t,$$

is a COGARCH(q, p) process with parameters A, $\mathbf{b}$, β and driving Lévy process M.

It can be shown that for $p = q = 1$ this definition is equivalent to the definition of the COGARCH(1,1) process given before. Brockwell et al. (2006) also give sufficient conditions for the existence of a strictly stationary solution $(V_t)_{t\geq 0}$ and its positivity, and show that $(V_t)_{t\geq 0}$ has the same autocorrelation structure as a CARMA$(p, q - 1)$ process. Hence the COGARCH(q, p) process allows a more flexible autocorrelation structure than the COGARCH(1,1) process, in particular, allowing for non-monotone autocovariance functions. A formula for the autocovariance function of the squared increments of $(G_t)_{t\geq 0}$ over non-overlapping intervals is also given.

6.4 Some estimation methods

6.4.1 Estimation of Ornstein–Uhlenbeck models

In this section we present some estimation methods for Lévy-driven Ornstein–Uhlenbeck processes.

A non-parametric estimator

We start with the non-parametric estimator of Jongbloed et al. (2005). For this, let $L = (L_t)_{t \in \mathbb{R}}$ be a subordinator without drift and with finite log-moment, and consider the stationary Lévy-driven Ornstein–Uhlenbeck process

$$V_t = e^{-t} \left(V_0 + \int_0^t e^s dL_s \right), \quad t \geq 0, \tag{6.42}$$

with $V_0 = \int_{-\infty}^0 e^s \, dL_s$. The aim is to estimate the stationary marginal distribution of V_t, non-parametrically based on observations $V_1, V_2, \ldots, V_n$. The method of Jongbloed et al. (2005) makes use of the fact that stationary solutions of Lévy-driven Ornstein–Uhlenbeck processes are self-decomposable, and since L was assumed to be a subordinator, V_t must be non-negative. A distribution μ is non-negative and selfdecomposable if and only if its characteristic function is of the form

$$\mathbb{R} \to \mathbb{C}, \quad z \mapsto \exp \left(\gamma z + \int_0^\infty (e^{izx} - 1) \frac{k(x)}{x} \, dx \right), \tag{6.43}$$

where $\gamma \geq 0$ and $k : (0, \infty) \to [0, \infty)$ is a decreasing right-continuous function such that $\int_0^1 k(x) \, dx < \infty$ and $\int_1^\infty x^{-1} k(x) dx < \infty$, see e.g. Sato (1999), Corollary 15.11 and Theorem 24.11. On the other hand, as follows for example from equation (4.17) in Barndorff–Nielsen and Shephard (2001b), the marginal stationary distribution V_t of the Lévy-driven Ornstein–Uhlenbeck process has characteristic function

$$\phi : \mathbb{R} \to \mathbb{C}, \quad z \mapsto \phi(z) := E^{iV_0 z} = \exp \left(\int_0^\infty (e^{izx} - 1) \frac{\nu_L((x, \infty))}{x} \, dx \right),$$

i.e. has the special form (6.43) with $\gamma = 0$ and $k(x) = \nu_L((x, \infty))$. The idea now is to take the empirical characteristic function based on observations $V_1, \ldots, V_n$ and then choose the decreasing function k from an appropriate class of functions that minimizes a certain distance between the empirical characteristic functions and the characteristic function defined by (6.43). To be more precise, define

$$\widetilde{\phi}_n(z) := \frac{1}{n} \sum_{j=0}^n e^{izV_j}, \quad z \in \mathbb{R}.$$

Then

$$\lim_{n \to \infty} \widetilde{\phi}_n(z) = \phi(z), \quad z \in \mathbb{R},$$

as a consequence of the strong mixing property and hence ergodicity of the process $(V_t)_{t \geq 0}$ as shown in Theorem 3.1 of Jongbloed et al. (2005). In Lemma 5.2 of the same paper, it is moreover shown that this limit is even uniform on compact subsets of $\mathbb{R}$, i.e. that

$$\lim_{n \to \infty} \sup_{z \in [-T,T]} |\widetilde{\phi}(z) - \phi(z)| = 0 \quad \text{a.s.} \quad \forall \, T > 0.$$

Since ϕ, as the characteristic function of a self-decomposable and hence infinitely divisible distribution, is nowhere zero on $\mathbb{R}$ and is continuous, it follows that $\widetilde{\phi}_n(z)$ is nowhere zero on $[-T,T]$ for large enough n, where $T > 0$ is fixed. It follows (e.g. Chung (2001, Section 7.6)), that there is a unique continuous function $\widetilde{g}_n : [-T,T] \to \mathbb{C}$, called the *distinguished logarithm of $\widetilde{\phi}_n$*, such that $\widetilde{g}_n(0) = 0$ and $\exp(\widetilde{g}_n(z)) = \widetilde{\phi}_n(z)$ for all $z \in [-T,T]$.

Now, let $T > 0$ be fixed and $w : [-T,T] \to [0,\infty)$ be an even, Lebesgue-integrable weight-function, which will be fixed. Moreover, define

$$K := \{k : (0,\infty) \to [0,\infty) : k \text{ decreasing and right-continuous with}$$

$$\int_0^1 k(x)\,dx < \infty \quad \text{and} \quad \int_1^\infty \frac{k(x)}{x}\,dx < \infty\},$$

and for $R \in K$ fixed, define

$$K_R := \{k \in K : k(x) \le R(x) \quad \forall \, x \in (0,\infty)\}.$$

Then as shown in Theorem 4.5 of Jongbloed et al. (2005), the *cumulant M-estimator*

$$\widehat{k}_n := \operatorname{argmin}_{k \in K_R} \int_{-T}^T \left| \int_0^\infty (e^{izx} - 1) \frac{k(x)}{x}\,dx - \widetilde{g}_n(z) \right|^2 w(z)\,dz \quad (6.44)$$

exists and is unique. The following is Theorem 5.3 of Jongbloed et al. (2005) which provides strong consistency of the cumulant M-estimator. We refer to the original paper for the proof.

Theorem 6.25 *Let $T > 0$ be fixed, w a weight-function as above and let $R \in K$. Let L be a subordinator without drift, finite log-moment and Lévy measure ν_L such that $x \mapsto \nu_L((x,\infty)) \in K_R$. Let V be a stationary version of (6.42) and let $\widehat{k}_n$ be the cumulant M-estimator as defined in (6.44), based on V_0, $V_1, \ldots, V_n$. Then*

$$\lim_{n \to \infty} \int_0^1 \left| \widehat{k}_n(z) - \nu_L((z,\infty)) \right| dz + \int_1^\infty \frac{|\widehat{k}_n(z) - \nu_L((z,\infty))|}{z}\,dz = 0 \quad \text{a.s.}$$

Estimating the parameter λ for the subordinator-driven CAR(1) process

For the stationary subordinator-driven Ornstein–Uhlenbeck process defined by (6.5) with $\lambda > 0$, Brockwell, Davis, and Yang (2007) consider inference based on observations of $(V_t)_{0 \le t \le T}$ at times $0, h, 2h, \ldots, Nh$, where $h > 0$ and

$N = \lfloor T/h \rfloor$. Defining $V_n^{(h)} = V_{nh}, n = 0, 1, 2, \ldots, N$, they estimate the coefficient λ using the estimator, also used by Jongbloed et al. (2005),

$$\hat{\lambda}_N = -h^{-1} \log \hat{\psi}_N, \qquad (6.45)$$

where

$$\hat{\psi}_N = \min_{1 \le n \le N} \frac{V_n^{(h)}}{V_{n-1}^{(h)}}. \qquad (6.46)$$

The estimator (6.46) is the estimator proposed by Davis and McCormick (1989) for AR(1) processes driven by non-negative noise, applied here to the sampled process

$$V_n^{(h)} = \psi V_{n-1}^{(h)} + Z_n,$$

where $\psi = e^{-\lambda h}$ and $(Z_n)_{n \in \mathbb{N}}$ is the non-negative i.i.d. sequence,

$$Z_n = \int_{(n-1)h}^{nh} e^{-\lambda(nh-u)} dL(u),$$

and transformed to give the estimator (6.45) of λ. Jongbloed et al. (2005) showed the weak consistency of $\hat{\lambda}_N$ as $N \to \infty$ with h fixed. Under the assumption that the distribution function, F of Z_n satisfies $F(0) = 0$ and is regularly varying at 0 with exponent α, i.e. if

$$\lim_{t \downarrow 0} \frac{F(tx)}{F(t)} = x^\alpha, \text{ for all } x > 0,$$

and if Z_n has finite β-moment for some $\beta > \alpha$, Brockwell et al. (2007) use the results of Davis and McCormick (1989) for AR(1) processes driven by non-negative noise to show that $\hat{\lambda}_N$ is strongly consistent and that $(\hat{\lambda}_N - \lambda)$ (suitably scaled) has the asymptotic Weibull distribution,

$$G_\alpha(x) = (1 - e^{-x^\alpha}) I_{[0,\infty)}(x).$$

(In particular, the conditions are satisfied if L is a gamma process.) In general we have the following theorem.

Theorem 6.26 *For the h-spaced observations $\{V_{nh}, n = 0, 1, \ldots, N\}$ of the stationary subordinator-driven CAR(1) process (6.5) with $\lambda, h > 0$ fixed, if the distribution function F of Z_n satisfies $F(0) = 0$ and if F is regularly varying at zero with exponent α, and if Z_n has finite β-moment for some $\beta > \alpha$, then $\hat{\lambda}_N \to \lambda$ almost surely as $N \to \infty$ and*

$$\lim_{N \to \infty} P[(-h)e^{-\lambda h} (EV_h^\alpha)^{1/\alpha} (\hat{\lambda}_N - \lambda)/F^{-1}(N^{-1}) \le x] = G_\alpha(x).$$

The following proposition provides some insight into the estimator $\hat{\lambda}_N$, suggests a generalized form for non-uniformly spaced observations and establishes strong consistency of both estimators when the observation interval is fixed and the maximum spacing of the observations goes to zero.

Proposition 6.27 *If the driving subordinator has zero drift (which is the case if the point 0 belongs to the closure of the support of $L(1)$), then for each fixed t, with probability 1,*

$$\lambda = \lim_{h \downarrow 0} \frac{\log V_t - \log V_{t+h}}{h}$$

and for any fixed $T > 0$,

$$\lambda = \sup_{0 \le s < t \le T} \frac{\log V_s - \log V_t}{t - s}.$$

Thus λ is determined almost surely by any small segment of continuously observed sample-path of V.

Proof. See Brockwell et al. (2007). □

If observations are available only at times $\{nh : n = 0, 1, 2, \ldots, N = \lfloor T/h \rfloor\}$ Proposition 6.27 suggests the estimator

$$\hat{\lambda}_T^{(h)} = \sup_{0 \le n < \lfloor T/h \rfloor} \frac{\log V_{nh} - \log V_{(n+1)h}}{h}.$$

which is precisely the estimator (6.45) discussed above. The analogous estimator, based on closely but irregularly spaced observations at times $t_1, t_2, \ldots, t_N$ such that $0 \le t_1 < t_2 < \cdots < t_N \le T$ is

$$\hat{\lambda}(T) = \sup_n \frac{\log V_{t_n} - \log V_{t_{n+1}}}{t_{n+1} - t_n}.$$

Proposition 6.27 implies that both estimators converge almost surely to λ as the maximum spacing between successive observations converges to zero.

Recovering the sample-path of L

In the idealized case when λ is known and the entire sample-path $(V_t)_{t \in [0,T]}$ is available, recovery of the sample-path of the driving Lévy process is straightforward, using the integrated form of (6.5). Thus

$$L_t = V_t - V_0 + \lambda \int_0^t V_u \, du.$$

From the sample path of L we could then extract i.i.d. samples of the increments of L on the intervals $((n-1)h, nh]$,

$$\Delta L_n^{(h)} := L_{nh} - L_{(n-1)h} = V_{nh} - V_{(n-1)h} + \lambda \int_{(n-1)h}^{nh} V_u \, du,$$

and use them to suggest an appropriate model for the background driving Lévy process.

In practice, however, λ must be estimated and the integral in the last equation replaced by an approximating sum. Replacing the CAR(1) coefficient by its estimator and the integral by a trapezoidal approximation, we obtain the estimated increments,

$$\Delta \hat{L}_n^{(h)} = V_n^{(h)} - V_{n-1}^{(h)} + \frac{1}{2}\hat{\lambda}_N h(V_n^{(h)} + V_{n-1}^{(h)}).$$

Brockwell et al. (2007) illustrate the effectiveness of the estimator $\hat{\lambda}$ and its use in recovering the increments of L in the case of a simulated gamma-driven CAR(1) process.

Some other estimators

Let us mention some other estimators for Lévy-driven Ornstein–Uhlenbeck processes, without attempting to give a complete survey here. Spiliopoulos (2009) gives a generalized method of moments estimator for the parameter λ in model (6.5), based on its autocorrelation function, and shows asymptotic normality of this estimator under certain conditions. This method is similar to the one presented in Section 6.4.3 for the COGARCH(1,1) estimator, and could in principle be applied also to get an estimator for λ in the volatility model of Barndorff–Nielsen and Shephard (2001b). We shall not go into further details here. Barndorff–Nielsen and Shephard (2001b) suggest further estimators for their model, among them a maximum likelihood approach when the stationary distribution is inverse Gaussian. A Markov chain methodology for Bayesian inference for this volatility model is given by Roberts, Papaspiliopoulos, and Dellaportas (2004).

6.4.2 Estimation of CARMA processes

As in the discussion of non-negative subordinator-driven Ornstein–Uhlenbeck models, Brockwell, Davis, and Yang (2011) consider the analogous questions for more general second-order Lévy-driven CARMA models, focusing in particular on the subordinator-driven CARMA(2,1) case and basing the inference on closely and uniformly spaced observations. The goal is to estimate the CARMA coefficients and to use these estimates to recover an approximation to the realization of the BDLP. A CARMA(2,1) model of this type was used by Todorov (2011) to represent the daily realized volatility of the Deutsch Mark/ U.S. dollar exchange rate.

It is assumed that observations are available of $\{Y_{nh}, n = 0, 1, 2, \ldots\}$ where

$(Y_t)_{t \geq 0}$ is the CARMA process defined by (6.13) and (6.14), $a(z)$ has distinct zeroes each with strictly negative real part and h is small.

Under these assumptions we know from Remark 6.8 that Y_t has a representation as a sum of dependent and possibly complex-valued CAR(1) processes, namely

$$Y_t = \sum_{r=1}^{p} Y_t^{(r)},$$

where

$$Y_t^{(r)} = \int_{-\infty}^{t} \alpha_r e^{\lambda_r (t-u)} dL_u \qquad (6.47)$$

and $\alpha_r = b(\lambda_r)/a'(\lambda_r)$. If L is a subordinator and the kernel $\sum_{r=1}^{p} \alpha_r e^{\lambda_r t}$ of Y is non-negative then Y itself is non-negative. A necessary and sufficient condition for the kernel of the CARMA(2,1) process to be non-negative is that the zeroes λ_1 and λ_2 of $a(z)$ are real and that $b_0 \geq \min(|\lambda_i|)$.

The sampled process $(Y_{nh})_{n \in \mathbb{Z}}$ is the sum of the sampled CAR(1) processes $Y_t^{(r)}$, $r = 1, \ldots, p$, each of which is a discrete-time AR(1) process satisfying

$$Y_{nh}^{(r)} = e^{\lambda_r h} Y_{(n-1)h} + \alpha_r \int_{(n-1)h}^{nh} e^{\lambda_r (nh-u)} dL_u.$$

Adding these equations, we see that the sampled process is a *weak* ARMA$(p, p-1)$ process satisfying

$$\phi(B)Y_{nh} = \theta(B)W_n,$$

where

$$\phi(z) := \prod_{r=1}^{p} (1 - e^{\lambda_r h} z) =: 1 - \phi_1 z - \cdots - \phi_p z^p,$$

$\theta(z)$ is a polynomial of degree $p-1$ and $(W_n)_{n \in \mathbb{Z}}$ is an uncorrelated sequence with constant mean and variance. W is not an independent sequence except in the case when L is Brownian motion; hence the use of the term *weak ARMA process* to describe $(Y_{nh})_{n \in \mathbb{Z}}$.

Under the assumption that there is no aliasing problem, i.e. that the covariance structure of the sampled process $(Y_{nh})_{n=0,1,2...}$ uniquely determines the parameters of the CARMA process $(Y_t)_{t \in \mathbb{R}}$ (this is certainly the case for the CARMA(2,1) process with real autoregressive roots λ_1 and λ_2) an almost surely consistent estimator (as $T \to \infty$ with h fixed) of the CARMA parameter vector $\beta_0 := (a_1, \ldots, a_p, b_0, \ldots, b_{q-1})'$ can be obtained by using the Kalman recursions to compute, for any specified β, the minimum mean-squared-error linear predictors $P(Y_{kh})$ of Y_{kh} in terms of $Y_0, \ldots, Y_{(k-1)h}$, $k = 1, \ldots, N = [T/h]$. (For details see Brockwell and Davis (1991, p. 477).) The least squares estimator $\hat{\beta}$ is then found by numerically minimizing the sum of squares,

$\sum_{k=1}^{N} \epsilon_k^2(\boldsymbol{\beta})$, of the one-step prediction errors,

$$\epsilon_k(\boldsymbol{\beta}) := Y_{kh} - P(Y_{kh}),$$

with respect to $\boldsymbol{\beta}$. For a discussion of the aliasing problem for processes with rational spectral density see Hansen and Sargent (1983). The use of the Kalman recursions here follows that of Jones (1981) who used them to maximize the Gaussian likelihood in order to fit a continuous-time autoregression with observational noise to irregularly-spaced observations. The almost sure consistency of $\hat{\beta}$ and, under further conditions, its asymptotic normality as $T \to \infty$ with h fixed, follows from asymptotic results of Francq and Zakoian (1998) for weak ARMA processes.

The asymptotic distribution of the maximum likelihood estimator $\hat{\boldsymbol{\beta}}_{MLE}$ for the corresponding Gaussian CARMA process observed *continuously* on the interval $[0, T]$ was found by Pham (1977) to be

$$\sqrt{T}(\hat{\boldsymbol{\beta}}_{MLE} - \boldsymbol{\beta}) \Rightarrow N(\mathbf{0}, \mathbf{M}^{-1}),$$

where $\mathbf{M}$ is the matrix with components, $M_{j,k}, \, j, k = 1, \ldots, p + q$, and

$$M_{jk} = \frac{1}{2\pi} \int_{-\infty}^{\infty} \left\{ \frac{\partial}{\partial \beta_j} \frac{a(i\omega)}{b(i\omega)} \right\} \left\{ \frac{\partial}{\partial \beta_k} \frac{a(-i\omega)}{b(-i\omega)} \right\} \left| \frac{b(i\omega)}{a(i\omega)} \right|^2 d\omega.$$

Empirically this is found to be a useful approximation when h is small, even in the non-Gaussian case.

Recovering the sample-path of L

In the idealized case of continuously observed $(Y_t)_{t \in \mathbb{R}}$, and under the additional assumption that the zeroes of the polynomial $b(z)$ all have strictly negative real parts, the $q \times 1$ vector $\mathbf{X}_t^{(q)}$, whose components are the first q components of the complete state-vector $\mathbf{X}_t$, can be found from

$$\mathbf{X}_t^{(q)} = \int_{-\infty}^{t} e^{B(t-u)} \mathbf{e}_q Y(u) du,$$

or

$$\mathbf{X}_t^{(q)} = e^{Bt} \mathbf{X}_0^{(q)} + \int_0^t e^{B(t-u)} \mathbf{e}_q Y(u) du, \tag{6.48}$$

where

$$\mathbf{B} = \begin{bmatrix} 0 & 1 & 0 & \cdots & 0 \\ 0 & 0 & 1 & \cdots & 0 \\ \vdots & \vdots & \vdots & \ddots & \vdots \\ 0 & 0 & 0 & \cdots & 1 \\ -b_0 & -b_1 & -b_2 & \cdots & -b_{q-1} \end{bmatrix} \quad \text{and} \quad \mathbf{e}_q = \begin{bmatrix} 0 \\ 0 \\ \vdots \\ 0 \\ 1 \end{bmatrix}.$$

The $(j + 1)^{\text{st}}$ component of $\mathbf{X}_t$, $j = q, \ldots, p - 1$, is obtained as the jth derivative with respect to t of the first component of $\mathbf{X}_t^{(q)}$.

If now we define the *canonical* state vector by

$$\mathbf{Y}_t = b(\Lambda)R^{-1}\mathbf{X}_t,\qquad(6.49)$$

where Λ is the diagonal matrix of zeroes, $\lambda_1, \ldots, \lambda_p$, of $a(z)$ and R is the matrix with columns equal to the corresponding right eigenvectors of A, namely $R = [\lambda_j^{i-1}]_{i,j=1}^{p}$, then the components of the canonical state vector $\mathbf{Y}_t$ are equal to the CAR(1) components $Y_t^{(r)}$ in (6.47) whose sum is Y_t.

From any one of the component CAR(1) processes the sample-path of the BDLP can be reconstructed, as in the discussion of the Ornstein–Uhlenbeck case, from

$$L(t) = \frac{a'(\lambda_r)}{b(\lambda_r)}\left[Y^{(r)}(t) - Y^{(r)}(0) - \lambda_r\int_0^t Y^{(r)}(s)ds\right], \ r = 1, \ldots, p.$$

In practice, of course, we do not observe Y continuously, the coefficients of the polynomials $a(z)$ and $b(z)$ must be estimated, and the integrals appearing in the exact solution must be replaced by approximating sums.

Example 6.28 (The CARMA(2,1) Process). In this case $q = 1$ so that $\mathbf{X}_t^{(q)}$ has only one component which we shall denote by X_t. Equation (6.48) takes the form

$$X_t = X_0 e^{-b_0 t} + \int_0^t e^{-b_0(t-u)}Y(u)du\qquad(6.50)$$

and the second component of the complete state-vector $\mathbf{X}_t$ is therefore

$$X_t' := DX_t = -b_0 X_t + Y(t).$$

Given X_0 and $(Y_t)_{0 \le t \le T}$, these equations determine $(\mathbf{X}_t)_{0 \le t \le T}$. The canonical state vector $\mathbf{Y}_t$ is then given by (6.49) and $(L_t)_{0 \le t \le T}$ can be determined from either of its two CAR(1) components as described above. In the CARMA(2,1) case a more direct calculation starting from the second component of the state-equation,

$$dL_t = dX_t' + a_1 X_t'dt + a_2 X_t dt,$$

and using the above expressions for X_t and X_t' leads to the relation,

$$dL_t = dY_t + (a_1 - b_0)Y_t dt + (b_0^2 - a_1 b_0 + a_2)X_t dt.$$

From this relation we obtain a simple approximation to the increment $L_{nh} - L_{(n-1)h}$, namely

$$\frac{h}{2}\left[(a_1 - b_0)(Y_{nh} + Y_{(n-1)h}) + (b_0^2 - a_1 b_0 + a_2)(X_{nh} + X_{(n-1)h})\right]$$

$$+ Y_{nh} - Y_{(n-1)h},$$

where the observations Y_{nh} are used to replace the integral in (6.50) by an approximating sum and hence to construct a realization of X at times $h, 2h, \ldots$. It is clear from (6.50), since $b_0 > 0$, that the effect of an assumed value for X_0 on the estimated increments becomes negligible for n sufficiently large.

Brockwell et al. (2011), following Todorov (2011), fit a CARMA(2, 1) model with real autoregressive roots to a realized volatility series of the Deutsch Mark/ U.S.Dollar exchange rate from December 1st, 1986 through June 30th, 1999. They use least squares to estimate the coefficients of $a(z)$ and $b(z)$, then choose EL_1 and $\text{Var}(L_1)$ to match the sample-mean and sample variance of the data. With these estimates EX_0 can be calculated and substituted for X_0 in (6.50). Sums based on the observations Y_{nh} are used to replace the integral in (6.50) by an approximating sum and hence to construct a realization of $\mathbf{X}$ at times $h, 2h, \ldots$ and a corresponding realization of the canonical state process $\mathbf{Y}$. The corresponding Lévy increments are then estimated as for the Ornstein–Uhlenbeck process considered earlier.

Applying this technique to the Deutsch Mark/ U.S.Dollar exchange rate series and comparing the empirical distribution function of the estimated Lévy increments with the distribution function of the Lévy increments of competing models, each with the same mean and variance as the empirical distribution, permits a comparison between the models. Of those considered by Brockwell et al. (2011) for this particular data set, the gamma process appeared to be the most appropriate.

6.4.3 Method of moment estimation for the COGARCH model

In this section we consider a COGARCH(1,1) process $(G_t)_{t \geq 0}$ as defined in Section 6.3.4 and give a generalized method of moment estimator for its parameters. We assume that we have equidistant observations $G_0, G_r, G_{2r}, \ldots, G_{nr}$ of a stationary COGARCH(1,1) process with parameters $\beta > 0$, $\lambda > 0$ and $\delta \in (0, 1)$, and for notational simplicity assume $r = 1$ and write

$$Y_t := G_t - G_{t-1}, \quad t \in \mathbb{N}.$$

We further assume that the driving Lévy process $M = (M_t)_{t \geq 0}$ satisfies

$$EM_1 = 0, \quad \text{Var}(M_1) = 1, \quad EM_1^4 < \infty, \quad \int_{\mathbb{R}} x^3 \, \nu_M(dx) = 0, \quad (6.51)$$

and that the Gaussian variance $A_M \in [0, 1)$ of M is known. The latter is satisfied e.g. if one assumes that M is a pure jump Lévy process, in which case $A_M = 0$. Without this assumption, the proposed estimator is not able to identify all parameters. The condition $\int_{\mathbb{R}} x^3 \, \nu_M(dx)$ simplifies calculations considerably, and $EM_1^4 < \infty$ is needed since we want G_t^2 to have finite variance.

We assume that the parameters β, λ, δ are unknown, but such that

$$\Psi_\xi(2) = 2\log\delta + \int_{\mathbb{R}} \left(\lambda^2\delta^{-2}y^4 + 2\lambda\delta^{-1}y^2\right) \nu_M(dy) < 0. \qquad (6.52)$$

We want to estimate the parameters β, λ and δ and define

$$\varphi := \lambda\delta^{-1} \quad \text{and} \quad \tau := -\log\delta.$$

This will be achieved using a "generalized method of moment" estimator. The first step is to express certain quantities in terms of their parameters, which is done in the next proposition:

Proposition 6.29 *Let $(G_t)_{t\geq0}$ be a COGARCH(1,1) process with stationary volatility, driven by M, and suppose that conditions (6.51) and (6.52) hold. Let $Y_t := G_t - G_{t-1}$ for $t \in \mathbb{N}$. Then $EY_1^4 < \infty$ and*

$$EY_1 = 0, \quad \mu := E(Y_1^2) = \frac{\beta}{|\Psi_\xi(1)|} \quad \text{and} \quad \text{Cov}(Y_t, Y_{t+h}) = 0$$

for every $t, h \in \mathbb{N}$. Further, with

$$p := |\Psi_\xi(1)|,$$

$$\gamma(0) := \text{Var}(Y_1^2) = EY_1^4 - (EY_1^2)^2, \quad \rho(h) := \text{Corr}(Y_t^2, Y_{t+h}^2)$$

and

$$k := \frac{\beta^2}{p^3\gamma(0)}(2\tau\varphi^{-1} + 2A_M - 1)\left(2|\Psi_\xi^{-1}(2)| - p^{-1}\right)\left(1 - e^{-p}\right)(e^p - 1)$$

we have

$$\begin{aligned}
\gamma(0) &= 6\frac{\beta^2}{p^3}(2\tau\varphi^{-1} + 2A_M - 1)(2|\Psi_\xi^{-1}(2)| - p^{-1})\left(p - 1 + e^{-p}\right) \\
&\quad + \frac{2\beta^2}{\varphi^2}\left(2|\Psi_\xi^{-1}(2)| - p^{-1}\right) + \frac{3\beta^2}{p^2} \quad \text{and} \\
\rho(h) &= ke^{-hp}, \quad h \in \mathbb{N}. \qquad (6.53)
\end{aligned}$$

In particular, the autocovariance structure of $(Y_t)_{t\in\mathbb{N}}$ is that of an ARMA(1,1) process.

The proof of this proposition follows by lengthy calculations using properties of the stochastic integral, and we refer to Haug, Klüppelberg, Lindner, and Zapp (2007) for details. The following theorem shows that the parameters β, φ and τ can be recovered from $\mu, \gamma(0), k, p$ as defined in the previous proposition. This will then lead to estimators for these parameters, by giving a generalized method of moment estimators for $\mu, \gamma(0), k$ and p.

Theorem 6.30 *Let the assumptions of Proposition 6.29 be satisfied, and let μ, $\gamma(0)$, k and p be as in Proposition 6.29. Define further*

$$Q_1 \ := \ \gamma(0) - 2\mu^2 - 6\frac{1 - p - e^{-p}}{(1 - e^p)(1 - e^{-p})} \, k\gamma(0) \quad and \quad (6.54)$$

$$Q_2 \ := \ \frac{2k\gamma(0)p}{Q_1(e^p - 1)(1 - e^{-p})}. \tag{6.55}$$

Then $Q_1, Q_2 > 0$, and the parameters β, φ and τ are uniquely determined by $\mu, \gamma(0), k$ and p and are given by the formulas

$$\beta \ := \ p\mu, \tag{6.56}$$

$$\varphi \ := \ p\sqrt{1 + Q_2} - p,, \tag{6.57}$$

$$\tau \ := \ p\sqrt{1 + Q_2}(1 - A_M) + pA_M = p + \varphi(1 - A_M). \tag{6.58}$$

Proof. This can be shown by solving the equations for $\mu, \gamma(0), k$ and p obtained in Proposition 6.29 for the parameters β, φ and τ. Details can be found in Theorem 3.1 of Haug et al. (2007). $\square$

It remains to estimate $\mu, \gamma(0), k$ and p. So, given observations $G_0, G_1, \ldots, G_n$, and hence $Y_1, \ldots, Y_n$, define estimators for μ and the autocovariance function $\gamma(h) = \text{Cov}(Y_t^2, Y_{t+h}^2)$ by

$$\widehat{\mu} := \frac{1}{n}\sum_{t=1}^{n} Y_i^2, \quad \widehat{\gamma}_n(h) := \frac{1}{n}\sum_{i=1}^{n-h}(Y_{i+h}^2 - \widehat{\mu})(Y_i^2 - \widehat{\mu}), \quad h = 0, 1, \ldots, n-h.$$

The empirical autocorrelation function is then given by $\widehat{\rho}(h) = \widehat{\gamma}(h)/\widehat{\gamma}(0)$, $h \geq 1$. Taking logarithms in (6.53) gives $\log \rho(h) = \log k - ph$, which suggests estimating k and p by a regression. For fixed $h_{\max} \geq 2$, this leads to the following estimators $\widehat{p}$ and $\widehat{k}$:

$$\widehat{p} \ := \ -\frac{\sum_{h=1}^{h_{\max}}(\log \widehat{\rho}(h) - \overline{\log \widehat{\rho}})(h - \frac{h_{\max}+1}{2})}{\sum_{h=1}^{h_{\max}}(h - \frac{h_{\max}+1}{2})^2}, \tag{6.59}$$

$$\widehat{k} \ := \ \exp\left\{\overline{\log \widehat{\rho}} + \frac{h_{\max} + 1}{2}\widehat{p}\right) \tag{6.60}$$

with $\overline{\log \widehat{\rho}} := h_{\max}^{-1} \sum_{h=1}^{h_{\max}} \log \widehat{\rho}(h)$. Finally, the *generalized method of moment estimators* $\widehat{\beta}, \widehat{\varphi}$ and $\widehat{\tau}$ for β, φ and τ are obtained by replacing $\mu, \widehat{\gamma}(0), p$ and k in formulas (6.54) – (6.58) by $\widehat{\mu}, \widehat{\gamma}(0), \widehat{p}$ and $\widehat{k}$, provided that the estimators $\widehat{Q}_1, \widehat{Q}_2$ and $\widehat{p}$ are strictly positive. If one of them fails to be strictly positive, let $(\widehat{\beta}, \widehat{\varphi}, \widehat{\tau}) := (0, 0, 0)$. The latter may happen, since although $p, Q_1, Q_2 > 0$, this does not necessarily imply the same for the corresponding estimators. However, since $\widehat{p}, \widehat{Q}_1$ and $\widehat{Q}_2$ converge to the corresponding properties as $n \to \infty$, giving in particular consistency, this exceptional case will not happen for large sample sizes. The asymptotic properties of the estimator are described in more detail in the following theorem of Haug et al. (2007):

Theorem 6.31 *Let $h_{\max} \in \mathbb{N}\backslash\{1\}$ be fixed and consider observations $Y_1, \ldots, Y_n$. Suppose that the conditions of Proposition 6.29 are satisfied. Then the generalized method of moments estimator $(\widehat{\beta}, \widehat{\varphi}, \widehat{\tau})$ as described above is strongly consistent, i.e.*

$$\lim_{n\to\infty} (\widehat{\beta}, \widehat{\varphi}, \widehat{\tau}) = (\beta, \varphi, \tau) \quad \text{a.s.}$$

Moreover if there is $\varepsilon > 0$ such that $E|M_1|^{8+\varepsilon} < \infty$ and $\Psi_\xi(4 + \varepsilon/2) < 0$, then $(\widehat{\beta}, \widehat{\varphi}, \widehat{\tau})$ is asymptotically normal, i.e. there is a 3×3-covariance matrix Σ such that

$$\sqrt{n}(\widehat{\beta} - \beta, \widehat{\varphi} - \varphi, \widehat{\tau} - \tau)' \xrightarrow{d} N_3(\mathbf{0}, \Sigma), \quad n \to \infty,$$

where $N_3(\mathbf{0}, \Sigma)$ denotes a three-dimensional centered normal distribution with covariance matrix Σ.

It is lengthy to give a semi-explicit formula for Σ. It involves infinite sums of covariances of the type $\text{Cov}(Y_1^2 Y_{1+k}^2, Y_{1+j}^2 Y_{1+l}^2)$ and $\text{Cov}(Y_1^2, Y_{1+j}^2 Y_{1+l}^2)$. See Haug et al. (2007) for details.

Outline of Proof. Under the given conditions, $(\sigma_t^2)_{t\geq0}$ is α-mixing with geometrically decreasing mixing constants, i.e. there are constants $C, a > 0$ such that

$$\sup_{A\in\mathcal{F}_u^-, B\in\mathcal{F}_{u+t}^+} |P(A \cap B) - P(A)P(B)| \leq Ce^{-at} \quad \forall\, u, t \geq 0,$$

where $\mathcal{F}_u^-$ denotes the σ-algebra generated by $(\sigma_s^2)_{0\leq s\leq u}$ and $\mathcal{F}_{u+t}^+$ the σ-algebra generated by $(\sigma_s^2)_{s\geq u+t}$. This is proved in Proposition 3.4 of Fasen (2010), who extends results of Masuda (2004) for Lévy-driven Ornstein–Uhlenbeck processes. In Theorem 3.5 of Haug et al. (2007) it is then shown that this implies α-mixing of $(Y_n)_{n\in\mathbb{N}}$ with geometrically decreasing mixing rate. Strong consistency of the estimators then follows from Birkhoff's ergodic theorem, and the asymptotic normality by a standard central limit theorem (e.g. Ibragimov and Linnik (1971, Theorem 18.5.3)) for α-mixing sequences with appropriate moment conditions. For details we refer to Haug et al. (2007). To see that the moment condition $E|G_1|^{8+\varepsilon}$ used there is satisfied, observe that $E|M_1|^{8+\varepsilon}$ and $\Psi_\xi(4 + \varepsilon/2) < 0$ imply $E\sigma_t^{8+\varepsilon} < \infty$ and hence $E|G_1|^{8+\varepsilon} = E|\int_0^t \sigma_{s-}\, dM_s|^{8+\varepsilon} < \infty$ since M is a Lévy process and (σ_{t-}) previsible with the corresponding moment condition. Similar moment calculations are given in Behme (2011). $\square$

The generalized method of moment estimator clearly suffers from the fact that finiteness of the fourth or $(8 + \delta)$-moment are assumed, which is hardly met in practice for financial data. Also, moment estimators are often not efficient and equally spaced observations are assumed. Hence it would be desirable to investigate other estimators and their properties. Two such estimation methods were suggested by Maller, Müller, and Szimayer (2008) and Müller (2010), respectively. Maller et al. (2008) assume observations $G_{t_1}, G_{t_2}, \ldots, G_{t_n}$, not necessarily equidistantly spaced, and develop a pseudo-maximum likelihood estimator based on these observations. They further show that every COGA-RCH(1,1) process is a limit in probability of GARCH(1,1) processes and that the pseudo-log-likelihood function is like the pseudo-log-likelihood function for GARCH(1,1) processes. The estimator is shown to perform well in a simulation study and used to fit a COGARCH(1,1) model to the ASX200 index of the Australian Stock exchange. Müller (2010) provides a Markov Chain Monte Carlo estimation procedure for the COGARCH(1,1) model, when the driving Lévy process is a compound Poisson process. We refer to Müller (2010) for further details.

Finally, let us take up again the statistical non-equivalence of Nelson's diffusion limit and the discrete time GARCH processes, as mentioned in Section 6.3.4. Since, unlike Nelson's diffusion limit, the COGARCH(1,1) process is driven by only one source of randomness, there seems hope that statistical inference for the COGARCH(1,1) process and GARCH(1,1) processes is statistically equivalent. However, as recently shown by Buchmann and Müller (2012), this is not the case in general. We refer to Buchmann and Müller (2012) for further details.

Acknowledgments

We would like to thank Anita Behme and Jens-Peter Kreiß for careful reading of the paper and many valuable comments. Support of Peter Brockwell's work by National Science Foundation Grant DMS-0744058 is gratefully acknowledged, as is support from an NTH-grant of the state of Lower Saxony.

References

Alsmeyer, G., Iksanov, A., & Rösler, U. (2009). On distributional properties of perpetuities. *J. Theoret. Probab.*, *22*, 666–682.

Applebaum, D. (2004). *Lévy Processes and Stochastic Calculus*. Cambridge: Cambridge University Press.

Barndorff–Nielsen, O. E., & Shephard, N. (2001a). Modelling by Lévy processes for financial econometrics. In O. E. Barndorff–Nielsen, T. Mikosch, & S. Resnick (Eds.), *Lévy Processes* (pp. 283–318). Boston: Birkhäuser.

Barndorff–Nielsen, O. E., & Shephard, N. (2001b). Non-Gaussian Ornstein-Uhlenbeck-based models and some of their uses in financial economics. *J. R. Stat. Soc. Ser. B Stat. Methodol.*, *63*, 167–241.

Behme, A. (2011). Distributional properties of stationary solutions of $dV_t = V_{t-}dU_t + dL_t$ with Lévy noise. *Adv. Appl. Probab.*, *43*, 688–711.

Behme, A., Lindner, A., & Maller, R. (2011). Stationary solutions of the stochastic differential equation $dV_t = V_{t-}dU_t + dL_t$ with Lévy noise. *Stochastic Process. Appl.*, *121*, 91–108.

Bertoin, J. (1996). *Lévy Processes*. Cambridge: Cambridge University Press.

Bingham, N. H., Goldie, C. M., & Teugels, J. L. (1989). *Regular Variation*. Cambridge: Cambridge University Press.

Bollerslev, T. (1986). Generalized autoregressive conditional heteroskedasticity. *Journal of Econometrics*, *31*, 307–327.

Brockwell, P. J. (2001). Lévy-driven CARMA processes. *Ann. Inst. Stat. Math.*, *53*, 113–124.

Brockwell, P. J., Chadraa, E., & Lindner, A. (2006). Continuous-time GARCH processes. *Ann. Appl. Probab.*, *16*, 790–826.

Brockwell, P. J., & Davis, R. A. (1991). *Time Series: Theory and Methods* (Second ed.). New York: Springer.

Brockwell, P. J., & Davis, R. A. (2001). Discussion of "Non-Gaussian Ornstein–Uhlenbeck based models and some of their uses in financial economics," by O.E. Barndorff–Nielsen and N. Shephard. *J. Royal Statist. Soc. Ser. B Stat. Methodol.*, *63*, 218–219.

Brockwell, P. J., Davis, R. A., & Yang, Y. (2007). Estimation for non-negative Lévy-driven Ornstein–Uhlenbeck processes. *J. Appl. Probab.*, *44*, 977–989.

Brockwell, P. J., Davis, R. A., & Yang, Y. (2011). Estimation for non-negative Lévy-driven CARMA processes. *J. Bus. Econom. Statist.*, *29*, 250–259.

Brockwell, P. J., & Lindner, A. (2009). Existence and uniqueness of stationary Lévy-driven CARMA processes. *Stoch. Process. Appl.*, *119*, 2660–2681.

Brockwell, P. J., Resnick, S. I., & Tweedie, R. L. (1982). Storage processes with general release rule and additive inputs. *Adv. Appl. Prob.*, *14*, 392–433.

Buchmann, B., & Müller, G. (2012). Limit experiments of GARCH. *Bernoulli*, *18*, 64–99.

Carmona, P., Petit, F., & Yor, M. (1997). On the distribution and asymptotic results for exponential functionals of Lévy processes. In M. Yor (Ed.), *Exponential Functionals and Principal Values Related to Brownian Motion* (pp. 73–130). Rev. Mat. Iberoamericana, Madrid.

Carmona, P., Petit, F., & Yor, M. (2001). Exponential functionals of Lévy processes. In O. E. Barndorff–Nielsen, T. Mikosch, & S. Resnick (Eds.), *Lévy Processes* (pp. 41–55). Boston: Birkhäuser.

Çinlar, E., & Pinsky, M. (1972). On dams with general input and a general release rule. *J. Appl. Prob.*, *9*, 422–429.

Chung, K. L. (2001). *A Course in Probability Theory* (Third ed.). San Diego: Academic Press.

Davis, R., & McCormick, W. (1989). Estimation for first-order autoregressive processes with positive and bounded innovations. *Stoch. Process. Appl.*, *31*, 237–250.

de Haan, L., & Karandikar, R. L. (1989). Embedding a stochastic difference equation in a continuous-time process. *Stoch. Process. Appl.*, *32*, 225–235.

Doney, R., & Maller, R. A. (2002). Stability and attraction to normality for Lévy processes at zero and at infinity. *J. Theoret. Probab.*, *15*, 751–792.

Einstein, A. (1905). Über die von der molekularkinetischen Theorie der Wärme geforderte Bewegung von in ruhenden Flüssigkeiten suspendierten Teilchen. *Annalen der Physik*, *322*, 549–560.

Engle, R. F. (1982). Autoregressive conditional heteroscedasticity with estimates of the variance of united kingdom inflation. *Econometrica*, *50*, 987–1008.

Erickson, K. B., & Maller, R. A. (2005). Generalised Ornstein-Uhlenbeck processes and the convergence of Lévy integrals. In M. Émery, M. Ledoux, & M. Yor (Eds.), *Séminaire de Probabilités XXXVIII. Lecture Notes in Math.* (Vol. 1857, pp. 70–94). Berlin: Springer.

Fasen, V. (2010). Asymptotic results for sample autocovariance functions and extremes of generalized Orstein–Uhlenbeck processes. *Bernoulli, 16,* 51–79.

Fasen, V., Klüppelberg, C., & Lindner, A. (2006). Extremal behavior of stochastic volatility models. In A. Shiryaev, M. D. R. Grossinho, P. Oliviera, & M. Esquivel (Eds.), *Stochastic Finance* (pp. 107–155). New York: Springer.

Francq, C., & Zakoian, J.-M. (1998). Estimating linear representations of non-linear processes. *J. Statistical Planning and Inference, 68,* 145–165.

Goldie, C. (1991). Implicit renewal theory and tails of solutions of random equations. *Ann. Appl. Probab., 1,* 126–166.

Hansen, L. P., & Sargent, T. P. (1983). The dimensionality of the aliasing problem in models with rational spectral densities. *Econometrics, 51,* 377–387.

Haug, S., Klüppelberg, C., Lindner, A., & Zapp, M. (2007). Method of moment estimation in the COGARCH(1,1) model. *The Econometrics Journal, 10,* 320–341.

Ibragimov, I. A., & Linnik, Y. V. (1971). *Independent and Stationary Sequences of Random Variables.* Wolters-Noordhoff Publishing, Groningen.

Jacod, J. (1985). Grossissement de filtration et processus d'Ornstein–Uhlenbeck généralisés. In T. Jeulin & M. Yor (Eds.), *Séminaire de Calcul Stochastique, Paris 1982/83. Lecture Notes in Math.* (Vol. 1118, pp. 36–44). Berlin: Springer.

Jaschke, S. (2003). A note on the inhomogeneous linear stochastic differential equation. *Insurance Math. Econom., 32,* 461–464.

Jones, R. H. (1981). Fitting a continuous-time autoregression to discrete data. In D. F. Findley (Ed.), *Applied Time Series Analysis II* (pp. 651–682). Academic Press, New York.

Jongbloed, G., van der Meulen, F. H., & van der Vaart, A. W. (2005). Nonparametric inference for Lévy-driven Ornstein–Uhlenbeck processes. *Bernoulli, 11,* 759–791.

Kesten, H. (1973). Random difference equations and renewal theory for products of random matrices. *Acta Math., 131,* 207–228.

Klüppelberg, C., Lindner, A., & Maller, R. (2004). Stationarity and second order behaviour of discrete and continuous-time GARCH(1,1) processes. *J. Appl. Probab., 41,* 601–622.

Klüppelberg, C., Lindner, A., & Maller, R. (2006). Continuous time volatility modelling: COGARCH versus Ornstein–Uhlenbeck models. In Y. Kabanov, R. Liptser, & J. Stoyanov (Eds.), *The Shiryaev Festschrift: From Stochastic Calculus to Mathematical Finance* (pp. 393–419). Berlin: Springer.

Kyprianou, A. E. (2006). *Introductory Lectures on Fluctuations of Lévy Processes with Applications*. Berlin: Springer.

Le Cam, L. (1986). *Asymptotic Methods in Statistical Decision Theory*. New York: Springer.

Lindner, A., & Maller, R. (2005). Lévy integrals and the stationarity of generalised Ornstein–Uhlenbeck processes. *Stoch. Process. Appl.*, *115*, 1701–1722.

Lindner, A., & Sato, K. (2009). Continuity properties and infinite divisibility of stationary solutions of some generalized Ornstein–Uhlenbeck processes. *Ann. Probab.*, *37*, 250–274.

Maller, R. A., Müller, G., & Szimayer, A. (2008). GARCH modelling in continuous-time for irregularly spaced time series data. *Bernoulli*, *14*, 519–542.

Masuda, H. (2004). On multidimensional Ornstein–Uhlenbeck processes driven by a general Lévy process. *Bernoulli*, *10*, 97–120.

Müller, G. (2010). MCMC estimation of the COGARCH(1,1) model. *Journal of Financial Econometrics*, *8*, 481–510.

Nelson, D. B. (1990). ARCH models as diffusion approximations. *Journal of Econometrics*, *45*, 7–38.

Paulsen, J. (1993). Risk theory in a stochastic economic environment. *Stoch. Process. Appl.*, *46*, 327–361.

Pham, D.-T. (1977). Estimation of parameters of a continuous-time Gaussian stationary process with rational spectral density. *Biometrika*, *64*, 385–399.

Protter, P. E. (2005). *Stochastic Integration and Differential Equations* (Second ed.). Berlin: Springer. (Version 2.1)

Roberts, G. O., Papaspiliopoulos, O., & Dellaportas, P. (2004). Markov chain Monte Carlo methodology for Bayesian inference for non-Gaussian Ornstein–Uhlenbeck stochastic volatility processes. *J. Royal Statist. Soc. Ser. B Stat. Methodol.*, *66*, 369–393.

Samorodnitsky, G. (2004). *Private communication*.

Sato, K. (1999). *Lévy Processes and Infinitely Divisible Distributions*. Cambridge: Cambridge University Press.

Sato, K., & Yamazato, M. (1983). Stationary processes of Ornstein-Uhlenbeck type. In *Probability Theory and Mathematical Statistics. Proceedings of the Fourth USSR-Japan-Symposium, held at Tbilisi, USSR, August 23–29, 1982. Lecture Notes in Math.* (Vol. 1021, pp. 541–551). Berlin: Springer.

Sato, K., & Yamazato, M. (1984). Operator-selfdecomposable distributions as limit distributions of processes of Ornstein–Uhlenbeck type. *Stoch. Process. Appl.*, *17*, 73–100.

Spiliopoulos, K. (2009). Method of moments estimation for Ornstein–Uhlenbeck processes driven by general Lévy process. *Annales de l'I.S.U.P.*, *53*, 3–19.

Todorov, V. (2011) Econometric analysis of jump-driven stochastic volatility models. *J. Econometrics*, *160*, 12–21.

Tsai, H., & Chan, K. S. (2005). A note on non-negative continuous-time processes. *J. Royal Statist. Soc. Ser. B Stat. Methodol.*, *67*, 589–597.

Uhlenbeck, G. E., & Ornstein, L. S. (1930). On the theory of the Brownian motion. *Physical Review*, *36*, 823–841.

Wang, Y. (2002). Asymptotic nonequivalence of GARCH models and diffusions. *Ann. Statist.*, *30*, 754–783.

Watanabe, T. (2009). *Private communication.*

Wolfe, S. J. (1982). On a continuous analogue of the stochastic difference equation $x_n = \rho x_{n-1} + b_n$. *Stoch. Process. Appl.*, *12*, 301–312.

Yoeurp, C., & Yor, M. (1977). *Espace orthogonal à une semimartingale: applications.* (Unpublished)

CHAPTER 7

Parameter estimation for multiscale diffusions: An overview

Grigorios A. Pavliotis, Yvo Pokern and Andrew M. Stuart

Department of Mathematics, Imperial College London
South Kensington Campus, London SW7 2AZ, United Kingdom
and
Department of Statistical Science, University College London
Gower Street, London WC1E 6BT, United Kingdom
and
Mathematics Department, University of Warwick
Coventry CV4 7AL , United Kingdom

7.1 Introduction

There are many applications where it is desirable to fit reduced stochastic descriptions (e.g. SDEs) to data. These include molecular dynamics (Schlick (2000), Frenkel and Smit (2002)), atmosphere/ocean science (Majda and Kramer (1999)), cellular biology (Alberts et al. (2002)) and econometrics (Dacorogna, Gençay, Müller, Olsen, and Pictet (2001)). The data arising in these problems often has a multiscale character and may not be compatible with the desired diffusion at small scales (see Givon, Kupferman, and Stuart (2004), Majda, Timofeyev, and Vanden-Eijnden (1999), Kepler and Elston (2001), Zhang, Mykland, and Aït-Sahalia (2005) and Olhede, Sykulski, and Pavliotis (2009)). The question then arises as to how to optimally employ such data to find a useful diffusion approximation.

The types of data available and the pertinent scientific questions depend on the particular field of application. While this chapter is about multiscale phenomena *common* to the above fields of application, we detail the type of data available and the pertinent scientific questions for the example of molecular dynamics.

Molecular dynamics data usually arises from large scale simulation of a high-dimensional Hamiltonian dynamical system, which stems from an approximation of molecular processes by, essentially, classical mechanics. The simulations can be deterministic or stochastic in nature, and applied interest usually focuses on some chemically interesting coordinates, the reaction coordinates, which are of much lower dimension than the simulated system. Such data typically evolves on a large range of timescales from fast and small vibrations of the distance between neighbouring atoms joined by a chemical bond (so-called bond-length vibrations) with characteristic timescale $t \approx 10^{-13}$s to large-scale conformational changes, like the folding of a protein molecule on timescales of at least $t \approx 10^{-6}$s. This creates an extremely challenging computational problem. See Schlick (2000) for an accessible overview of this application area.

Molecular dynamics data can be available at inter-observation times as low as $t \approx 10^{-15}$s but because the data may itself be deterministic it is clear that successfully fitting a stochastic model at those timescales is unlikely. At slightly larger timescales, fits to SDEs are routinely attempted and fitting the special class of hypoelliptic SDEs can be advantageous, as it allows for some smoothness (i.e. the paths being of greater regularity than that e.g of Brownian motion) of the input path as well as imposing physically meaningful structures, like that of a damped-driven Hamiltonian system.

Furthermore, as the diffusivity is most affected by information from small timescales, it is interesting to note that in non-parametric drift estimation, local time (or, more generally, the empirical measure) can be an almost sufficient statistic, so that time-ordering of the data is not relevant and hence, drift estimation performed in this way will be less affected by inconsistencies at small timescales.

It may also be advantageous to model the separation in timescales between e.g. bond-length vibrations and large scale conformational changes explicitly by a system of SDEs operating at different timescales. In the limit of infinite separation of these timescales, effective SDEs for the slow process can be derived through the mathematical techniques of averaging and homogenization for diffusions.

If the fitted SDEs are of convenient type, it is then possible to glean information of applied interest, concerning e.g. effective energy barriers to a conformational transition, relative weights of transition paths, number and importance of metastable states, etc.

We illustrate the issues arising from multiscale data, first through studying some illustrative examples in Section 7.2, including a toy-example from molecular dynamics, and then more generally in the context of averaging and homogenization for diffusions in Section 7.3. In Section 7.4 we show how subsampling may be used to remove some of the problems arising from multiscale data. Sections 7.5 and 7.6 treat the use of hypoelliptic diffusions and ideas stemming from non-parametric drift estimation, respectively.

The material in this overview is based, to a large extent, on the papers of Papaspiliopoulos, Pokern, Roberts, and Stuart (2009), Papavasiliou, Pavliotis, and Stuart (2009), Pavliotis and Stuart (2007), Pokern, Stuart, and Vanden-Eijnden (2009), and Pokern, Stuart, and Wiberg (2009). We have placed the material in a common framework, aiming to highlight the interconnections in this work. The details, however, are in the original papers, including the proofs where we do not provide them.

7.2 Illustrative examples

In this section we start with four examples to illustrate the primary issue arising in this context. To understand these examples it is necessary to understand the concept of the quadratic variation process for a diffusion. Consider the stochastic differential equation (SDE)

$$\frac{dz}{dt} = h(z) + \gamma(z)\frac{dW}{dt}, \quad z(0) = z_0. \tag{7.1}$$

Here $z(t) \in \mathcal{Z}$ with $\mathcal{Z} = \mathbb{R}^d$ or $\mathbb{T}^d$, the d-dimensional torus*. We assume that h, γ are Lipschitz on $\mathcal{Z}$. To make the notion of solution precise we let $\mathcal{F}_t$ denote the filtration generated by $\{W(s)\}_{0 \leq s \leq t}$ and define $z(t)$ to be the unique $\mathcal{F}_t$-adapted process which is a semimartingale defined via the integral equation

$$z(t) = z_0 + \int_0^t h(z(s))ds + \int_0^t \gamma(z(s))dW(s). \tag{7.2}$$

The stochastic integral is interpreted in the Itô sense. It is an $\mathcal{F}_t$−martingale and we write

$$m(t) = \int_0^t \gamma(z(s))dW(s). \tag{7.3}$$

A matrix valued process $Q(t)$ is increasing if $Q(t) - Q(s)$ is non-negative for all $t \geq s \geq 0$. The quadratic variation process of z, namely $\langle z \rangle_t := Q(t)$ is defined as the unique adapted, increasing and continuous process for which

$$m(t)m(t)^T - Q(t)$$

is an $\mathcal{F}_t$−martingale, see Da Prato and Zabczyk (1992) for a definition. The quadratic variation is non-zero precisely because of the lack of regularity of sample paths of diffusion processes. It is given by the expression

$$Q(t) = \int_0^t \gamma(z(s))\gamma(z(s))^T ds.$$

---

* See Appendix 2 for a definition.

It is possible to give a corresponding differential statement, namely that

$$\lim_{\tau \to 0} \sum_{i=0}^{n-1} \left(z(t_{i+1}) - z(t_i) \right) \left(z(t_{i+1}) - z(t_i) \right)^T = Q(T) \quad \text{in prob.} \qquad (7.4)$$

where the ordered times $t_0 = 0 < t_1 < \ldots < t_{n-1} < t_n = T$ on $[0, T]$ are such that their largest distance $\tau = \max_{i=1,\ldots n} (t_i - t_{i-1})$ decreases like $\mathcal{O}(1/n)$. An easily accessible treatment for one-dimensional continuous local martingales including this result (Theorems 2.3.1 and 2.3.8) is given in Durrett (1996). The issue of the required decay of τ is treated more carefully in Marcus and Rosen (2006).

If γ is constant then

$$Q(t) = t \gamma \gamma^T.$$

Notice that, for ordinary differential equations (ODEs) where $\gamma \equiv 0$ the quadratic variation is zero. The definition we have given here may be generalized from solutions of SDEs to Itô processes where the drift depends upon the past history of $z(t)$. In this fashion, it is possible to talk about the quadratic variation associated with a single component of a system of SDEs in several dimensions.

7.2.1 Example 1. SDE from ODE

This example is taken from Melbourne and Stuart (2011).

Consider the scale-separated system of ODEs

$$
\begin{aligned}
\frac{dx}{dt} &= \frac{1}{\epsilon} f_0(y) + f_1(x, y), \quad x(0) = \xi, \\
\frac{dy}{dt} &= \frac{1}{\epsilon^2} g_0(y), \quad y(0) = \eta,
\end{aligned}
$$

where $x \in \mathbb{R}^d$. We make some technical assumptions on y (detailed in Melbourne and Stuart (2011)) which imply that it is mixing with invariant measure μ. We assume that

$$\mathbb{E}^\mu f_0(y) = 0.$$

This ensures that the first term on the right-hand side of the equation for x gives rise to a well-defined limit as $\epsilon \to 0$. In fact this term will be responsible for the creation of white noise in the limiting equation for x. The technical assumptions imply that

$$\frac{1}{\epsilon} \int_0^t f_0(y(s)) ds \Rightarrow \sqrt{2\Sigma} W(t)$$

for some covariance matrix $\Sigma \in \mathbb{R}^{d \times d}$, and W a standard d-dimensional Brownian motion. Here, $\Rightarrow$ denotes weak convergence in $C([0, T], \mathbb{R}^d)$ and we use uppercase letters for effective quantities (like diffusivity) arising from averaging or homogenization here and throughout the remainder of the chapter.

Intuitively, this weak convergence follows because y has correlation decay with timescale $1/\epsilon^2$ so that $\frac{1}{\epsilon} f_0(y)$ has an autocorrelation function which approximates a Dirac delta distribution.

Now define

$$F(x) = \mathbb{E}^\mu f_1(x, y),$$

which will become the mean drift in the limiting equation for x.

Theorem 7.1 *(Melbourne and Stuart (2011)) Let $\eta \sim \mu$. Then, under some technical conditions on the fast process y, $x \Rightarrow X$ in $C([0, T], \mathbb{R}^d)$ as $\epsilon \to 0$, where*

$$\frac{dX}{dt} = F(X) + \sqrt{2\Sigma}\frac{dW}{dt}, \quad X(0) = \xi.$$

The important point that we wish to illustrate with this example is that the limit of X, and x itself, have vastly different properties at small scales. In particular, the quadratic variations differ:

$$\langle x \rangle_t = 0; \quad \langle X \rangle_t = 2\Sigma t.$$

Any parameter estimation procedure which attempts to fit an SDE in X to data generated by the ODE in x will have to confront this issue. Specifically, any parameter estimation procedure which sees small scales in the data will have the potential to *incorrectly* identify an appropriate SDE fit to the data.

7.2.2 Example 2. Smoluchowski from Langevin

The situation arising in the previous example can also occur when considering scale-separated SDEs. We illustrate this with a physically interesting example taken from Papavasiliou et al. (2009): consider the *Langevin equation* for $x \in \mathbb{R}^d$:

$$\epsilon^2 \frac{d^2 x}{dt^2} + \frac{dx}{dt} + \nabla V(x) = \sqrt{2\sigma}\frac{dW}{dt}.$$

As a first order system this is

$$\frac{dx}{dt} = \frac{1}{\epsilon} y,$$

$$\frac{dy}{dt} = -\frac{1}{\epsilon}\nabla V(x) - \frac{1}{\epsilon^2} y + \sqrt{\frac{2\sigma}{\epsilon^2}}\frac{dW}{dt}.$$

Using the method of homogenization the following may be proved:

Theorem 7.2 *(Pavliotis and Stuart (2008)) As $\epsilon \to 0$, $x \Rightarrow X$ in $C([0, T], \mathbb{R}^d)$ where X is the solution of the Smoluchowski equation* [†]

$$\frac{dX}{dt} = -\nabla V(X) + \sqrt{2\Sigma}\,\frac{dW}{dt}.$$

Thus, similarly to the previous example,

$$\langle x \rangle_t = 0; \quad \langle X \rangle_t = 2\Sigma t.$$

Again, any parameter estimation procedure for the SDE in X, and which sees small scales in the data x, will have the potential to incorrectly identify an appropriate homogenized SDE fit to the data.

7.2.3 Example 3. Butane

The two previous examples both possessed a known effective equation which may be untypical of many practical applications. In this example, an effective equation is not known; neither is the range of timescales at which such an equation would be approximately valid. This can, however, be assessed empirically to some extent, which is typical of practical applications of multiscale diffusions. The data presented in this example is taken from Pokern (2006). We consider a classical molecular dynamics model for butane. The model comprises the positions x and momenta of four (extended) atoms interacting with one another through various two, three and four body interactions, such as bond angle, bond stretch and dihedral angle interactions all combined in a potential $V(x)$. The equations of motion are

$$M\frac{d^2 x}{dt^2} + \gamma M\frac{dx}{dt} + \nabla V(x) = \sqrt{2\gamma k_B T M}\frac{dB}{dt}. \tag{7.5}$$

The choice $\gamma = 0$ gives deterministic dynamics, whilst for $\gamma > 0$ stochastic dynamics are obtained. A typical configuration of the molecule is shown in Figure 7.1. The *dihedral angle* ϕ is the angle formed by intersecting the planes passing through the first three atoms and through the last three atoms respectively. The molecule undergoes conformational changes which can be seen in changes in the dihedral angle. This is shown in Figure 7.2 which exhibits a time series for the dihedral angle, as well as its histogram, for $\gamma > 0$. Clearly the dihedral angle has three preferred values, corresponding to three different molecular conformations. Furthermore, the transitions between these states is reminiscent of thermally activated motion in a three well potential. This fact concerning the time series remains true even when $\gamma = 0$.

[†] In molecular dynamics, this equation would be termed *Brownian Dynamics*, whereas in the statistical literature it is sometimes called the *Langevin equation*.

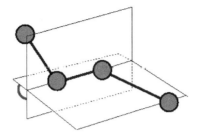

Figure 7.1 *The butane molecule*

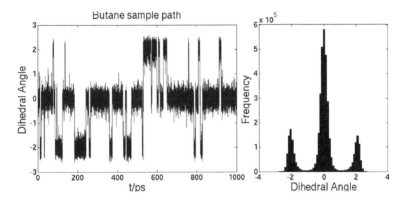

Figure 7.2 *The dihedral angle (time series and histogram)*

It is hence natural to try and fit an SDE to the dihedral angle, and we consider an SDE of the form

$$\frac{d\Phi}{dt} = -\Psi'(\Phi) + \sigma\frac{dW}{dt}, \tag{7.6a}$$

$$\Psi(\phi) = \sum_{j=1}^{5} \theta_j \Big(\cos(\phi)\Big)^j, \tag{7.6b}$$

where $\{\theta_j\}_{j=1}^{5}$ and σ are the parameters to be estimated. However, the fit to such an SDE is highly sensitive to the rate at which the data is sampled, as shown in Figure 7.3. Here the diffusion coefficient is estimated exploiting (7.4), and the maximum likelihood principle is used to estimate θ.

The behaviour of the estimator at different scales is caused by the fact that the data is again incompatible with the diffusion approximation at small scales. The true dihedral angle ϕ has zero quadratic variation because it is a function of the positions of x only, and not the momenta $M\dot{x}$; only the momenta are directly forced by noise. Thus $\langle\phi\rangle_t = 0$. In contrast, for the model (7.6), we have

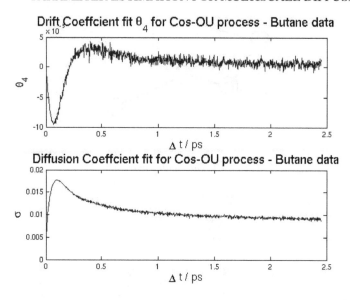

Figure 7.3 *Parameter fits to θ_4 and σ from (7.6) given data from Figure 7.2.*

$\langle \Phi \rangle_t = \sigma^2 t$. Notice that the estimated σ in Figure 7.3 tends to zero as the sampling rate increases, reflecting the smoothness of the data at small scales. When the invariant measure is to be preserved, a direct link between the maximum likelihood estimator for the drift to the maximum likelihood estimator for the diffusion arises for this diffusion, see Pavliotis and Stuart (2007). Therefore, the drift estimator is inconsistent between different scales of the data, too.

The situation is even more complex in the case where data is taken from (7.5) with $\gamma = 0$, a Hamiltonian ODE – see Figure 7.4 for a typical time series and Figure 7.5 for estimated diffusivities. Again the data is inconsistent at small scales, leading to estimates of drift and diffusion which tend to zero. But at intermediate scales oscillations caused by bond length stretches between atoms cause inflated, large diffusion coefficient estimates – a resonance effect.

7.2.4 *Example 4. Thermal motion in a multiscale potential*

All of the previous examples have the property that the quadratic variation of the data at finest scales is zero, and this is incompatible with the assumed diffusion. Here we present an example (taken from Pavliotis and Stuart (2007)) where, at small scales the quadratic variation of the data is *larger* than that of the desired model to be fitted to the data.

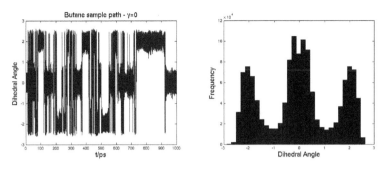

Figure 7.4 *The dihedral angle time series for* $\gamma = 0$

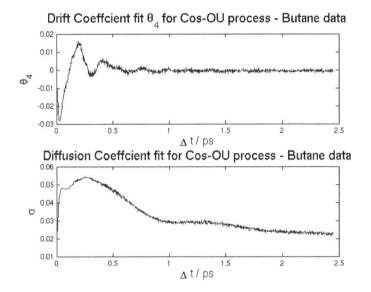

Figure 7.5 *Parameter fits to* σ *from* (7.6) *given data from* (7.5) *with* $\gamma = 0$

Consider the equation

$$\frac{dx}{dt} = -K^\epsilon(x)\nabla V^\epsilon(x) + \sqrt{2K^\epsilon(x)}\,\frac{dW}{dt}, \qquad (7.7)$$

where $x \in \mathbb{R}^d$. Here, for K, p $1-$periodic functions we define

$$K^\epsilon(x) = K(x/\epsilon),$$
$$V^\epsilon(x) = V(x) + p(x/\epsilon).$$

We also assume that K is symmetric positive-definite, so that its square-root

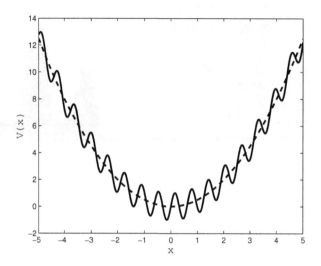

Figure 7.6 $V^\epsilon(x)$ and $V(x)$.

is defined, and divergence-free in the sense that each row of the matrix K is a vector with zero divergence.

By applying standard techniques from homogenization theory it is possible to prove the following theorem:

Theorem 7.3 *(Pavliotis and Stuart (2008)) There exist matrices $\overline{K}$ and K^* such that as $\epsilon \to 0$, $x \Rightarrow X$ in $C([0,T], \mathbb{R}^d)$, where X satisfies the equation*

$$\frac{dX}{dt} = -\overline{K}\nabla V(X) + \sqrt{2\overline{K}}\,\frac{dW}{dt}. \tag{7.8}$$

Furthermore

$$\lim_{\epsilon \to 0} \int_0^T \langle x\rangle_t dt = K^\star T, \qquad \int_0^T \langle X\rangle_t dt = \overline{K}T; \quad \overline{K} < K^\star.$$

Here, $\overline{K} < K^\star$ means that $\left(K^\star - \overline{K}\right)$ is positive definite. A typical potential is shown in Figure 7.6. Notice that the rapid oscillations persist at $\mathcal{O}(1)$ in the limit $\epsilon \to 0$. Thus, although the oscillatory part of the potential, p, is not present in the homogenized equation, its effect is felt in slowing down the diffusion process, as the particle must cross the energy barriers caused by this oscillation.

Once again we expect that parameter estimation which sees the small scales will fail. However, the situation differs from the three previous examples: here the quadratic variation of the data x is *larger* than that of the desired homogenized model X.

7.3 Averaging and homogenization

7.3.1 Orientation

In this section we probe the phenomenon exhibited in the preceding examples by means of the study of scale-separated SDEs. These provide us with a set of model problems which can be used to study the issues arising when there is a mismatch between the statistical model and the data at small scales. The last example in the preceding section illustrates such a model problem. Many coupled systems for (x, y) contain a parameter $\epsilon \ll 1$ which characterizes scale-separation. If y evolves more quickly than x, then it can sometimes be eliminated to produce an equation for $X \approx x$ alone. Two important situations where this arises are *averaging* and *homogenization*. We will be interested in fitting parameters in an effective averaged or homogenized equation for X, given data x from the coupled system for x, y. All proofs from this section may be found in the paper by Papavasiliou et al. (2009).

7.3.2 Set-up

In the following we set $\mathcal{X} = \mathbb{T}^l$ and $\mathcal{Y} = \mathbb{T}^{d-l}$. Let $\varphi_\xi^t(y)$ denote the Markov process which solves the SDE

$$\frac{d}{dt}\left(\varphi_\xi^t(y)\right) = g_0\left(\xi, \varphi_\xi^t(y)\right) + \beta\left(\xi, \varphi_\xi^t(y)\right)\frac{dV}{dt}, \quad \varphi_\xi^0(y) = y. \qquad (7.9)$$

Here $\xi \in \mathcal{X}$ is a fixed parameter and, for each $t \geq 0$, $\varphi_\xi^t(y) \in \mathcal{Y}$, $g_0 : \mathcal{X} \times \mathcal{Y} \to \mathbb{R}^{d-l}$, $\beta : \mathcal{X} \times \mathcal{Y} \to \mathbb{R}^{(d-l)\times m}$ and V is a standard $m-$dimensional Brownian motion. The generator of the process is

$$\mathcal{G}_0(\xi) = g_0(\xi, y) \cdot \nabla_y + \frac{1}{2}B(\xi, y) : \nabla_y\nabla_y \qquad (7.10)$$

with $B(\xi, y) := \beta(\xi, y)\beta(\xi, y)^T$ and equipped with periodic boundary conditions and : denotes the matrix inner product, see Appendix 2. Notice that $\mathcal{G}_0$ is a differential operator in y alone.

Our interest is in data generated by the projection onto the x coordinate of systems of SDEs for (x, y) in $\mathcal{X} \times \mathcal{Y}$. In particular, for U, a standard Brownian motion in $\mathbb{R}^n$, we will consider either of the following coupled systems of

SDEs:

$$\frac{dx}{dt} = f_1(x,y) + \alpha_0(x,y)\frac{dU}{dt} + \alpha_1(x,y)\frac{dV}{dt}, \qquad (7.11\text{a})$$

$$\frac{dy}{dt} = \frac{1}{\epsilon}g_0(x,y) + \frac{1}{\sqrt{\epsilon}}\beta(x,y)\frac{dV}{dt}; \qquad (7.11\text{b})$$

or the SDEs

$$\frac{dx}{dt} = \frac{1}{\epsilon}f_0(x,y) + f_1(x,y) + \alpha_0(x,y)\frac{dU}{dt} + \alpha_1(x,y)\frac{dV}{dt}, \qquad (7.12\text{a})$$

$$\frac{dy}{dt} = \frac{1}{\epsilon^2}g_0(x,y) + \frac{1}{\epsilon}g_1(x,y) + \frac{1}{\epsilon}\beta(x,y)\frac{dV}{dt}. \qquad (7.12\text{b})$$

Here $f_i : \mathcal{X} \times \mathcal{Y} \to \mathbb{R}^l, \alpha_0 : \mathcal{X} \times \mathcal{Y} \to \mathbb{R}^{l\times n}, \alpha_1 : \mathcal{X} \times \mathcal{Y} \to \mathbb{R}^{l\times m}$, $g_1 : \mathcal{X} \times \mathcal{Y} \to \mathbb{R}^{d-l}$ and g_0, β are as above. Note that in (7.11) (resp. (7.12)) the equation for y with x frozen has solution $\varphi_x^{t/\epsilon}(y(0))$ (resp. $\varphi_x^{t/\epsilon^2}(y(0))$ with $g_1 = 0$). Of course x is not frozen, but since it evolves much more slowly than y, intuition based on freezing x and considering the process (7.9) is useful. In addition to the generator $\mathcal{G}_0$, we also define the operator $\mathcal{G}_1$ as follows:

$$\mathcal{G}_1 = f_0 \cdot \nabla_x + g_1 \cdot \nabla_y + C : \nabla_y\nabla_x,$$

where the matrix-valued function C is defined as

$$C(x,y) = \alpha_1(x,y)\beta(x,y)^T.$$

Assumptions 7.4 • *All the functions f_i, g_i, α_i and β are C^∞ on the torus $\mathbb{T}^d$.*
• *The equation*

$$-\mathcal{G}_0^*(\xi)\rho(y;\xi) = 0, \quad \int_\mathcal{Y}\rho(y;\xi)dy = 1$$

has a unique non-negative solution $\rho(y;\xi) \in L^1(\mathcal{Y})$ for every $\xi \in \mathcal{X}$; furthermore $\rho(y;\xi)$ is C^∞ in y and ξ. Here as throughout, we use $\cdot^*$ to denote the adjoint operator.
• *Define the weighted Hilbert space $L^2_\rho(\mathcal{Y})$ with inner-product*

$$\langle a,b\rangle_\rho := \int_\mathcal{Y}\rho(y;\xi)a(y)b(y)dy.$$

The Poisson equation

$$-\mathcal{G}_0(\xi)\Theta(y;\xi) = h(y;\xi), \quad \int_\mathcal{Y}\rho(y;\xi)\Theta(y;\xi)dy = 0$$

has a unique solution $\Theta(y;\xi) \in L^2_\rho(\mathcal{Y})$, provided that

$$\int_\mathcal{Y}\rho(y;\xi)h(y;\xi)dy = 0.$$

- If $h(y; \xi)$ and all its derivatives with respect to y, ξ are uniformly bounded in $\mathcal{X} \times \mathcal{Y}$ then the same is true of Θ solving the Poisson equation above.

The second assumption is an ergodicity assumption. It implies the existence of an invariant measure $\mu_\xi(dy) = \rho(y; \xi)dy$ for $\varphi_\xi^t(\cdot)$. From the Birkhoff ergodic theorem it follows that, for μ–almost all $y \in \mathcal{Y}$,

$$\lim_{T \to \infty} \frac{1}{T} \int_0^T \phi(\xi, \varphi_\xi^t(y))dt = \int_\mathcal{Y} \phi(\xi, y)\rho(y; x)dy.$$

The averaging and homogenization theorems we now state arise from the calculation of appropriate averages against the measure $\mu_x(dy)$.

7.3.3 Averaging

Starting from model (7.11), we define $F(x)$ by

$$F(x) = \int_\mathcal{Y} f_1(x, y)\rho(y; x)dy$$

and $\Sigma(x)$ to be the matrix satisfying

$$2\Sigma(x) = \int_\mathcal{Y} \left(\alpha_0(x, y)\alpha_0(x, y)^T + \alpha_1(x, y)\alpha_1(x, y)^T \right)\rho(y; x)dy.$$

We note that $\Sigma(x)$ is positive semidefinite and hence its square root is well-defined.

Theorem 7.5 (*Papavasiliou et al. (2009)*) *Let Assumptions 7.4 hold. Then $x \Rightarrow X$ in $C([0, T], \mathcal{X})$ where X solves the SDE*

$$\frac{dX}{dt} = F(X) + \sqrt{2\Sigma(X)}\frac{dW}{dt} \tag{7.13}$$

with W a standard l–dimensional Brownian motion.

7.3.4 Homogenization

In order for the equations (7.12) to produce a sensible limit as $\epsilon \to 0$ it is necessary to impose a condition on f_0. Specifically we assume the following which, roughly, says that $f_0(x, y)$ averages to zero against the empirical measure of the fast y process.

Assumptions 7.6

$$\int_\mathcal{Y} \rho(y; x)f_0(x, y)dy = 0.$$

Let $\Phi(x, y) \in L^2_\rho(\mathcal{Y})$ be the unique solution of the equation

$$-\mathcal{G}_0\Phi(y; x) = f_0(x, y), \quad \int_\mathcal{Y} \rho(y; x)\Phi(y; x)dy = 0.$$

This has a unique solution by Assumptions 7.4 and 7.6 (by the Fredholm Alternative, see Evans (1998) or a presentation in context in Pavliotis and Stuart (2008)). Define

$$F(x) = F_0(x) + F_1(x)$$

where

$$F_0(x) = \int_\mathcal{Y} \left((\nabla_x\Phi f_0)(x, y) + (\nabla_y\Phi g_1)(x, y) \right.$$
$$\left. + (\alpha_1\beta^T : \nabla_y\nabla_x\Phi)(x, y) \right)\rho(y; x)dy,$$

$$F_1(x) = \int_\mathcal{Y} f_1(x, y)\rho(y; x)dy.$$

Also define $\Sigma(x)$ to be the matrix satisfying

$$2\Sigma(x) = A_1(x) + A_2(x)$$

where

$$A_1(x) = \int_\mathcal{Y} \left((\nabla_y\Phi\beta + \alpha_1)(\nabla_y\Phi\beta + \alpha_1)^T \right)(x, y)\rho(y; x)dy,$$

$$A_2(x) = \int_\mathcal{Y} \alpha_0(x, y)\alpha_0(x, y)^T \rho(y; x)dy.$$

By construction $\Sigma(x)$ is positive semidefinite and so its square root is well-defined.

Theorem 7.7 *(Papavasiliou et al. (2009)) Let Assumptions 7.4, 7.6 hold. Then $x \Rightarrow X$ in $C([0, T], \mathcal{X})$ where X solves the SDE*

$$\frac{dX}{dt} = F(X) + \sqrt{2\Sigma(X)}\frac{dW}{dt} \tag{7.14}$$

with W a standard l−dimensional Brownian motion.

7.3.5 Parameter estimation

A statistical approach to multiscale data $\{x(t)\}_{t\in[0,T]}$ might consist of simply using equations of the form

$$\frac{dX}{dt} = F(X; \theta) + \sqrt{2\Sigma(X)}\frac{dW}{dt}. \tag{7.15}$$

(which is just (7.13) or (7.14) but with an unknown parameter θ and we assume $\Sigma(X)$ is uniformly positive definite on $\mathcal{X}$) to fit multiscale data that may not necessarily arise from that diffusion, so that the diffusion is a good description only at some timescales.

If we assume that the data is actually generated by the particular multiscale systems (7.11) or (7.12) we can analyze how classical maximum likelihood estimators behave in the presence of multiscale data. Naturally, this only covers one particular instance of model-misspecification due to the presence of multiscale data but it has the advantage of being amenable to rigorous analysis.

Suppose that the actual drift compatible with the data is given by $F(X) = F(X; \theta_0)$. We ask whether it is possible to correctly identify $\theta = \theta_0$ by finding the *maximum likelihood estimator* (MLE) when using a statistical model of the form (7.15), but given data from (7.11) or (7.12). We assume that (7.15) is ergodic with invariant measure $\pi(x)$ at $\theta = \theta_0$. This enables us to probe directly the question of how parameter estimators function when the desired model-fit is incompatible with the data at small scales.

Given data $\{z(t)\}_{t \in [0,T]}$, application of the Girsanov theorem shows that the log likelihood for θ satisfying (7.15) is given by

$$\mathcal{L}(\theta; z) = \int_0^T \langle F(z; \theta), dz \rangle_{\Sigma^{-1}(z)} - \frac{1}{2} \int_0^T |F(z; \theta)|^2_{\Sigma^{-1}(z)} dt, \qquad (7.16)$$

where

$$\langle r_1, r_2 \rangle_{\Sigma(z)^{-1}} = \frac{1}{2} \langle r_1, \Sigma(z)^{-1} r_2 \rangle.$$

The MLE is a random variable given by

$$\hat{\theta} = \mathrm{argmax}_\theta \mathcal{L}(\theta; z).$$

Before analyzing the situation which arises when data and model are incompatible, we first recap the situation that occurs when data is taken from the model used to fit the data, in order to facilitate comparison. The following theorem shows how the log likelihood behaves, for large T, when the data is generated by the model used to fit the data itself.

Theorem 7.8 *(Papavasiliou et al. (2009)) Assume that (7.15) is ergodic with invariant density $\pi(X)$ at $\theta = \theta_0$, and that $\{X(t)\}_{t \in [0,T]}$ is a sample path of (7.15) with $\theta = \theta_0$. Then*

$$\lim_{T \to \infty} \frac{2}{T} \mathcal{L}(\theta; X) = \int_y |F(Z; \theta_0)|^2_{\Sigma^{-1}(Z)} \pi(Z) dZ$$

$$- \int_y |F(Z; \theta) - F(Z; \theta_0)|^2_{\Sigma^{-1}(Z)} \pi(Z) dZ,$$

where convergence takes place in $L^2(W)$ (square integrable random variables on the probability space for the Brownian motion W) and is almost sure wrt.

the initial condition $X(0)$. *The above expression is maximized by choosing* $\hat{\theta} = \theta_0$.

We make three observations: (i) for large T the likelihood is asymptotically independent of the particular sample path chosen — it depends only on the invariant measure; (ii) as a consequence we see that, asymptotically, time-ordering of the data is irrelevant to drift parameter estimation — this is something we will exploit in our non-parametric estimation in Section 7.6; (iii) the large T expression also shows that choosing data from the model which is to be fitted leads to the correct estimation of drift parameters, in the limit $T \to \infty$.

In the following we make:

Assumptions 7.9 *Equation* (7.11) *(resp.* (7.12)) *is ergodic with invariant measure* $\rho^\epsilon(x, y)dxdy$. *This measure converges weakly to the measure* $\pi(x)\rho(y; x)$ $dxdy$ *where* $\rho(y; x)$ *is the invariant density of the fast process* (7.9) *and* $\pi(x)$ *is the invariant density for* (7.13) *(resp.* (7.14)).

This assumption may be verified under mild assumptions on the drift and diffusion coefficients of the SDEs.

We now ask what happens when the MLE for the averaged equation (7.15) is confronted with data from the original multiscale equation (7.11). The following result explains what happens if the estimator sees the small scales of the data and shows that, in the averaging scenario, there is no problem arising from the incompatibility. Specifically the large T and small ϵ limit of the log-likelihood with multiscale data converges to the likelihood arising with data taken from the statistical model itself.

Theorem 7.10 *(Papavasiliou et al.* (2009)) *Let Assumptions 7.4 and 7.9 hold. Let* $\{x(t)\}_{t\in[0,T]}$ *be a sample path of* (7.11) *and* $X(t)$ *a sample path of* (7.15) *at* $\theta = \theta_0$.

$$\lim_{\epsilon \to 0} \lim_{T \to \infty} \frac{1}{T}\mathcal{L}(\theta; x) = \lim_{T \to \infty} \frac{1}{T}\mathcal{L}(\theta; X),$$

where convergence takes place in the same sense as in Theorem 7.8.

We now ask what happens when the MLE for the homogenized equation (7.15) is confronted with data from the original multiscale equation (7.12). In contrast to the situation with averaging, here there is a problem arising from the incompatibility at small scales. Specifically the large T and small ϵ limit of the log-likelihood with multiscale data differs from the likelihood arising with data taken from the statistical model at the correct parameter value.

In order to state the theorem we introduce the Poisson equation

$$-\mathcal{G}_0\Gamma = \langle F(x; \theta), f_0(x, y)\rangle_{\Sigma^{-1}(x)}, \quad \int_y \rho(y; \xi)\Gamma(y; x)dy = 0 \qquad (7.17)$$

which has a unique solution $\Gamma(y; \xi) \in L^2_\rho(\mathcal{Y})$ (as for Φ, by the Fredholm Alternative). Note that

$$\Gamma = \langle F(x; \theta), \Phi(x, y) \rangle_{\Sigma^{-1}(x)}.$$

Then define

$$E = \int_{\mathcal{X} \times \mathcal{Y}} \Big(\mathcal{G}_1 \Gamma(x, y) - \langle F(x; \theta), (\mathcal{G}_1 \Phi(x, y)) \rangle_{\Sigma^{-1}(x)} \Big) \pi(x) \rho(y; x) dx dy.$$

Theorem 7.11   *Let Assumptions 7.4, 7.6 and 7.9 hold. Let $\{x(t)\}_{t \in [0,T]}$ be a sample path of (7.12) and $X(t)$ a sample path of (7.15) at $\theta = \theta_0$. Then*

$$\lim_{\epsilon \to 0} \lim_{T \to \infty} \frac{1}{T} \mathcal{L}(\theta; x) = \lim_{T \to \infty} \frac{1}{T} \mathcal{L}(\theta; X) + E,$$

where convergence is in the sense given in Theorem 7.8 and the order in which limits are taken is, of course, crucial.

This theorem shows that the correct limit of the log likelihood is not obtained unless $\mathcal{G}_1$ is a differential operator in y only, in which case we recover the averaging situation covered in the Theorem 7.5. The paper Papavasiliou et al. (2009) contains examples in which E can be calculated explicitly. These examples demonstrate that E is non-zero and leads to a bias. This bias indicates that fitting multiscale data to an effective homogenized model equation can lead to incorrect identification of parameters if the multiscale data is interrogated at the fastest scales. We now investigate methods designed to overcome this problem.

7.4 Subsampling

In the previous section we demonstrated that, in the situation where homogenization pertains, using classical MLE on multiscale data may result in convergent estimates of the homogenized coefficients, but the estimated homogenized coefficients can be incorrect!

In this section we illustrate the first of three ideas which can be useful in overcoming the fact that data may be incompatible with the desired diffusion at small scales. In other words, the basic idea is to use subsampling of the data, at an appropriate rate, to ensure that the data is interrogated on a scale where it "behaves like" data from the homogenized equation. This section is based on Pavliotis and Stuart (2007). Similar ideas relating to the role of subsampling are encountered in the market microstructure noise models discussed in Chapter 2.

We present results of an analysis for the special case of linear dependence of the drift on the unknown parameter θ, i.e. we assume that the vector field has the form $F(z; \theta) = \theta F(z)$ for some scalar $\theta \in \mathbb{R}$. This simplifies the results considerably, but the observation that subsampling at the correct rate leads to correct estimates of the homogenized coefficients is valid more generally; see Papavasiliou et al. (2009).

In the numerical experiments that we will present, the data will be in discrete time: it will be in the form of a sequence $z = \{z_n\}_{n=0}^N$ which we will view as approximating a diffusion process, whose parameters we wish to estimate, at time increment δ. The maximum likelihood estimator derived from (7.16) gives

$$\hat{\theta} = \frac{\int_0^T \langle F(z), dz\rangle_{\sigma(z)}}{\int_0^T |F(z)|^2_{\sigma(z)} dt}.$$

A natural discrete time analogue of this estimator, which we will use in this paper, is

$$\hat{\theta}_{N,\delta}(z) = \frac{\sum_{n=0}^{N-1} \langle F(z_n), z_{n+1} - z_n\rangle_{\sigma(z_n)}}{\sum_{n=0}^{N-1} \delta |F(z_n)|^2_{\sigma(z_n)}}. \tag{7.18}$$

Although we concentrated in the previous section on drift parameter estimation, in numerical experiments presented here we will also investigate the estimation of the diffusion coefficient. Specifically, in the case where Σ is constant, given a discrete time-series $\{z_n\}_{n=0}^N$, we estimate Σ by

$$\hat{\Sigma}_{N,\delta}(z) = \frac{1}{2T} \sum_{j=0}^{N-1} (z_{j+1} - z_j)(z_{j+1} - z_j)^T, \tag{7.19}$$

where $z_j = z(j\delta)$ and $N = \lfloor \frac{T}{\delta} \rfloor$. This is derived from equation (7.4).

For our numerical investigations we revisit Example 4 in one dimension. Consider the equation

$$\frac{dx}{dt} = -\alpha \nabla V^\epsilon(x) + \sqrt{2\sigma} \frac{dW}{dt}. \tag{7.20}$$

Here $V^\epsilon(x) = V(x) + p(x/\epsilon)$ and p is 1−periodic. To write this in the form to which homogenization applies notice that setting $y = x/\epsilon$ we obtain

$$\frac{dx}{dt} = -\alpha \nabla V(x) - \frac{\alpha}{\epsilon} \nabla p(y) + \sqrt{2\sigma} \frac{dW}{dt},$$

$$\frac{dy}{dt} = -\frac{\alpha}{\epsilon} \nabla V(x) - \frac{\alpha}{\epsilon^2} \nabla p(y) + \sqrt{\frac{2\sigma}{\epsilon^2}} \frac{dW}{dt}.$$

This is now a specific case of (7.12).

Theorem 7.3 shows that the homogenized equation is

$$\frac{dX}{dt} = -\theta \nabla V(X) + \sqrt{2\Sigma} \frac{dW}{dt}. \tag{7.21}$$

Furthermore, the theorem shows that

$$\theta < \alpha, \quad \Sigma < \sigma.$$

In fact $\theta/\Sigma = \alpha/\sigma$. It may be shown that Σ is exponentially small in $\sigma \to 0$ Campillo and Pitnitski (2002). Thus the relative discrepancy between the original and homogenized diffusion coefficients is enormous in the small diffusion limit.

The numerical experiments that we now describe concern the case where

$$V(x) = \frac{1}{2}x^2, \quad p(y) = \cos(y).$$

The experiments are conducted in the following way. We generate the data by simulating the multiscale process x using a time-step Δt which is small compared to ϵ^2 so that the data is a fully resolved approximation of the multiscale process. We then use this data in the estimators (7.18), (7.19) which are based on a homogenized model. We study two cases: in the first we take data sampled at time-step $\delta = \Delta t$ so that the data is high frequency relative to the small scales in the equation; we anticipate that this scenario should be close to that covered by the theory in the previous section where we take continuous time data as input. We then show what happens if we subsample the data and take a step δ which is comparable to ϵ; as the fast process has time-scale ϵ^2 the hope is that, on the scale ϵ, which is long compared with ϵ^2, the data will "look like" that of the homogenized process.

The time-interval used is $t \in [0, 10^4]$ and the data is generated with time-step $\Delta t = 5 \cdot 10^{-4}$. Figure 7.7 shows the maximum likelihood and quadratic variation estimators (7.18) and (7.19), for the drift and diffusion coefficients respectively, with data $z = x$ at the fine-scale $\delta = \Delta t$. The figure clearly shows that the estimators fail to correctly identify coefficients in the homogenized equation (7.14) for X when employing multiscale data x. Indeed the estimator finds α and σ, from the unhomogenized equation, rather than θ and Σ. Hence it overestimates.

Figure 7.8 shows that, if subsampling is used, this problem may be overcome by interrogating the data at scale ϵ; this is shown for $\epsilon = 0.1$ and using the time interval $t \in [0, 2 \cdot 10^4]$ and again generating data with a timestep of $\Delta t = 5 \cdot 10^{-4}$, by choosing $\delta = 256 \times \Delta t$, $512 \times \Delta t$ and $\delta = 1024 \times \Delta t$ and showing that, with these choices, the correct parameters are estimated for both drift and diffusion, uniformly over a wide range of σ.

The following theorem justifies the observation that subsampling, at an appropriate rate, results in correct estimators.

Theorem 7.12 *(Papavasiliou et al. (2009)) Consider the parameter estimators (7.18) and (7.19) for drift and diffusion parameters in the statistical model (7.21). Define $x = \{x(n\delta)\}_{n=0}^{N-1}$ where $\{x(t)\}$ is a sample path of (7.20).*

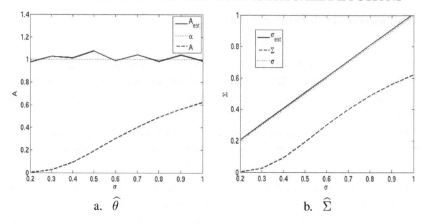

a. $\widehat{\theta}$ b. $\widehat{\Sigma}$

Figure 7.7 $\widehat{\theta}$ and $\widehat{\Sigma}$ vs σ for $\alpha = 1$, $\epsilon = 0.1$

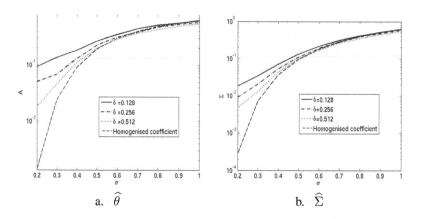

a. $\widehat{\theta}$ b. $\widehat{\Sigma}$

Figure 7.8 $\widehat{\theta}$ and $\widehat{\Sigma}$ vs σ for $a = 1$, $\epsilon = 0.1$. $\delta \in \{0.128, 0.256, 0.512\}$

- *Let $\delta = \epsilon^{\alpha}$ with $\alpha \in (0, 1)$, $N = \lfloor \epsilon^{-\gamma} \rfloor, \gamma > \alpha$. Then*

$$\lim_{\epsilon \to 0} \hat{\theta}_{N,\delta}(x) = \theta \quad \text{in distribution.}$$

- *Fix $T = N\delta$ with $\delta = \epsilon^{\alpha}$ with $\alpha \in (0, 1)$. Then*

$$\lim_{\epsilon \to 0} \widehat{\Sigma}_{N,\delta}(x) = \Sigma \quad \text{in distribution.}$$

7.5 Hypoelliptic diffusions

In this section we return to the Butane molecule considered in Section 7.2.3. In that example we showed problems arising from trying to fit a scalar diffusion to the raw time series data from the dihedral angle. Amongst several problems with the attempted fit, we highlighted the fact that the data came from a time series with zero quadratic variation, derived as a nonlinear function of the time-series $x(t)$ from (7.5), whilst the equation (7.6) which we attempted to fit had non-zero quadratic variation. In this section we show how we may attempt to overcome this problem by fitting a *hypoelliptic diffusion process* to the dihedral angle data. Technical details can be found in the paper Pokern, Stuart, and Wiberg (2009).

Specifically we attempt to fit to the data $\{\phi_n\}_{n=0}^N$, where observations are made at small but fixed inter-sample time δ so that $\phi_n = \phi(n\delta)$, a model of the form

$$\frac{d^2q}{dt^2} + \gamma\frac{dq}{dt} - V'(q) = \sigma\frac{dW}{dt}, \qquad (7.22a)$$

$$V(q) = \sum_{j=1}^{5} \frac{\theta_j}{j}\Big(\cos(q)\Big)^j, \qquad (7.22b)$$

q here plays the same role as Φ in (7.6): it describes the dihedral angle in a postulated lower dimensional stochastic fit to data derived from a higher di-mensional dynamical model. The statistical task at hand is then to use the data ϕ_n to infer the value of the parameters $\{\theta_j\}_{j=1}^5$ as well as γ and σ.

This task is problematic because

1. observations of only ϕ rather than ϕ and its time derivative $\frac{d\phi}{dt}$ are assumed to be available, so a missing data problem is to be dealt with;

2. the two components of the process, namely ϕ and $\frac{d\phi}{dt}$ have different smooth-ness rendering straightforward statistical models ill-conditioned.

Velocities for ϕ may be available in practice (molecular dynamics codes can certainly produce such output if desired) but it may be preferable to ignore them — such data may be incompatible with SDE models.

It should be emphasized that both the missing data aspect (1. above) and the different degrees of differentiability (2. above) are serious problems *regardless* of whether a frequentist or a Bayesian approach is developed. For simplicity, we will explain these problems further using the maximum likelihood principle and only later introduce a full (Bayesian) estimation algorithm.

In order to better understand these problems we rewrite (7.22) as a damped-driven Hamiltonian system as follows:

$$dq = pdt, \tag{7.23a}$$

$$dp = \left(-\gamma p - \sum_{j=1}^{5} \theta_j \sin(q) \cos^{j-1}(q) \right) dt + \sigma dB, \tag{7.23b}$$

where we have used the new variable $p(t) = \frac{dq}{dt}$. To understand how the two problems 1. and 2. enumerated above come about, consider one of the most widespread discrete time approximations to this SDE, the Euler-Maruyama approximation:

$$Q_{i+1} = Q_i + \delta P_i \tag{7.24a}$$

$$P_{i+1} = P_i - \left(\gamma P_i + \sum_{j=1}^{5} \theta_j \sin(Q_i) \cos^{j-1}(Q_i) \right) \delta + \sigma \sqrt{\delta} \xi_i \tag{7.24b}$$

where $\xi_i \sim \mathcal{N}(0,1)$ is a sequence of iid. standard normal random variables and we have used capital variable names to indicate that this is the discretised version of a continuous time system. It is possible to estimate all desired parameters using this model, by first using (7.24a) to obtain $P_i = \frac{1}{\delta}(Q_{i+1} - Q_i)$ and then estimating σ from the quadratic variation of the path $\{P_i\}_{i=0}^{N-1}$ and the drift parameters can then be estimated by applying the maximum likelihood principle to (7.24b).

Using this approximation to estimate σ given the data $\{\phi_n\}_{n=0}^{N}$ for $\{Q_i\}_{i=0}^{N}$ and no data for $\{P_i\}_{i=0}^{N}$ leads to gross mis-estimation. In fact for the simpler example of stochastic growth

$$dq = pdt$$
$$dp = \sigma dB$$

it is straightforward to show that in the limit of infinitely many observations $N \to \infty$ (both in the case when δ is fixed and in the case when $T = N\delta$ is fixed!) the maximum likelihood estimator $\hat{\sigma}$ converges to an incorrect estimate almost surely:

$$\hat{\sigma}^2 \longrightarrow \frac{2}{3}\sigma^2 \quad \text{a.s.}$$

This failure can be traced back to the fact that (7.24) effectively uses numerical differentiation of the time series Q_i to solve the missing data problem, i.e. to estimate P_i. This approximation neglects noise contributions of order $\mathcal{O}(\delta^{\frac{3}{2}})$ in (7.24a) which are of the same order as the contributions obtained via numerical differentiation of Q_i.

Understanding the source of the error suggests that replacing the Euler-Muruyama scheme with a higher order discretization scheme that propagates noise to both rows of the equation (7.23) results in successful estimators for σ. One such scheme is given by

$$\begin{bmatrix} Q_{i+1} \\ P_{i+1} \end{bmatrix} - \begin{bmatrix} Q_i \\ P_i \end{bmatrix} + \delta \begin{bmatrix} P_i \\ \sum_{j=1}^{5} \theta_j \sin(Q_i) \cos^{j-1}(Q_i)) - \gamma P_i \end{bmatrix} + \sigma\sqrt{\delta} R \begin{bmatrix} \xi_1 \\ \xi_2 \end{bmatrix}$$
(7.25)

where the matrix R is given as

$$R = \begin{bmatrix} \frac{\delta}{\sqrt{12}} & \frac{\delta}{2} \\ 0 & 1 \end{bmatrix}$$
(7.26)

and ξ_1 and ξ_2 are again independent standard normal random variables. Generally, Itô-Taylor expansions of sufficiently high order should be used to propagate noise to all components of the process.

The approximation (7.25) can be used not only to infer σ but also to infer the missing component of the path.

Finally, it remains to estimate the drift parameters $\{\theta_j\}_{j=1}^{5}$ and γ and it turns out that the approximation (7.25) yields results with a large bias that does not decay as δ decreases or the observation time T increases. In fact, for the simpler case of a harmonic oscillator

$$\begin{cases} dq &= pdt \\ dp &= -\theta qdt - \gamma pdt + \sigma dB. \end{cases}$$
(7.27)

it is possible to compute by Itô-Taylor-expansion that the maximum likelihood estimator for θ and γ based on an analogous model to (7.25) satisfies:

$$\mathbb{E}\hat{\theta} = \frac{1}{4}\theta + \mathcal{O}(\delta)$$

$$\mathbb{E}\hat{\gamma} = \frac{1}{4}\gamma + \mathcal{O}(\delta).$$

This can be traced back to the fact that such an estimator assumes the drift parameters in the first row of (7.27) to be known exactly whereas the discrete time path only satisfies (7.25) (or the analogous model for the harmonic oscillator) approximately. The ill-conditioning of the inverse of the matrix R introduced in (7.26) as $\delta \to 0$ causes small errors to be amplified to $\mathcal{O}(1)$ deviations in the drift parameter estimates. Using the Euler-Maruyama approximation (7.24) instead delivers satisfactory results.

Having used a maximum likelihood framework to highlight both the fact that the missing data problem adds significant difficulty and that the different degrees of differentiability of ϕ and $\frac{d\phi}{dt}$ produce ill-conditioning, we now proceed to a Bayesian algorithm to infer the missing data $\{P_j\}_{j=0}^{N}$ as well as the diffusion and drift parameters σ and $\{\theta_j\}_{j=1}^{5}$ and γ. The sequential Gibbs algorithm suggested to produce approximate samples $\gamma^{(i)}$, $\{\theta_j^{(i)}\}_{j=1}^{5}$, $\sigma^{(i)}$ and $P_n^{(i)}$ indexed by $i = 1, 2, \ldots$ thus reads as follows:

1. Sample $\theta^{(i+1)}, \gamma^{(i+1)}$ from $\mathbb{P}(\{\theta_j\}_{j=1}^5, \gamma | \{Q_j\}_{j=0}^N, \{P_j^{(i)}\}_{j=0}^N, \sigma^{(i)})$ using (7.24).

2. Sample $\sigma^{(i+1)}$ from $\mathbb{P}(\sigma | \{Q_j\}_{j=0}^N, \{P_j^{(i)}\}_{j=0}^N, \{\theta_j^{(i+1)}\}_{j=1}^5, \gamma^{(i+1)})$ using (7.25).

3. Sample $\{P_j^{(i+1)}\}_{j=0}^N$ from $\mathbb{P}(\{P_j\}_{j=0}^N | \{Q_j\}_{j=0}^N, \{\theta_j^{(i+1)}\}_{j=1}^5, \gamma^{(i+1)}, \sigma^{(i+1)})$ using (7.25).

Note that we have omitted the initialization stage and that sampling the missing path, stage 3, is simplified by the fact that p only ever enters the SDE linearly, so that a direct Gaussian sampler can be used. Stage 1 is also Gaussian, whereas stage 2 is not, and an MCMC method, for example, can be used.

The status of Gibbs samplers, such as this one, combining different approximate likelihoods, especially in the presence of ill-conditioning which renders some approximations unsuitable for some of the estimation tasks, is not yet theoretically understood. In the particular case of the SDE being fitted, the method has been subjected to very careful numerical studies detailed in Pokern, Stuart, and Wiberg (2009) and found to be convergent.

Finally, the method can be applied to the data given as Example 3 in Section 7.2.3 and Figure 7.9 shows posterior mean parameter estimates as a function of sampling interval δ.

We have shown in this section how to extend parameter estimation from the elliptic case (7.6) to the hypoelliptic case (7.22) and we have highlighted how to do this in the case of missing velocities. This is useful e.g. when neglecting the velocities is viewed as a means to decrease the fitting process' sensitivity to incompatibility between the model and the data at the short timescales. Still, Figure 7.9 shows that the fitted drift and diffusion parameters are far from independent of the timescale on which we look at the data. It is natural to ask why this is so. Since the dihedral angle is a nonlinear transformation of the Cartesian coordinates in (7.5) and hence (as a brief application of the Itô formula readily shows) will exhibit multiplicative rather than additive noise, it will *not* be well-described by a hypoelliptic diffusion with constant diffusivity. It is this problem with model fit that results in the timescale dependence of fitted parameters evidenced in Figure 7.9.

7.6 Non-parametric drift estimation

Theorem 7.8 illustrates the fact that, for large times, drift parameter estimation does not see path properties of the data, but rather just sees the invariant measure. This suggests an approach to drift parameter estimation which exploits this property directly and uses only the empirical measure of the data, thereby

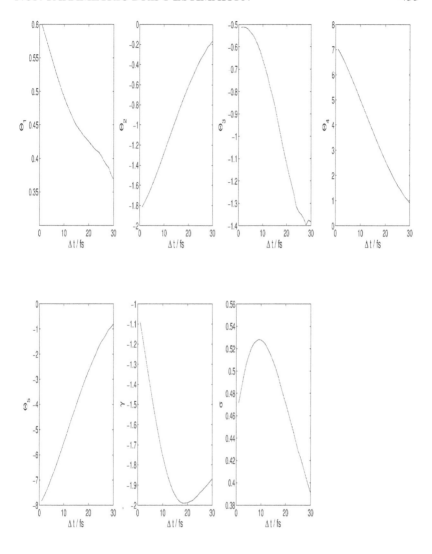

Figure 7.9 *Mean posterior estimates for drift and diffusion parameters using Butane data as a function of sampling interval δ*

avoiding issues relating to incompatibility at small scales. These ideas are pursued in Pokern, Stuart, and Vanden-Eijnden (2009) and then, building on this in a Bayesian context, in Papaspiliopoulos et al. (2009).

Here we illustrate the basic ideas in the context of the one-dimensional equation

$$\frac{dx}{dt} = -V'(x) + \frac{dW}{dt}. \tag{7.28}$$

In Pokern, Stuart, and Vanden-Eijnden (2009) we treat the multidimensional case, matrix diffusion coefficients and the second order Langevin equation. We note for later use that the SDE (7.28) has invariant measure with density

$$\rho \propto \exp(-2V), \tag{7.29}$$

provided $\exp(-2V) \in L^1(\mathbb{R})$.

Our strategy is a non-parametric one. We write down the log likelihood for this equation which, from (7.16), has the form

$$\mathcal{L}(V;x) = -\int_0^T V'(x)dx - \frac{1}{2}\int_0^T |V'(x)|^2 dt. \tag{7.30}$$

Notice that now we view the log likelihood as being a functional of the unknown drift with potential V. This reflects our non-parametric stance. Applying the Itô formula to $V(x(t))$ we deduce that

$$-\int_0^T V'(x)dx = \frac{1}{2}\int_0^T V''(x)dt + \Big(V(x(0)) - V(x(T))\Big).$$

Thus we obtain

$$\mathcal{L}(V;x) = V(x(0)) - V(x(T)) + \frac{1}{2}\int_0^T \Big(V''(x) - |V'(x)|^2\Big)\,dt. \tag{7.31}$$

With the goal of expressing the likelihood independently of small-timescale structure of the data we now express the log likelihood in terms of the local time L_T^a of the process. Recall that the local time L_T^a measures the time spent at a up to time T, so L_T^a/T is proportional to an empirical density function for the path history up to time T. Note that use of the local time L_T^a of the process indeed removes any time-ordering in the data and thus makes drift estimation independent of the dynamical information contained in the data. Since time-ordered data at small scales is at the root of the problems we are confronting in this paper, taking this point of view is likely to be beneficial.

To make the notion of local time as a scaled empirical density rigorous, consider Theorem 2.11.7 in Durrett (1996) which states that for x being a 1-d continuous semimartingale with local time L_T^a and quadratic variation process $\langle x \rangle_t$, the following identity holds for any Borel-measurable, bounded function g:

$$\int_{-\infty}^{\infty} L_T^a g(a)da = \int_0^T g(x_s)d\langle x \rangle_s. \tag{7.32}$$

Note that for the process (7.28) we have

$$dt = d\langle x \rangle_t$$

so that

$$\int_{-\infty}^{\infty} L_t^a g(a) da = \int_0^t g(x_s) ds.$$

In terms of the local time we have

$$\mathcal{L}(V; x) = -\left(V(x(T)) - V(x(0))\right) + \frac{1}{2} \int_{\mathbb{R}} \left(V''(a) - |V'(a)|^2\right) L_T^a da.$$
(7.33)

To make the mathematical structure of this functional more apparent, we re-express the likelihood in terms of the drift function $b = -V'$. To do this, we first introduce the signed indicator function

$$\tilde{\chi}(a; X_0, X_T) = \begin{cases} 1 & \text{if } X_0 < a < X_T \\ -1 & \text{if } X_T < a < X_0 \\ 0 & \text{otherwise} \end{cases}.$$

The expression for the likelihood then takes the following form:

$$\mathcal{L}(V; x) = -\frac{1}{2} \int_{\mathbb{R}} \left(\left(|V'(a)|^2 - V''(a)\right) L_T^a + 2\tilde{\chi}(a; X_0, X_T) V'(a)\right) da.$$

Having eliminated V by expressing everything in terms of its derivatives, we replace those derivatives by the drift function b as planned. We abuse notation and now write $\mathcal{L}$ as a functional of b instead of V:

$$\mathcal{L}(b; x) = -\frac{1}{2} \int_{\mathbb{R}} \left(b^2(a) L_T^a + b'(a) L_T^a - 2\tilde{\chi}(a; X_0, X_T) b(a)\right) da. \quad (7.34)$$

We would like to apply the likelihood principle to $\mathcal{L}(b; x)$ to estimate b. **Purely formally**, seeking a critical point of the functional (7.34) is possible, i.e. one asks that its functional derivative with respect to b be zero:

$$\frac{\delta \mathcal{L}(b; x)}{\delta b} \overset{!}{=} 0$$

To carry this out, integrate by parts to obtain

$$\mathcal{L}(b; x) = -\frac{1}{2} \int_{\mathbb{R}} \left(b^2(a) L_T^a - b(a) L_T'(a) - 2\tilde{\chi}(a; X_0, X_T) b(a)\right) da.$$

Expand this expression at $b(\epsilon) = b + \epsilon u$ where u is an arbitrary smooth function to compute the functional derivative:

$$\frac{\delta \mathcal{L}(b; x)}{\delta b} u = \lim_{\epsilon \to 0} \frac{1}{\epsilon} \left(\mathcal{L}(b + \epsilon u; x) - \mathcal{L}(b; x)\right)$$

Equating the functional derivative to zero yields the **formal** maximum likelihood estimate

$$\hat{b} = \frac{L_T'}{2L_T} - \frac{\tilde{\chi}(\cdot; X_0, X_T)}{L_T}. \qquad (7.35)$$

Note the derivative of the local time figures prominently in this estimate – however, it is not defined since L_T is not differentiable. It can be shown that, in one dimension, the local time L_t^a is jointly continuous in (t, a), but that it is not in general differentiable; it is only α-Hölder continuous up to but excluding exponent $\alpha = 1/2$. Therefore, the likelihood functional in (7.34) would not be expected to be bounded above and the estimate (7.35) is *not* a proper maximiser of the likelihood.

To get an idea of why this is, consider the case where local time is replaced by a Brownian bridge (which has essentially the same Hölder regularity as local time) where it is possible to show the following theorem (see Appendix 1):

Theorem 7.13 *Let $w \in C([0, 1], \mathbb{R})$ be a realisation of the standard Brownian bridge on $[0, 1]$. Then with probability one, the functional*

$$I(b; w) = -\frac{1}{2} \int_0^1 \left(b^2(s)w(s) + b'(s)w(s) \right) ds$$

is not bounded above for $b \in H^1([0, 1])$.

If $\mathcal{L}(b; x)$ given by (7.34) is not bounded above, then application of the maximum likelihood principle will fail. To remedy this problem, several options are available. One can introduce a parametrization $b(x, \theta)$ for $\theta \in \Theta \subset \mathbb{R}^m$ for some finite m with the attendant problems of choosing a set of basis functions that make the parameters well-conditioned and easy to interpret. Alternatively, one can work with a mollified version of the local time, $\tilde{L}_T^a$, which is smooth enough to ensure existence of a maximizer of the likelihood functional. Finally, it is possible to use Tikhonov regularization and then, taking this further, to adopt a Bayesian framework and use a prior to ensure sufficient regularity. In Section 7.6.1, we will investigate the use of mollified local time in detail and in Section 7.6.2 we briefly introduce the Bayesian non-parametric approach.

7.6.1 Mollified local time

To start using mollified local time, we proceed in three steps adopting a traditional regularization and truncation approach. Firstly, in (7.34) we replace L_T^a by a mollification $\tilde{L}_T^a$ which is assumed to be compactly supported and nonnegative just like the original local time. Additionally, we assume that it has Sobolev regularity $L_T \in H^1(\mathbb{R})$. Secondly, we integrate by parts exploiting the smoothness and compact support of $\tilde{L}_T^a$:

$$\mathcal{L}(b; x) \approx -\frac{1}{2} \int_{\mathbb{R}} \left(b^2(a)\tilde{L}_T^a - b(a)\left(\tilde{L}_T^a\right)' - 2\tilde{\chi}(a; X_0, X_T)b(a) \right) da$$
$$(7.36)$$

Thirdly, we restrict attention to a bounded open interval $U \subset \mathbb{R}$ on the real line which is chosen such that

$$\exists \epsilon > 0 \, \forall a \in U : \tilde{L}_T^a > \epsilon. \tag{7.37}$$

This leads to the final approximation of $\mathcal{L}(b; x)$ by the following functional:

$$\tilde{\mathcal{L}}(b; x) = -\frac{1}{2} \int_U \left(b^2(a) \tilde{L}_T^a - b(a) \left(\tilde{L}_T^a \right)' - 2\tilde{\chi}(a; X_0, X_T) b(a) \right) da \tag{7.38}$$

This functional is quadratic in b and it is straightforward to prove the following theorem:

Theorem 7.14 *The functional $\tilde{\mathcal{L}}(b; x)$ is almost surely bounded above on $b \in L^2(U)$ and its maximum is attained at*

$$\hat{b} = \frac{1}{2} \left(\log \tilde{L}_T^a \right)' - \frac{\tilde{\chi}(a; X_0, X_T)}{\tilde{L}_T^a} \tag{7.39}$$

Proof. We rewrite the mollified likelihood functional by completing the square as follows:

$$\tilde{\mathcal{L}}(b; x) = -\frac{1}{2} \int_U \left[\left(b - \frac{\left(\tilde{L}_T^a \right)'}{2\tilde{L}_T^a} - \frac{\tilde{\chi}(a; X_0, X_T)}{\tilde{L}_T^a} \right)^2 \tilde{L}_T^a \tag{7.40} \right.$$

$$\left. - \left(\frac{\left(\tilde{L}_T^a \right)'}{2\tilde{L}_T^a} + \frac{\tilde{\chi}(a; X_0, X_T)}{\tilde{L}_T^a} \right)^2 \tilde{L}_T^a \right] da \tag{7.41}$$

Observe that the first summand in the integrand is always non-negative. To avoid potentially subtracting two infinite terms from each other, we verify that the second summand in the integrand has a finite integral:

$$\int_U \left(\frac{\left(\tilde{L}_T^a \right)'}{2\tilde{L}_T^a} + \frac{\tilde{\chi}(a; X_0, X_T)}{\tilde{L}_T^a} \right)^2 \tilde{L}_T^a da$$

$$= \int_U \frac{1}{\tilde{L}_T^a} \left(\frac{\left(\tilde{L}_T^a \right)'}{2} + \tilde{\chi}(a; X_0, X_T) \right)^2 da$$

$$\leq \frac{1}{\epsilon} \int_U \left(\frac{1}{2} \left(\tilde{L}_T^a \right)' + \tilde{\chi}(a; X_0, X_T) \right)^2 da < \infty$$

where we have used condition (7.37) in the penultimate inequality and the fact that U is compact and that $\tilde{L}_T$ is smooth (and hence it and its derivative are square-integrable on U). All that remains is to read off the maximizer (7.39) from the brackets in (7.40).   □

The last term in (7.39) is integrable on U (since $\tilde{L}$ is bounded away from zero on U and U is bounded), the integrated version of that MLE thus reads

$$\widehat{V}(a) = -\frac{1}{2}\log \tilde{L}_T^a + \int_{\inf(U)}^a \frac{\tilde{\chi}(s; X_0, X_T)}{\tilde{L}_T^s}\,ds, \quad a \in U.$$

Furthermore, we expect that $\tilde{L}_T^a$ scales like $\mathcal{O}(T)$. Thus the first term gives the dominant term in the estimator for large T. Retaining only the first term gives the approximation

$$\widehat{V}(a) = -\frac{1}{2}\log \tilde{L}_T^a.$$

If we make the reasonable assumption that $\frac{1}{T}\tilde{L}_T^a \to \rho(a)$ as $T \to \infty$ where ρ is the invariant measure for the process we deduce that, for this approximation,

$$\lim_{T \to \infty} \widehat{V}(a) = -\frac{1}{2}\log \rho(a).$$

But the invariant density is given by (7.29) and so we deduce that, under these reasonable assumptions,

$$\lim_{T \to \infty} \widehat{V}(a) = V(a)$$

as expected.

7.6.2 Regularized likelihood functional

Another way to regularize the functional (7.33) is to add a penalty function to the logarithm of the likelihood to obtain

$$\mathcal{L}_p(b; x) = \mathcal{L}(b; x) - \|b - b_0\|_H^2.$$

Here H is a suitable Hilbert subspace and we call b_0 the centre of regularization. We refer to this procedure as Tikhonov regularization; in the case $b_0 = 0$ it coincides with the standard usage – see Kaipio and Somersalo (2005). If the additional term $\|b - b_0\|_H^2$ is chosen to penalize roughness it is possible to ensure that the combined logarithm has a unique maximum. We will outline that in the sequel that since the penalization is quadratic, this may be linked to the introduction of a Gaussian prior. The posterior will be seen to be Gaussian, too, because the likelihood is quadratic in b, and all densities and probabilities arising can be given a fully rigorous interpretation. We leave technical and implementation details to Papaspiliopoulos, Pokern, Roberts, and Stuart (2011) and Papaspiliopoulos et al. (2009) and merely outline the key calculations first concentrating on the viewpoint of the Tikhonov regularization $\mathcal{L}_p$ of the likelihood functional $\mathcal{L}$, then introducing the Bayesian viewpoint at the end.

Tikhonov regularization

We first state a theorem to show that regularization using the Hilbert space norm

$$\|b\|_H^2 = 2 \int_{\mathbb{R}} b(a)^2 + (b'(a))^2 + (b''(a))^2 \, da$$

on the Sobolev space $H = H^2(\mathbb{R})$ is indeed possible:

Theorem 7.15 *For any fixed $c > 0$, the functional*

$$\mathcal{L}_p(b; x) = \mathcal{L}(b; x) - c\|b - b_0\|_{H^2(\mathbb{R})}^2$$

is almost surely bounded above on $b \in H^2(\mathbb{R})$ for any $b_0 \in H^2(\mathbb{R})$.

Proof. The proof follows along the same lines as the proof of Theorem 7.17 which will be given in full. □

Showing that a maximizer exists and that it is the solution of the accompanying Euler-Lagrange equations requires a more detailed analysis which is easier to carry out in the periodic setting. Thus, we henceforth consider Itô SDEs with constant diffusivity on the circle parametrized by $[0, 2\pi]$,

$$dx = b(x)dt + dW, \quad x(0) = x_0 \quad \text{on } [0, 2\pi].$$

The state space of the diffusion process is now compact. Also note that using the circle as a state space introduces another linear term into $\mathcal{L}(b; x)$ so that we now have

$$\mathcal{L}(b; x) = -\frac{1}{2} \int_0^{2\pi} \left(b^2(a)L_T^a + b'(a)L_T^a - 2(M + \tilde{\chi}(a; X_0, X_T))b(a) \right) da,$$

$$(7.42)$$

where $M \in \mathbb{Z}$ is the winding number of the process x, i.e. the number of times x has gone around the circle in $[0, T]$.

Let us now consider the Tikhonov regularization of $\mathcal{L}$ by the H^2-seminorm as follows:

$$\mathcal{L}_p(b; x) = \mathcal{L}(b; x) - \frac{1}{2} |b - b_0|_{H^2}^2 \qquad b \in H_{\text{per}}^2([0, 2\pi]), \qquad (7.43)$$

where $H_{\text{per}}^2([0, 2\pi])$ refers to the Sobolev space of twice weakly differentiable periodic functions on $[0, 2\pi]$ and $|b|_{H^2} = \int_0^{2\pi} (b''(a))^2 \, da$. Note that the prefactor $\frac{1}{2}$ in front of the seminorm can be replaced by an arbitrary positive constant, allowing an adjustment of the strength of regularization. To analyze this regularized functional, we separate off its quadratic terms by introducing the bilinear form $q(u, v)$ defined for $u, v \in H^2_{\text{per}}([0, 2\pi])$ as follows:

$$q(u, v) = \frac{1}{2} \int_0^{2\pi} \Delta u(a) \Delta v(a) + u(a) L^a_T v(a) \, da, \qquad (7.44)$$

where we denote second derivatives by the Laplace operator, $\Delta = \frac{d^2}{da^2}$. Important properties of this bilinear form are given in the following Lemma whose proof is slightly technical and can be found in Papaspiliopoulos et al. (2009).

Lemma 7.16 *If the local time L_T is not identically zero on $[0, 2\pi]$, then the form q, defined in (7.44), is a continuous, coercive, symmetric bilinear form, i.e. there are constants $\alpha, C \in \mathbb{R}_+$ which may depend on L_T but not on u, v such that the following relations hold:*

$$\alpha \|u\|^2_{H^2} \leq q(u, u) \quad \forall u \in H^2_{\text{per}}([0, 2\pi]) \qquad (7.45)$$

$$q(u, v) \leq C \|u\|_{H^2} \|v\|_{H^2} \quad \forall u, v \in H^2_{\text{per}}([0, 2\pi])$$

$$q(u, v) = q(v, u) \quad \forall u, v \in H^2_{\text{per}}([0, 2\pi])$$

We now state an analogous theorem to Theorem 7.14:

Theorem 7.17 *The functional $\mathcal{L}_p(b)$ defined in (7.43) is almost surely bounded above on $b \in H^2_{\text{per}}([0, 2\pi])$ and its maximum is attained at $\widehat{b} \in H^2_{\text{per}}([0, 2\pi])$ which is given by the unique weak solution of the boundary value problem*

$$\left(\Delta^2 + L_T \right) \widehat{b} = \frac{1}{2} L'_T + M + \tilde{\chi}(\cdot; X_0, X_T) + \Delta^2 b_0. \qquad (7.46)$$

Proof. We present a heuristic calculation first that simply proceeds by completing the square and reading off the answer. We then indicate how this can be approached rigorously. To simplify notation we assume that the centre of regularization is identically zero: $b_0 = 0$.

$$\mathcal{L}_p(b; x) = -\frac{1}{2} \int_0^{2\pi} \left(b^2(a) L^a_T + (\Delta b(a))^2 + b'(a) L^a_T \right.$$

$$\left. -2 \left(\tilde{\chi}(a; X_0, X_T) + M \right) b(a) \right) da$$

To simplify the notation we drop the arguments. To formally derive the maximizer we repeatedly integrate by parts, pretending that the local time is sufficiently regular.

$$\mathcal{L}_p = -\frac{1}{2} \int \left(b(\Delta^2 + L_T) b - b(L'_T + 2(\tilde{\chi} + M)) \right) da$$

We now introduce the abbreviations

$$\mathcal{D} = \Delta^2 + L_T \qquad (7.47)$$

$$c = -\frac{1}{2}L_T' \quad (\tilde{\chi} + M) \qquad (7.48)$$

and the following notational convention for the square root of the operator $\mathcal{D}$ and its inverse:

$$\left| \mathcal{D}^{\frac{1}{2}} u \right|^2 = \langle u, \mathcal{D} u \rangle \qquad u \in H^2_{\text{per}}([0, 2\pi])$$

$$\left| \mathcal{D}^{-\frac{1}{2}} u \right|^2 = \langle u, \mathcal{D}^{-1} u \rangle \qquad u \in H^2_{\text{per}}([0, 2\pi]),$$

where $\langle \cdot, \cdot \rangle$ denotes the inner product on the Hilbert space $L^2([0, 2\pi])$. These conventions are intended to make the following calculation more transparent, so we rewrite $\mathcal{L}_p$ first of all:

$$\mathcal{L}_p(b; x) = -\frac{1}{2} \langle b, \mathcal{D} b \rangle - \langle b, c \rangle$$

Now we complete the square

$$\mathcal{L}_p(b; x) = -\frac{1}{2} \left| \mathcal{D}^{\frac{1}{2}} \left(b + \mathcal{D}^{-1} c \right) \right|^2 + \frac{1}{2} \left| \mathcal{D}^{-\frac{1}{2}} c \right|^2$$

and note that the maximizer of the regularized functional can be seen to be given by

$$\hat{b} = -\mathcal{D}^{-1} c.$$

This is identical to (7.46) which can be seen by inserting the terms from (7.47) and (7.48); boundedness from above is also apparent.

We rigorously establish boundedness from above (again in the case $b_0 = 0$) and leave the rest of the proofs to Papaspiliopoulos et al. (2009). To do this, we first rewrite the Tikhonov-regularized likelihood using the quadratic form (7.44) as follows:

$$\mathcal{L}_p(b; x) = \frac{1}{2} \int_0^{2\pi} \left(-b'(a) L_T^a + 2 \left(M + \tilde{\chi}(a; X_0, X_T) \right) b(a) \right) da - q(b, b)$$

Now bound the linear term in $\mathcal{L}_p$ as follows:

$$\left| \int_0^{2\pi} b'(a) L_T^a - 2b(a) (M + \tilde{\chi}(a; X_0, X_T)) da \right|$$

$$\leq \int_0^{2\pi} \epsilon_1 \left(b'(a) \right)^2 + \frac{1}{\epsilon_1} \left(L_T^a \right)^2 + \epsilon_2 b^2(a) + \frac{1}{\epsilon_2} \left(M + \tilde{\chi}(a; X_0, X_T) \right)^2 da$$

which holds for any $\epsilon_1, \epsilon_2 \in (0, \infty)$. Now choose $\epsilon = \epsilon_1 = \epsilon_2 < \alpha$ where

α is given by (7.45) from Lemma 7.16. Exploit coercivity (i.e. (7.45)) in the following way:

$$\mathcal{L}_p(b) \leq -q(b,b) + \epsilon\|b\|^2_{H^1} + \frac{1}{\epsilon}\|L_T\|^2_{L^2} + \frac{1}{\epsilon}\|M + \tilde{\chi}(\cdot; X_0, X_T)\|^2_{L^2}$$
$$\leq -(\alpha - \epsilon)\|b\|^2_{H^2} + \text{const.}$$

The case $b_0 \neq 0$ presents mainly notational complications, whereas deriving the PDE (7.46) requires a little variational calculus and showing existence and uniqueness of its solutions is an application of standard PDE theory given in Papaspiliopoulos et al. (2009). □

Bayesian viewpoint

In this subsection we show that the Tikhonov regularization given above underpins the adoption of a Bayesian framework and we briefly outline some speculation concerning the limit $T \to \infty$.

To introduce a prior, we decompose the space $H^2_{\text{per}}([0, 2\pi])$, into the direct sum of the (one-dimensional) space of constant functions and the space of Sobolev functions with average zero (denoted by $\dot{H}^2_{\text{per}}([0, 2\pi])$):

$$H^2_{\text{per}}([0, 2\pi]) = \{\alpha \mathbf{1} | \alpha \in \mathbb{R}\} \bigoplus \dot{H}^2_{\text{per}}([0, 2\pi]),$$

where we use $\mathbf{1}$ to denote the constant function with value one. We now define the prior measure as the product measure found from Lebesgue measure on the space of constant functions and the Gaussian measure $\mathcal{N}(b_0, A)$ on the space $\dot{H}^2_{\text{per}}([0, 2\pi])$ where $A = \left(-\frac{d^2}{da^2}\right)^{-2}$ subject to periodic boundary conditions:

$$p_0 = \lambda \otimes \mathcal{N}(b_0, A).$$

This is the prior measure. Note that as this is a degenerate Gaussian, we are using an improper prior. Purely formally, this prior can be written as a density with respect to (non-existing) Lebesgue measure on $H^2_{\text{per}}([0, 2\pi])$:

$$p_0(b) \sim \exp\left(-\frac{1}{2}\int_0^{2\pi}(b - b_0)(a)(\Delta^2(b - b_0))(a)da\right). \tag{7.49}$$

Having defined a prior we now use the Radon Nikodym derivative $\mathcal{L}(b; x)$ as the likelihood just as before. The posterior measure then follows the Bayes formula:

$$\mathbb{P}\left(b|\{x_t\}_{t=0}^{T}\right) \propto \mathbb{P}(\{x_t\}_{t=0}^{T}|b)p_0(b)$$

$$= \exp\left(-\frac{1}{2}\int_{0}^{2\pi} L_T^a b'^2(a) - b(a)\left(L_T^a\right)' \quad 2b(a)\left(M + \tilde{\chi}(a; X_0, X_T)\right) da \right.$$

$$\left. -\frac{1}{2}\int_{0}^{2\pi}(b(a)-b_0(a))\Delta^2(b(a)-b_0(a))da\right)$$

where M again corresponds to the number of times the path $\{x_s\}_{s=0}^{T}$ winds around the circle. Note that this is just the straightforward product of the likelihood (7.42) with the prior measure (7.49). We simplify this expression by considering the case $b_0 = 0$ and by dropping the arguments:

$$\mathbb{P}\left(b|\{x_t\}_{t=0}^{T}\right) \propto \exp\left(-\frac{1}{2}\int_{0}^{2\pi} |\Delta b|^2 + L_T b^2 - b\left(L_T' + 2M + 2\tilde{\chi}\right) da\right).$$

Formally completing the square in the exponent as in the proof of Theorem 7.17 one finds that this posterior measure is again a Gaussian with formal density

$$\mathbb{P}\left(b|\{x_t\}_{t=0}^{T}\right) \sim \exp\left(-\frac{1}{2}\left|\mathcal{D}^{-\frac{1}{2}}\left(b+\mathcal{D}^{-1}c\right)\right|^2 + \frac{1}{2}\left|\mathcal{D}^{-\frac{1}{2}}c\right|^2\right),$$

where we have used the abbreviations (7.47) and (7.48) to shorten notation. Its mean is given by (7.46) and its covariance is

$$C_o = \left(\Delta^2 + L_T\right)^{-1}. \qquad (7.50)$$

This establishes the usual connection between regularization of the likelihood and the mean of an appropriate Bayesian posterior. It is possible to prove existence and robustness of these measures against small errors in the local time, including stability of a numerical implementation, see Papaspiliopoulos et al. (2009) for details.

Finally, let us rewrite (7.46) in a suggestive form:

$$\hat{b} = \left(\Delta^2 + L_T\right)^{-1}\left(\frac{1}{2}L_T' + M + \tilde{\chi}(\cdot; X_0, X_T) + \Delta^2 b_0\right). \qquad (7.51)$$

Heuristically we expect that L_T is $\mathcal{O}(T)$ for large T, and since Δ^2 is $\mathcal{O}(1)$ on low frequencies we expect that equation (7.50) defines a small operator, at least on low frequencies, and that the covariance of the posterior tends to the zero operator as $T \to \infty$. Likewise the mean, given by (7.47), will approach

$$\frac{1}{2}(\log L_T^a)'$$

for large T. Note that we have a similar result for the maximizers of the likelihood for regularized local times, see (7.39).

In summary, these heuristics indicate that the Bayesian framework should be amenable to a posterior consistency result with the posterior measure converging to a Dirac distribution centred at the true drift function.

7.7 Conclusions and further work

In this overview we have illustrated the following points:

- At small scales, data is often incompatible with the diffusion process that we wish to fit.
- In Section 7.3 we saw that
  1. this situation can be understood in the context of fitting averaged/homogenized equations to multiscale data,
  2. in the *averaging* situation fine-scale data produces the *correct* averaged equation,
  3. in the *homogenization* situation fine-scale data produces an *incorrect* homogenized equation.
- In Section 7.4 we saw that
  1. to estimate the drift and diffusion coefficients accurately in the homogenization scenario it is necessary to *subsample*,
  2. there is an optimal subsampling rate, between the two characteristic timescales of the multiscale data,
  3. the optimal subsampling rate may differ for different parameters.
- In Section 7.5 we observed that in the case where the data is smooth at small scales a useful approach can be to fit hypoelliptic diffusions; such models are often also dictated by physical considerations.
- In Section 7.6 we observed that when fitting the drift, another approach is to use estimators which do not see time-ordering of the data and use, instead, the *local time* (or empirical measure) of the data.

There are many open questions for further investigation:

Section 7.3: How to identify multiscale character from time-series?

Section 7.4:

1. If subsampling is used, then what is the optimal subsampling rate?
2. Is subsampling at random helpful?
3. Is it possible to optimize the data available by combining shifts?

4. How to estimate diffusion coefficients from low frequency data?

Section 7.5:

1. A theoretical understanding of the conditions under which hybrid Gibbs samplers, using different approximate likelihoods for different parts of the sampling problem, yield approximately correct samples from the true posterior is yet to be attained.

2. While the recipe described in this section extends to higher order hypoellipticity, the method is still to be tested in this region.

Section 7.6:

1. How to obtain estimates of the local time L_T^a for all a which are good in a suitable norm, e.g. L^2?

2. What is the convergence behaviour as $T \to \infty$ for the mollified maximum likelihood and the Bayesian estimators?

3. How to extend the Bayesian approach to higher dimensions where the empirical measure is even less regular than in the one dimensional case?

Acknowledgements We gratefully acknowledge crucial contributions from all our co-authors on work we cited in this chapter.

7.8 Appendix 1

Let us consider the random functional

$$\mathcal{I}(b; w) = \int_0^1 b^2(x)w(x) + b'(x)w(x)dx. \qquad (7.52)$$

where $b(\cdot) \in H^1(0,1)$ and $w(x)$ is a standard Brownian bridge. We claim that this functional is not bounded below and state this as a theorem:

Theorem 7.18 *There almost surely exists a sequence $b^{(n)}(\cdot) \in H^1(0,1)$ such that*

$$\lim_{n \to \infty} \mathcal{I}(b^{(n)}; w) = -\infty \quad a.s.$$

Proof. For the Brownian bridge we have the representation

$$w(x) = \sum_{i=1}^{\infty} \frac{\sin(i\pi x)}{i} \xi_i \qquad (7.53)$$

where the $\{\xi_i\}_{i=1}^{\infty}$ are a sequence of iid normal $\mathcal{N}(0,1)$ random variables. This series converges in $L^2(\Omega; L^2((0,1), \mathbb{R}))$ and almost surely in $C([0,1], \mathbb{R})$, see Kahane (1985).

Now consider the following sequence of functions $b^{(n)}$:

$$b^{(n)}(x) = \sum_{i=1}^{n} \frac{\xi_i}{i} \cos(i\pi x). \qquad (7.54)$$

We think of a fixed realization $\omega \in \Omega$ of (7.53) for the time being and note that $\{w(x) : x \in [0,1]\}$ is almost surely bounded in $L^\infty((0,1), \mathbb{R})$, so if there exists a $C > 0$ (which may depend on $\{\xi_i\}_{i=0}^\infty$) such that

$$\|b^{(n)}\|_{L^2} < C \quad \forall n \in \mathbb{N} \qquad (7.55)$$

the first integral in (7.52) will stay finite. By Parseval's identity, it is clear that for the sequence of functional (7.54) this will be the case if the coefficients $\frac{\xi_i}{i}$ are square-summable.

Computing the second summand in (7.52) is straightforward, since the series terminates due to orthogonality:

$$\int_0^1 \left(\sum_{i=1}^\infty \frac{\sin(i\pi x)}{i} \xi_i \right) \cdot \left(\sum_{j=1}^n \frac{\xi_j}{j} \cos(j\pi x) \right)' dx = -\frac{\pi}{2} \sum_{j=1}^n \frac{\xi_j^2}{j}.$$

It can now be seen that (7.52) is unbounded from below if the following two conditions are fulfilled:

$$\lim_{n \to \infty} \sum_{j=1}^n \frac{1}{j} \xi_j^2 = \infty \qquad (7.56)$$

$$\lim_{n \to \infty} \sum_{j=1}^n \frac{1}{j^2} \xi_j^2 < \infty \qquad (7.57)$$

We finally allow ω to vary and seek to establish that the conditions (7.56) and (7.57) are almost surely fulfilled. To do this, first note that the random variables being summed are independent. Thus, by the Kolmogorov 0-1 law the probability for convergence is either zero or one. We proceed by applying Kolmogorov's Three-Series Theorem (Theorem 12.5 in Williams (1991)) to each of the three sequences to establish (7.56) and (7.57).

We start by treating (7.56). Denote by $X_j \mid^K$ the truncation of the random variable for some $K > 0$ in the sense:

$$X_j \mid^K (\omega) = \begin{cases} X_j(\omega) & \text{if } |X_j(\omega)| \le K \\ 0 & \text{if } |X_j(\omega)| > K \end{cases}.$$

To abbreviate notation, define the following two sequences of random variables:

$$X_j = \frac{1}{j}\xi_j^2$$

$$Y_j = \frac{1}{j^2}\xi_j^2$$

Now consider the summability of expected values for the sequence X_j: since ξ_j^2 follows a χ-squared distribution with one degree of freedom, its expected value is one. For the truncated variable $X_j \mid^K$, for any $K > 0$, there will be some j^* so that for all $j \geq j^*$ we have that

$$\mathbb{E}(X_j \mid^K) = \mathbb{E}\left[\frac{1}{j}\left(\xi^2 \mid^{jK}\right)\right] > \frac{1}{2j}$$

Therefore, the expected value summation fails as follows:

$$\sum_{j=1}^{\infty} \mathbb{E}(X_j \mid^K) = \sum_{j=1}^{\infty} \frac{1}{j}\mathbb{E}\left(\xi^2 \mid^{jK}\right)$$

$$\geq \sum_{j=j^*}^{\infty} \frac{1}{2j} = \infty$$

Therefore, the series $\sum_{j=1}^{\infty} X_j$ diverges to infinity almost surely, thus (7.56) is established.

Now let us establish (7.57) using the Three-series theorem. First check the summability of the expected values:

$$\sum_{j=1}^{\infty} \mathbb{E}(Y_j \mid^K) \leq \sum_{j=1}^{\infty} \mathbb{E}Y_j = \sum_{j=1}^{\infty} \frac{1}{j^2} < \infty$$

Now let us establish the summability of the variances:

$$\sum_{j=1}^{\infty} \text{Var}(Y_n \mid^K) \leq \sum_{j=1}^{\infty} \text{Var}Y_n$$

$$= \sum_{j=1}^{\infty} \frac{1}{j^4}\text{Var}\xi_j^2$$

$$= 2\sum_{j=1}^{\infty} \frac{1}{j^4} < \infty$$

where we used that ξ_j^2 follows a χ-squared distribution with one degree of freedom and hence has variance $\text{Var}\xi_j^2 = 2$. Finally, to establish the summability of the tail probabilities we use the following argument for any $K > 0$:

$$\sum_{j=1}^{\infty} P(|Y_j| > K) \leq \sum_{j=1}^{\infty} \frac{1}{K} \mathbb{E}|Y_j|$$

$$\leq \frac{1}{K} \sum_{j=1}^{\infty} \frac{1}{j^2} < \infty$$

where we have used the Markov inequality and the previous calculation of the expected value of $Y_j = |Y_j|$.

To put everything together, let us reconsider the functional $I_B[b]$:

$$I_B[b^{(n)}] = \int_0^1 \left(b^{(n)}\right)^2 (x)w(x) + \left(b^{(n)}\right)' (x)w(x)dx$$

$$\leq \left(\sup_{x\in[0,1]} w(x)\right) \int_0^1 \left(b^{(n)}\right)^2 (x)dx - \frac{\pi}{2} \sum_{j=1}^{n} \frac{1}{j}\xi_j^2$$

$$\leq \left(\sup_{x\in[0,1]} w(x)\right) \frac{1}{2}\sum_{j=1}^{n} X_j - \frac{\pi}{2} \sum_{j=1}^{n} Y_j$$

Now use the almost surely true convergence and divergence statements (7.56) and (7.57) to conclude:

$$\lim_{n\to\infty} I_B[b^{(n)}] = -\infty \quad \text{a.s.}$$

□

7.9 Appendix 2

Torus

We denote by $\mathbb{T}^d$ the d-dimensional torus. We parameterise the torus by d variables $z_i \in [0, 2\pi]$ where we identify the end points 0 and 2π so as to obtain periodicity in each direction z_i.

Matrix inner product

Given two matrices $A, B \in \mathbb{R}^{n\times m}$ we define their inner product as

$$A : B = \sum_{i=1}^{n}\sum_{j=1}^{m} A_{i,j} B_{i,j}.$$

This defines a positive-definit symmetric bilinear form on $\mathbb{R}^{n \times m}$ and turns this space into an inner product space (also known as a finite dimensional Hilbert space).

References

Alberts, B., Johnson, A., Lewis, J., Raff, M., Roberts, K., & Walter, P. (2002). *Molecular Biology of the Cell*. New York: Garland Science.

Campillo, F., & Pitnitski, A. (2002). Effective diffusion in vanishing viscosity. In *Nonlinear Partial Differential Equations and Their Applications* (pp. 133–145). Amsterdam: North-Holland. (France Seminar, Vol. XIV (Paris 1997/1998), volume 31 of Stud. Math. Appl.)

Da Prato, G., & Zabczyk, J. (1992). *Stochastic Equations in Infinite Dimensions*. Cambridge: Cambridge University Press.

Dacorogna, M. M., Gençay, R., Müller, U., Olsen, R. B., & Pictet, O. (2001). *An Introduction to High-Frequency Finance*. San Diego: Academic Press.

Durrett, R. (1996). *Stochastic Calculus - A Practical Introduction*. London: CRC Press.

Evans, L. C. (1998). *Partial Differential Equations*. American Mathematical Society.

Frenkel, D., & Smit, B. (2002). *Understanding Molecular Simulation. From algorithms to applications*. London: Academic Press.

Givon, D., Kupferman, R., & Stuart, A. M. (2004). Extracting macroscopic dynamics: model problems and algorithms. *Nonlinearity*, *17*, R55–R127.

Kahane, J.-P. (1985). *Some Random Series of Functions*. Cambridge: Cambridge University Press.

Kaipio, J., & Somersalo, E. (2005). *Statistical and Computational Inverse Problems*. Berlin: Springer.

Kepler, T., & Elston, T. (2001). Stochasticity in transcriptional regulation: origins, consequences, and mathematical representations. *Biophys. J.*, *81*, 3116–3136.

Majda, A. J., & Kramer, P. R. (1999). Simplified models for turbulent diffusion: Theory, numerical modelling and physical phenomena. *Physics Reports*, *314*(4-5), 237–574.

Majda, A. J., Timofeyev, I., & Vanden-Eijnden, E. (1999). Models for stochastic climate prediction. *Proc. Natl. Acad. Sci. USA*, *96*(26), 14687–14691.

Marcus, M. B., & Rosen, J. (2006). *Markov Processes, Gaussian Processes, and Local Times*. Cambridge: Cambridge University Press.

Melbourne, I., & Stuart, A. M. (2011). A note on diffusion limits of chaotic skew-product flows. *Nonlinearity, 24*, 1361-1367.

Olhede, S. C., Sykulski, A., & Pavliotis, G. A. (2009). Frequency domain estimation of integrated volatility for Itô processes in the presence of market-microstructure noise. *SIAM J. Multiscale Model. Simul., 8*, 393–427.

Papaspiliopoulos, O., Pokern, Y., Roberts, G. O., & Stuart, A. M. (2009). *Nonparametric Bayesian drift estimation for one-dimensional diffusion processes* (Tech. Rep.). CRiSM report no. 09-29, Warwick.

Papaspiliopoulos, O., Pokern, Y., Roberts, G. O., & Stuart, A. M. (2011). Nonparametric estimation of diffusions: a differential equations approach. (Submitted, available from http://www.econ.upf.edu/~omiros/papers/submission_arxiv.pdf)

Papavasiliou, A., Pavliotis, G. A., & Stuart, A. M. (2009). Maximum likelihood drift estimation for multiscale diffusions. *Stochastic Process. Appl., 119*, 3173–3210.

Pavliotis, G. A., & Stuart, A. M. (2007). Parameter estimation for multiscale diffusions. *J. Stat. Phys., 127*, 741–781.

Pavliotis, G. A., & Stuart, A. M. (2008). *Multiscale Methods: Averaging and Homogenization*. New York: Springer.

Pokern, Y. (2006). *Fitting stochastic differential equations to molecular dynamics data*. Unpublished doctoral dissertation, Warwick University.

Pokern, Y., Stuart, A. M., & Vanden-Eijnden, E. (2009). Remarks on drift estimation for diffusion processes. *Multiscale Modeling & Simulation, 8*, 69-95.

Pokern, Y., Stuart, A. M., & Wiberg, P. (2009). Parameter estimation for partially observed hypoelliptic diffusions. *J. Roy. Statist. Soc. B, 71*, 49–73.

Schlick, T. (2000). *Molecular Modeling and Simulation – an Interdisciplinary Guide*. New York: Springer.

Williams, D. (1991). *Probability with Martingales*. Cambridge: Cambridge University Press.

Zhang, L., Mykland, P. A., & Aït-Sahalia, Y. (2005). A tale of two time scales: Determining integrated volatility with noisy high frequency data. *J. Amer. Statistical Assoc., 100*, 1394-1411.

Index

A

α-mixing, 50
ADD, 169
Approximate quadratic variation, 197
AQVT, *see* Asymptotic quadratic variation of time
ARMA process, 391
ARV, 161
Assumption (H), 213
Assumption (H'), 214
Assumption (K), 270
Assumption (K'), 281
Assumption (L), 291
Assumption (SH), 218
Assumption (SL), 295
Assumptions [A1] – [A6], 342
Assumptions 7.4, 440
Assumptions 7.6, 441
Assumptions 7.9, 444
Asymptotic decoupling delay, 169
Asymptotic quadratic variation of time, 147
Average realized volatility, 161
Averaging, method of 167, 441
    limit theorem, 441
    maximum likelihood estimator, 444

B

Bernstein-type inequality, 353
Bipower variation, 217
Black–Scholes–Merton differential equations, 134
Black–Scholes–Merton formula, 115–116
Black–Scholes model, 83
Block approximation, 168
Blumenthal–Getoor index, 287

J

K

L

For Product Safety Concerns and Information please contact our
EU representative GPSR@taylorandfrancis.com Taylor & Francis
Verlag GmbH, Kaufingerstraße 24, 80331 München, Germany